T0271854

Practical Mathematical Cryptography

Practical Mathematical Cryptography provides a clear and accessible introduction to practical mathematical cryptography.

Cryptography, both as a science and as practice, lies at the intersection of mathematics and the science of computation, and the presentation emphasises the essential mathematical nature of the computations and arguments involved in cryptography.

Cryptography is also a practical science, and the book shows how modern cryptography solves important practical problems in the real world, developing the theory and practice of cryptography from the basics to secure messaging and voting.

The presentation provides a unified and consistent treatment of the most important cryptographic topics, from the initial design and analysis of basic cryptographic schemes towards applications.

Features

- Builds from theory toward practical applications
- Suitable as the main text for a mathematical cryptography course
- Focus on secure messaging and voting systems.

Kristian Gjøsteen is a professor of mathematical cryptography at NTNU – Norwegian University of Science and Technology. Gjøsteen has worked on cryptographic voting, electronic identification, privacy, public key encryption and key exchange.

Chapman & Hall/CRC Cryptography and Network Security Series

Series Editors: Douglas R. Stinson and Jonathan Katz

Secret History: The Story of Cryptology, Second Edition
Craig P. Bauer

Data Science for Mathematicians
Nathan Carter

Discrete Explorations
Craig P. Bauer

Cryptography: Theory and Practice, Fourth Edition
Douglas R. Stinson and Mary P. Paterson

Cryptology: Classical and Modern, Second Edition
Richard Klima and Neil Sigmon

Group Theoretic Cryptography
Maria Isabel Gonzalez Vasco and Rainer Steinwandt

Advances of DNA Computing in Cryptography
Suyel Namasudra and Ganesh Chandra Deka

Mathematical Foundations of Public Key Cryptography
Xiaoyun Wang, Guangwu Xu, Mingqiang Wang, Xianmeng Meng

Guide to Pairing-Based Cryptography
Nadia El Mrabet and Marc Joye

Techniques for Designing and Analyzing Algorithms
Douglas R. Stinson

Practical Mathematical Cryptography
Kristian Gjøsteen

For more information about the series: https://www.crcpress.com/Chapman--HallCRC-Cryptography-and-Network-Security-Series/book-series/CHCRYNETSEC

Practical Mathematical Cryptography

Kristian Gjøsteen

Norwegian University of Science and Technology, Norway

CRC Press
Taylor & Francis Group
Boca Raton London New York

CRC Press is an imprint of the
Taylor & Francis Group, an **informa** business

A CHAPMAN & HALL BOOK

First edition published 2023

by CRC Press
6000 Broken Sound Parkway NW, Suite 300, Boca Raton, FL 33487-2742

and by CRC Press
4 Park Square, Milton Park, Abingdon, Oxon, OX14 4RN

CRC Press is an imprint of Taylor & Francis Group, LLC

ISBN: 978-0-367-71085-9 (hbk)
ISBN: 978-0-367-71119-1 (pbk)
ISBN: 978-1-003-14942-2 (ebk)

DOI: 10.1201/9781003149422

Typeset in Latin Modern font
by KnowledgeWorks Global Ltd.

Publisher's note: This book has been prepared from camera-ready copy provided by the authors.

Contents

Preface

Cryptography as a practice is more than two millennia old. Two people trying to talk without being overheard is the original cryptographic problem. And until recently, that was all of cryptography. This is obviously an extremely important problem, and cryptographic histories are full of stories of how cryptography, and in particular cryptanalysis (attacking cryptographic systems), has played a pivotal role in historical events.

But modern cryptography is about communication in a very general sense of the word in the presence of powerful adversaries. It is about much more than plain communication between Alice and Bob and now includes human activities such as finance and voting.

This book is intended as an introduction to practical mathematical cryptography. Cryptography, both as a science and as a practice, lies at the intersection of mathematics and the science of computation. By saying mathematical cryptography, we emphasise the essential mathematical nature of the computations and arguments involved in cryptography. The word practical emphasises that our goal is to understand how modern cryptography solves important problems in the real world. While this book covers many theoretical topics, the topics included are motivated in practical needs and understanding.

Motivating problems   A far loftier goal for this book is to enable the reader to use modern cryptography in the real world to help defend things we care about, both abstract and concrete. In order to help structure the material with this goal in mind, we have chosen two important practical topics as our motivating problems: secure messaging and cryptographic voting schemes.

Secure messaging is the original cryptographic problem. The modern, connected society would simply not work without practical secure messaging. Modern messaging schemes are very different objects than those of any previous time, in terms of what kind of functionality can be achieved and in terms of what kind of security can be achieved. Functionality-wise, the main achievement is the ability to establish secure messaging between parties that have not previously established a shared secret. Security-wise, we can now defend against powerful adversaries that steal users' secret keys.

Elections are vital for modern, free societies. Consequently, we have to care about the security of the voting systems used to hold elections. Cryptographic voting systems use the techniques of modern cryptography to improve elections, both by enabling new functionality such as remote voting but also by allowing new kinds of security, such as letting voters verify that their vote was counted correctly without compromising privacy.

Overview of the book   This book is divided into four parts, where the final part discusses our motivating problems in detail. The first three parts discuss the mathematical cryptography, both practical and theoretical, that is needed to understand these applications and their solutions.

The first part of the book is an informal introduction to basic cryptography. It covers the traditional focus of cryptography, which is confidentiality, integrity and authenticity. The emphasis is on describing the cryptographic functionality we want, and explain the rudiments of cryptanalysis, how to attack cryptographic constructions, both through purely cryptographic attacks and through involved mathematical computations. The question of exactly what security means for cryptographic systems is discussed in some detail, though informally. In addition to covering the basics of cryptography, the goal of this section is to give the reader an understanding of how to attack a cryptographic system, and a feel for the adversarial nature that lies at the heart of cryptography. This is necessary since it is difficult to understand how to defend without knowing both how to attack and that defense is necessary.

The second part introduces the modern practice of provable security by showing how it applies to the cryptographic constructions introduced and analysed in the first part. The idea is to use techniques from computational complexity theory to relate the security of cryptosystems to the hardness of the underlying computational problems. A significant part of this work involves precisely defining what security actually means, which turns out to be a highly non-trivial question. The emphasis is on studying the theory of these constructions from a practical point of view and using concrete constructions as examples for the theory developed. In this part, cryptanalysis plays a much smaller role, though defining security is hardly possible without a clear instinct for what constitutes a practical attack on a system.

The third part continues discussing the practice of provable security, but now the object are not the traditional cryptographic constructions covering confidentiality, integrity and authenticity, but rather more general cryptographic protocols. We cover three fundamental topics in cryptography: key exchange, zero knowledge and multiparty computation. They are the fundamental building blocks of modern practical cryptography, without which we could not do anything.

The future   At the time of writing, quantum computers are a potential threat to almost all deployed cryptography. Nobody knows when, or even if, quantum computers will ever become practical, but at the moment, nobody who takes cryptographic practice seriously can afford to assume that quantum computers will not someday become practical. This book covers the very basics of quantum computations, but the goal is not to give a general introduction to quantum computation. Instead, the goal is to motivate the need for so-called quantum-safe cryptography and some techniques for quantum algorithms for solving computational problems related to cryptography. With respect to quantum-safe cryptography, we have decided to cover mainly lattice-based cryptography. There are other important forms of plausible quantum-safe cryptography not covered by this book. As future research either conclusively proves that sufficiently large and reliable quantum computers can be built, or somehow demonstrates that practical quantum computations seem infeasible, this book will be updated. Either the cryptography that is not quantum-safe will be removed, or quantum-safe cryptography will only be discussed to the extent that it provides interesting new cryptographic functionality.

The practical viewpoint of this book is definitely not intended to disparage the theory of cryptography. As in any field, theoretical study has time and again provided crucial tools and understanding required for practical cryptography. As our practical viewpoint has guided the selection of topics from cryptographic theory, a large number of very interesting topics have been left out, and some of them will certainly turn out to be of greater practical interest in the future, since theory has a habit of turning practical every now and then. It is to be hoped that the current selection of topics will provide the reader with the skills required to deal with new topics, but we expect any future version of this book to contain a different set of topics, again guided by a future practical point of view.

Using the book for teaching There are a number of ways to use this book in teaching. From a cryptographic point of view, the first part covers roughly the material expected of an introductory course in cryptography. It is suitable for students with some background in abstract algebra and probability theory. Some exercises require programming skills, which will be suitable for students interested in computations. The first section is structured such that concepts from abstract algebra can be introduced in a natural way for students that need it, but this book is not a self-contained exposition for such students.

The material in the final three parts can be structured in many ways. One option is to teach the material as a single large course in provable security, or one basic course covering the second part and an advanced course covering the third and fourth parts. Another option is to teach the second part together with the material on key exchange and secure messaging as a first course in provable security, and then teach the material on zero knowledge, multiparty computation and cryptographic voting as a second course in provable security.

Acknowledgements: This book would not have been possible without my family's great patience, love and faith. I also owe a great debt of gratitude to my students, colleagues, coauthors and mentors, who have taught me so many things and helped me in so many ways.

CHAPTER 1

Symmetric Cryptography

We begin with the original cryptographic problem: Alice wants to send messages to Bob via some communications channel. Eve has access to the channel, and she may eavesdrop on and tamper with anything sent over the channel.

Alice does not want Eve to be able to eavesdrop on her messages, sometimes called *plaintext*. She wants to communicate *confidentially*. When Bob receives a message seemingly from Alice, then Bob wants to be sure that Alice sent the message and that it was not tampered with. Alice and Bob want *integrity*.

Alice and Bob share a secret, called a *key*. Cryptography where the sender and the receiver (or the *honest users*) have the same key is called *symmetric cryptography*, where *symmetry* refers to the symmetry of knowledge.

Our first task is to define what a *symmetric cryptosystem* is and discuss informally what the *security requirements* are for such cryptosystems.

To illustrate standard *attacks*, we shall first study some historic cryptosystems and how those systems can be attacked. The exposition alternates between discussing the cryptosystems and discussing interesting attacks that apply to them. Note that this chapter is no history of cryptography.

Eventually we describe several constructions that provide confidentiality against eavesdroppers. These constructions do not provide integrity, nor do they in practice provide confidentiality if Eve tampers with ciphertexts.

Providing integrity is in some sense orthogonal to providing confidentiality against eavesdroppers, but how to combine the constructions described in Sections 1.2 and 1.3 into cryptosystems providing both integrity and confidentiality even when Eve tampers with the ciphertexts has some pitfalls.

While modern high-level constructions are discussed in this chapter, low-level constructions are out of scope for this book. This explains why this chapter defines what a block cipher is and gives an informal explanation of what it means for a block cipher to be secure but does not contain a single example of a modern block cipher.

1.1 DEFINITIONS

We begin with the definition of a symmetric cryptosystem. We want to capture the functionality Alice and Bob use: Alice wants Bob to learn a plaintext. She will send something; we shall call it a *ciphertext* or *encryption*, from which Bob learns the plaintext. Creating the ciphertext is called *encryption*. Learning the plaintext from the ciphertext is called *decryption*. In order to make this possible, Alice and Bob have a shared secret called a *key*.

Definition 1.1. A *symmetric cryptosystem* consists of a set $\mathfrak{K}$ of *keys*, a set $\mathfrak{P}$ of *plaintexts*, a set $\mathfrak{C}$ of *ciphertexts*, and two algorithms:

- an *encryption* algorithm $\mathcal{E}$ that on input of a key and a plaintext outputs a ciphertext; and

- a *decryption* algorithm $\mathcal{D}$ that on input of a key and a ciphertext outputs either a plaintext or the special symbol $\perp$ (indicating an invalid ciphertext).

For any key k and any plaintext m, we have that $\mathcal{D}(k, \mathcal{E}(k, m)) = m$.

The plaintext set $\mathfrak{P}$ will usually be a set of finite sequences (strings) of letters from an *alphabet*.

We shall assume that the key Alice and Bob share has been chosen uniformly at random from the set of keys.

1.2 CONFIDENTIALITY AGAINST EAVESDROPPERS

The situation we shall now consider has Alice sending messages to Bob while Eve eavesdrops. Eve wants to understand what Alice is saying to Bob.

Our discussion involves historic cryptosystems because this gives us a gentle introduction to the basic concepts in cryptography and provides some insight into important attack strategies. The presentation in this section alternates between describing a cryptosystem and describing how to attack that cryptosystem until we reach systems that will provide confidentiality.

Informally. A symmetric cryptosystem provides *confidentiality* if it is – without knowledge of the key – hard to learn anything at all about the decryption of a ciphertext from the ciphertext itself, except possibly the length of the decryption.

Remark. It cannot be emphasised strongly enough that cryptography does not try to hide the length of the plaintext. The reason is that this would be prohibitively expensive. However, this means that applications where the length of messages must do so themselves, perhaps by using fixed-length encodings.

1.2.1 Shift Cipher

The *shift cipher* is also known as the *Cæsar cipher*.

We first give our alphabet G a group structure. There is a natural bijection between the English alphabet $\{A, B, C, \ldots, Z\}$ and the group $\mathbb{Z}_{26}^{+}$, given by $0 \leftrightarrow A$, $1 \leftrightarrow B$, etc. We add F and G by applying the bijection to get 5 and 6, adding them to 11, and then applying the inverse bijection to get L.

Remark. Unless explicitly said otherwise, for any finite group we discuss, there is a canonical representation for group elements, and we always use this representation. This is important. If group elements have multiple representations, the representation could contain information about more than just the group element, such as how the group element was computed.

The *plaintext* m is a sequence of letters $m_1 m_2 \ldots m_l$ from the alphabet. The *key* is an element k from G. We *encrypt* the message by adding the key to each letter, that is, the ith ciphertext letter is

$$c_i = m_i + k, \qquad 1 \le i \le l. \tag{1.1}$$

The ciphertext c is the sequence of letters $c_1 c_2 \ldots c_l$.

To *decrypt* a ciphertext $c = c_1 \ldots c_l$, we subtract the key from each ciphertext letter, that is, the ith plaintext letter is

$$m_i = c_i - k, \qquad 1 \le i \le l.$$

Example 1.1. The *shift cipher* based on a group G is the following: The set of keys is the group, $\mathfrak{K} = G$. The plaintext set is the set of strings of group elements, $\mathfrak{P} = \cup_l G^l$. The ciphertext set is the same, $\mathfrak{C} = \cup_l G^l$.

- The *encryption* algorithm $\mathcal{E}$ takes as input a key $k \in G$ and a tuple of group elements $m_1 m_2 \ldots m_l \in G^l$ and computes the ciphertext $c_1 c_2 \ldots c_l \in G^l$ as

$$c_i \leftarrow m_i + k, \qquad 1 \le i \le l.$$

- The *decryption* algorithm $\mathcal{D}$ takes as input a key $k \in G$ and a tuple of group elements $c_1 c_2 \ldots c_l \in G^l$ and computes the message $m_1 m_2 \ldots m_l \in G^l$ as

$$m_i \leftarrow c_i - k, \qquad 1 \le i \le l.$$

Exercise 1.1. Show that the shift cipher $(\mathfrak{K}, \mathfrak{P}, \mathfrak{C}, \mathcal{E}, \mathcal{D})$ is a symmetric cryptosystem. Implement the two algorithms $\mathcal{E}$ and $\mathcal{D}$ for the English alphabet.

Exercise 1.2. How many different keys are there for the shift cipher when the alphabet has 26 elements?

Example 1.2. To encrypt BABOONSAREFUNNY using the shift cipher with the key $k = \text{D}$ (D corresponds to the number 3), we do the following computations:

B	A	B	O	O	N	S	A	R	E	F	U	N	N	Y
+D	+D	+D	+D	+D	+D	+D	+D	+D	+D	+D	+D	+D	+D	+D
↓	↓	↓	↓	↓	↓	↓	↓	↓	↓	↓	↓	↓	↓	↓
E	D	E	R	R	Q	V	D	U	H	I	X	Q	Q	B

The ciphertext is EDERRQVDUHIXQQB.

Attack: Exhaustive Search   The easiest attack on the shift cipher is an *exhaustive search* for the key, or a *brute force attack*. We make two assumptions: only one key will give a reasonable decryption, and we will be able to recognise that reasonable decryption. Both assumptions are usually true.

   If there are few keys, we can decrypt with all possible keys in reasonable time. The correct key will be the one that gives a reasonable decryption.

Exercise 1.3. Find all the possible decryptions of HGHUUT. How many are English words? What about the possible decryptions of MBQ?

Exercise 1.4. Choose a random key for the shift cipher and encrypt a message. Let someone else decrypt the ciphertext without telling them the key.

1.2.2   Affine Cipher

Over a ring, the equation $Y = k_1 X + k_2$ has a unique solution if k_1 is invertible in the ring. We shall use this fact to construct an *affine cipher*.

   We give our alphabet R a ring structure, $\mathbb{Z}_{26}$. We add as before. We multiply F and G by applying the bijection to get 5 and 6, multiplying them to get 30, which is 4 modulo 26, and then applying the inverse bijection to get E.

Example 1.3. The *affine cipher* based on a ring R is the following: The set of keys is $\mathfrak{K} = R^* \times R$, the plaintext set is the set of strings of ring elements, $\mathfrak{P} = \cup_l R^l$, and the ciphertext set is the same, $\mathfrak{C} = \mathfrak{P}$.

- The encryption algorithm $\mathcal{E}$ takes as input a key $(k_1, k_2) \in R^* \times R$ and a tuple of ring elements $m_1 m_2 \ldots m_l \in R^l$ and computes the ciphertext $c_1 c_2 \ldots c_l \in R^l$ as

$$c_i = k_1 m_i + k_2, \qquad 1 \le i \le l. \tag{1.2}$$

- The decryption algorithm $\mathcal{D}$ takes as input a key $(k_1, k_2) \in R^* \times R$ and a tuple of ring elements $c_1 c_2 \ldots c_l \in R^l$ and computes the message

$$m_1 m_2 \ldots m_l \in R^l \text{ as}$$

$$m_i = k_1^{-1}(c_i - k_2), \qquad 1 \le i \le l.$$

E | *Exercise* 1.5. Show that the affine cipher $(\mathfrak{K}, \mathfrak{P}, \mathfrak{C}, \mathcal{E}, \mathcal{D})$ is a symmetric cryptosystem. Implement the two algorithms $\mathcal{E}$ and $\mathcal{D}$ for the English alphabet.

E | *Exercise* 1.6. How many different keys are there for the affine cipher when the alphabet has 26 elements?

Attack: Known Plaintext When the attacker knows the plaintext corresponding to a piece of ciphertext, that is called *known plaintext*.

Suppose Eve knows that Alice always begins her messages with HI. One ciphertext starts with the letters UB. Eve knows that Alice used the affine cipher, which means that equation (1.2) was used to encrypt H to U and I to B. She gets the following two equations:

$$\begin{aligned} \text{U} &= k_1 \text{H} + k_2, \\ \text{B} &= k_1 \text{I} + k_2. \end{aligned} \qquad (1.3)$$

This is a linear system of equations with two equations and two unknowns. As long as the difference $\text{H} - \text{I}$ is invertible in the ring (which it is), we can solve the system and recover the key (k_1, k_2).

E | *Exercise* 1.7. Solve the system of equations given in (1.3) to find the key.

E | *Exercise* 1.8. Choose a random key for the affine cipher and encrypt a message. Let someone else decrypt the ciphertext without knowing the key, but give them some known plaintext.

E | *Exercise* 1.9. Describe a known plaintext attack for the shift cipher.

1.2.3 Substitution Cipher

The formulas (1.1) and (1.2) define bijections on the alphabet. We can generalise these schemes by using any bijection or *permutation* on our alphabet.

E | *Example* 1.4. The *substitution cipher* on an alphabet S is the following: The set of keys is the set of permutations on S. The plaintext set is the set of strings of set elements, $\mathfrak{P} = \cup_l S^l$. The ciphertext set is the same, $\mathfrak{C} = \cup_l S^l$.

- The encryption algorithm $\mathcal{E}$ takes as input a key π and a tuple of set elements $m_1 m_2 \ldots m_l \in S^l$ and computes a tuple $c_1 c_2 \ldots c_l \in S^l$ as

$$c_i = \pi(m_i), \qquad 1 \le i \le l. \qquad (1.4)$$

The output is $c_1 c_2 \ldots c_l$.

- The decryption algorithm $\mathcal{D}$ takes as input a key π and a tuple $c = c_1 \ldots c_l \in S^l$ and computes a tuple $m_1 m_2 \ldots m_l \in S^l$ as

$$m_i = \pi^{-1}(c_i), \qquad 1 \leq i \leq l.$$

[E] *Exercise* 1.10. Show that the substitution cipher $(\mathfrak{K}, \mathfrak{P}, \mathfrak{C}, \mathcal{E}, \mathcal{D})$ is a symmetric cryptosystem. Implement two algorithms $\mathcal{E}$ and $\mathcal{D}$ for the English alphabet.

[E] *Exercise* 1.11. How many different keys are there for the substitution cipher when the alphabet has 26 elements?

[E] *Exercise* 1.12. Explain how we can recover part of the key (a *partial key*) from known plaintext, but not necessarily the full key.

Attack: Frequency Analysis   Known plaintext will reveal part of the key. But there are stronger attacks on the substitution cipher, based on the number of times the various ciphertext letters appear.

If the permutation takes the plaintext letter A to the ciphertext letter Z, the number of Z's in the ciphertext will be the same as the number of A's in the plaintext. This means that the relative frequencies of the ciphertext letters will be the same as the relative frequencies of the plaintext letters, up to permutation. *Frequency analysis* exploiting relative frequencies in the plaintext (inferred) and ciphertext (known) to deduce the plaintext.

Most long English texts have the same relative frequency of the various letters. This means that for encryptions of long English texts, the relative frequencies of ciphertext letters is a simple permutation of the relative frequencies of letters in English text. It is easy to match plaintext letters and ciphertext letters and thereby recover the key and thus the plaintext.

For texts of moderate length, the relative frequencies of the less common letters will vary, and we cannot reliably match plaintext letters to ciphertext letters. But some letters, E in particular, are so common in English that they will be the most frequent letter in most texts, even for fairly short ones.

[E] *Exercise* 1.13. Gather a collection of English texts of varying topics and lengths. Compute the frequency distributions. Use these distributions to estimate how long a text must be before we can expect to identify with reasonable certainty (a) E, (b) the five most frequent letters and (c) the ten most frequent letters.

This means that even though we cannot reliably match every plaintext letter to every ciphertext letter, we can match a few plaintext letters to a few ciphertext letters. This gives us a partial key and partial decryption.

Now we pretend that this partial decryption is a crossword puzzle and guess some plaintext words that fit with the partial decryption. We treat these guesses as known plaintext and recover more of the key. This gives us more of the partial decryption. If the new partial decryption does not make sense or is impossible, we guessed wrong. We backtrack and guess again.

When the new partial decryption makes sense, we probably guessed right. Again, we treat the new partial decryption as a crossword puzzle. We repeat this process of *guessing* and *verifying* until we have the complete decryption.

E *Exercise* 1.14. Choose a random key for the substitution cipher and encrypt a long message. Let someone else decrypt the ciphertext without knowing the key. You may give them some known plaintext.

1.2.4 Towards Block Ciphers

One counter to frequency analysis is to permute pairs of letters, that is, our permutation acts on the set S of pairs of letters, not the alphabet.

E *Exercise* 1.15. For a substitution cipher based on permutations on pairs, write down carefully what the three sets $\mathfrak{K}, \mathfrak{P}, \mathfrak{C}$ are, and implement the two algorithms $\mathcal{E}$ and $\mathcal{D}$. Show that $(\mathfrak{K}, \mathfrak{P}, \mathfrak{C}, \mathcal{E}, \mathcal{D})$ is a symmetric cryptosystem.

Unfortunately, the frequencies of pairs are uneven, which means that frequency analysis still works, although it is less effective. A permutation on triples of letters would be better, but still not perfect.

Even better would be L-tuples. The number L is called the *block length*. Unfortunately, representing a random permutation over a large set is impractical. Merely writing down a permutation requires at least $\log_2(|S|^L!) \approx |S|^L(\ln|S|^L - 1)/\ln 2$ binary digits.

One idea would be to use not a random permutation, but instead use a random member of some smaller family of permutations.

E *Example* 1.5. The *Hill cipher* is an example of such a family of permutations, the permutations given by invertible matrices. We give our alphabet R a ring structure, say $\mathbb{Z}_{26}$. We denote an L-tuple of letters as $\mathbf{m} \in R^L$. An invertible $L \times L$ matrix $\mathbf{K}$ acts on L-tuples through matrix multiplication, denoted by $\mathbf{Km}$.

The plaintext m is a sequence of L-tuples of letters $\mathbf{m}_1\mathbf{m}_2\ldots\mathbf{m}_l$. The key is an invertible $L \times L$ matrix $\mathbf{K}$. We encrypt the message using the formula

$$\mathbf{c}_i = \mathbf{Km}_i, \qquad 1 \le i \le l.$$

The ciphertext c is the sequence of L-tuples $\mathbf{c}_1\mathbf{c}_2\ldots\mathbf{c}_l$.

To decrypt a ciphertext $c = \mathbf{c}_1\ldots\mathbf{c}_l$, we compute the ith plaintext tuple using the formula

$$\mathbf{m}_i = \mathbf{K}^{-1}\mathbf{c}_i, \qquad 1 \le i \le l.$$

E | *Exercise* 1.16. The above is an informal description. Write down carefully what the three sets $\mathfrak{K}, \mathfrak{P}, \mathfrak{C}$ are, and implement the two algorithms $\mathcal{E}$ and $\mathcal{D}$. Show that $(\mathfrak{K}, \mathfrak{P}, \mathfrak{C}, \mathcal{E}, \mathcal{D})$ is a symmetric cryptosystem.

E | *Exercise* 1.17. How many different keys are there for the Hill cipher with block length 2 when the alphabet has 29 elements?

E | *Exercise* 1.18. How many blocks of ciphertext-plaintext correspondences do you need to recover $\mathbf{K}$ with reasonable probability, when the block length is 2 and the alphabet has 29 elements? For the purposes of this exercise, you may assume that the known plaintext consists of random letters from the alphabet.

E | *Example* 1.6. To encrypt ABABOONISFUNNY using Hill cipher encryption with the key $\mathbf{K} = \begin{pmatrix} B & D \\ A & F \end{pmatrix}$, we do the following computations:

AB	AB	OO	NI	SF	UN	NY
|	|	|	|	|	|	|
K·	K·	K·	K·	K·	K·	K·
↓	↓	↓	↓	↓	↓	↓
DF	DF	ES	LO	HZ	HN	HQ

The ciphertext is then DFDFESLOHZHNHQ. Observe the repetition in the first two plaintext blocks reflecting in the ciphertext, while the final three blocks all start with H, which is not correlated to any particular plaintext feature.

E | *Exercise* 1.19. Choose a random key for the Hill cipher and encrypt a message. Let someone else decrypt it without knowing the key. Let them have some known plaintext.

E | *Example* 1.7. A second example of a family of permutations is the *Pohlig-Hellman exponentiation cipher*. This time, we do not consider tuples of letters but rather a very large alphabet. We give our large alphabet G the structure of a cyclic group of order n.

The plaintext m is a sequence of group elements $m_1 m_2 \ldots m_l$. The key is an integer k between 0 and n that is relatively prime to n. We encrypt the message using the formula

$$c_i = k m_i, \qquad 1 \leq i \leq l.$$

The ciphertext c is the sequence of group elements $c_1 c_2 \ldots c_l$.

To decrypt a ciphertext $c = c_1 c_2 \ldots c_l$, we compute the i plaintext tuple using the formula

$$m_i = k^{-1} c_i, \qquad 1 \leq i \leq l,$$

where the inverse k^{-1} of k is computed modulo n.

Note that the expression km_i where k is an integer and m_i is a group element is different from the expression km_i when k and m_i are elements in a ring. The former denotes $m_i + m_i + \cdots + m_i$, where the sum contains k terms.

E *Exercise* 1.20. The above is an informal description. Write down carefully what the three sets $\mathfrak{K}, \mathfrak{P}, \mathfrak{C}$ are, and implement the two algorithms $\mathcal{E}$ and $\mathcal{D}$. Show that $(\mathfrak{K}, \mathfrak{P}, \mathfrak{C}, \mathcal{E}, \mathcal{D})$ is a symmetric cryptosystem.

It is generally believed that if the group G is carefully chosen, it is hard to find the key, even with known or chosen plaintext.

Attack: Distinguishing The permutations used in the Hill cipher (linear, invertible maps) are very different from most permutations. For any two L-tuples $\mathbf{m}, \mathbf{m}'$ of letters from the alphabet, an invertible matrix $\mathbf{K}$ satisfies

$$\mathbf{Km} + \mathbf{Km}' = \mathbf{K}(\mathbf{m} + \mathbf{m}').$$

The same observation holds for the permutations used in the Pohlig-Hellman cipher. For any two elements $m, m' \in G$, we get that

$$km + km' = k(m + m').$$

Most permutations do not satisfy these equations. It is easy to *distinguish* the Hill cipher and Pohlig-Hellman cipher permutations from the random permutations. We can also use this fact to make deductions about plaintexts based on ciphertext properties.

E *Exercise* 1.21. Consider the Hill cipher. Suppose c, c' and c'' are ciphertexts such that $c_i + c_i' = c_i''$ for one or more indexes i. What can you say about the corresponding plaintexts?

1.2.5 Block Ciphers

We will now work with a set S, usually the set of L-tuples of letters.

D **Definition 1.2.** A *block cipher* is a pair of maps $\pi, \pi^{-1} : \mathfrak{K} \times S \to S$ such that for all $k \in \mathfrak{K}$ and $s \in S$ we have that

$$\pi(k, \pi^{-1}(k, s)) = s \text{ and } \pi^{-1}(k, \pi(k, s)) = s.$$

That is, a block cipher is a family of permutations on S indexed by $\mathfrak{K}$.

E *Exercise* 1.22. The Hill cipher is based on a block cipher. Describe the sets $\mathfrak{K}, S$ and the functions π, π^{-1} for the block cipher.

E *Exercise* 1.23. The Pohlig-Hellman cipher is based on a block cipher. Describe the sets $\mathfrak{K}, S$ and the functions π, π^{-1} for the block cipher.

Despite the name, a block cipher by itself is not a cryptosystem. But we can easily construct a cryptosystem based on a block cipher.

E *Example* 1.8. The plaintext m is a sequence of elements $m_1 m_2 \ldots m_l$ from the set S. The key is an element k in $\mathfrak{K}$. We encrypt the message using

$$c_i = \pi(k, m_i), \qquad 1 \leq i \leq l.$$

The ciphertext c is the sequence of set elements $c_1 c_2 \ldots c_l$.
    We decrypt a ciphertext $c = c_1 \ldots c_l$, using

$$m_i = \pi^{-1}(k, c_i), \qquad 1 \leq i \leq l.$$

E *Exercise* 1.24. The above is an informal description of a block cipher used in *electronic code book (ECB) mode*. Write down carefully what the three sets $\mathfrak{K}$, $\mathfrak{P}$, $\mathfrak{C}$ are, and implement the two algorithms $\mathcal{E}$ and $\mathcal{D}$. Use the block cipher from Exercise 1.16. Show that $(\mathfrak{K}, \mathfrak{P}, \mathfrak{C}, \mathcal{E}, \mathcal{D})$ is a symmetric cryptosystem.

Our hope is that this scheme will be both practical and as secure as the substitution cipher when used on L-tuples. But as we have seen, the Hill cipher is easy to attack.

Informally. A block cipher is *secure* if it is hard to distinguish a pair of randomly chosen inverse permutations (π, π^{-1}) from the pair of inverse permutations $(\pi(k, \cdot), \pi^{-1}(k, \cdot))$, when k has been sampled from the uniform distribution on $\mathfrak{K}$.
    The distinguisher will only see the function values at various points. The distinguisher will never see the permutation π or the key k.

The idea is that if it is hard to distinguish the block cipher's permutations from average permutations, we may as well use the block cipher with a random key instead of a random permutation. If there is an attack on the block cipher cryptosystem that does not work on the substitution cipher, that will be one way to distinguish the block cipher.
    We note that for a block cipher to be secure, the key set must be very large. Otherwise, one can recognise the block cipher permutations with high probability by enumerating all the keys and observing how the corresponding permutation affects one or two elements of the set.

Sketch: Feistel Ciphers   How to construct secure block ciphers is out of scope for this book. However, we shall very briefly discuss one popular design for block ciphers, the *Feistel cipher*. There are other design strategies, and we have very good block ciphers.

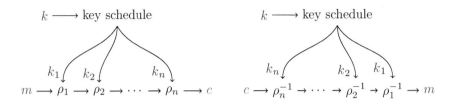

FIGURE 1.1 Typical high-level block cipher design.

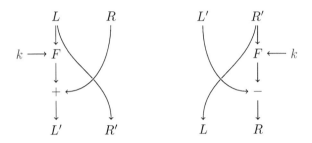

FIGURE 1.2 The Feistel round and its inverse.

Block ciphers are typically built by repeatedly applying one or more simple block ciphers called *rounds*. A single round will be very easy to break, but the composition of sufficiently many rounds may be hard to break. Given the rounds $\rho_1, \rho_2, \ldots, \rho_n$ with corresponding inverse rounds, we get a block cipher π by composition:

$$\pi(k, m) = \rho_n(k_n, \cdots \rho_2(k_2, \rho_1(k_1, m)) \cdots) \qquad \text{and}$$
$$\pi^{-1}(k, m) = \rho_1^{-1}(k, \cdots \rho_{n-1}^{-1}(k_{n-1}, \rho_n^{-1}(k_n, m)) \cdots)$$

The *round keys* $k_1, k_2, \ldots, k_n$ are derived from the key k. Using the same key for each round is problematic. Using independent keys such that $k = (k_1, k_2, \ldots, k_n)$ leads to impractically large keys. Instead, a *key schedule* is usually used to derive round keys from the block cipher key. This high-level design is shown in Figure 1.1. We note that for good ciphers, the key schedule and the rounds are highly dependent on each other.

One convenient way to design rounds is the *Feistel round*. The construction assumes that $S = G \times G$ for some finite group G. It uses a *round function* $F : \mathcal{K} \times G \to G$ to construct the permutation

$$\rho(k, (L, R)) = (R, L + F(k, R)),$$

illustrated in Figure 1.2. The inverse permutation is

$$\rho^{-1}(k, (L', R')) = (R' - F(k, L'), L').$$

Choosing a good round function is difficult, especially when the goal is to find a round function that can be computed very quickly and that does not require many rounds. Again, this problem is out of scope for this book.

Padding Schemes The plaintext set for the cryptosystem from Exercise 1.24 is the set of finite sequences of elements from S, where each set element is typically an L-tuple of letters from the alphabet. In other words, the plaintext set is the set of letter sequences whose length is divisible by L.

But when L is large, it is unreasonable to expect message lengths to be a multiple of L. We usually need to encrypt arbitrary sequences of letters. Since we need to decrypt correctly, we cannot just append some fixed letter until the sequence length is a multiple of L.

We extend a cryptosystem to accept sequences of any length by applying a suitable injective function before encryption.

D **Definition 1.3.** Let $\mathfrak{P}$ and $\mathfrak{P}'$ be sets. A *padding scheme for $\mathfrak{P}$ and $\mathfrak{P}'$* consists of two functions $\iota : \mathfrak{P} \to \mathfrak{P}'$ and $\lambda : \mathfrak{P}' \to \mathfrak{P} \cup \{\bot\}$ satisfying

$$\lambda(\iota(m)) = m \text{ for all } m \in \mathfrak{P}.$$

E *Exercise* 1.25. Suppose you have a cryptosystem $(\mathfrak{K}, \mathfrak{P}', \mathfrak{C}, \mathcal{E}', \mathcal{D}')$ and a padding scheme (ι, λ) for $\mathfrak{P}$ and $\mathfrak{P}'$. Based on the padding scheme and the cryptosystem, construct a new cryptosystem $(\mathfrak{K}, \mathfrak{P}, \mathfrak{C}, \mathcal{E}, \mathcal{D})$. Show that it is indeed a cryptosystem.

Typically, the alphabet is $\{0, 1\}$ and the set is $S = \{0, 1\}^L$, bit strings of length L. The plaintext set $\mathfrak{P}'$ will then be bit strings of length divisible by L.

One padding scheme is the following: We first add one 1-bit, then we add 0-bits until the total length is divisible by L. If the block size L is 8, the bit string 10 10 1 will become 10 10 11 00. If the block size L is 5, the bit string 01 01 0 becomes 01 01 01 00 00.

To remove the padding, we remove up to $L - 1$ trailing 0-bits and exactly one 1-bit. If the block size L is 8, the string 01 01 01 01 01 becomes the bit string 01 01 01 01 0. If the block size L is 5, the string 01 01 0 becomes 01 0, while the string 10 10 10 00 00 cannot be decoded and therefore becomes $\bot$.

If decoding fails as in the last example, we have a *padding error*. Padding errors have subtle effects on the security of cryptographic protocols.

Attack: Block Repetitions We make two observations about ECB mode from Exercise 1.24.

- Repetition of plaintext blocks will cause repetitions of ciphertext blocks.

- If we encrypt the same message twice, we get the same ciphertext.

Both of these observations allow an eavesdropper to learn something about the message from the ciphertext and thereby break confidentiality. Both observations are independent of which block cipher we use. The problem is not with the concept of block cipher but with how we use the block cipher.

Exercise 1.26. Suppose a message has been encrypted with ECB mode and a block cipher with block length 4, and that you know that the message is

SELL THE HOUSE NOW DO NOT SELL THE CABIN

or

SELL THE HOUSE AND EVERYTHING ELSE NOW.

(Ignore spaces and punctuation.) Explain how you can trivially decide which message is the decryption by looking at the ciphertext.

1.2.6   A Correct Use of a Block Cipher

We now let our block cipher operate on a set G with a group structure. Few permutations on G respect the group operation, nor should the block cipher.

Example 1.9. The plaintext m is a sequence of elements $m_1 m_2 \ldots m_l$ from the group G. The key is an element k in $\mathfrak{K}$. We encrypt the message elementwise by first choosing a random group element c_0 and then using the formula

$$c_i = \pi(k, m_i + c_{i-1}), \qquad 1 \leq i \leq l.$$

The ciphertext c is the sequence of set elements $c_0 c_1 c_2 \ldots c_l$.
    We decrypt a ciphertext $c = c_0 c_1 \ldots c_l$ using the formula

$$m_i = \pi^{-1}(k, c_i) - c_{i-1}, \qquad 1 \leq i \leq l.$$

Exercise 1.27. The above is an informal description of a block cipher used in *cipherblock chaining (CBC) mode*. See also Figure 1.3. Write down carefully what the three sets $\mathfrak{K}, \mathfrak{P}, \mathfrak{C}$ are, and implement the two algorithms $\mathcal{E}$ and $\mathcal{D}$. Show that $(\mathfrak{K}, \mathfrak{P}, \mathfrak{C}, \mathcal{E}, \mathcal{D})$ is a symmetric cryptosystem.

What happens when we encrypt is that we start at a random group element. This element is added to the first message block, which is then permuted, resulting in essentially a random-looking group element. This element is added to the second message block, which is again permuted, resulting in essentially a second random-looking group element. This process continues, producing a ciphertext that consists of a tuple of random-looking group elements.

Remark. The initial group element c_0 is often called an *initialisation vector*.

We will prove a precise variant of the following statement in Chapter 7.

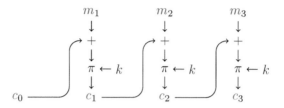

FIGURE 1.3 Cipherblock chaining (CBC) mode encryption diagram using the block cipher (π, π^{-1}). To get the decryption diagram, reverse the direction of the vertical arrows, replace $+$ by $-$ and π by π^{-1}.

Informally. *A secure block cipher used in CBC mode provides confidentiality against eavesdroppers.*

E *Exercise* 1.28. Consider the attacks discussed in previous sections. Explain why they fail against a secure block cipher used in CBC mode.

E *Exercise* 1.29. CBC mode is insecure if the initialisation vector c_0 is predictable. Suppose the group G is $\mathbb{Z}_{2^{128}}^+$. We play the following game.

1. You get an encryption $c^* = c_0^* c_1^*$ of a known plaintext $m^* \in \mathbb{Z}_{2^{128}}^+$.

2. You choose $m \in \mathbb{Z}_{2^{128}}^+$.

3. You are given an encryption c of $m + \Delta$, $\Delta \in \{0, 1\}$, where the initial random element c_0 was not chosen at random, but rather as c_1^*. That is, $c = c_0 c_1$ with $c_0 = c_1^*$.

Show how you can play this game to determine Δ by choosing m carefully.

1.2.7 Vigenère Cipher

Frequency analysis worked against the substitution cipher from Section 1.2.3. A different approach to preventing frequency analysis might be to encrypt different plaintext letters with different substitution ciphers.

The idea is that the frequencies produced by the encryption will be the average of the frequencies produced by different substitution ciphers, which should tend towards a uniform distribution, thus blocking frequency analysis.

E *Example* 1.10. Again, we let our alphabet be a group G, such as $\mathbb{Z}_{26}^+$.

The plaintext m is a sequence of elements $m_1 m_2 \ldots m_l$ from the group G. The key is a sequence of elements $k_1, k_2, k_3, \ldots, k_l$ from G. We encrypt the message elementwise with the formula

$$c_i = m_i + k_j, \text{ where } 1 \le i \le l, 1 \le j \le l \text{ and } j \equiv i \pmod{l}.$$

We decrypt a ciphertext $c = c_1 c_2 \ldots c_l$ using

$$m_i = c_i - k_j, \text{ where } 1 \leq i \leq l, 1 \leq j \leq l \text{ and } j \equiv i \pmod{l}.$$

E *Exercise* 1.30. The above is an informal description of the *Vigenère cipher*. Write down carefully what the three sets $\mathfrak{K}, \mathfrak{P}, \mathfrak{C}$ are, and implement the two algorithms $\mathcal{E}$ and $\mathcal{D}$. Show that $(\mathfrak{K}, \mathfrak{P}, \mathfrak{C}, \mathcal{E}, \mathcal{D})$ is a symmetric cryptosystem.

E *Example* 1.11. To encrypt BABOONSAREFUNNY using the Vigenère cipher with the key $k = $ JAPE, we do the computations:

B	A	B	O	O	N	S	A	R	E	F	U	N	N	Y
|	|	|	|	|	|	|	|	|	|	|	|	|	|	|
+J	+A	+P	+E	+J	+A	+P	+E	+J	+A	+P	+E	+J	+A	+P
↓	↓	↓	↓	↓	↓	↓	↓	↓	↓	↓	↓	↓	↓	↓
K	A	Q	S	X	N	H	E	A	E	U	Y	W	N	N

The ciphertext is KAQSXNHEAEUYWNN.

E *Exercise* 1.31. Choose a random key for the Vigenère cipher and encrypt a message. Let someone else decrypt it without knowing the key. Let them have some known plaintext.

Attack: Frequency Analysis Again   There is an easy attack against the Vigenère cipher if we have known plaintext. If we subtract the known plaintext from the corresponding ciphertext, we get the key repeated over and over. However, there are stronger attacks based on frequency analysis.

We begin with an English text and create a subsequence of letters by starting at the ith letter and then adding every lth letter. Such subsequences tend to have the same frequency distribution as the entire text.

E *Exercise* 1.32. In Exercise 1.13, you gathered a collection of English texts. Extract subsequences as above and compute the frequency distributions. Compare these distributions to the frequency distribution of the whole text, and estimate how long subsequences must be before we can expect to identify with reasonable certainty (a) E and (b) the five most frequent letters.

What happens to such subsequences when we encrypt the text with the Vigenère cipher using a key of length l? The letters in the subsequence are encrypted by adding the same letter from the key to it. In other words, the subsequence is encrypted using a shift cipher.

Earlier, we attacked the shift cipher by exhaustive search, but recognising subsequences of English text is more difficult than recognising English text. A better approach is to use frequency analysis. We know that E will likely be the most common letter in the plaintext subsequence, which corresponds to the most common letter in the ciphertext subsequence.

To recover a key of length l, all we have to do is run l frequency analysis attacks against the shift cipher (as in Exercise 1.9).

E | *Exercise* 1.33. Choose a random key for the Vigenère cipher and encrypt a sufficiently long message. Let someone else decrypt it without knowing the key. Let them know the key length.

One issue remains: We do not know the key length. The simplest approach is to try every possible length, beginning with $l = 1$. When we have the wrong key length, the attack will fail to produce a sensible decryption.

If everything has to be done by hand, there are faster ways to determine the key length. The oldest method relies on repetitions in the plaintext affecting the ciphertext. A newer method uses the so-called *index of coincidence*.

E | *Exercise* 1.34. Redo Exercise 1.33, but use the index of coincidence to determine the key length.

1.2.8   One-Time Pad

Suppose the key letter for the Vigenère cipher is sampled independently from the uniform distribution. Suppose also that the key is at least as long as the message and used for only one message. When used like this, the Vigenère cipher is known as the *one-time pad*.

E | *Example* 1.12. Again, our alphabet is a group G, such as $\mathbb{Z}_{26}^+$.

The single plaintext m is a sequence of l group elements $m_1 m_2 \dots m_l \in G^l$. The key k is a sequence of l group elements $k_1 k_2 \dots k_l \in G^l$. We encrypt the message elementwise using the formula

$$c_i = m_i + k_i, \text{ where } 1 \leq i \leq l.$$

The ciphertext c is the sequence of l group elements $c_1 c_2 \dots c_l$.

To decrypt a ciphertext $c = c_1 c_2 \dots c_l$, we compute the ith plaintext element using the formula

$$m_i = c_i - k_i, \text{ where } 1 \leq i \leq l.$$

E | *Exercise* 1.35. The above is an informal description of the *one-time pad*. Write down carefully what the three sets $\mathfrak{K}$, $\mathfrak{P}$, $\mathfrak{C}$ are, and implement the two algorithms $\mathcal{E}$ and $\mathcal{D}$. Show that $(\mathfrak{K}, \mathfrak{P}, \mathfrak{C}, \mathcal{E}, \mathcal{D})$ is a symmetric cryptosystem.

We shall prove that Eve's information about the message after she saw the ciphertext is the same as the information she had before she saw the ciphertext.

Eve has some information about Alice's plaintext, even before Alice sends the ciphertext. We model this information as a probability space M, which means that Eve in some sense assigns a probability for Alice's plaintext being any particular message m. We denote this probability by $\Pr[M = m]$.

Likewise, we model Eve's information about the ciphertext as a probability space C. Again, Eve assigns a probability to the likelihood of seeing any given ciphertext. We denote that probability by $\Pr[C = c]$. We can also consider, from Eve's point of view, the conditional probability of seeing a particular ciphertext, given a particular plaintext. We denote that conditional probability by $\Pr[C = c \mid M = m]$.

When Eve sees the ciphertext, there is no longer any uncertainty. Now, however, her information about the message may have changed. We denote the probability of Alice's message being a specific message conditioned on the observed ciphertext by $\Pr[M = m \mid C = c]$.

Theorem 1.1. *If the above scheme has been used to encrypt exactly one message, then*

$$\Pr[M = m_1 m_2 \dots m_l \mid C = c_1 c_2 \dots c_l] = \Pr[M = m_1 m_2 \dots m_l].$$

Proof. The assumption on the key means that from Eve's point of view before seeing the ciphertext, the key letters are uniformly and independently distributed, which is expressed as the statement

$$\Pr[K = k_1 k_2 \dots k_l] = |G|^{-l}.$$

Again, K is a random variable describing Eve's knowledge about the key.

We first compute the probabilities

$$\Pr[C = c_1 c_2 \dots c_l \mid M = m_1 m_2 \dots m_l]$$
$$= \Pr[K = (c_1 - m_1)(c_2 - m_2) \dots (c_l - m_l)] = |G|^{-l}$$

and

$$\Pr[C = c] = \sum_m \Pr[C = c \mid M = m]\Pr[M = m]$$
$$= |G|^{-l} \sum_m \Pr[M = m] = |G|^{-l}.$$

Then we compute the a posteriori probability

$$\Pr[M = m \mid C = c] = \frac{\Pr[M = m \wedge C = c]}{\Pr[C = c]} = \frac{\Pr[C = c \mid M = m]\Pr[M = m]}{\Pr[C = c]}$$
$$= \Pr[M = m],$$

which completes the proof. $\qquad \square$

The a posteriori probability is the same as the a priori probability, which means that Eve has learned nothing new by observing the ciphertext. We have proved that the one-time pad provides confidentiality against eavesdroppers.

Informally. *If the key is used to encrypt only once, the one-time pad provides confidentiality against eavesdroppers.*

Note that this claim is *unconditional*, unlike the corresponding claim for CBC mode in Section 1.2.6, which is conditional on the use of a secure block cipher. This is good. Cryptosystems with the property implied by Theorem 1.1 are often called *perfect*. Unfortunately, we must note that the one-time pad is impractical in almost every application.

E| *Exercise* 1.36. The one-time pad is not secure if the key is used more than once. Choose a random key of sufficient length for the one-time pad and encrypt two different messages using the same key. Let someone else decrypt them without knowing the key. To ease decryption, the messages should consist of mostly long words.

1.2.9 Stream Ciphers

Stream ciphers cover a wide range of design approaches, but we shall consider only the notion of *synchronious* or *additive* stream ciphers. The idea is that a *key stream generator* expands a key and an *initialisation vector* into something that looks like a one-time pad key, which is then used to encrypt the message.

D| **Definition 1.4.** A *key stream generator* is a function $f : \mathfrak{K} \times \mathfrak{V} \to G^N$.

Informally. A key stream generator is *secure* if it is hard to distinguish the function values $f(k, iv_1), f(k, iv_2), \ldots, f(k, iv_n)$ from random values when k has been chosen uniformly at random from $\mathfrak{K}$ and the values $iv_1, iv_2, \ldots, iv_n$ have been chosen uniformly at random from $\mathfrak{V}$.

E| *Example* 1.13. Again, our alphabet is a group G, such as $\mathbb{Z}_{26}^+$.
The plaintext m is a sequence of group elements $m_1 m_2 \ldots m_l$ of length $l \leq N$. The key k is an element in $\mathfrak{K}$. We encrypt the message by first choosing iv uniformly at random from $\mathfrak{V}$, then computing the l first elements $z_1 z_2 \ldots z_l$ of $f(k, iv) = z_1 z_2 \ldots z_N$. We encrypt the plaintext using

$$w_i = m_i + z_i, \text{ where } 1 \leq i \leq l.$$

The ciphertext c is the pair $(iv, w_1 w_2 \ldots w_l)$.
To decrypt a ciphertext $c = (iv, w_1 w_2 \ldots w_l)$, we first compute the l first elements $z_1 z_2 \ldots z_l$ of $f(k, iv)$ and then use

$$m_i = w_i - z_i, \text{ where } 1 \leq i \leq l.$$

Remark. We shall denote the first l elements of $f(k, iv)$ by $f(k, iv, l)$.

E *Exercise* 1.37. The above is an informal description of a *stream cipher*. Write down carefully what the three sets $\mathfrak{K}, \mathfrak{P}, \mathfrak{C}$ are, and implement the two algorithms $\mathcal{E}$ and $\mathcal{D}$. Show that $(\mathfrak{K}, \mathfrak{P}, \mathfrak{C}, \mathcal{E}, \mathcal{D})$ is a symmetric cryptosystem.

Informally. *A stream cipher using a secure key stream generator provides confidentiality against eavesdroppers.*

Key Stream Generators from Block Ciphers   Let $\pi, \pi^{-1} : \mathfrak{K} \times G \to G$ be a block cipher. Suppose the set $\mathfrak{V} \times \{1, 2, \ldots, N\}$ is a subset of the set of group elements of G. We shall use this block cipher to construct two key stream generators, $\mathrm{OFB}_\pi : \mathfrak{K} \times G \to G^N$ and $\mathrm{CTR}_\pi : \mathfrak{K} \times \mathfrak{V} \to G^N$.

E *Example* 1.14. For any $iv \in G$ and $k \in \mathfrak{K}$, the *output feedback (OFB) mode* is $\mathrm{OFB}_\pi(k, iv) = z_1 z_2 \ldots z_N$, with

$$z_1 = \pi(k, iv) \text{ and } z_i = \pi(k, z_{i-1}), \text{ where } 2 \leq i \leq N.$$

E *Example* 1.15. For any $iv \in \mathfrak{V}$ and $k \in \mathfrak{K}$, the *counter (CTR) mode* is $\mathrm{CTR}_\pi(k, iv) = z_1 z_2 \ldots z_N$, with

$$z_i = \pi(k, (iv, i)), \text{ where } 1 \leq i \leq N.$$

We shall discuss this statement further in Chapter 7.

Informally. *Output feedback mode* OFB_π *and counter mode* CTR_π *using a secure block cipher are secure key stream generators.*

Note that counter mode is very easy to parallelise and can therefore be made very fast. Output feedback mode is inherently unparallelisable.

1.3   INTEGRITY

In this section, we shall consider the situation where Eve controls the communications channel between Alice and Bob. Eve's goal is to tamper with the messages sent by Alice so that Bob receives a different message, *without noticing*. For the moment, we shall not care about confidentiality.

The main tool we shall use is the *message authentication code* (MAC), where Alice adds an *authentication tag* to her message that allows Bob the verify that the message is unchanged.

D **Definition 1.5.** *A simple message authentication code* (MAC) *is a function* $\mu : \mathfrak{K} \times \mathfrak{P} \to \mathfrak{T}$.

Informally. *A simple message authentication code is* secure *if it is hard to guess the function value* $\mu(k, m)$ *for any* m, *even after seeing the values*

$\mu(k, m_1), \mu(k, m_2), \ldots, \mu(k, m_n)$ for any $m_1, m_2, \ldots, m_n \in \mathfrak{P}$. Here, k has been chosen uniformly at random from $\mathfrak{K}$, and $m \neq m_i$ for $i = 1, 2, \ldots, n$.

When Alice wants to send a message m to Bob, she sends the pair $(m, \mu(k, m))$. When Bob receives the pair (m', t), he checks that $\mu(k, m') = t$. If so, he accepts that m came from Alice. Otherwise, he discards the message.

Informally. *A secure simple message authentication code used as above provides integrity.*

Note that nothing prevents Eve from *replaying* old messages by sending them to Bob. We shall discuss such attacks further in Section 7.4.

One of the most popular MAC constructions – HMAC – is based on so-called *hash functions*, which we return to in Sections 4.2 and 7.5. We shall discuss two constructions of message authentication codes.

1.3.1 Polynomial Evaluation MACs

We begin our discussion with a *one-time* polynomial-evaluation MAC. This MAC is insecure if it is used on more than one message. Such a MAC is impractical, so we shall also discuss how a block cipher can be used to make a more practical variant.

Example 1.16. Our alphabet is a prime finite field $\mathbb{F}$ with p elements.

The plaintext m is a sequence of field elements $m_1 m_2 \ldots m_l$. The key (k_1, k_2) is a pair of field elements. The function OTPE is computed as

$$\text{OTPE}(k_1, k_2, m) = k_2 + k_1^{l+1} + \sum_{i=1}^{l} m_i k_1^i. \tag{1.5}$$

Exercise 1.38. The above is an informal description of a *polynomial evaluation MAC*. Write down carefully what the three sets $\mathfrak{K}$, $\mathfrak{P}$, $\mathfrak{T}$ are, and implement an algorithm computing the function OTPE. Show that OTPE is a message authentication code.

We shall first prove the following statement.

Theorem 1.2. *Let m, m' be two messages of length at most l. Let $t, t' \in \mathbb{F}$. The probability that $\text{OTPE}(k_1, k_2, m') = t'$ given that $\text{OTPE}(k_1, k_2, m) = t$ is at most $(l+1)/p$.*

Proof. For simplicity, we shall assume that both messages have length exactly l. We want to compute the probability

$$\Pr[\text{OTPE}(k_1, k_2, m') = t' \mid \text{OTPE}(k_1, k_2, m) = t].$$

We can do that by computing how many pairs (k_1, k_2) satisfy both

$$t = k_2 + k_1^{l+1} + \sum_{i=1}^{l} m_i k_1^i \text{ and} \tag{1.6}$$

$$t' = k_2 + k_1^{l+1} + \sum_{i=1}^{l} m_i' k_1^i. \tag{1.7}$$

For every value of k_1, there is exactly one value of k_2 that satisfies (1.6), so there are exactly p pairs that we need to consider.

Combining the two equations, we get that k_1 must satisfy the equation

$$t' - t = \sum_{i=1}^{l} (m_i' - m_i) k_1^i.$$

A solution to this equation is a zero of a polynomial equation of degree at most l, which means that there are at most l solutions.

Out of p possible keys (k_1, k_2) satisfying (1.6), there are at most l pairs that also satisfy (1.7). A bound on the probability follows.

When the messages have different lengths, the exact same argument applies, but the polynomial we consider has degree at most $l + 1$. □

Note that this means that when Alice sends a single message m to Bob with the tag $t = \text{OTPE}(k_1, k_2, m)$, and Bob receives the message $m' \neq m$ and the tag t', the probability that Bob accepts that message as coming from Alice is at most $(l + 1)/p$. This proves the following claim.

Informally. *If the key is used to create a MAC tag only once, the one-time polynomial evaluation MAC is a secure message authentication code.*

Exercise 1.39. Show that the scheme is not secure if we

(a) replace k_1^i with k_1^{i-1} in the sum in (1.5);

(b) remove the term k_2 from (1.5); or

(c) remove the term k_1^{l+1} from (1.5).

The above construction is just a one-time MAC, which is usually impractical. Looking at the construction, we can see that in some sense, it is just a polynomial evaluated at a secret encrypted with the shift cipher. Perhaps using better encryption could work? Indeed, a block cipher will do.

Example 1.17. Let $\pi, \pi^{-1} : \mathcal{K}' \times S \to S$ be a block cipher with $\mathbb{F} \subseteq S$. We construct an alternative polynomial evaluation MAC $\text{PE}_\pi : \mathbb{F} \times \mathcal{K}' \times \cup_l \mathbb{F}^l \to S$

that can be used more than once, using

$$\text{PE}_\pi(k_1, k_2, m) = \pi(k_2, k_1^{l+1} + \sum_{i=1}^{l} m_i k_1^i).$$

E *Exercise* 1.40. The above is an informal description. Write down carefully what the three sets $\mathfrak{K}$, $\mathfrak{P}$, $\mathfrak{T}$ are, and implement an algorithm computing the function PE_π. Show that PE_π is a message authentication code.

1.3.2 Block-Cipher-Based MACs

One simple MAC based on a block cipher $\pi, \pi^{-1} : \mathfrak{K} \times G \to G$ is *CBC-MAC*, which is somewhat similar to cipherblock chaining mode from Section 1.2.6.

E *Example* 1.18. Fix some element $t_0 \in G$. The plaintext m is a sequence of group elements $m_1 m_2 \dots m_l$. The key k is an element of $\mathfrak{K}$. Let

$$t_i = \pi(k, t_{i-1} + m_i), \text{ where } 1 \le i \le l.$$

Then $\text{CBC-MAC}(k, m) = t_l$.

E *Exercise* 1.41. The above is an informal description. Write down carefully what the three sets $\mathfrak{K}$, $\mathfrak{P}$, $\mathfrak{T}$ are, and implement an algorithm computing the function CBC-MAC. Show that CBC-MAC is a message authentication code.

This MAC is not secure if messages of different lengths are allowed.

E *Exercise* 1.42. Find an attack against this MAC when messages of different lengths are allowed.

There are many secure variants of this MAC. One variant is based on a block cipher $\pi, \pi^{-1} : \mathfrak{K} \times G \to G$ over a group G. The idea is to modify the final plaintext block with an unpredictable value.

The plaintext m is a sequence of group elements $m_1 m_2 \dots m_l$. The key k is an element of $\mathfrak{K}$. Let $t_0 = 0$, $h = \pi(k, 0)$ and

$$t_i = \pi(k, t_{i-1} + m_i), \text{ where } 1 \le i \le l - 1.$$

Then $\text{CBC-MAC}'(k, m) = \pi(k, h + t_{l-1} + m_l)$.

Informally. *With a secure block cipher, $\text{CBC-MAC}'$ is a secure MAC.*

1.4 CONFIDENTIALITY AND INTEGRITY

Finally, we consider the situation where Eve controls the communications channel between Alice and Bob. Eve's goal is both to read Alice's messages to Bob and tamper with them.

We shall combine the secure cryptosystems we saw in Section 1.2 with the message authentication codes we saw in Section 1.3 into cryptosystems that provide both confidentiality and integrity.

Informally. A symmetric cryptosystem provides *integrity* if it is hard to create a ciphertext that decrypts to anything other than $\perp$ without knowledge of the key.

First, we consider a general principle in cryptography (part of *crypto hygiene*): *Never use the same key for two different things.* This means that we should use different keys for the cryptosystem and the MAC when we combine them. That leaves us with three obvious ways of combining a cryptosystem with a MAC: Encrypt the message, then authenticate the ciphertext; or authenticate the message, then encrypt the message and the tag; or encrypt the message and authenticate the message separately.

- Encrypt-then-MAC: $w = \mathcal{E}(k_e, m)$; $t = \mu(k_m, w)$; $c = (w, t)$.
- MAC-then-encrypt: $t = \mu(k_m, m)$; $c = \mathcal{E}(k_e, (m, t))$.
- Encrypt-and-MAC: $w = \mathcal{E}(k_e, m)$; $t = \mu(k_m, m)$; $c = (w, t)$.

In all three cases, the decryption algorithm verifies the MAC, and if the verification fails, the output of the decryption algorithm is $\perp$.

Encrypt-and-MAC is, in general, insecure. If you encrypt the same message twice, you will always get the same tag, something that Eve will notice. Therefore, it fails confidentiality.

MAC-then-encrypt is often secure, but there are special cases where it is not secure. In particular, such schemes will often fail when combined with padding schemes.

Encrypt-then-MAC is always secure, and therefore the best choice.

Informally. *A cryptosystem that provides confidentiality against eavesdroppers and a secure message authentication code combined using Encrypt-then-MAC provides both confidentiality and integrity.*

Remark. Practice has shown that Encrypt-then-MAC seems to be hard to do right. In practice, there is also a need to protect the integrity of more than the encrypted message, so-called *associated data*, which is actually non-trivial to get right. Modern practice has therefore moved towards newer security notions. We discuss this in Chapter 7.

Note that still nothing prevents Eve from *replaying* old ciphertexts by sending them to Bob.

1.5   THE KEY DISTRIBUTION PROBLEM

In many situations, it is easy for Alice and Bob to agree on a shared secret, simply by meeting in a secure location and generating a secret by flipping coins or throwing dice. But Alice does not just want to talk to Bob, she wants to talk to many people. Establishing shared secrets suddenly becomes more difficult, and so is managing the shared secrets. The problem becomes even more difficult. Alice has a habit of finding new people to talk to all the time.

It is also worthwhile recalling that even though we use Alice and Bob consistently, people almost never use cryptography. It is devices that use cryptography, and how do those devices establish shared secrets?

One possible approach is to use a trusted third party. If everyone has a shared secret with a trusted third party, Alice and Bob can request a shared secret from the trusted third party, who will then send them a shared key, encrypted under their respective shared secrets.

Such an approach is feasible in some situations, but the trusted third party would be a major liability if it ever turned out not to be trusted. Also, such an approach is not very flexible. Which means that we need to search for a better approach.

CHAPTER 2

Key Exchange and Diffie-Hellman

In the previous chapter, we discussed how Alice and Bob could use a shared secret to protect their messages. Establishing a shared secret is a prerequisite for using the theory of symmetric cryptography to communicate securely over insecure channels. Traditionally, this was done by meeting in person, using couriers or relying on trusted third parties. But as networks grow and the number of connections increases, the traditional approaches become impractical or introduce unpleasant trust assumptions.

Therefore, Alice and Bob want to establish a shared secret by communicating via their usual communications channel, even though Eve has access to the channel and may eavesdrop on anything sent over the channel. This process is known as *key exchange*.

The Diffie-Hellman protocol is our first example of a key exchange protocol. Conjecturally, if the underlying mathematical structure is carefully chosen, running the protocol requires Alice and Bob to do relatively little work to establish a shared value, but the eavesdropper Eve will have to do infeasibly much work to deduce the shared value. In other words, the shared value will remain a secret known (fully) only to Alice and Bob.

This chapter is an introduction to this protocol and the study of its security, through its mathematical foundations, namely finite cyclic groups. Not every finite cyclic group is suitable for use in the Diffie-Hellman protocol. By studying the problem of breaking attacking the Diffie-Hellman protocol, we shall find necessary requirements for a cyclic group structure to be suitable. We shall arrive at two plausible families of cyclic groups based on finite fields and elliptic curves. To find suitable group structures, we will also need new mathematical tools to distinguish prime numbers from composite numbers.

2.1   THE DIFFIE-HELLMAN PROTOCOL

The *Diffie-Hellman protocol* is a *key exchange protocol* that allows Alice and Bob to establish a shared value. Alice sends the first message in the protocol and is the *initiator*. Bob is the *responder* since he responds to Alice's message. We want the shared value to be *secret*, that is, the eavesdropper Eve should not know the shared value.

Informally. A *key exchange* protocol is a two-party protocol between an *initiator* and a *responder*. The players eventually output either a *shared secret* or *session key* or ⊥.

Informally. A key exchange protocol is *secure* if someone who sees the protocol messages but does not actually participate in the protocol, cannot learn the agreed-upon shared secret.

The protocol is based on a finite cyclic group, and the study of its security turns out to involve the study of an interesting computational problem in finite cyclic groups, computing so-called *discrete logarithms*.

Before we can describe the Diffie-Hellman protocol, we must establish the underlying abstract mathematical structure. We begin with a bit of notation.

Exponentiation in a group is defined as

$$x^a = \underbrace{x \cdot x \cdots x}_{a \text{ terms}}$$

for any group element x and any integer $a > 0$. Then we define x^0 to be the identity element and $x^a = (x^{-1})^{-a}$ when $a < 0$. If we use additive notation for the group, we write $a \cdot x$.

Example 2.1. Let G be a finite cyclic group of order n, and let g be a generator. The Diffie-Hellman protocol works as follows (see also Figure 2.1).

The group G, the group order n and the generator g are public and fixed.

- Alice samples a from the uniform distribution on $\{0, 1, 2, \ldots, n-1\}$. She computes $x = g^a$ and sends x to Bob.
- Bob receives x. He samples b from the uniform distribution on $\{0, 1, 2, \ldots, n-1\}$, computes $y = g^b$ and $z_B = x^b$, and sends y to Alice.
- Alice receives y. Alice computes $z_A = y^a$.

We shall use the notation $s \xleftarrow{r} S$ to denote that s is sampled from the uniform distribution on the set S.

Remark. The structures $\mathbb{Z}_n$ and $\mathbb{F}_p$, which we will use for many examples, are typically constructed as factor rings. As such, their elements are really cosets and should be denoted as $k + \langle n \rangle$ or $k + \langle p \rangle$. In principle, any integer in the coset can be used to represent the coset.

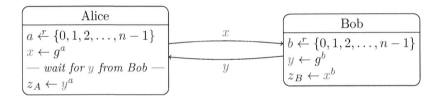

FIGURE 2.1 The Diffie-Hellman protocol for a group G of order n with generator g.

In the interest of brevity, we will simply write k instead of $k + \langle n \rangle$ and $k + \langle p \rangle$ whenever the meaning is clear. So in $\mathbb{Z}_{13}^+$ we will write $3 \cdot 10 = 4$, instead of the cumbersome $(3 + \langle 13 \rangle)(10 + \langle 13 \rangle) = (4 + \langle 13 \rangle)$.

When Alice and Bob transmit group elements, it is usually extremely important that they use some *canonical representation*.

Example 2.2. Alice and Bob use Diffie-Hellman with the group $G = \mathbb{Z}_{13}^+$ and generator $g = 2$. Note the additive notation.

Alice chooses $a = 3$ and computes $x = a \cdot g = 3 \cdot 2 = 6$. She sends 6 to Bob.

Bob receives 6. He chooses $b = 5$ and computes $y = b \cdot g = 5 \cdot 2 = 10$ and $z_B = b \cdot x = 5 \cdot 6 = 4$. He sends 10 to Alice.

Alice receives 10 from Bob. She computes $z_A = a \cdot y = 3 \cdot 10 = 4$.

Remark. The exponentiation coincides with ring multiplication for $\mathbb{Z}_n^+$. We shall later see that $\mathbb{Z}_n^+$ cannot, therefore, be used safely with Diffie-Hellman.

Example 2.3. Alice and Bob use Diffie-Hellman with the group $G = \mathbb{F}_{13}^*$ and generator $g = 2$.

Alice chooses $a = 3$ and computes $x = g^a = 2^3 = 8$. She sends 8 to Bob.

Bob receives 8. He chooses $b = 5$ and computes $y = g^b = 2^5 = 6$ and $z_B = x^b = 8^5 = 8$. He sends 6 to Alice.

Alice receives 6 from Bob. She computes $z_A = y^a = 6^3 = 8$.

Exercise 2.1. Agree on a cyclic group and a generator with someone else, then use the Diffie-Hellman protocol to agree on a secret.

The protocol is *complete*, in the sense that Alice and Bob arrive at a shared value. Because $z_A = y^a = (g^b)^a = (g^a)^b = x^b = z_B$, the Diffie-Hellman protocol establishes a single shared value, which we shall denote by z.

Next, we must consider how much work Alice and Bob must do to execute the protocol. The only non-trivial computations involved are the two *exponentiations* done by each of them. Alice and Bob can do their two exponentiations according to the definition using less than $2n$ group operations:

$$2^5 = 2 \cdot 2^4 = 4 \cdot 2^3 = 8 \cdot 2^2 = 16 \cdot 2 = 32.$$

However, there is a better way to do these computations.

Proposition 2.1. *For any $x \in G$ and integer $a > 0$, the group element x^a can be computed using at most $2 \log_2 a$ group operations.*

Proof. Since $x^{2^{i+1}} = (x^{2^i})^2$, we can compute the $l + 1$ group elements $x^{2^0}, x^{2^1}, x^{2^2}, \ldots, x^{2^l}$ using l group operations.

Writing a in binary we get

$$a = \sum_{i=0}^{l} a_i 2^i, \qquad a_i \in \{0, 1\} \qquad \text{and} \qquad \prod_{i=0}^{l} (x^{2^i})^{a_i} = x^{\sum_{i=0}^{l} a_i 2^i} = x^a.$$

Computing this product therefore requires at most l group operations.

We have done at most $2l$ group operations. If $a > 0$, we may assume that $a_l = 1$, and then we have that $l \leq \log_2 a$. The claim follows. $\square$

Example 2.4. We can compute 2^{13} in $\mathbb{Z}$ by observing that $13 = 2^0 + 2^2 + 2^3$, computing

$$2^{2^0} = 2 \qquad\qquad\qquad 2^{2^1} = (2^{2^0})^2 = 2^2 = 4$$
$$2^{2^2} = (2^{2^1})^2 = 4^2 = 16 \qquad\qquad 2^{2^3} = (2^{2^2})^2 = 16^2 = 256$$

and
$$2^{13} = 2^{2^0 + 2^2 + 2^3} = 2^{2^0} 2^{2^2} 2^{2^3} = 2 \cdot 16 \cdot 256 = 32 \cdot 256 = 8192.$$

This required only 5 multiplications, while using the definition would require 12 multiplications.

Example 2.5. We can compute 2^{257} in $\mathbb{F}_{19}^*$ by computing successively $2^2 = 4$, $2^{2^2} = 4^2 = 16$, $2^{2^3} = 16^2 = 9$, $2^{2^4} = 9^2 = 5$, $2^{2^5} = 5^2 = 6$, $2^{2^6} = 6^2 = 17$, $2^{2^7} = 17^2 = 15$ and $2^{2^8} = 15^2 = 16$, and finally

$$2^{257} = 2 \cdot 16 = 13.$$

Note that by choosing small coset representatives for the intermediate results, we avoid computing with large integers.

Exercise 2.2. The proof of Proposition 2.1 essentially describes an algorithm for computing x^a. Implement this algorithm and restate Proposition 2.1 as a statement about the algorithm's time complexity in terms of group operations, ignoring any other form of computation involved.

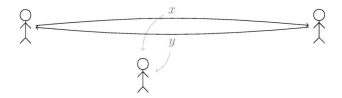

FIGURE 2.2 Eve is listening and learns x and y, but cannot see Alice' and Bob's secret values or computations. She already knows G, g and n.

E *Exercise* 2.3. Use the algorithm from Exercise 2.2 to compute (by hand) 5^{582} in the cyclic group $\mathbb{F}^*_{10007}$.

E *Exercise* 2.4. Note that we can define x^a for $a \geq 0$ using the rules that $x^0 = 1$ and

$$x^a = \begin{cases} (x^{a/2})^2 & \text{when } a \text{ is even, and} \\ (x^{(a-1)/2})^2 x & \text{when } a \text{ is odd.} \end{cases}$$

Use this fact to give an alternative proof of Proposition 2.1 leading to an alternative algorithm.

We have shown that as long as group operations can be computed in a reasonable time, Alice and Bob can execute the Diffie-Hellman protocol in a reasonable time, even when the group order is very large.

2.2 DISCRETE LOGARITHMS

Our goal is for Alice and Bob to establish a shared *secret*. We must consider what the eavesdropper Eve can do to learn the established shared value.

The Diffie-Hellman protocol should work even if Alice and Bob have had no previous communication, which is why the group G, the group order n and the generator g are public, and therefore known by Eve. When Alice and Bob run the protocol, Eve additionally learns x and y. She wants to know z.

D **Definition 2.1.** The *Diffie-Hellman problem* in a cyclic group G of order n with generator g is to find $z = g^{ab}$ given $x = g^a$ and $y = g^b$, when a and b have been sampled independently from the uniform distribution on $\{0, 1, \ldots, n-1\}$.

Solving the Diffie-Hellman problem is more or less the same as attacking the Diffie-Hellman key exchange protocol, as seen in Figure 2.2.

We must choose the cyclic group that Alice and Bob shall use. There are two considerations. Alice and Bob want to spend as little effort as possible to establish the shared secret, both with respect to computation and communication. This means that computing the group operation should be reasonably fast, and group elements should have a reasonably compact representation.

At the same time, Alice and Bob want the established shared value to be secret, in the sense that recovering it must require too much computational effort for Eve. This is true if and only if solving the Diffie-Hellman problem is computationally expensive.

If Eve can find a or b, then she can trivially compute $z = y^a = x^b$.

D | **Definition 2.2.** Let G be a finite cyclic group. The *discrete logarithm* of x to the base g is the smallest non-negative integer a such that $x = g^a$. We write $\log_g x = a$. The *discrete logarithm problem* in G is to find $\log_g x$ when x has been sampled from the uniform distribution on G.

It is generally believed, and there is evidence to suggest that this is the case, that if Eve can compute the shared value, she must be able to compute discrete logarithms as well.

Informally. *It is conjectured that solving the Diffie-Hellman problem in a group G is not (much) easier than computing discrete logarithms in G.*

Under this conjecture, the study of the security of the Diffie-Hellman protocol reduces to the study of how easy it is to compute discrete logarithms in the group we want to use. Our first result is that which generator we choose is not important for security.

E | *Exercise 2.5.* Let x be a group element of order m and a, b be integers. Prove that $x^a = x^b$ if and only if $a \equiv b \pmod{m}$.

T | **Proposition 2.2.** *Let g_1 and g_2 be generators, and suppose $\log_{g_1} g_2 = a$. Then $\log_{g_2} g_1 \equiv a^{-1} \pmod{n}$.*

Proof. It is given that $g_2 = g_1^a$, and since g_2 is a generator, there exists an integer b such that $g_1 = g_2^b$. We get that $g_1 = g_1^{ab}$. Exercise 2.5 then says that

$$ab \equiv 1 \pmod{n},$$

and our claim follows. □

E | *Exercise 2.6.* Show that if you can compute discrete logarithm with some generator g as base, then you can compute discrete logarithms with any generator as base, at roughly twice the cost.

E | *Exercise 2.7.* Any cyclic group G of order n is isomorphic to $\mathbb{Z}_n^+$. Let $\lambda : G \to \mathbb{Z}_n^+$ be a group isomorphism taking a generator g to $1 + \langle n \rangle$. Explain why computing discrete logarithms to the base g is essentially the same as computing the group isomorphism λ.

We now begin by establishing a level of effort that will certainly be sufficient to compute discrete logarithms, namely exhaustive search. This will lead to our first requirement for a group to be suitable for Diffie-Hellman.

T **Proposition 2.3.** *Let G be a cyclic group of order n. The discrete logarithm of a group element $x \in G$ can be computed using less than n group operations.*

Proof. Let g be a generator. Since $g^a g = g^{a+1}$, we can compute the elements of the sequence $g^0, g^1, g^2, \ldots, g^{n-1}$ using $n-1$ group operations. We can keep track of the discrete logarithm of each element as we compute it, by counting how many group operations we have done. Since g is a generator, x must an element in the sequence, and because of canonical representations we can recognise it when we reach it. The claim follows. □

The proof of the proposition implies an algorithm for computing discrete logarithms. In addition to group operations, it also compares group elements and counts how many group operations it has done. The group operations will dominate the computational effort. Therefore, we focus on the number of group operations when we measure the time complexity of the computation. This will be the case for all the algorithms in this section.

E *Exercise* 2.8. The proof of Proposition 2.3 essentially describes an algorithm for computing $\log_g x$. Implement this algorithm and restate Proposition 2.3 as a statement about the algorithm's time complexity, in terms of group operations, ignoring everything else.

E *Example* 2.6. Consider the group $G = \mathbb{F}_{13}^*$ with generator $g = 2$. We want to compute the discrete logarithm of $x = 6$.
    We compute $2^0 = 1$, $2^1 = 2$, $2^2 = 2 \cdot 2 = 4$, $2^3 = 4 \cdot 2 = 8$, $2^4 = 8 \cdot 2 = 3$ and $2^5 = 3 \cdot 2 = 6$. We find that $\log_2 6 = 5$.

E *Exercise* 2.9. Use the algorithm from Exercise 2.8 to compute (by hand) the discrete logarithm of 3972 to the base 5 in the group $\mathbb{F}_{10007}^*$.

The structure of our study of the discrete logarithm computation is to study various ways of solving or simplifying the computation. Every time we improve our ability to compute discrete logarithms in various groups, we better understand what kind of group Alice and Bob can use for Diffie-Hellman.
    Based on Proposition 2.3, we arrive at the following requirement.

Requirement 2.1. If n is the group order, n group operations must be an infeasible computation.

2.2.1 An Unsuitable Group

It is easy to find cyclic groups with large group order. If n is any large number, then $\mathbb{Z}_n^+$ is a cyclic group of order n. A natural generator is $1 + \langle n \rangle$, but the coset of any integer relatively prime to n will do. Since the group is written additively, the Diffie-Hellman protocol looks like:

- Alice samples $a \xleftarrow{r} \{0, 1, \ldots, n-1\}$, computes $x = ag$ and sends x.

- Bob receives x. He samples $b \xleftarrow{r} \{0, 1, \ldots, n-1\}$, computes $y = bg$ and $z_B = bx$, and sends y.
- Alice receives y. Alice computes $z_A = a \cdot y$.

We see that x is essentially a, so this group is unsuitable for Diffie-Hellman.

E *Exercise 2.10.* Show that the *Extended Euclidean algorithm* can be used to compute multiplicative inverses modulo n in time proportional to $\log_2 n$ integer divisions, multiplications and additions. Implement this algorithm.

E *Exercise 2.11.* Alice's x is essentially equal to a since we used $1 + \langle n \rangle$ as a generator. Use Proposition 2.2 and the previous exercise to show that any other generator would be equally insecure.

E *Exercise 2.12.* In Example 2.2, Alice and Bob use the group $G = \mathbb{Z}_{13}^+$ and generator $g = 2$. Alice sends $x = 6$ to Bob, and Bob replies with $y = 10$. Use only this information, together with the result from the previous exercise, to compute Alice's secret value a, Bob's secret value b and the shared secret z.

We see that a large group is necessary, but not sufficient for our purposes.

2.2.2 Pohlig-Hellman I

There are many ways to describe and analyse the first part of the Pohlig-Hellman algorithm for computing discrete logarithms. We shall rely on the algebraic structure of cyclic groups, and begin with the observation that computing a discrete logarithm in G is the same as computing the isomorphism from G to $\mathbb{Z}_n^+$ taking g to $1 + \langle n \rangle$. If G has a suitable group structure, we can use that to simplify the computation of the isomorphism.

Suppose for now that $n = n_1 n_2$ with $\gcd(n_1, n_2) = 1$. Define the two sets $H_1 = \{x^{n_2} \mid x \in G\}$ and $H_2 = \{x^{n_1} \mid x \in G\}$, and the maps $\pi_1 : G \to H_1$, $\pi_2 : G \to H_2$ and $\pi : G \to H_1 \times H_2$ by

$$\pi_1(x) = x^{n_2}, \quad \pi_2(x) = x^{n_1} \quad \text{and} \quad \pi(x) = (\pi_1(x), \pi_2(x)).$$

E *Exercise 2.13.* Prove that the sets H_1, H_2 are subgroups of G of order n_1, n_2, respectively.

E *Exercise 2.14.* Prove that the maps π_1, π_2 are well-defined, that they are group homomorphisms and that they are surjective.

E *Exercise 2.15.* Prove that the map π is a group isomorphism by giving an inverse homomorphism.

It is interesting to compare the above results with the situation for $\mathbb{Z}_n^+$. We begin by stating a version of the Chinese remainder theorem without proof.

T **Fact 2.4** (Chinese Remainder Theorem). *Let $n = n_1 n_2$ with $\gcd(n_1, n_2) = 1$.* *Then as rings*

$$\mathbb{Z}_n \simeq \mathbb{Z}_{n_1} \times \mathbb{Z}_{n_2},$$

and the unique ring isomorphism and its inverse are easy to compute.

The map $\mathbb{Z}_n^+ \to \mathbb{Z}_{n_1}^+ \times \mathbb{Z}_{n_2}^+$ corresponding to π above is given by $k + \langle n \rangle \mapsto (kn_2 + \langle n_1 \rangle, kn_1 + \langle n_2 \rangle)$. This map is a group isomorphism, but not a ring isomorphism, unlike the Chinese remainder theorem ring isomorphism, which takes $k + \langle n \rangle$ to $(k + \langle n_1 \rangle, k + \langle n_2 \rangle)$.

Let $\lambda : G \to \mathbb{Z}_n^+$, $\lambda_1 : H_1 \to \mathbb{Z}_{n_1}^+$ and $\lambda_2 : H_2 \to \mathbb{Z}_{n_2}^+$ be the group isomorphisms satisfying $\lambda(g) = 1 + \langle n \rangle$, $\lambda_1(\pi_1(g)) = 1 + \langle n_1 \rangle$ and $\lambda_2(\pi_2(g)) = 1 + \langle (n_2) \rangle$. Note that λ_1 and λ_2 are carefully chosen, and correspond to discrete logarithms to the bases $\pi_1(g)$ and $\pi_2(g)$.

T **Proposition 2.5.** *Denote by $\lambda_1 \times \lambda_2$ the group isomorphism taking (x_1, x_2) to $(\lambda_1(x_1), \lambda_2(x_2))$. We get a commutative diagram:*

$$
\begin{array}{ccc}
G & \xrightarrow{\ \ \lambda\ \ } & \mathbb{Z}_n^+ \\
{\scriptstyle \pi}\downarrow & & \uparrow{\scriptstyle CRT} \\
H_1 \times H_2 & \xrightarrow{\ \lambda_1 \times \lambda_2\ } & \mathbb{Z}_{n_1} \times \mathbb{Z}_{n_2}
\end{array}
$$

Proof. It is sufficient to consider what happens to a group generator. By definition, we have that $\lambda_1(\pi_1(g)) = 1 + \langle n_1 \rangle$ and $\lambda_2(\pi_2(g)) = 1 + \langle n_2 \rangle$. Since $\gcd(n_1, n_2) = 1$, the Chinese remainder theorem says that $\mathbb{Z}_{n_1}^+ \times \mathbb{Z}_{n_2}^+$ is isomorphic to $\mathbb{Z}_n^+$. Moreover, the ring isomorphism given by the Chinese remainder theorem takes the ring identity $(1 + \langle n_1 \rangle, 1 + \langle n_2 \rangle)$ to the ring identity $1 + \langle n \rangle$, which concludes the proof. □

T **Proposition 2.6.** *Let G be a cyclic group of order $n = n_1 n_2$, $\gcd(n_1, n_2) = 1$. Let H_1 and H_2 be the subgroups of order n_1 and n_2, respectively. The discrete logarithm of any group element $x \in G$ can be computed by computing one discrete logarithm in H_1 and one discrete logarithm in H_2, and also two exponentiations of x and two exponentiations of g.*

Proof. The diagram from Proposition 2.5 gives another way to compute λ, which in turn is equivalent to computing discrete logarithms in G.

Computing the maps π_1 and π_2 costs one exponentiation each, and we need to apply the maps to g and x in order to know what the λ_1 and λ_2 maps are. Computing discrete logarithms in H_1 and H_2 is the same as computing the maps λ_1 and λ_2. Computing the ring isomorphism described by the Chinese remainder theorem is cheap, so we can ignore that cost. □

E *Exercise* 2.16. The proof of Proposition 2.6 essentially describes an algorithm for computing discrete logarithms, using a different algorithm for computing discrete logarithms in subgroups. Implement this algorithm and restate Proposition 2.6 as a statement about the algorithm's time complexity (in terms of group operations and computing discrete logarithms in subgroups, ignoring any other form of computation involved).

E *Example* 2.7. Consider the group $G = \mathbb{Z}_{72}^{+}$ with generator $g = 1$, and let $x = 23$. We want to compute $\log_g x$ with the above algorithm.

Since $n = 72 = 8 \cdot 9$, we let $n_1 = 8$ and $n_2 = 9$.

We compute $\pi_1(g) = 9 \cdot 1 = 9$ and $\pi_1(x) = 9 \cdot 23 = 63$. We can now observe that $\pi_1(x) = 63 = 7 \cdot 9 = 7 \cdot \pi_1(g)$, so $\lambda_1(\pi_1(x)) = 7$.

Next, we compute $\pi_2(g) = 8 \cdot 1 = 8$ and $\pi_2(x) = 8 \cdot 23 = 40$. Again, we observe that $\pi_1(x) = 40 = 5 \cdot 8 = 5 \cdot \pi_2(g)$, so $\lambda_2(\pi_2(x)) = 5$.

Finally, the canonical CRT map takes $(7, 5)$ to 23, as expected.

E *Exercise* 2.17. Observe that $72 = 8 \cdot 9$. Use the algorithm from Exercise 2.16 to compute (by hand) $\log_5 70$ in the group $\mathbb{F}_{73}^{*}$.

An alternative, more computational approach to proving the theorem is to observe that if $\log_g x = a$ then

$$x^{n_2} = g^{an_2} = (g^{n_2})^a,$$

which by Exercise 2.5 means that $a \equiv \log_{g^{n_2}} x^{n_2} \pmod{n_1}$. In the same way, we get that $a \equiv \log_{g^{n_1}} x^{n_1} \pmod{n_2}$. It follows that if we can compute the discrete logarithms to the bases g^{n_1} and g^{n_2}, we can use the Chinese remainder theorem to recover the logarithm to the base g.

E *Example* 2.8. We redo Example 2.7.

We compute $n_2 \cdot g = 9 \cdot 1 = 9$ and $n_2 \cdot x = 9 \cdot 23 = 63$. We compute that $\log_9 63 = 7$, which means that $a \equiv 7 \pmod 8$.

Next, we compute $n_1 \cdot g = 8 \cdot 1 = 8$ and $n_1 \cdot x = 8 \cdot 23 = 40$. We compute that $\log_8 40 = 5$, which means that $a \equiv 5 \pmod 9$.

Finally, we use CRT to find that $a = 23$.

E *Exercise* 2.18. State a variant of Proposition 2.6 and prove it using the above computational approach.

T **Theorem 2.7** (Pohlig-Hellman I). *Let G be a cyclic group of order $n = \prod_{i=1}^{l} \ell_i^{r_i}$, where $\ell_i \neq \ell_j$ when $i \neq j$. The discrete logarithm of a group element $x \in G$ can be computed by computing one discrete logarithm for each subgroup of order $\ell_i^{r_i}$, l exponentiations of x, and l exponentiations of g.*

Proof. We apply Proposition 2.6 repeatedly and the theorem follows. □

E *Exercise* 2.19. The proof of Theorem 2.7 essentially describes an algorithm for computing discrete logarithms. Implement this algorithm and restate Theorem 2.7 as a statement about the algorithm's time complexity (in terms of group operations and computing discrete logarithms in subgroups, ignoring any other form of computation involved).

You may use the algorithm and statement from Exercise 2.16 as a subroutine and lemma, respectively.

We get the following requirement from Proposition 2.3 and Theorem 2.7.

Requirement 2.2. If $n = \prod_{i=1}^{l} \ell_i^{r_i}$ and $\ell_l^{r_l}$ is the largest prime power dividing n, then $\ell_l^{r_l}$ group operations must be an infeasible computation.

2.2.3 Pohlig-Hellman II

In the previous section, we saw how computing discrete logarithms can be reduced to computing discrete logarithms in subgroups whose orders are prime powers. We shall now see how computing discrete logarithms in a group whose order is a prime power can be reduced to computing discrete logarithms in a prime-ordered subgroup.

Suppose for the remainder of this section that the group order n is a prime power ℓ^r for some prime ℓ. Define the sets $H_i = \{x^{\ell^i} \mid x \in G\}$ for $0 \leq i \leq r$. Note that $H_r = \{1\}$. We shall study the maps $\pi_i : H_i \to H_{r-1}$ defined by

$$\pi_i(y) = y^{\ell^{r-i-1}}, \qquad 0 \leq i \leq r-1.$$

E *Exercise* 2.20. Prove that the sets $H_0, H_1, \ldots, H_r$ are subgroups of G such that $H_0 \supsetneq H_1 \supsetneq \cdots \supsetneq H_r$. Prove also that H_{r-1} is isomorphic to $\mathbb{Z}_\ell^+$.

E *Exercise* 2.21. Prove that the π_i maps are surjective group homomorphisms.

E *Exercise* 2.22. Show that if $y \in H_i$, then $y^{\ell^j} \in H_{i+j}$ for any $j \geq 0$. Show also that $\pi_{i+j}(y^{\ell^j}) = \pi_i(y)$.

E *Exercise* 2.23. Prove that the kernel of π_i is H_{i+1}.

T **Proposition 2.8.** *Let g be a generator for G, let $y \in H_i$ and let $a = \log_{\pi_0(g)} \pi_i(y)$. Then*

$$yg^{-\ell^i a} \in H_{i+1}.$$

Proof. Note first that $g^{\ell^i} \in H_i$, and that $\pi_i(y) = \pi_0(g)^a$. Using Exercise 2.22, we compute

$$\pi_i(yg^{-\ell^i a}) = \pi_i(y)\pi_i(g^{\ell^i})^{-a} = \pi_0(g)^a \pi_0(g)^{-a} = 1.$$

The claim now follows from Exercise 2.23.   ∎

This suggests a recursive algorithm for computing discrete logarithms.

T **Theorem 2.9** (Pohlig-Hellman II). *Let G be a cyclic group of order $n = \ell^r$, where ℓ is prime. The discrete logarithm of a group element $x \in G$ can be computed by computing r discrete logarithms in the subgroup of order ℓ (all to the same base), and also $r - 1$ exponentiations of group elements, $r - 1$ group operations and r exponentiation of g.*

Proof. We construct a sequence of group elements $y_0, y_1, \ldots, y_r$ and integers $a_0, a_1, \ldots, a_{r-1}$ as follows. We begin with $y_0 = x$, and compute

$$a_i = \log_{\pi_0(g)} \pi_i(y_i) \text{ and } y_{i+1} = y_i g^{-\ell^i a_i}, \text{ for } 0 \le i < r.$$

By Proposition 2.8, this sequence is well-defined and $y_r = 1$, so

$$1 = y_{r-1} g^{-\ell^{r-1} a_{r-1}} = \cdots = x g^{-\sum_{i=0}^{r-1} a_i \ell^i}.$$

Note that $0 \le a_i < \ell$ for $0 \le i < r$, which means that $0 \le \sum_{i=0}^{r-1} a_i \ell^i < \ell^r$. By Exercise 2.5, we get that

$$\log_g x = \sum_{i=0}^{r-1} a_i \ell^i,$$

Each iteration requires the computation of one discrete logarithm in the subgroup H_{r-1}, and also one evaluation of π_i (one exponentiation of a group element), one exponentiation of g and one group operation. The latter can be ignored for the final iteration, of course. There is a total of r iterations. We also have to compute $\pi_0(g)$ (one exponentiation of g) once. □

E *Exercise* 2.24. The proof of Theorem 2.9 essentially describes an algorithm for computing discrete logarithms. Implement this algorithm and restate Theorem 2.9 as a statement about the algorithm's time complexity (in terms of group operations and computing discrete logarithms in subgroups, ignoring any other form of computation involved).

E *Example* 2.9. Consider the group $G = \mathbb{Z}_{27}^+$ with generator $g = 1$, and let $x = 23$. We want to compute $\log_g x$.

Since $n = 27 = 3^3$, we find that the subgroup H_1 is all the multiples of 3, while $H_2 = \{0, 9, 18\}$ is the multiples of $9 = 3^2$. The map π_0 then maps $y \in G = H_0$ to $3^2 \cdot y \in H_2$, while π_1 maps $y \in H_1$ to $3 \cdot y \in H_2$.

First, we see that $\pi_0(g) = 9 \cdot 1 = 9$.

With $y_0 = x = 23$, we get $\pi_0(x) = 9 \cdot 23 = 18 = 2 \cdot 9$, so $a_0 = 2$.

We compute $y_1 = x_0 - a_0 \cdot g = 23 - 2 \cdot 1 = 21$. We then get $\pi_1(21) = 3 \cdot 21 = 9 = 1 \cdot 9$, so $a_1 = 1$.

Finally, we compute $y_2 = y_1 - (3a_1) \cdot g = 21 - (3 \cdot 1) \cdot 1 = 18 = 2 \cdot 9$, so $a_2 = 2$.

We get $\log_1 23 = 2 + 1 \cdot 3 + 2 \cdot 3^2 = 23$.

E| *Exercise 2.25.* Use the algorithm from Exercise 2.24 to compute the discrete logarithm of 85 to the base 11 in the group $\mathbb{Z}_{125}^+$. (Note that we are talking about the additive group here, where the group operation is addition modulo 125 and exponentiation is multiplication modulo 125.)

E| *Exercise 2.26.* Observe that $72 = 2^3 3^2$.

(a) In $\mathbb{F}_{73}^*$, 2 generates a subgroup of order 3^2. Use the algorithm from Exercise 2.24 to compute (by hand) the discrete logarithm of 64 to the base 2 in the subgroup of $\mathbb{F}_{73}^*$.

(b) In $\mathbb{F}_{73}^*$, 10 generates a subgroup of order 2^3. Use the algorithm from Exercise 2.24 to compute (by hand) the discrete logarithm of 27 to the base 10 in the subgroup of $\mathbb{F}_{73}^*$.

(c) Use the algorithm from Exercise 2.16 together with the algorithm from Exercise 2.24 to compute (by hand) the discrete logarithm of 70 to the base 5 in the group $\mathbb{F}_{73}^*$. You should use the results of the previous two exercises in your computation.

An alternative, more computational approach to proving Proposition 2.9 begins by writing $a = \sum a_i \ell^i$ and then observing that if

$$y = g^{\sum_{i=j}^{r-1} a_i \ell^i}, \qquad \text{then} \qquad y^{\ell^{j-1}} = g^{a_j \ell^{r-1}}.$$

E| *Exercise 2.27.* Give a proof of Theorem 2.9 using the computational approach.

E| *Example 2.10.* Consider the group $G = \mathbb{Z}_{27}^+$ with generator $g = 1$, and let $x = 23$. We want to compute $\log_g x$.
We write $\log_g x = a_0 + 3a_1 + 3^2 a_2$.
First, we compute $9 \cdot g = 9 \cdot 1 = 9$.
Let $y_0 = x = (a_0 + 3a_1 + 3^2 a_2) \cdot 1$. We compute $9 \cdot 23 = 18 = 2 \cdot 9$, which tells us that

$$9(a_0 + 3a_1 + 3^2 a_2) \equiv 9 \cdot 2 \pmod{27} \qquad \Rightarrow \qquad a_0 = 2.$$

We get $y_1 = y_0 - 2 \cdot 1 = 21 = (3a_1 + 3^2 a_2) \cdot 1$. We compute $3 \cdot y_1 = 3 \cdot 21 = 9 = 1 \cdot 9$, which tells us that

$$3(3a_1 + 3^2 a_2) \equiv 9 \cdot 1 \pmod{27} \qquad \Rightarrow \qquad a_1 = 1.$$

We get $y_2 = y_1 - (3 \cdot 1) \cdot 1 = 18 = (3^2 a_2) \cdot 1$. We compute $y_2 = 2 \cdot 9$, which tells us that
$$3^2 a_2 \equiv 9 \cdot 2 \pmod{27} \qquad \Rightarrow \qquad a_2 = 2.$$
We get that $\log_g 23 = 2 + 3 \cdot 1 + 3^2 \cdot 2 = 23$.

Proposition 2.3 with Theorems 2.7 and 2.9 give us this requirement.

Requirement 2.3. If $n = \prod_{i=1}^{l} \ell_i^{r_i}$ is the group order and ℓ_l is the largest prime dividing n, ℓ_l group operations must be an infeasible computation.

2.2.4 Shanks' Baby-step Giant-step

In this section, we shall improve on Proposition 2.3 by trading reduced computational effort for increased memory use. Let G be a cyclic group of order n. We must assume that there is some *total order* $\preceq$ on the group elements, such that the following two claims hold when $L \ll n$:

Sorting With effort comparable to computing L group operations, a list of pairs of group elements and integers $(x_1, a_1), (x_2, a_2), \ldots, (x_L, a_L)$ can be rearranged into a new list $(y_1, b_1), (y_2, b_2), \ldots, (y_L, b_L)$ satisfying $y_i \preceq y_j$ when $i \leq j$.

Searching With effort comparable to computing one group operation, we can decide if a group element x is present in a list of pairs of group elements and integers $(y_1, b_1), (y_2, b_2), \ldots, (y_L, b_L)$ satisfying $y_i \preceq y_j$ when $i \leq j$. If x is present, we also learn what its corresponding number b is.

The algorithm we are interested in begins with the observation that $\log_g x = bL + b'$, where $0 \leq b' < L$.

Ⓣ **Theorem 2.10** (Shanks algorithm). *Let G be a cyclic group of order n. For any positive integer $L \ll n$, the discrete logarithm of an element $x \in G$ can be computed using memory to hold L group elements and $L + \lceil n/L \rceil$ group operations.*

Proof. First, we construct a list of group elements with their discrete logarithms

$$(1,0), (g,1), (g^2, 2), (g^3, 3), \ldots, (g^{L-1}, L-1).$$

Computing this list requires less than L group operations and memory to hold L group elements. By assumption, we can sort the list into the list

$$(y_1, b_1), (y_2, b_2), (y_3, b_3), \ldots, (y_L, b_L)$$

using effort comparable to what it took to create the list. We can now quickly decide if the discrete logarithm of any group element is less than L, and if so, what its discrete logarithm is.

Recall that we can write $\log_g x = bL + b'$, where $0 \leq b' < L$. Therefore,

$$0 \leq \log_g x (g^{-L})^b < L.$$

We can find b and b' by computing successively

$$x, xg^{-L}, x(g^{-L})^2, x(g^{-L})^3, \ldots$$

and each time check if it is in our list. Computing the next element requires one group operation, while the check requires comparable effort. Before computing $\lceil n/L \rceil$ elements, we will find an element in our list. The claim follows. □

E Exercise 2.28. The proof of Theorem 2.10 essentially describes an algorithm for computing discrete logarithms. Implement this algorithm and restate Theorem 2.10 as a statement about the algorithm's time complexity (in terms of group operations, ignoring any other form of computation involved).

E Example 2.11. Consider the group $G = \mathbb{Z}_{29}^{+}$ with generator $g = 1$, and let $x = 23$. We want to compute $\log_g x$.

We first compute the list of group elements, choosing $L = 5$. Sorting is easy for this group.

$i \cdot g$	0	1	2	3	4
i	0	1	2	3	4

We also compute $-5 \cdot g = 24$.

We see that 23 is not in the table, so we compute $23 + 24 = 18$. This is not in the table, so we compute $18 + 24 = 13$. This is not in the table, so we compute $13 + 24 = 8$. Again, this is not in the table, so we compute $8 + 24 = 3$.

We find that 3 is in the table, with discrete logarithm 3. We have added $-5 \cdot g$ to 23 four times. It follows that

$$\log_g 23 = 4 \cdot 5 + 3.$$

E Exercise 2.29. What value of L minimises computational effort?

E Exercise 2.30. Using the algorithm from Exercise 2.28 to compute (by hand) the discrete logarithm of 51 to the base 2 in the group $\mathbb{F}_{59}^{*}$.

E Exercise 2.31. Decide the optimal value of L if we need to compute the discrete logarithm of k group elements $x_1, x_2, \ldots, x_k$.

E Exercise 2.32. Suppose we know that $\log_g x < k$. Show that for any $L > 0$ we can compute this discrete logarithm using at most $L + \lceil k/L \rceil$ group operations.

E Exercise 2.33. Suppose we know that $k_1 \leq \log_g x < k_2$. Show that for any $L > 0$ we can compute this discrete logarithm using at most $L + \lceil (k_2 - k_1)/L \rceil$ group operations.

Based on Theorems 2.7, 2.9 and 2.10, we get the following requirement.

Requirement 2.4. If n is the group order and ℓ is the largest prime dividing n, then $L + \lceil \ell/L \rceil$ group operations using memory for L group elements must be an infeasible computation.

2.2.5 Pollard's ρ Method

In this section, we shall consider a group G with a prime n number of elements. In the previous section, we saw that we could trivially recover the discrete logarithm of x given an equation of the form $x(g^{-L})^b = g^{b'}$. Pollard's ρ method will try to generate an equation of the form

$$g^a x^b = g^{a'} x^{b'}, \tag{2.1}$$

with $b \not\equiv b' \pmod{n}$, which gives us $\log_g x \equiv (a - a')(b' - b)^{-1} \pmod{n}$.

Pollard's ρ method relies on connecting two separate ideas with a conjecture to arrive at an algorithm for finding a relation like (2.1) above. The first idea is that in sequences of randomly chosen elements we will quickly see repetitions. The second idea is that in certain sequences we can quickly find cycles. The conjecture is that, with respect to repetitions, certain non-random sequences "look random". It is the repetitions in this sequence that will cause cycles, which we can quickly find and which will give us the required relation.

We begin with a sequence of elements chosen at random. What is the probability that there are no repetitions within the first L elements of the sequence?

Proposition 2.11. *Suppose we have a sequence of elements $s_1, s_2, s_3, \ldots$, the elements chosen independently and uniformly at random from a set S with n elements. Let E be the event that the L first elements are all distinct. Then*

$$1 - \frac{L(L-1)}{2n} \leq \Pr[E] \leq \exp\left(-\frac{L(L-1)}{2n}\right).$$

Proof. If $L \geq n$, then $\Pr[E] = 0$ and the claim holds. So assume $L < n$.

Choosing elements one after another, we see that we have n choices for the first element, $n - 1$ choices for the second element, $n - 2$ choices for the third element, and so forth. By independence and uniformity, we get that

$$\Pr[E] = (1 - 1/n)(1 - 2/n) \cdots (1 - (L-1)/n).$$

Note that $1 - \epsilon \leq \exp(-\epsilon)$ for any ϵ, so

$$\Pr[E] \leq e^{-1/n} e^{-2/n} \cdots e^{-(L-1)/n} = \exp\left(-\sum_{i=1}^{L-1} i/n\right) = \exp\left(-\frac{L(L-1)}{2n}\right).$$

For the lower bound, we consider the complementary event. Let F_i be the event that the ith chosen element coincides with at least one of the previous elements. Then $\Pr[F_i] \leq (i-1)/n$, and we get that

$$1 - \Pr[\neg E] = 1 - \Pr[F_1 \vee F_2 \vee \cdots \vee F_L]$$

$$\geq 1 - \sum_{i=1}^{L} \Pr[F_i] \geq 1 - \sum_{i=1}^{L} \frac{i-1}{n} \geq 1 - \frac{L(L-1)}{2n}. \qquad \square$$

The sequence we shall now consider is far from random, so this proposition and its proof do not apply. But many non-random sequences look random with respect to repetitions in the sequence, which means that the bounds in the proposition apply in practice. This will be sufficient for our purposes.

Now we shall consider the problem of finding cycles in the special kind of infinite sequences generated by a starting point $s_1 \in S$, a function $f : S \to S$ and the rule $s_{i+1} = f(s_i)$. The sequence eventually becomes cyclic (for some integers i, j, k, $s_{j+k} = s_j$ when $j \geq i$) when S is finite.

T **Proposition 2.12.** *Let $s_1, s_2, \ldots$ be the sequence defined by s_1 and the rule $s_{i+1} = f(s_i)$. Suppose k is the smallest integer such that $s_k = s_{k'}$ for some $k' < k$. Then distinct indexes i, j can be found such that $s_i = s_j$ using at most $3k$ evaluations of f.*

Proof. We consider the sequence $u_1, u_2, \ldots$ given by $u_j = s_{2j}$. Periodicity ensures that for some i, $u_i = s_i$, and that this i is at most k.

We can compute successively the pairs $(s_1, u_1), (s_2, u_2), \ldots$ using the rule

$$(s_{i+1}, u_{i+1}) = (f(s_i), f(f(u_i))).$$

We will notice when $s_i = u_i$, at which point we have found $s_i = s_{2i}$. Computing each new pair requires evaluating f three times. The claim follows. □

E *Exercise 2.34.* The proof of Proposition 2.12 essentially describes an algorithm for finding two integers. Implement this algorithm and restate Proposition 2.12 as a statement about the algorithm's time complexity (in terms of evaluations of the function f).

Now we are ready to construct a sequence that should allow us to find an equation like (2.1) and thereby compute discrete logarithms.

Our set will be the group G, of course. Suppose $\{S_1, S_2, S_3\}$ is a partition of G, that is, S_1, S_2 and S_3 are pairwise disjoint sets whose union is G, where the three subsets have approximately the same cardinality. Suppose also that it is easy to check which subset an element is in.

The sequence $y_1, y_2, \ldots$ is based on S_1, S_2, S_3, a generator g and a group element x. We let $y_1 = x$ and

$$y_{i+1} = \begin{cases} y_i g & y_i \in S_1, \\ y_i^2 & y_i \in S_2, \text{ or} \\ y_i x & y_i \in S_3. \end{cases} \tag{2.2}$$

We define a sequence of integer pairs $(a_1, b_1), (a_2, b_2), \ldots$ related to $y_1, y_2, \ldots$ by defining $(a_1, b_1) = (0, 1)$ and using the rule

$$(a_{i+1}, b_{i+1}) = \begin{cases} (a_i + 1 \bmod n, b_i) & y_i \in S_1, \\ (2a_i \bmod n, 2b_i \bmod n) & y_i \in S_2, \text{ or} \\ (a_i, b_i + 1 \bmod n) & y_i \in S_3. \end{cases} \tag{2.3}$$

E *Exercise* 2.35. Show that the above two sequences defined by (2.2) and (2.3) satisfy $y_i = g^{a_i} x^{b_i}$.

E *Example* 2.12. Consider $G = \mathbb{Z}_{31}^+$ with generator $g = 1$, $x = 7$ and the (randomly chosen) partition:

$$\begin{array}{c|l}
S_1 & 4, 9, 11, 14, 15, 16, 21, 24, 29, 30 \\
S_2 & 1, 6, 7, 8, 10, 18, 19, 20, 25, 27 \\
S_3 & 0, 2, 3, 5, 12, 13, 17, 22, 23, 26, 28
\end{array}$$

Given the starting point $y_1 = x = 7$, we get the sequences

i	1	2	3	4	5	6	7	8	9	10	11	12	13	14	15	16
j		1		2		3		4		5		6		7		8
y_i	7	14	15	16	17	24	25	19	7	14	15	16	17	24	25	19
	S_2	S_1	S_1	S_1	S_3	S_1	S_2	S_2	S_2	S_1	S_1	S_1	S_3	S_1	S_2	S_2
a_i	0	0	1	2	3	3	4	8	16	1	2	3	4	4	5	10
b_i	1	2	2	2	2	3	3	6	12	24	24	24	24	25	25	19

By inspection, we see that $y_9 = y_1 = 7$, while the algorithm from Exercise 2.34 finds the repetition $y_8 = y_{2 \cdot 8}$.

E *Exercise* 2.36. Consider the subgroup G of $\mathbb{F}_{107}^*$ with 53 elements. Use the generator $g = 11$ and $x = 85$. Also use the partition where $y \in S_i \Leftrightarrow y \equiv i - 1 \pmod 3$.

(a) Compute the 20 first terms of the sequences from (2.2) and (2.3).

(b) Find by inspection the first repetition in the sequence $y_1, y_2, \ldots$.

(c) Use the algorithm from Exercise 2.34 to find a repetition in this sequence.

E *Exercise* 2.37. Let $y_1, y_2, \ldots$ and $(a_1, b_1), (a_2, b_2), \ldots$ be the above sequences defined by (2.2) and (2.3) and suppose k is the smallest integer such that $y_k = y_{k'}$ for some $k' < k$. Prove that distinct indexes i, j along with corresponding pairs (a_i, b_i) and (a_j, b_j) can be found such that $y_i = y_j$ using at most $3k$ group operations and $6k$ additions and multiplications modulo n.

Let E be the event that the L first elements of $y_1, y_2, \ldots$ are all distinct, and let E' be the event that the L first pairs of $(a_1, b_1), (a_2, b_2), \ldots$ are all distinct. Then we can define the two functions

$$\theta(L, n) = \Pr[E] \qquad \text{and} \qquad \gamma(L, n) = 1 - \Pr[E'],$$

where the probability is taken over the choice of x.

T | **Theorem 2.13** (Pollard's rho method). *Let G be a cyclic group of order n. The discrete logarithm of x sample from the uniform distribution on G can be computed using $3L$ group operations and $6L + 3$ arithmetic operations modulo n, except with probability at most $\theta(L, n) + \gamma(L, n)$.*

Proof. Some element will appear twice among the L first elements of the sequence (2.2) except with with probability $\theta(L, n)$. The corresponding pairs (a_i, b_i) and (a_j, b_j) will be distinct except with probability $\gamma(L, n)$.

This repetition $y_i = y_j$ gives us an equation of the form (2.1), namely

$$g^{a_i} x^{b_i} = g^{a_j} x^{b_j}.$$

When $b_i \neq b_j$, we can compute the discrete logarithm of x, since n is prime.

By Exercise 2.37 we can find the indexes of the repetition, while at the same time keeping track of the sequence described by (2.3), using at most $3L$ group operations and $6L$ arithmetic operations modulo n. The claim follows. □

E | *Exercise 2.38.* The proof of Theorem 2.13 essentially describes an algorithm for computing discrete logarithms. Implement this algorithm and restate the claim in Theorem 2.13 as a statement about the algorithm's time complexity (in terms of group operations and arithmetic operations modulo n) and success probability (in terms of $\theta(L, n)$ and $\gamma(L, n)$).

E | *Example 2.13.* Continuing from Example 2.12, we use the algorithm from Exercise 2.37 to find a repetition in the sequence $y_1, y_2, \ldots$, and find that $y_{16} = y_8$. From the already computed values, we find that

$$8 \cdot 1 + 6 \cdot x = 10 \cdot 1 + 19 \cdot x \quad \Rightarrow \quad \log_g x \equiv \frac{8 - 10}{19 - 6} \equiv (-2)(12) \equiv 7 \pmod{31}.$$

E | *Exercise 2.39.* Use the algorithm from Exercise 2.38 to compute (by hand) the discrete logarithm of 85 to the base 11 in the group $\mathbb{F}_{107}^*$.

When we choose a reasonable partition $\{S_1, S_2, S_3\}$, it is plausible that claims similar to Proposition 2.11 should hold for the two sequences, and this seems to hold in practice. We phrase this in terms of a conjecture. Note that in this conjecture the probability is taken over the choice of the element x.

T | **Conjecture 2.14.** *For reasonable partitions $\{S_1, S_2, S_3\}$ we have*

$$\theta(L, n) \approx \exp\left(-\frac{L(L - 1)}{2n}\right) \quad and \quad \gamma(L, n) \approx \frac{L(L - 1)}{2n^2}.$$

Conjecture 2.14 and Theorems 2.7, 2.9 and 2.13 give us this requirement.

Requirement 2.5. If n is the group order and ℓ is the largest prime dividing n, then $\sqrt{\ell}$ group operations must be an infeasible computation.

2.3 PRIMALITY TESTING

Throughout this section, we shall take n to be a large odd integer.

We want to be able to distinguish prime integers from composite integers. Since a composite integer must have a proper divisor smaller than the square root of the integer, we can in principle check every possible divisor.

T | **Proposition 2.15.** *We can decide if n is prime using at most $\sqrt{n}$ divisions.*

Unfortunately, the algorithm implied by this proposition is useless for large integers, and therefore for our purposes. However, having divisors is just one difference in behaviour between composite and prime integers.

We shall identify subsets S of $\mathbb{Z}_n^*$ with three properties:

1. It is easy to check if an element is in S or not.

2. If n is prime, then $S = \mathbb{Z}_n^*$.

3. If n is composite, then S contains at most half of all the elements of $\mathbb{Z}_n^*$.

If our three properties are satisfied, we can recognise primes. First, we choose k elements from $\mathbb{Z}_n^*$ uniformly and independently at random. Then we check if all of these elements are in the subset S. If any element is not in the subset, then n is composite. Otherwise, we conclude that n *may be prime*.

If one or more of the elements are not in the subset, we have proved that S is a proper subset, from which it follows that n is composite. Any element that is not in the subset is a *witness* for the fact that n is not prime.

The probability that all of the random elements lie inside S is $(|S|/|\mathbb{Z}_n^*|)^k$. For moderately large k, if $|S|/|\mathbb{Z}_n^*| < 1/2$, this probability will be very small. It follows logically that if all of the random elements lie inside S, it is most likely that n is prime. Note that *it is not certain that n is prime*, just likely.

2.3.1 Fermat's Test

Our set is $G_n = \{x \in \mathbb{Z}_n^* \mid x^{n-1} = 1\}$, and it is a group. It is easy to test for membership and $G_n = \mathbb{Z}_n^*$ by Fermat's little theorem.

T | **Theorem 2.16.** *If n is prime, then for any integer a not divisible by n,*

$$a^{n-1} \equiv 1 \pmod{n}. \tag{2.4}$$

E | *Exercise* 2.40. Show that G_n is a subgroup of $\mathbb{Z}_n^*$. Also, $G_n = \mathbb{Z}_n^*$ if n is prime.

E *Exercise* 2.41. Show that there is an algorithm that can decide if an element $x \in \mathbb{Z}_n^*$ is in G_n or not using at most one exponentiation.

We have now shown that our subset G_n satisfies the first and second property. Furthermore, if G_n is a proper subgroup, then it will contain at most half of all the elements of $\mathbb{Z}_n^*$.

E *Exercise* 2.42. Exercise 2.41 together with the strategy described above informally describes an algorithm that decides if the subgroup G_n is a proper subgroup of $\mathbb{Z}_n^*$. Implement this algorithm. Formulate and prove a statement about the algorithm's time complexity (in terms of exponentiations) and success probability (the probability that the algorithm decides correctly).

Note the careful phrasing. The exercise only says that the algorithm can decide if G_n is likely to be a proper subgroup, not if n is likely to be prime. Of course, if G_n is a proper subgroup, we know that n is composite, so we can attempt to use this to prove that numbers are composites.

E *Example* 2.14. Consider $n = 55$. First, we try 34 and compute that

$$34^{54} \equiv 1 \pmod{55}.$$

This means that 34 does not disprove that 55 is prime. Next, we try 21 and compute that
$$21^{54} \equiv 1 \pmod{55}.$$

Again, this means that 21 does not disprove that 55 is prime. Next, we try 27 and compute that
$$27^{54} \equiv 9 \pmod{55},$$

which proves that 55 is composite. Note that we did not factor 55.

E *Exercise* 2.43. Use the algorithm from Exercise 2.42 to prove two of the three numbers 767, 13457 and 83693 composite.

Unfortunately, our strategy will fail for some composite numbers.

D **Definition 2.3.** A *Carmichael number* is a composite n such that $G_n = \mathbb{Z}_n^*$.

The algorithm from Exercise 2.42 does not distinguish prime numbers, but rather prime and Carmichael numbers.

E *Exercise* 2.44. Use the algorithm from Exercise 2.42 to prove one of the two numbers 294409 and 951253 composite. What about the other number?

Carmichael numbers not only exist, but there are infinitely many of them. Fortunately, Carmichael numbers are rare, in the sense that we are probably unlikely to run into one in most cryptographic applications. This test might be good enough for our purposes. But we can do better.

2.3.2 Soloway-Strassen Test

The Soloway-Strassen test relies on a simple way to decide if a number is a square modulo a prime. Before we can explain the test, we require a bit of number theory.

D | **Definition 2.4.** Let p be a prime. For any integer a, the *Legendre symbol* is

$$\left(\frac{a}{p}\right) = \begin{cases} 1 & \text{if there exists } b \not\equiv 0 \pmod{p} \text{ such that } b^2 \equiv a \pmod{p}, \\ 0 & p \text{ divides } a, \text{ and} \\ -1 & \text{otherwise.} \end{cases}$$

Let n be an integer with prime factorisation $n = \prod_i p_i^{r_i}$. For any integer a, the *Jacobi symbol* is

$$\left(\frac{a}{n}\right) = \prod_i \left(\frac{a}{p_i}\right)^{r_i}.$$

If n is prime, the Jacobi symbol coincides with the Legendre symbol. We can consider the Legendre symbol to be a map from $\mathbb{F}_p^*$ to $\{\pm 1\}$. The inverse image of 1 contains all the square group elements. Likewise, we can consider the Jacobi symbol $\left(\frac{\cdot}{n}\right)$ to be a map from $\mathbb{Z}_n^*$ to $\{\pm 1\}$.

E | *Exercise 2.45.* Show that $H_n \subseteq \mathbb{Z}_n^*$ is a subgroup, when

$$H_n = \left\{ x \in \mathbb{Z}_n^* \mid x^{\frac{n-1}{2}} = \left(\frac{x}{n}\right) \right\}$$

We have defined our subset. We now need to be able to decide if an element lies in H_n or not, which means that we need to be able to compute the Jacobi symbol quickly. The properties of the Jacobi symbol allow us to do so.

T | **Fact 2.17.** *Let n be an odd integer and let a, b be integers that are relatively prime to n. For the final point, a should also be odd. Then the following hold:*

1. $\left(\frac{a}{n}\right) = \left(\frac{b}{n}\right)$ *if $a \equiv b \pmod{n}$.*

2. $\left(\frac{ab}{n}\right) = \left(\frac{a}{n}\right)\left(\frac{b}{n}\right).$

3. $\left(\frac{2}{n}\right) = \begin{cases} 1 & \text{if } n \text{ is congruent to 1 or 7 modulo 8,} \\ -1 & \text{otherwise.} \end{cases}$

4. $\left(\frac{a}{n}\right) = \begin{cases} -\left(\frac{n}{a}\right) & \text{if } n \text{ and } a \text{ are both congruent to 3 modulo 4,} \\ \left(\frac{n}{a}\right) & \text{otherwise.} \end{cases}$

E | *Exercise 2.46.* With the above properties of the Jacobi symbol, show that there is an algorithm that quickly computes the Jacobi symbol. Implement it.

E *Example* 2.15. We want to compute the Jacobi symbol $\left(\frac{34}{55}\right)$. With the Jacobi properties used marked above the equal signs, we get

$$\left(\frac{34}{55}\right) \overset{2.}{=} \left(\frac{2}{55}\right)\left(\frac{17}{55}\right) \overset{3.}{=} \left(\frac{17}{55}\right) \overset{4.}{=} \left(\frac{55}{17}\right) \overset{1.}{=} \left(\frac{4}{17}\right) \overset{3.}{=} 1.$$

E *Example* 2.16. We want to compute the Jacobi symbol $\left(\frac{34}{385}\right)$. We get

$$\left(\frac{34}{385}\right) \overset{2.}{=} \left(\frac{2}{385}\right)\left(\frac{17}{385}\right) \overset{3.}{=} \left(\frac{17}{385}\right) \overset{4.}{=} \left(\frac{385}{17}\right) \overset{1.}{=} \left(\frac{11}{17}\right) \overset{4.}{=} \left(\frac{17}{11}\right)$$

$$\overset{1.}{=} \left(\frac{6}{11}\right) \overset{2.}{=} \left(\frac{2}{11}\right)\left(\frac{3}{11}\right) \overset{3.}{=} -\left(\frac{3}{11}\right) \overset{4.}{=} \left(\frac{11}{3}\right) \overset{1.}{=} \left(\frac{2}{3}\right) \overset{3.}{=} -1.$$

E *Exercise* 2.47. Compute the Jacobi symbol $\left(\frac{7}{294409}\right)$.

The next theorem shows that $H_n = \mathbb{Z}_n^*$ when n is prime.

T **Theorem 2.18.** *Let n be an odd prime and let $a \in \mathbb{Z}_n^*$. Then*

$$\left(\frac{a}{n}\right) = a^{\frac{n-1}{2}}. \tag{2.5}$$

Proof. Recall that the Legendre symbol of a group element is 1 if and only if that element is a square in the group.

When n is prime, $\mathbb{Z}_n^*$ is cyclic of order $n-1$. Let g be any generator. When the group order $n-1$ is even, the squares are the elements of the form g^{2i} for integers i, while the non-squares are of the form g^{2i+1}. We see that

$$(g^{2i})^{\frac{n-1}{2}} = (g^{n-1})^i = 1 \quad \text{and} \quad (g^{2i+1})^{\frac{n-1}{2}} = (g^{n-1})^i g^{\frac{n-1}{2}} = -1,$$

which proves the theorem. ∎

To get an algorithm that decides if a number is prime or not, it remains to show that whenever n is composite, H_n is a proper subgroup of $\mathbb{Z}_n^*$.

T **Theorem 2.19.** *Let n be an odd composite number. Then there exists $a \in \mathbb{Z}_n^*$ such that (2.5) does not hold.*

Proof. We shall construct a when n is square-free, and when it is not.

First, suppose there is a prime p such that p^2 divides n. Let $n_2 = n/p$, and let a be the integer $1 + n_2$, which is not congruent to 1 modulo n but is congruent to 1 modulo n_2 and p. The Jacobi symbol is

$$\left(\frac{a}{n}\right) = \left(\frac{a}{p}\right)\left(\frac{a}{n_2}\right) = \left(\frac{1}{p}\right)\left(\frac{1}{n_2}\right) = 1.$$

The binomial theorem says that

$$a^p \equiv (1 + n_2)^p \equiv \sum_{i=0}^{p} \binom{p}{i} n_2^i \pmod{n}.$$

Every term in the sum will be divisible by n except the first term (which is 1) and possibly the second term. But p divides $\binom{p}{1}$, so n divides the second term too, and a has order p modulo n. Then since p does not divide $n - 1$, it will follow that $a^{(n-1)/2}$ is not congruent to 1 modulo n.

Second, suppose no square of any prime divides n. Let p be a prime dividing n and let $n_2 = n/p$. Let also b be relatively prime to n and not a square modulo p. The Chinese remainder theorem gives us a such that

$$a \equiv 1 \pmod{n_2} \qquad \text{and} \qquad a \equiv b \pmod{p}.$$

We have $a^{(n-1)/2} \not\equiv -1 \pmod{n}$ by construction, but the Jacobi symbol of a is

$$\left(\frac{a}{n}\right) = \left(\frac{b}{p}\right)\left(\frac{1}{n_2}\right) = (-1) \cdot 1 = -1.$$

In either case, $a + \langle n \rangle$ does not satisfy (2.5). □

Exercise 2.48. Exercise 2.45, Exercise 2.46 and our strategy can be combined into an algorithm that decides if the subgroup H_n is a proper subgroup of $\mathbb{Z}_n^*$. Implement this algorithm and formulate and prove a statement about the algorithm's time complexity (in terms of exponentiations) and success probability.

We use Exercise 2.48 to decide if H_n is a proper subgroup of $\mathbb{Z}_n^*$. By Theorems 2.18 and 2.19, we know that $H \neq \mathbb{Z}_n^*$ if and only if n is composite. In other words, we have an efficient way to decide if a number is prime or not.

Example 2.17. Consider $n = 55$. First, we try 34 and compute that

$$34^{27} \equiv 34 \not\equiv \pm 1 \pmod{55},$$

which proves that 55 is composite. Note that we have not factored 55.

Example 2.18. Consider $n = 385$. First, we try 309 and compute that

$$309^{192} \equiv 1 \equiv \left(\frac{309}{385}\right) \pmod{385}.$$

So 309 does not disprove that 385 is prime. Next, we try 34 and compute that

$$34^{192} \equiv 1 \not\equiv \left(\frac{34}{385} \right) \pmod{385},$$

which proves that 385 is composite. Note that we have not factored 385.

E | *Exercise* 2.49. Use the algorithm from Exercise 2.48 to prove the two numbers 294409 and 951253 composite.

E | *Exercise* 2.50. Choose several random numbers between 10^6 and 10^7. Use the algorithm from Exercise 2.48 to decide primality for each number with reasonable certainty. For the composites, how many random numbers do you need to test before disproving their primality.

Redo the exercise for random numbers between 10^{20} and 10^{21}.

Are the results surprising? Should they be?

2.4 FINITE FIELDS

The non-zero elements of a finite field with q form an abelian group under multiplication, and we denote this group by $\mathbb{F}_q^*$. The question we shall now investigate is if this group is suitable for use in Diffie-Hellman.

E | *Exercise* 2.51. Prove that the group $\mathbb{F}_q^*$ is cyclic.

It turns out (for reasons that we shall not investigate) that computing discrete logarithms in extension fields is easier than in prime fields of comparable size. Therefore, we restrict our study to prime fields.

Requirement 2.5 says that the order of any group we use in Diffie-Hellman should be divisible by a large prime. In other words, we need a large prime q such that $q - 1$ is divisible by a large prime. Before we go on to discuss how we can find suitable primes, we first discuss the details of the group operation.

2.4.1 Group Operation

Let p be a prime. Mathematically, the finite field $\mathbb{F}_p$ consists of a set of p elements, among which are 0 and 1, and two binary operations $+$ and $\cdot$.

The field is isomorphic to the factor ring $\mathbb{Z}/\langle p \rangle$, where it is easy to compute. To add, subtract or multiply, we simply add, subtract or multiply representatives of the cosets to get new representatives. To divide by $\xi + \langle p \rangle$, we first find an inverse ζ of ξ modulo p, then multiply by $\zeta + \langle p \rangle$.

This is unproblematic mathematically, but computationally it is awkward because the size of the representatives tends to grow, making arithmetic slow. However, we know that two integers represent the same coset if and only if they are congruent modulo p. Which means that after we do arithmetic operations on the representatives, we may divide the new representative by p and use the remainder as our representative for the result coset.

In practice, we usually represent the field elements using the integers $\{0, 1, \ldots, p-1\}$ and do arithmetic as integer arithmetic followed by taking the remainder after division by p. A group operation then costs two arithmetic operations. Finding an inverse in the group is somewhat more expensive.

Exercise 2.52. Find reasonable algorithms for addition, subtraction, multiplication, division. We can use the Extended Euclidian algorithm to compute inverses. Implement any of these that are missing in your favourite programming language. Compare these algorithms in terms of *runtime*.

2.4.2 Finding Suitable Primes

The prime number theorem says that in any given range of integers, primes are fairly common (inversely proportional to the number of digits). Since we can efficiently recognise primes, we can fairly quickly find large primes simply by choosing random large numbers until we find a prime.

By Requirement 2.5, we need a cyclic group such that the group order is divisible by a large prime. This means that we are not just looking for primes, we are looking for a prime p such that $p - 1$ is divisible by a large prime.

We can do this by first choosing a sufficiently large prime ℓ and then choosing random numbers k until $2k\ell + 1$ is prime. In practice, this algorithm performs as well as a search for an arbitrary prime.

Example 2.19. Testing integers sequentially starting at 2000, we find that $\ell = 2003$ is prime. We then test multiples starting with 101 and find that when $k = 103$ the number $p = 2k\ell + 1 = 412619$ is prime.

Sometimes, we want the $p-1$ to be twice a prime ℓ. In this case, p is called a *safe prime* and ℓ is called a *Sophie-Germain prime*. This time, the need for ℓ and $2\ell + 1$ to be prime simultaneously means that we need to look at many more candidates before we find a suitable prime. This will be slow, but there are techniques to speed up the search.

Example 2.20. Again, testing integers sequentially starting at 2003, we find that $\ell = 2039$ is prime at the same time as $p = 2\ell + 1 = 4079$.

Exercise 2.53. About one third of all candidates will be divisible by 3. Checking that a number is not divisible by 3 is much faster than using the primality testing algorithms from Section 2.3. Expand on this idea and explain how we can use so-called trial division by small primes to exclude most candidate primes before finally using the algorithms from Section 2.3.

Exercise 2.54. The Sieve of Eratosthenes can be used to quickly find the first small primes. Adapt the sieving idea to quickly find the most likely prime candidates from an integer range.

2.4.3 Index Calculus

All of the algorithms from Section 2.2 will work for $\mathbb{F}_p^*$. But it turns out that we can do very much better. We shall develop the ideas of index calculus in a general setting, and then show how the properties of prime fields give us a more efficient algorithm for computing discrete logarithms.

We begin with an observation about abstract cyclic groups, which is an extension of (2.1). Let G be a cyclic group of order n. Let g be some generator and let x be a group element. Suppose we have ν pairs of integers $(r_1, t_1), (r_2, t_2), \ldots, (r_\nu, t_\nu)$ and integers $\alpha_1, \alpha_2, \ldots, \alpha_\nu$ (not all congruent to zero modulo n) such that

$$\prod_{i=1}^{\nu} \left(x^{t_i} g^{r_i} \right)^{\alpha_i} = 1. \tag{2.6}$$

This will give us the equation

$$x^{\sum_i \alpha_i t_i} g^{\sum_i \alpha_i r_i} = 1,$$

which is of the same form as (2.1). As long as $\sum_i \alpha_i t_i$ is invertible modulo n, we can recover $\log_g x$ from the equation using $2\nu + 2$ arithmetic operations ($\nu + 1$ multiplications, ν additions and one inversion).

We first consider the case when the group order n is prime.

Example 2.21. Consider the prime $p = 1019$ with the elements $g = 3$, generating a subgroup G of $\mathbb{F}_p^*$ of order 509. Let $x = 11$.

With the relations

$$y_1 = g^{112} x^{239} = 576 \qquad y_2 = g^{477} x^{274} = 70 \qquad y_3 = g^{378} x^{248} = 180$$

$$y_4 = g^{80} x^{66} = 42 \qquad y_5 = g^{331} x^{488} = 720$$

and integers $\alpha_1 = 145$, $\alpha_2 = 436$, $\alpha_3 = 72$, $\alpha_4 = 73$ and $\alpha_5 = 1$, we get that

$$y_1^{145} x_2^{436} y_3^{72} y_4^{73} y_5 = 1.$$

This means that modulo 509 we get

$$\log_g x \equiv -\frac{145 \cdot 112 + 436 \cdot 477 + 72 \cdot 378 + 73 \cdot 80 + 331}{145 \cdot 239 + 436 \cdot 274 + 72 \cdot 180 + 73 \cdot 66 + 488} \equiv -\frac{45}{149} \equiv 191.$$

A quick calculation proves that this is the correct discrete logarithm.

E *Exercise* 2.55. Suppose n is prime. Let

$$y_i = x^{t_i} g^{r_i}, \; i = 1, 2, \ldots, \nu.$$

Suppose further that coefficients $\alpha_1, \alpha_2, \ldots, \alpha_\nu$, depending only on the group elements $y_1, y_2, \ldots, y_\nu$, satisfy (2.6). If the coefficients $r_1, r_2, \ldots, r_\nu$, $t_1, t_2, \ldots, t_\nu$ have been chosen uniformly at random, show that $\sum_i \alpha_i t_i$ is divisible by n with probability $1/n$.

T **Proposition 2.20.** *Suppose n is prime and that $\bar{\ell}_1, \bar{\ell}_2, \ldots, \bar{\ell}_l$ are elements of G. Given $l + 1$ distinct, non-trivial relations of the form*

$$y_i = \prod_{j=1}^{l} \bar{\ell}_j^{s_{ij}}, \qquad i = 1, 2, \ldots, l + 1,$$

we can compute $\alpha_1, \alpha_2, \ldots, \alpha_{l+1}$ satisfying $\prod_{i=1}^{l+1} y_i = 1$ using at most $(l+1)^3$ arithmetic operations.

Proof. Let $\mathbf{S}$ be the $(l+1) \times l$ matrix (s_{ij}), where each of the relations defines one row. If we consider $\mathbf{S}$ as a matrix over $\mathbb{F}_n$, it has rank at most l, so there exists a non-zero vector $\boldsymbol{\alpha}$ such that $\boldsymbol{\alpha}\mathbf{S} = \mathbf{0}$. Given such a vector, we get

$$\prod_{i=1}^{l+1} y_i^{\alpha_i} = \prod_{i=1}^{l+1} \left(\prod_{j=1}^{l} \bar{\ell}_j^{s_{ij}} \right)^{\alpha_i} = \prod_{j=1}^{l} \bar{\ell}_j^{\sum_{i=1}^{l+1} \alpha_i s_{ij}} = 1.$$

Gaussian elimination will find a vector in the kernel of $\mathbf{S}$ using at most $(l+1)^3$ arithmetic operations. □

E *Example* 2.22. Continuing with the discrete logarithm problem and relations from Example 2.21, we have the elements 2, 3, 5 and 7 in the subgroup of $\mathbb{F}_p^*$ and the relations:

$$y_1 = 576 = 2^6 \cdot 3^2 \qquad y_2 = 70 = 2 \cdot 5 \cdot 7 \qquad y_3 = 180 = 2^2 \cdot 3^2 \cdot 5$$
$$y_4 = 42 = 2 \cdot 3 \cdot 7 \qquad y_5 = 720 = 2^4 \cdot 3^2 \cdot 5$$

This gives us the matrix

$$\mathbf{S} = \begin{pmatrix} 6 & 2 & 0 & 0 \\ 1 & 0 & 1 & 1 \\ 2 & 2 & 1 & 0 \\ 1 & 1 & 0 & 1 \\ 4 & 2 & 1 & 0 \end{pmatrix}.$$

The Gaussian elimination in $\mathbb{F}_{509}$ proceeds as follows:

$$
\left(\begin{array}{cccc|cccc}
6 & 2 & 0 & 0 & 1 & & & \\
1 & 0 & 1 & 1 & & 1 & & \\
2 & 2 & 1 & 0 & & & 1 & \\
1 & 1 & 0 & 1 & & & & 1 \\
4 & 2 & 1 & 0 & & & & & 1
\end{array}\right)
\sim
\left(\begin{array}{cccc|cccc}
6 & 2 & 0 & 0 & 1 & & & \\
 & 339 & 1 & 1 & 424 & 1 & & \\
 & 171 & 1 & 0 & 339 & & 1 & \\
 & 340 & 0 & 1 & 424 & & & 1 \\
 & 340 & 1 & 0 & 169 & & & & 1
\end{array}\right)
$$

$$
\sim \cdots \sim
\left(\begin{array}{cccc|ccccc}
6 & 2 & 0 & 0 & 1 & & & & \\
 & 339 & 1 & 1 & 424 & 1 & & & \\
 & & 5 & 4 & 508 & 4 & 1 & & \\
 & & 205 & 458 & 204 & 305 & 1 & \\
 & & & 145 & 436 & 72 & 73 & 1
\end{array}\right).
$$

We find $(\alpha_1, \alpha_2, \alpha_3, \alpha_4, \alpha_5)$ in the bottom row of the right half of the matrix.

Proposition 2.21. *Suppose n is prime and that $\bar{\ell}_1, \bar{\ell}_2, \ldots, \bar{\ell}_l$ are elements of G. Suppose further that for a group element y chosen uniformly at random from G we can find a relation of the form*

$$
y = \prod_{j=1}^{l} \bar{\ell}_j^{s_j} \tag{2.7}
$$

with probability σ, using at most τ_s arithmetic operations.

Then, except with probability $1/n$, we can compute $\log_g x$ using an expected

$$
\sigma^{-1}(l+1)(\tau_s + 2\chi) + (l+1)^3 + 2l + 3
$$

arithmetic operations, where χ is the number of arithmetic operations required for an exponentiation in G.

Proof. We choose random elements of the form $x^t g^r$ and for each element use at most τ_s arithmetic operations to try to find a relation of the form (2.7). When we have found $l+1$ relations, we shall apply Proposition 2.20 to compute a relation among the random group elements.

The coefficients s_j in each relation will be independent of the exact random numbers t, r chosen, satisfying the requirements of Exercise 2.55, which then allows us to compute the discrete logarithm except with probability $1/n$.

We expect to find $l+1$ relations after choosing $\sigma^{-1}(l+1)$ random group elements. Sampling a random element costs 2χ arithmetic operations, and trying to find relations costs τ_s arithmetic operations per group element tested. The linear algebra will require $(l+1)^3$ arithmetic operations, and computing $\log_g x$ will cost at most $2l+3$ arithmetic operations. The claim follows. □

If n is a prime power q^k then Proposition 2.20 does not apply. If q is small, the algorithms from Section 2.2 suffice to compute the discrete logarithm quickly. The case where q is not small can be handled as well, but since this is almost never the case in cryptography, we do not discuss the details.

Finally, we consider the case of a more general group order.

Proposition 2.22. *Let* $n = n_1 n_2$, *such that* n_1 *is a large prime that does not divide* n_2, *let* g *be a generator of* G, *let* $\bar{\ell}_1, \bar{\ell}_2, \ldots, \bar{\ell}_l$ *be elements of* G, *and let* H *be the subgroup of* G *of order* n_1. *Also, let* d *be an inverse of* n_2 *modulo* n_1.

Suppose that for a group element y *chosen uniformly at random from* G, *we can find a relation of the form*

$$y = \prod_{j=1}^{l} \bar{\ell}_j^{s_j}$$

with probability σ, *using at most* τ_s *arithmetic operations.*

Then for a group element y' *chosen uniformly at random from* H, *we can find a relation of the form*

$$y' = \prod_{j=1}^{l} (\bar{\ell}_j^{n_2 d})^{s_j}$$

with probability σ, *using at most* $\tau_s + \chi$ *arithmetic operations, where* χ *is the number of arithmetic operations required for an exponentiation in* G.

Proof. If y' has been chosen uniformly at random from H, and b has been chosen uniformly at random from $\{0, 1, \ldots, n_2 - 1\}$, then $y = y'g^{n_1 b}$ has been chosen uniformly at random from G.

With probability σ we can find $s_1, s_2, \ldots, s_l$ such that $y = \prod_{j=1}^{l} \bar{\ell}_j^{s_j}$. Then

$$y' = y^{n_2 d} = \prod_{j=1}^{l} \bar{\ell}_j^{s_j n_2 d} = \prod_{j=1}^{l} (\bar{\ell}_j^{n_2 d})^{s_j}.$$

We have a relation of the required form, and we succeed with probability σ.

Finding the relation requires τ_s arithmetic operations. Generating y' requires χ arithmetic operations. □

This result says that if we can find relations in the big group, we can move that relation into a subgroup. By Proposition 2.21, we can then compute discrete logarithms in the subgroup. Theorem 2.7 now applies. In fact, we can do even better by reusing relations that we find in the big group for each of the subgroups, and not finding new relations for each subgroup.

Now we let $G = \mathbb{F}_p^*$. We want a way to find relations of the form (2.7) for randomly chosen group elements. Let $\ell_1, \ell_2, \ldots, \ell_l$ be the l smallest primes (listed in order, so that $\ell_1 = 2$, $\ell_2 = 3$, etc.), and let $\bar{\ell}_1, \bar{\ell}_2, \ldots, \bar{\ell}_l$ be the corresponding group elements in $\mathbb{F}_p^*$. Define the sets

$$P = \left\{ \prod_{j=1}^{l} \ell_j^{s_j} < p \,\middle|\, s_1, s_2, \ldots, s_l \geq 0 \right\} \quad \text{and} \quad \bar{P} = \{k + \langle p \rangle \mid k \in P\}.$$

The numbers in P are often called *smooth*. For any $y \in \mathbb{F}_p^*$ we can choose a smallest non-negative coset representative and test if $y \in \bar{P}$ by trial division.

E *Exercise 2.56.* Show that we can decide if a group element is in $\bar{P}$ or not using at most $\tau_s = \lfloor l + \log_2 p \rfloor$ integer divisions.

E *Exercise 2.57.* Show that we can use the algorithm implied by the previous exercise to find relations of the form (2.7) for a fraction $\sigma = |\bar{P}|/(p-1)$ of all elements in $\mathbb{F}_p^*$ using at most $\tau_s = \lfloor l + \log_2 p \rfloor$ arithmetic operations.

E *Exercise 2.58.* A group operation in $\mathbb{F}_p^*$ requires two arithmetic operations (one multiplication and one division). Use Proposition 2.1 to show that we can compute an exponentiation (with an exponent smaller than the group order) in $\mathbb{F}_p^*$ using at most $4 \log_2 p$ arithmetic operations.

Example 2.23. Consider $p = 272003$ and the subgroup of $\mathbb{F}_p^*$ of order 307, generated by $g = 231904$, and the group element $x = 4644$. We want to compute $\log_g x$. (Note that $p = 2 \cdot 443 \cdot 307 + 1$.)

We find that $h = 255754$ has order $2 \cdot 443$ modulo p. We choose to use the small primes 2, 3, 5, 7 and 11. There are 1647 integers between 1 and p who have no other prime factors, so we expect to find the 6 relations we need after about 1000 tries. A computer-aided search finds:

$$
\begin{array}{ll}
g^{238} x^4 h^{20} = 1089 & = 3^2 \cdot 11^2 \\
g^{217} x^{264} h^{142} = 180075 & = 3 \cdot 5^2 \cdot 7^4 \\
g^{206} x^{305} h^{355} = 128 & = 2^7 \\
g^{259} x^{261} h^{464} = 26244 & = 2^2 3^8 \\
g^{303} x^{231} h^{64} = 4158 & = 2 \cdot 3^3 \cdot 7 \cdot 11 \\
g^{106} x^{125} h^{793} = 13122 & = 2 \cdot 3^8
\end{array}
$$

We get the matrix

$$
S = \begin{pmatrix}
0 & 2 & 0 & 0 & 2 \\
0 & 1 & 2 & 4 & 0 \\
7 & 0 & 0 & 0 & 0 \\
2 & 8 & 0 & 0 & 0 \\
1 & 3 & 0 & 1 & 1 \\
1 & 8 & 0 & 0 & 0
\end{pmatrix}.
$$

With $\alpha = (0, 0, 1, -7, 0, 7)$ we get $\alpha S = 0$ because $128 \cdot 26244^{-7} \cdot 13122^7 = 1$. We then compute

$$\log_g x \equiv -\frac{206 - 7 \cdot 259 + 7 \cdot 106}{305 - 7 \cdot 261 + 7 \cdot 125} \equiv 11 \pmod{307}.$$

For simplicity, we consider a single large prime divisor q of $p - 1$ such that q^2 does not divide $p - 1$. With the above results, Proposition 2.21 and Exercises 2.56 and 2.58 say that we can compute logarithms modulo q using

$$\sigma^{-1}(l+1)(l + \log_2 p + 8 \log_2 p) + (l+1)^3 + 2l + 3$$

arithmetic operations, where $\sigma = |\bar{P}|/(p-1)$. We expect l to be much larger than $\log_2 p$, so we get an approximate cost

$$\sigma^{-1}l^2 + l^3. \tag{2.8}$$

Making l larg will make the fraction σ large, but that will also increase the cost of generating relations and the number of relations we need. Making l too large may increase the total cost, so we need to find a good value.

We must estimate $\sigma^{-1} = (p-1)/|P| \approx p/|P|$. We begin by taking logarithms in the equation $\prod_j \ell_j^{s_j} < p$ and observing that the number of integers in P is the same as the number of non-negative integer solutions to the inequality

$$\sum_{j=1}^{l} s_j \log \ell_j < \log p.$$

Observing that small primes have approximately the same logarithm, we can instead count the number of non-negative integer solutions to the inequality

$$\sum_{j=1}^{l} s_j < \frac{\log p}{\log \ell_l}.$$

Letting $u = \log p / \log \ell_l$, we get that

$$|P| \approx \binom{\lfloor u \rfloor + l}{l} = \frac{(\lfloor u \rfloor + l)!}{\lfloor u \rfloor! l!}.$$

Taking logarithms, using Stirling's approximation and $\lfloor u \rfloor \approx u$, we get that

$$\log |P| \approx (u+l)\log(u+l) - (u+l) - (u\log u - u) - (l\log l - l).$$

If we additionally assume that u is much smaller than l, we get

$$\log |P| \approx (u+l)\log l - u - l - u\log u + u - l\log l + l = u\log l - u\log u.$$

The prime number theorem says that $l \approx \ell_l/\log \ell_l$, so we get that

$$\begin{aligned}
\log \sigma^{-1} &\approx \log p - \log |P| \approx u\log \ell_l - u\log l + u\log u \\
&= u\log \ell_l - u\log \ell_l + u\log\log \ell_l + u\log\log p - u\log\log \ell_l \\
&= \frac{\log p}{\log \ell_l}\log\log p.
\end{aligned}$$

We get the rough estimate

$$\sigma^{-1} \approx \exp\left(\frac{\log p}{\log \ell_l} \log \log p\right).$$

The prime number theorem says that

$$l \approx \ell_l / \log \ell_l = \exp\left(\log \ell_l - \log \log \ell_l\right),$$

which means that (2.8) gives us an approximate cost of

$$\exp\left(\frac{\log p \log \log p}{\log \ell_l} + 2 \log \ell_l - 2 \log \log \ell_l\right) + \exp\left(3 \log \ell_l - 3 \log \log \ell_l\right).$$

Ignoring the $\log \log \ell_l$ terms, we want to choose ℓ_l so as to make the sum

$$\exp\left(\frac{\log p \log \log p}{\log \ell_l} + 2 \log \ell_l\right) + \exp\left(3 \log \ell_l\right)$$

as small as possible. The sum is dominated by the exponential with the biggest exponent. The second exponent increases monotonously with increasing ℓ_l. Since the first exponent has its minimum $\sqrt{8}\sqrt{\log p \log \log p}$ at

$$\log \ell_l = \sqrt{\log p \log \log p / 2}$$

where the second exponent takes the value $\sqrt{9/2}\sqrt{\log p \log \log p}$, we see that taking the larger value for $\log \ell_l$ should approximate a minimum. In particular, by using this value we should be able to compute discrete logarithms using approximately

$$\exp\left(\sqrt{8}\sqrt{\log p \log \log p}\right)$$

arithmetic operations.

We have arrived at the following requirement.

Requirement 2.6. If G is any subgroup of $\mathbb{F}_p^*$, then $\exp(\sqrt{8 \log p \log \log p})$ arithmetic operations must be an infeasible computation.

There are much better algorithms for computing discrete logarithms in finite fields. While we shall not study these algorithms, we note that the above requirement is not the final requirement.

E | *Exercise* 2.59. Suppose we want to compute a number (say hundreds) of discrete logarithms in a prime-order subgroup of $\mathbb{F}_p^*$. That is, suppose we have $x_1, \ldots, x_\nu \in G$, $G \subseteq \mathbb{F}_p$, and want to compute $\log_g x_1, \ldots, \log_g x_\nu$.

Show that given ν relations

$$g^{v_i} \prod_i x_i^{u_i} = 1,$$

you can compute the discrete logarithms. Show that the above index calculus algorithms can be adapted to find such relations. Show that this computation is much faster than computing the ν discrete logarithms separately.

Hint: Be optimistic about linear independence when needed.

2.5 ELLIPTIC CURVES

Elliptic curves have been studied for a long time by number theorists and a rich and varied theory has been developed. We are interested in elliptic curves because the points on an elliptic curve over a finite field forms a group that is suitable for use in cryptography.

From a mathematical point of view, studying elliptic curves over any field is interesting. From a cryptographic point of view, our groups come from elliptic curves over finite fields, which must therefore be our main interest. To simplify our presentation, we shall restrict ourselves to elliptic curves of a special form defined over prime fields. We note that essentially all of the theory we discuss works equally well for elliptic curves defined over other fields, though sometimes with minor modifications.

Even though we only discuss curves over finite prime fields, it is still convenient to use drawings of curves over the real numbers to illustrate ideas.

We begin by considering the algebraic curve C defined over the field $\mathbb{F}_p$, p a large prime, given by the polynomial equation

$$Y^2 = X^3 + AX + B, \qquad A, B \in \mathbb{F}_p. \tag{2.9}$$

The *points* on the curve are the points in the affine plane that satisfy the curve equation. However, we cannot restrict the coordinates of the points to be elements of $\mathbb{F}_p$. We fix an algebraic closure $\bar{\mathbb{F}}$ of $\mathbb{F}_p$ and consider the points on the curve to be all the pairs $(x, y) \in \bar{\mathbb{F}}^2$ satisfying the curve equation.

A point on a curve with coordinates in $\mathbb{F}_p$ is $\mathbb{F}_p$-*rational* or just *rational*.

Example 2.24. Consider the curve over $\mathbb{F}_{13}$ defined by $Y^2 = X^3 + X + 2$. The points in $\mathbb{F}_p^2$ satisfying the equation are (observe that $11 = -2$, $8 = -5$, etc.)

$$(1, 2), (1, 11), (2, 5), (2, 8), (6, 4), (6, 9), (7, 1), (7, 12), (9, 5), (9, 8), (12, 0).$$

A curve is *smooth* if its partial derivatives never vanish all at the same time for points on the curve.

Example 2.25. Consider the curve over $\mathbb{F}_{13}$ defined by $Y^2 = X^3 + X + 3$. The point $(2, 0)$ is on the curve. Since both partial derivatives vanish at $(2, 0)$, there is no well-defined tangent at that point.

The slope of a line passing through two distinct points on the curve is well-defined. Over a finite field, the tangent line can no longer be defined in terms of limits, but we can still use formal derivatives to define tangent lines.

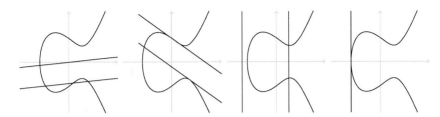

FIGURE 2.3 Intersection of lines and elliptic curves. From left to right: three distinct intersection points, possibly with complex Y-coordinates; tangent lines with one or two distinct intersection points; two distinct intersection points, possibly with complex Y-coordinates; and a tangent line with one intersection point.

E | *Exercise* 2.60. Show that a smooth curve has a well-defined, unique tangent line at any point on the curve.

T | **Proposition 2.23.** *An algebraic curve defined by* (2.9) *is smooth if and only if the polynomial* $X^3 + AX + B$ *has three distinct zeros.*

Proof. We first compute the partial derivative with respect to Y to get

$$2Y = 0.$$

Thus, both partial derivatives can only vanish on the X-axis.

It follows that the X-coordinate of a point where both partial derivatives vanish will be a zero of both $X^3 + AX + B$ and its derivative $3X^2 + A$. The only way a polynomial and its derivative can have a common zero is if the polynomial has a double or triple zero. We may conclude that the curve is smooth if and only if the polynomial $X^3 + AX + B$ has three distinct zeros. $\square$

T | **Fact 2.24.** *A polynomial* $aX^3 + bX^2 + cX + d$ *has three distinct zeros if and only if its* discriminant $b^2c^2 - 4ac^3 - 4b^3d - 27a^2d^2 + 18abcd$ *is non-zero.*

In general, an elliptic curve is a smooth cubic curve with one chosen point, but we shall restrict attention to curves of the form we have already discussed.

D | **Definition 2.5.** *An* elliptic curve E *over the field* $\mathbb{F}_p$ *is a cubic curve over* $\mathbb{F}_p$ *given by* (2.9) *with* $4A^3 + 27B^2 \neq 0$.

E | *Example* 2.26. Over $\mathbb{F}_{13}$, the curve defined by $Y^2 = X^3 + X + 2$ is an elliptic curve, while the curve defined by $Y^2 = X^3 + X + 3$ is not.

We need to study the points of intersection between elliptic curves and straight lines. The situation can be illustrated over the real numbers as in Figure 2.3. Lines can intersect the curve in zero, one, two or three points. This is untidy, and we want a better understanding of what is going on.

T **Proposition 2.25.** *Let E be an elliptic curve defined by $Y^2 = X^3 + AX + B$. If L is a line defined by $Y = \alpha X + \beta$, then the X-coordinates of the points of intersection are the zeros of the polynomial*

$$X^3 - \alpha^2 X^2 + (A - 2\alpha\beta)X + B - \beta^2. \tag{2.10}$$

If L is a line defined by $X = \beta$, then the Y-coordinates of the points of intersection are the zeros of the polynomial

$$Y^2 - \beta^3 - A\beta - B. \tag{2.11}$$

Proof. We begin with a line L of the form $Y = \alpha X + \beta$. We want to compute the intersection of this line and an elliptic curve defined by $Y^2 = X^3 + AX + B$. Using the line equation to eliminate Y from the curve equation, we find

$$\alpha^2 X^2 + 2\alpha\beta X + \beta^2 = X^3 + AX + B.$$

The solutions to this equation corresponds to the zeros of the cubic polynomial (2.10). If x is any zero of this polynomial, we know that the point $(x, \alpha x + \beta)$ is on both the line and the curve and therefore a point of intersection.

Next, we consider a line L' of the form $X = \beta$. This time, we use the line equation to eliminate X from the curve equation and get $Y^2 = \beta^3 + A\beta + B$. The solutions to this equation corresponds to the zeros of the quadratic polynomial (2.11). If y is any zero of this polynomial, we know that the point (β, y) is on both the line and the curve, and therefore a point of intersection. □

E *Example 2.27.* Consider the elliptic curve over $\mathbb{F}_{13}$ defined by $Y^2 = X^3 + X + 2$, and the lines $X = 2$, $X = 3$, $X = 12$, $Y = X + 3$ and $Y = X + 4$ and $Y = 2X$.

The line $X = 2$ gives us the equation $Y^2 = 2^3 + 2 + 2 = 12$. We see that $5^2 = 25 = 12$, so the zeros of this polynomial equation are $Y = \pm 5$, giving us the intersection points $(2, 5)$ and $(2, -5) = (2, 8)$.

The line $X = 3$ gives us the equation $Y^2 = 3^3 + 3 + 2 = 6$, but 6 is not a square in $\mathbb{F}_{13}$, so the line does not intersect the curve in any rational points.

The line $X = 12$ gives us the equation $Y^2 = 12^3 + 12 + 2 = 0$, so 0 is a double zero of the polynomial. This gives us a single intersection point $(12, 0)$.

The line $Y = X + 3$ gives us the equation $X^2 + 6X + 9 = X^3 + X + 2$, or $X^3 - X^2 - 5X - 7 = 0$. A quick search tells us that the left hand side is $(X - 2)(X - 6)^2$, which means that there are two intersection points, $(2, 5)$ and $(6, 9)$, where $(6, 9)$ is a double zero and the line is a tangent.

The line $Y = X + 4$ gives us the equation $X^2 + 8X + 3 = X^3 + X + 2$, or $X^3 - X^2 - 7X - 1 = 0$. A quick computation tells us that this equation has no rational zeros (solutions in $\mathbb{F}_p$).

The line $Y = 2X$ gives us the equation $4X^2 = X^3 + X + 2$, or $X^3 - 4X^2 + X + 2 = 0$. The left hand side factors as $(X - 1)(X - 7)(X - 9)$, which gives us the intersection points $(1, 2)$, $(7, 1)$ and $(9, 5)$.

T **Fact 2.26.** *Let $\mathbb{F}$ be a field and let $\bar{\mathbb{F}}$ be an algebraic closure of $\mathbb{F}$. Counting multiplicities, a polynomial of degree d over $\mathbb{F}$ has d zeros in $\bar{\mathbb{F}}$.*

D **Definition 2.6.** Let E be an elliptic curve, P a point on E and L a line intersecting E in P. The *multiplicity* of the point of intersection P is the multiplicity of the corresponding zero of the polynomials in (2.10) or (2.11).

E *Exercise* 2.61. Let E be an elliptic curve, P a point on E and L a line intersecting E in P. Show that the multiplicity of the intersection point is greater than 1 if and only if L is tangent to E at P.

E *Exercise* 2.62. Let E be an elliptic curve. Show that E has 0, 1 or 3 rational points with vertical tangents.

E *Example* 2.28. The elliptic curve over $\mathbb{F}_{13}$ defined by $Y^2 = X^3 + X + 2$ has just one rational point with vertical tangent, namely $(12, 0)$.
   The elliptic curve over $\mathbb{F}_{13}$ defined by $Y^2 = X^3 + 2X + 6$ has three rational points with vertical tangents, namely $(3, 0)$, $(4, 0)$ and $(6, 0)$.
   The elliptic curve over $\mathbb{F}_{13}$ defined by $Y^2 = X^3 + 2X + 4$ has no rational points with vertical tangents.

   Returning to Figure 2.3, in the left drawing we see two lines, one of which has three real points of intersection. In this case, the polynomial (2.10) has three real solutions. The second line seemingly has only one point of intersection. In this case, (2.10) has only one real zero. But it also has two complex zeros, and these zeros correspond to points of intersection with complex coordinates. Over a general field, all three points of intersection may have X-coordinates in an extension field.

   The second drawing in Figure 2.3 again has two lines, one of which has two points of intersection and one which has a single point of intersection. The corresponding cubic polynomial (2.10) does not have complex zeros, but double or triple zeros. The intersection point has multiplicity equal to the multiplicity of the corresponding zero. We do not have three distinct points of intersection, but if we count multiplicity, we have three points of intersection.

   In the third drawing in Figure 2.3, by considering complex coordinates, we see that both lines have two distinct points of intersection. In the fourth drawing, by counting multiplicity, we get two points of intersection. But we never get three points of intersection, because (2.11) is a degree two polynomial.

   It is very inconvenient that vertical lines have only two points of intersection, while non-vertical lines have three points of intersection. It would have been nice if every line intersected the curve in three points, counting multiplicity. The proper solution to this issue lies in projective geometry, but in this text we shall choose a much simpler solution.

We declare that one extra point $\mathcal{O}$ exists with the following properties:

$\mathcal{O}$-1 The point $\mathcal{O}$ does not lie in the plane (hence has no coordinates), but lies on the curve and is a rational point.

$\mathcal{O}$-2 Any line of the form $X = \beta$ intersects the curve in $\mathcal{O}$ with multiplicity one. No line of the form $Y = \alpha X + \beta$ intersects the curve in $\mathcal{O}$.

$\mathcal{O}$-3 The curve has a tangent line in $\mathcal{O}$, and that line intersects the curve only in $\mathcal{O}$, with multiplicity three.

The special point $\mathcal{O}$ is often called the *point at infinity*.

As we saw in the discussion of Figure 2.3, considering coordinates in an algebraic closure and the notion of multiplicity somewhat simplifies the situation with regard to intersection points. Now that we have introduced the point at infinity, we get the following nice results.

Proposition 2.27. *Let E be an elliptic curve defined over $\mathbb{F}_p$, and let L be a line. Counting multiplicities, L intersects E in exactly three points. If two of the intersection points are rational, then the third point is also rational.*

Proof. The first claim follows from Proposition 2.25 and the properties of $\mathcal{O}$.

If the two points are both $\mathcal{O}$, then by $\mathcal{O}$-2 and $\mathcal{O}$-3 the line must be the tangent line to E at $\mathcal{O}$, which means that the third point of intersection is again $\mathcal{O}$, which is rational.

If only one point is $\mathcal{O}$, then by $\mathcal{O}$-2 the line takes the form $X = \beta$. By (2.11), the other two points of intersection must be $(\beta, \pm y)$ for some $y \in \bar{\mathbb{F}}_p$. It is then clear that either both or none of these points are rational.

Finally, suppose neither of the two rational points is $\mathcal{O}$. We consider two cases. If the line is of the form $X = \beta$, then by $\mathcal{O}$-2 the third point of intersection is $\mathcal{O}$, which is rational.

Otherwise, the line is of the form $Y = \alpha X + \beta$. If the line passes through two distinct rational points, we know that $\alpha, \beta \in \mathbb{F}_p$. If the multiplicity of the intersection point is 2 or greater, Exercise 2.61 says that the line is a tangent, which means that $\alpha, \beta \in \mathbb{F}_p$.

This means that the polynomial (2.10) has coefficients in $\mathbb{F}_p$. Also, two of its zeros lie in $\mathbb{F}_p$, which means that the third zero also is in $\mathbb{F}_p$. This means that the X-coordinate of the third point of intersection lies in $\mathbb{F}_p$, and since it lies on the line $Y = \alpha X + \beta$, the third point of intersection is rational. □

Proposition 2.28. *Let P, Q be points (not necessarily distinct) on an elliptic curve E. Then there exists a unique line L and a unique point R such that P, Q and R are the points of intersection of E and L.*

Proof. If $P = Q = \mathcal{O}$ then $\mathcal{O}$-2 and $\mathcal{O}$-3 say that $R = \mathcal{O}$.

If P and Q are distinct points and one of them is $\mathcal{O}$, then the vertical line through the other point is determined by that point. By Proposition 2.27, this line intersects the curve in one more point, though not necessarily distinct.

If P and Q are distinct points and neither of them is $\mathcal{O}$, the line through the points is uniquely determined. By Proposition 2.27, this line intersects the curve in one more point, though not necessarily distinct.

If $P = Q \neq \mathcal{O}$, then the line we are looking for must be a tangent line. By Exercise 2.60 the tangent line is unique, and by Proposition 2.27 this line intersects the curve in one more point, though not necessarily distinct.   □

Example 2.29. Consider the elliptic curve over $\mathbb{F}_{13}$ defined by $Y^2 = X^3+X+2$.

The points $(1,2)$ and $(7,1)$ both lie on the curve, and uniquely determines the line $Y = 2X$. The line also intersects the curve in a third point $(9,5)$.

The point $(6,9)$ lies on the curve. The line $Y = X+3$ is the unique tangent to the curve at $(6,9)$. The line also intersects the curve in the point $(2,5)$.

The points $\mathcal{O}$ and $(2,5)$ both line on the curve. The line $X = 2$ is, by $\mathcal{O}$-2, the unique line determined by the two points. The line also intersects the curve in a third point $(2,-5)$.

We conclude by precisely defining the points on an elliptic curve are.

Definition 2.7. The *points* on an elliptic curve E over $\mathbb{F}_p$ defined by

$$Y^2 = X^3 + AX + B, \qquad 4A^3 + 27B^2 \neq 0,$$

are the points with coordinates in the algebraic closure $\bar{\mathbb{F}}$ of $\mathbb{F}_p$ satisfying the curve equation, plus the special point $\mathcal{O}$:

$$E(\bar{\mathbb{F}}) = \{(x,y) \in \bar{\mathbb{F}}^2 \mid y^2 = x^3 + Ax + B\} \cup \{\mathcal{O}\}.$$

The $\mathbb{F}_p$-*rational* (or just rational) points on E are the points with coordinates in $\mathbb{F}_p$, plus the special point $\mathcal{O}$:

$$E(\mathbb{F}_p) = \{(x,y) \in \mathbb{F}_p^2 \mid y^2 = x^3 + Ax + B\} \cup \{\mathcal{O}\}.$$

Example 2.30. Consider the elliptic curve over $\mathbb{F}_{13}$ defined by $Y^2 = X^3+X+2$. The rational points on the curve are

$$\mathcal{O}, (1,2), (1,11), (2,5), (2,8), (6,4), (6,9), (7,1), (7,12), (9,5), (9,8), (12,0).$$

Except for $\mathcal{O}$, this is exactly as in Example 2.24

The rational points on an elliptic curve form a group. We are interested in how many rational points an elliptic curve has.

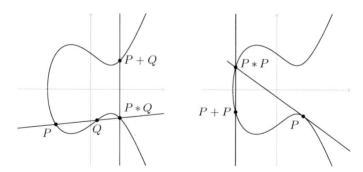

FIGURE 2.4 The group operation on elliptic curves. Addition of distinct points is shown on the left, point doubling is shown on the right.

Fact 2.29 (Hasse's theorem). *Let E be an elliptic curve defined over $\mathbb{F}_p$. The number of rational points on E is*

$$|E(\mathbb{F}_p)| = p + 1 - t, \qquad |t| \le 2\sqrt{p}.$$

Example 2.31. Consider the elliptic curve over $\mathbb{F}_{13}$ defined by $Y^2 = X^3 + X + 2$. We saw in Example 2.30 that there are 12 rational points on this curve, including $\mathcal{O}$. We get that $12 = 13 + 1 - 2$, so in this case $t = 2$.

Example 2.32. Consider the elliptic curve over $\mathbb{F}_{13}$ defined by $Y^2 = X^3 + 4$. An exhaustive computation shows that there are 21 rational points on this curve, including $\mathcal{O}$. We get that $21 = 13 + 1 - (-7)$, so in this case $t = -7$.

2.5.1 Group Operation

We are now ready to turn the set of points on an elliptic curve into a group. We begin by defining a binary operation on the points of the elliptic curve, and then define the actual group operation in terms of the binary operation, as shown in Figure 2.4.

Definition 2.8. We define two binary operations $*$ and $+$ on an elliptic curve E as: $P * Q$ is the point given by Proposition 2.28, and $P + Q = (P * Q) * \mathcal{O}$.

Example 2.33. Consider the elliptic curve over $\mathbb{F}_{13}$ defined by $Y^2 = X^3 + X + 2$.

Since the unique line through $(1, 2)$ and $(7, 1)$ intersects the curve in a third point $(9, 5)$, we find that $(1, 2) * (7, 1) = (9, 5)$.

The line through $(9, 5)$ and $\mathcal{O}$ intersects the curve in $(9, -5)$, so according to the definition $(1, 2) + (7, 1) = ((1, 2) * (7, 1)) * \mathcal{O} = (9, -5)$.

The unique line through $(6, 9)$ that is tangent to the curve intersects the curve in the point $(2, 5)$, so $(6, 9) * (6, 9) = (2, 5)$.

The line through $(2, 5)$ and $\mathcal{O}$ intersects the curve in $(2, -5)$, so $(6, 9) + (6, 9) = (2, -5)$.

The unique line through $(2, 5)$ and $(2, -5)$ intersects the curve in $\mathcal{O}$. The line through $\mathcal{O}$ and $\mathcal{O}$ is the tangent at $\mathcal{O}$, which intersects only there with multiplicity 3. In other words, $(2, 5) + (2, -5) = \mathcal{O}$.

We shall show that there exists an identity element for $+$, there exists inverses for $+$ and that it is commutative. We shall not show that $+$ is associative. There are many ways to do so, but they are either tedious or advanced.

E *Exercise* 2.63. Let E be an elliptic curve and let P, Q be points on E. Show that $P * Q = Q * P$, and consequently that $P + Q = Q + P$.

T **Proposition 2.30.** *Let E be an elliptic curve and let P be a point on E. Then $P + \mathcal{O} = \mathcal{O} + P = P$.*

Proof. First, suppose $P = \mathcal{O}$. By $\mathcal{O}$-2 and $\mathcal{O}$-3, we see that the third point of intersection identified by Proposition 2.28 must be $\mathcal{O}$. It follows that $\mathcal{O} * \mathcal{O} = \mathcal{O}$ and that $\mathcal{O} + \mathcal{O} = \mathcal{O}$.

Next, suppose $P \neq \mathcal{O}$. By $\mathcal{O}$-2 the line through $\mathcal{O}$ and P intersects the curve in some point $Q \neq \mathcal{O}$ and $P * \mathcal{O} = Q$. Therefore, the line through P and Q has $\mathcal{O}$ as its third point of intersection, which means that $P + \mathcal{O} = P$. □

T **Proposition 2.31.** *Let E be an elliptic curve and let $P = (x, y)$ be a point on E. Let $Q = (x, -y)$. Then Q is also on the curve, $P * \mathcal{O} = Q$ and $P + Q = \mathcal{O}$.*

Proof. It is immediately clear that if P is on the curve, then so is Q.

If $y = 0$, then $P = Q$ and the tangent in that point is a vertical line, which intersects the curve in $\mathcal{O}$ by $\mathcal{O}$-2, so $P * Q = \mathcal{O}$ and $P + Q = \mathcal{O}$.

If $y \neq 0$, the line through P and Q is vertical, so by $\mathcal{O}$-2 intersects the curve in $\mathcal{O}$. It follows that $P * Q = \mathcal{O}$, that $P + Q = \mathcal{O}$ and that $P * \mathcal{O} = Q$. □

T **Fact 2.32.** *Let E be an elliptic curve and let P, Q, R be points on E. Then $(P + Q) + R = P + (Q + R)$.*

T **Theorem 2.33.** *Let E be an elliptic curve. The set of points on E is a commutative group under $+$. The set of rational points is a subgroup.*

Proof. The fact that the set of points is a group follows from Propositions 2.30 and 2.31 and Fact 2.32. Commutativity follows from Exercise 2.63.

That the rational points form a subgroup follows from Proposition 2.28. □

We conclude by developing explicit formulas for computing $P + Q$.

By Proposition 2.30, if either point to be added is $\mathcal{O}$, the answer is the other point. Otherwise, we may assume that $P = (x_1, y_1)$ and $Q = (x_2, y_2)$.

By Proposition 2.31, if $x_1 = x_2$ and $y_1 = -y_2$, then the answer is $\mathcal{O}$.

Otherwise, we need to find the third point of intersection of the line through P and Q. We begin by finding the slope α of the line, after which the constant term is $\beta = y_1 - \alpha x_1$. If $x_1 = x_2$, then $P = Q$, and we must use the tangent. The tangent line has slope

$$\alpha = \frac{3x_1^2 + A}{2y_1}.$$

If $x_1 \neq x_2$, then the slope is

$$\alpha = \frac{y_2 - y_1}{x_2 - x_1}.$$

The X-coordinates of the intersection points of this line and the elliptic curve are zeros of (2.10). We know two of the zeros, namely x_1 and x_2, and we need to find the third zero x_3. The monic polynomial in (2.10) should be equal to the polynomial $(X - x_1)(X - x_2)(X - x_3)$. Comparing the coefficients of the X^2 term, we get that $-\alpha^2 = -x_1 - x_2 - x_3$, or

$$x_3 = \alpha^2 - x_1 - x_2.$$

The Y-coordinate of the third point is $\alpha x_3 + y_1 - \alpha x_1 = \alpha(x_3 - x_1) + y_1$. By Proposition 2.31, the Y-coordinate of $P + Q$ is the negative of this, which is

$$y_3 = \alpha(x_1 - x_3) - y_1.$$

We summarise the above in the following result.

Proposition 2.34. *Let E be an elliptic curve and P, Q be points on E.*

- *If $P = \mathcal{O}$, then $P + Q = Q$.*

- *If $Q = \mathcal{O}$, then $P + Q = P$.*

If neither point is $\mathcal{O}$, then let $P = (x_1, y_1)$ and $Q = (x_2, y_2)$.

- *If $x_1 = x_2$ and $y_1 = -y_2$, then $P + Q = \mathcal{O}$.*

- *Otherwise, $P + Q = (x_3, y_3)$ with $x_3 = \alpha^2 - x_1 - x_2$ and $y_3 = \alpha(x_1 - x_3) - y_1$, where*

$$\alpha = \begin{cases} \frac{3x_1^2 + A}{2y_1} & x_1 = x_2, \\ \frac{y_2 - y_1}{x_2 - x_1} & x_1 \neq x_2. \end{cases}$$

Example 2.34. Consider the elliptic curve over $\mathbb{F}_{13}$ defined by $Y^2 = X^3 + X + 2$. We want to add $(1, 2)$ and $(7, 1)$. We compute the slope as $-1/6 = 2$, and then $x_3 = 2^2 - 1 - 7 = 9$ and $y_3 = 2(1 - 9) - 2 = 8$, that is

$$(1, 2) + (7, 1) = (9, -5).$$

We want to add $(6, 9)$ to itself. We compute the slope as $(3 \cdot 6^2 + 1)/(2 \cdot 9) = 1$, and then $x_3 = 1^2 - 6 - 6 = 2$ and $y_3 = 6 - 2 - 9 = -5$, that is

$$(6, 9) + (6, 9) = (2, -5).$$

The addition operation is written additively, so the *exponentiation* we have studied in Sections 2.1 and 2.2 is now called *point multiplication*

$$aP = \underbrace{P + P + \cdots + P}_{a \text{ terms}}.$$

Even though the notation we have chosen for the elliptic curve group operation suggests addition and not multiplication, we use the words from Definition 2.2 and say that the discrete logarithm of a point Q to the base P is the smallest non-negative integer a such that $Q = aP$. We write $\log_P Q = a$.

E | *Exercise* 2.64. Redo Exercise 2.2 for point multiplications. Is anything different? What about Exercise 2.4?

2.5.2 Finding Suitable Curves

We have shown that the points on an elliptic curve form a commutative group, and the set of rational points is a finite commutative group. As we saw in Section 2.2, we need a cyclic group whose order is divisible by a large prime.

E | *Exercise* 2.65. Show that an elliptic curve has 0, 1 or 3 rational points of order 2.

Hint: Use Exercise 2.62.

E | *Exercise* 2.66. Let E be the curve defined by $Y^2 = X^3 + 12$ over the field $\mathbb{F}_{13}$. Show that E is not cyclic.

The group of rational points is not cyclic in general, but there must be a large cyclic subgroup.

T | **Fact 2.35.** *Let E be an elliptic curve defined over $\mathbb{F}_p$. Then there exists n_1, n_2, where n_1 divides both n_2 and $p - 1$, such that*

$$E(\mathbb{F}_p) \simeq \mathbb{Z}_{n_1}^+ \times \mathbb{Z}_{n_2}^+.$$

But a large cyclic subgroup is not sufficient for our purposes, we also need to know that its order is divisible by a large prime.

T | **Proposition 2.36.** *Let p be a prime congruent to 2 modulo 3, and let E be an elliptic curve defined by $Y^2 = X^3 + B$ over $\mathbb{F}_p$. Then $|E(\mathbb{F}_p)| = p + 1$.*

Proof. Since $p \equiv 2 \pmod 3$, 3 is invertible modulo $p-1$. Then the map $\zeta \mapsto \zeta^3$ is invertible. If $k \equiv 3^{-1} \pmod{p - 1}$, then $\zeta \mapsto \zeta^k$ is the inverse map.

For any value γ, the equation $\gamma = X^3 + B$ therefore has the unique solution $(\gamma - B)^k$ in $\mathbb{F}_p$. We conclude that for every possible Y-coordinate y,

$$((y^2 - B)^k, y)$$

is a point on the curve, and it is the only point with Y-coordinate y. There are p such points and these are all the points with coordinates, which when counting $\mathcal{O}$ makes for $p + 1$ points on the curve. $\qquad\square$

Unfortunately, curves of the form $Y^2 = X^3 + B$ are so-called *supersingular*, which are less suitable for our purposes than ordinary elliptic curves.

For a randomly chosen curve, the number of points is close to evenly distributed within the range given by Hasse's theorem. Since a curve with a prime (or close to prime) number of points would be reasonable, if we can determine the number of rational points on a curve, we could easily find a suitable curve.

We shall briefly sketch Schoof's algorithm for counting points on an elliptic curve. The story begins with the *Frobenius map*. Let E be an elliptic curve defined over $\mathbb{F}_p$ and define a map $\phi : E(\bar{\mathbb{F}}) \to E(\bar{\mathbb{F}})$ by

$$\phi(P) = \begin{cases} \mathcal{O} & P = \mathcal{O}, \text{ and} \\ (x^p, y^p) & P = (x, y). \end{cases}$$

E | *Exercise* 2.67. Show that ϕ is a well-defined map.

E | *Exercise* 2.68. Show that ϕ is a group homomorphism.

The fixed field of the map $\bar{\mathbb{F}} \to \bar{\mathbb{F}}$ given by $\alpha \mapsto \alpha^p$ is $\mathbb{F}_p$. It follows that the set of fixed points of ϕ is the set of rational points on E.

Hasse's theorem (Fact 2.29) says that there is a number t such that the number of rational points is $p + 1 - t$. It turns out that this number is closely related to the Frobenius map.

T | **Fact 2.37.** *Let E be an elliptic curve defined over $\mathbb{F}_p$, and let t be the integer such that the number of rational points on E is $p + 1 - t$. Then*

$$\phi^2(P) - t\phi(P) + pP = \mathcal{O} \tag{2.12}$$

for any point $P \in E(\bar{\mathbb{F}})$.

The action of the Frobenius map on points of small order will be important.

D | **Definition 2.9.** Let ℓ be an integer, $\ell > 0$. The set of ℓ-torsion points is

$$E[\ell] = \{P \in E(\bar{\mathbb{F}}) \mid \ell P = \mathcal{O}\}.$$

If $P \in E[\ell]$, then (2.12) becomes

$$\phi^2(P) - t_\ell \phi(P) + pP = \mathcal{O},$$

where $t_\ell = t \bmod \ell$. Reordering terms, we get

$$\phi^2(P) + pP = t_\ell \phi(P).$$

This is nothing more than a discrete logarithm problem, but note that the subgroup generated by $\phi(P)$ has order ℓ. If ℓ is not too big and computations on ℓ-torsion points are not too expensive, it will be possible to recover t_ℓ.

Note that $t \equiv t_\ell \pmod{\ell}$. If we can find t_ℓ for many small primes ℓ whose product is greater than $4\sqrt{p}$, we can recover t and thereby the group order.

It remains to show that we can find ℓ-torsion points and compute with them.

Fact 2.38. *Let E be an elliptic curve. There exists an efficiently computable sequence of polynomials (called* division polynomials*) $\psi_1(X,Y), \psi_2(X,Y), \ldots$ such that*

- *for any $P = (x,y) \in E(\bar{\mathbb{F}})$, $\psi_\ell(x,y) = 0$ if and only if $P \in E[\ell]$; and*

- *for odd ℓ, $\psi_\ell(X,Y)$ is a polynomial in X only, and has degree $(\ell^2 - 1)/2$.*

Fact 2.39. *Let E be an elliptic curve over $\mathbb{F}_p$. If p does not divide ℓ, then $E[\ell] \simeq \mathbb{Z}_\ell^+ \times \mathbb{Z}_\ell^+$.*

We now have a polynomial that characterises all the X-coordinates of the ℓ-torsion points. Suppose that ℓ is an odd prime and smaller than p.

We can decide if $\psi_\ell(X)$ has rational zeros by computing $\gcd(X^p - X, \psi_\ell(X))$. It must either have 0, $(\ell - 1)/2$ or $(\ell^2 - 1)/2$ rational zeros. Since each X-coordinate gives rise to two points, for the latter two cases we immediately know that $p + 1 - t$ is congruent to 0 modulo ℓ or ℓ^2, respectively.

More usually, $\psi_\ell(X)$ will not have any rational zeros. But it will usually not be irreducible. Let $f(X)$ be an irreducible factor of $\psi_\ell(X)$ of degree d. Then by constructing the extension field

$$\mathbb{F}_{p^d} \simeq \mathbb{F}_p[X]/\langle f(X) \rangle,$$

we know that the element $x \in \mathbb{F}_{p^d}$ corresponding to $X + \langle f(X) \rangle$ is a zero of $f(X)$ and hence of $\psi_\ell(X)$ and hence the X-coordinate of an ℓ-torsion point.

We now have the X-coordinate of an ℓ-torsion point, but we do not have a Y-coordinate. We can find a Y-coordinate by computing a square root of $x^3 + Ax + B$. Note that sometimes, we have to go to yet another field extension to find a square root, but this is unproblematic.

Factoring $\psi_\ell(X)$ is possible but costly. Instead of finding an irreducible factor of $\psi_\ell(X)$, we can compute in the factor ring

$$\mathbb{F}_p[X]/\langle \psi_\ell(X) \rangle.$$

If $\psi_\ell(X)$ factors into distinct irreducibles as $f_1(X) f_2(X) \cdots f_l(X)$, then

$$\mathbb{F}_p[X]/\langle \psi_\ell(X) \rangle \simeq \mathbb{F}_p[X]/\langle f_1(X) \rangle \times \cdots \times \mathbb{F}_p[X]/\langle f_l(X) \rangle.$$

Computing in this ring is just a simultaneous computation in every possible field extension where we could have found X-coordinates for ℓ-torsion points.

Since our computations involve divisions, and our ring contains non-invertible elements, the computation may not be possible to complete. However, when we cannot find an inverse of some element, we find a divisor of $\psi_\ell(X)$. If this happens, we simply restart the computation using one of the factors of $\psi_\ell(X)$. Eventually, we must either complete the computation or find an irreducible factor of $\psi_\ell(X)$.

Going into an extension of these rings to find the Y-coordinate of the ℓ-points is also unproblematic.

The above algorithm sketch works but is very inefficient and mostly impractical. A more careful algorithm will work significantly faster.

The most important improvement to the algorithm is that for some small primes it is relatively easy to find an irreducible factor of $\psi_\ell(X)$. Since this factor has much smaller degree, we can work in a small field extension where arithmetic is much faster than in $\mathbb{F}_p[X]/\langle\psi_\ell(X)\rangle$. The resulting algorithm is significantly faster.

The upshot is that there are feasible algorithms that can compute the number of points on an elliptic curve. We will not have to count the number of points of many curves until we find one with a suitable number of points.

2.5.3 Discrete Logarithms

In the group $\mathbb{F}_p^*$ the group operation requires two arithmetic operations (one integer multiplication and one integer division), while finding inverses is much more costly (using the extended Euclidian algorithm). Note that division in a finite field is usually done by multiplying with inverses.

In the group $E(\mathbb{F}_p)$, Proposition 2.34 says that adding distinct non-inverse points requires one inversion, three multiplications and six additions. Adding a point to itself requires one inversion, two multiplications by small constants, four multiplications, one addition of a constant and four additions.

At first glance, it would seem odd to consider the elliptic curve group, since the group operation there is much more complicated than the group operation in $\mathbb{F}_p^*$. But there is one more variable to consider: the size of the underlying field. Recall that we choose the size of the group such that the discrete logarithm problem in the group is sufficiently difficult.

The algorithms in Section 2.2 work for any group. For $\mathbb{F}_p^*$ we also have much faster index calculus methods. For most elliptic curves, there are no equivalents of the small primes, so there are no useful index calculus methods.

We briefly describe some results on elliptic curve discrete logarithms.

- For so-called *anomalous elliptic* curves, curves defined over $\mathbb{F}_p$ with p elements, there are very efficient algorithms for computing discrete logarithms. These curves are completely unsuitable for use in cryptography.

- For certain subgroups G, H of $E(\bar{\mathbb{F}})$, there are so-called *bilinear maps* (called *pairings*) $e : G \times H \to \mathbb{F}_{p^d}^*$ satisfying

$$e(aP, bQ) = e(P, Q)^{ab}.$$

When the field extension degree d is small, these maps can be computed easily and can sometimes be used to move a discrete logarithm problem from an elliptic curve into a finite field. Since index calculus methods can be used in finite fields, curves with low extension degree must be defined over larger finite fields, making arithmetic slower.

One class of curves with extension degree 2 is the *supersingular curves*.

Note that bilinear maps can sometimes be used constructively in cryptography. But unless we need easily computable bilinear maps, curves with such maps are not useful for cryptography.

- For many other small families of elliptic curves, there are algorithms for computing discrete logarithms faster than the methods from Section 2.2.

For most elliptic curves, the best algorithms for computing discrete logarithms are those from Section 2.2. Compared to $\mathbb{F}_p^*$, our elliptic curves can therefore be defined over much smaller fields where arithmetic is much faster, so even if we have to do more arithmetic operations, each group operation may be faster.

We should mention that elliptic curves over non-prime finite fields have been studied extensively. In certain cases, such fields may be more convenient than prime fields, but there have been many more advances in computing discrete logarithms for such curves. Usually, the advantage gained is outweighed by the uncertainty.

2.6   ACTIVE ATTACKS

The only attackers we have considered so far are eavesdroppers. As we discussed in Chapter 1, an attacker that can eavesdrop can very often also tamper with communications. We often call such attackers *active*. And a large number of applications actually need integrity more than they need confidentiality.

As we see in Figure 2.5, Diffie-Hellman on its own will not be secure in practice. We shall consider how Diffie-Hellman can be secured later.

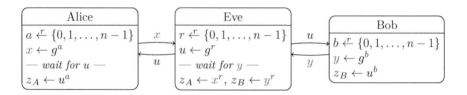

FIGURE 2.5 A man-in-the-middle attack on the Diffie-Hellman protocol, where Eve establishes separate shared secrets with Alice and Bob, who do not notice the attack.

CHAPTER 3

Public Key Encryption

We shall now return to the original cryptographic problem: Alice wants to send messages to Bob via some channel. Eve has access to the channel, and she may eavesdrop on anything sent over the channel. Alice does not want Eve to know the content of her messages to Bob.

The solution we have arrived at so far is for Alice and Bob to first run a key exchange protocol (such as Diffie-Hellman from Chapter 2) to establish a shared secret. Then Alice can use the shared secret to encrypt her messages using symmetric cryptography as discussed in Chapter 1. For some channels, such as ordinary mail, this is very inconvenient, since each message may take a long time to arrive, and even longer before it is read and acted upon.

Of course, Alice and Bob could establish a shared secret and then use that secret from then on. But Alice may need to talk to many people, not just Bob. Sharing secrets and managing the shared secrets may be inconvenient.

So Alice wants to be able to send a single encrypted message to Bob or one of her other correspondents. She does not want to manage shared secrets with every correspondent, but she may be willing to manage public information, just as she is already managing names, phone numbers and adresses.

The answer to Alice's problem is public key encryption, where we want anyone to be able to encrypt, but the ability to decrypt should be restricted. To achieve this we shall have *two* keys, one key for encryption and a different key for decryption. Hopfully we can make the encryption key public, allowing anyone to encrypt, while we keep the decryption key secret and thus restrict the decryption ability to those who know the secret. The two keys must be connected in some way, of course, which means that we cannot just sample keys from the uniform distribution on some simple set.

We begin by studying encryption schemes based on Diffie-Hellman. We then study a family of public key encryption schemes based on arithmetic modulo products of large primes and the famous RSA cryptosystem. The best general attack on these schemes is to factor integers, so we study factoring algorithms, just as we studied the computation of discrete logarithms.

DOI: 10.1201/9781003149422-3

We shall also study several public key cryptosystems essentially based on lattices. For two of the systems, lattices are not required for describing the scheme (relying instead on linear algebra and polynomial arithmetic). Instead, lattices are required for the analysis of the schemes. Again, we need to study algorithms for solving the related lattice problems in order to understand the security of these schemes.

3.1 DEFINITIONS

Alice, Bob and a many others want to be able to send confidential messages to each other. For various reasons, using a key exchange protocol to establish a shared secret every time they want to send messages is impractical. Furthermore, Alice does not want to manage a long-term secret for each correspondent. She is willing to manage public information for each correspondent.

In this situation, what is needed is public key encryption.

Definition 3.1. A *public key encryption* scheme $(\mathcal{K}, \mathcal{E}, \mathcal{D})$ consists of:

- The *key generation* algorithm $\mathcal{K}$ takes no input and outputs an *encryption key* ek and a *decryption key* dk. To each encryption key ek there is an associated message set $\mathfrak{M}_{ek}$.

- The *encryption* algorithm $\mathcal{E}$ takes as input an encryption key ek and a message $m \in \mathfrak{M}_{ek}$ and outputs a ciphertext c.

- The *decryption* algorithm $\mathcal{D}$ takes as input a decryption key dk and a ciphertext c and outputs either a message m or the special symbol $\perp$ indicating decryption failure.

For any key pair (ek, dk) output by $\mathcal{K}$ and any message $m \in \mathfrak{M}_{ek}$

$$\mathcal{D}(dk, \mathcal{E}(ek, m)) = m.$$

Just as for symmetric encryption, the users of a system require confidentiality and some sense of integrity. However, since the encryption key is known anyone can create a ciphertext, so the informal notion of integrity does not work for public key encryption. Instead, we have a notion of non-malleability where it should be hard to modify a ciphertext in a predictable way. We return to these notions in Chapter 8.

Informally. A public key encryption scheme provides *confidentiality* if it is hard to learn anything at all about the decryption of a ciphertext from the ciphertext itself, possibly except the length of the decryption.

Informally. A public key encryption scheme is *non-malleable* if it is hard to create a new ciphertext based on a given ciphertext such that the decryption of the new ciphertext is not $\perp$, but predictably related to the decryption of the given ciphertext.

3.2 SCHEMES BASED ON DIFFIE-HELLMAN

We shall develop a public key encryption scheme based on the Diffie-Hellman protocol. The initial situation is that Alice, Carol and Bob will use the Diffe-Hellman protocol to establish a shared secret, and then send the message encrypted using a symmetric encryption scheme.

Let G be a finite cyclic group of order n and let g be a generator. Let $(G, \mathfrak{P}, \mathfrak{C}, \mathcal{E}_s, \mathcal{D}_s)$ be a symmetric cryptosystem. Note that G is the key set for the symmetric cryptosystem.

When Alice wants to send a message $m_A \in \mathfrak{P}$ to Bob, she uses Diffie-Hellman to establish a shared secret, encrypts her message using the symmetric cryptosystem and sends the ciphertext to Bob. Bob decrypts the ciphertext.

1. Alice chooses a number r_A uniformly at random from the set $\{0, 1, 2, \ldots, n-1\}$. She computes $x_A = g^{r_A}$ and sends x_A to Bob.

2. Bob receives x_A from Alice. He chooses a number b_A uniformly at random from the set $\{0, 1, \ldots, n-1\}$, computes $y_A = g^{b_A}$ and $z_A = x_A^{b_A}$.

3. Alice receives y_A from Bob. Alice computes $z_A = y_A^{r_A}$, encrypts the message as $w \leftarrow \mathcal{E}_s(z_A, m_A)$, and sends w to Bob.

Figure 3.1 illustrates how Alice and Carol send messages to Bob.

But a variation on this topic is possible. Before anything happens, Bob executes part of the Diffie-Hellman protocol. He samples a random number b and computes $y = g^b$. Whenever someone contacts him, he immediately responds with y. When he receives the symmetric ciphertext, he computes the shared secret and decryptes the ciphertext. This variation is illustrated in the middle part of Figure 3.1. It is possible to prove that Bob does not lose any security by doing this, which we return to in Chapter 8.

Repeatedly sending the same value y is wasteful. Bob therefore announces what he is doing and publishes the value y. When Alice and Carol want to send a message to Bob, they already know y. Therefore, they do not have to wait to receive it. Instead, they can immediately compute the shared secret and encrypt the message. This final situation is shown in the bottom of Figure 3.1.

What has happened is that we have turned the Diffie-Hellman protocol combined with a symmetric cryptosystem into a public key encryption scheme.

Example 3.1. The public key encryption scheme $(\mathcal{K}, \mathcal{E}, \mathcal{D})$ is based on a cyclic group G of order n with generator g and a symmetric cryptosystem $(G, \mathfrak{P}, \mathfrak{C}, \mathcal{E}_s, \mathcal{D}_s)$. The message set associated to any ek is $\mathfrak{P}$.

- The *key generation* algorithm $\mathcal{K}$ samples $b \xleftarrow{r} \{0, 1, 2, \ldots, n-1\}$, computes $y = g^b$ and outputs $ek = y$ and $dk = b$.

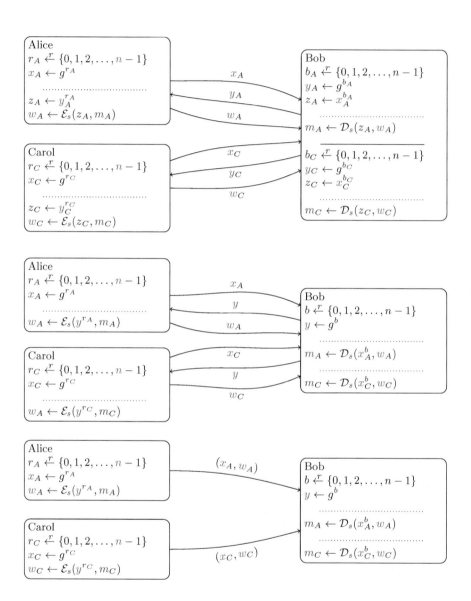

FIGURE 3.1 From Diffie-Hellman to public key encryption scheme. (Top) Alice and Carol use Diffie-Hellman to establish shared secrets with Bob and send him encrypted messages. (Middle) Bob uses a single random number for all the Diffie-Hellman protocol runs. (Bottom) Bob publishes his Diffie-Hellman message. Alice and Carol complete the Diffie-Hellman protocol to establish a shared secret and send Bob encrypted messages.

- The *encryption* algorithm $\mathcal{E}$ takes as input an encryption key y and a message $m \in \mathfrak{P}$. It samples $r \xleftarrow{r} \{0, 1, 2, \ldots, n-1\}$ and computes $x = g^r$, $z = y^r$ and $w = \mathcal{E}_s(z, m)$. It outputs the ciphertext $c = (x, w)$.

- The *decryption* algorithm $\mathcal{D}$ takes as input a decryption key b and a ciphertext $c = (x, w)$. It computes $z = x^b$ and $m = \mathcal{D}_s(z, w)$. If $\mathcal{D}_s$ outputs the special symbol $\bot$, then $\mathcal{D}$ outputs $\bot$, otherwise m.

E| *Exercise* 3.1. The above is an informal description of a public key encryption scheme. Implement the three algorithms $\mathcal{K}$, $\mathcal{E}$ and $\mathcal{D}$. Show that the triple $(\mathcal{K}, \mathcal{E}, \mathcal{D})$ is a public key encryption scheme.

Informally. *If it is hard to break the Diffie-Hellman protocol based on G, and $(G, \mathfrak{P}, \mathfrak{C}, \mathcal{E}_s, \mathcal{D}_s)$ provides confidentiality and integrity, then the above public key encryption scheme provides confidentiality and non-malleability.*

3.2.1 ElGamal

The security of the above scheme depends both on the security of the Diffie-Hellman protocol and the security of the symmetric cryptosystem.

In order to illustrate certain attacks, it is convenient to consider a variant of the above public key encryption scheme that uses an extremely simple symmetric cryptosystem, namely a variant of the shift cipher from Section 1.2.1 given by $(G, G, G, \mathcal{E}_s, \mathcal{D}_s)$, where $\mathcal{E}_s(k, m) = mk$ and $\mathcal{D}_s(k, w) = wk^{-1}$.

The result is the ElGamal public key encryption scheme.

E| *Example* 3.2. Let G be a cyclic group of order n with generator g. The *ElGamal* public key cryptosystem works as follows:

- The *key generation* algorithm is as in Example 3.1.

- The *encryption* algorithm $\mathcal{E}$ takes as input an encryption key y and a message $m \in G$. It samples $r \xleftarrow{r} \{0, 1, 2, \ldots, n-1\}$ and computes $x = g^r$ and $w = my^r$. It outputs the ciphertext $c = (x, w)$.

- The *decryption* algorithm $\mathcal{D}$ takes as input a decryption key b and a ciphertext $c = (x, w)$. It computes $m = wx^{-b}$ and outputs m.

E| *Example* 3.3. Consider the group $\mathbb{Z}_{23}^+$ with generator $g = 5$.

Bob chooses the random number $b = 13$ and computes $y = 13 \cdot 5 = 19$.

Alice wants to encrypt the message $m = 7$. She knows Bob's encryption key $y = 19$. She chooses the random number $r = 5$ and computes $x = 5 \cdot 5 = 2$ and $w = 7 + 5 \cdot 19 = 10$. The ElGamal ciphertext is $c = (2, 10)$.

Bob computes $10 + (-13) \cdot 2 = 7$.

E *Example* 3.4. Consider the group $\mathbb{F}_{23}^*$ with generator $g = 5$.

Bob chooses the random number $b = 13$ and computes $y = 5^{13} = 21$.

Alice wants to encrypt the message $m = 7$. She knows Bob's encryption key $y = 21$. She chooses the random number $r = 5$ and computes $x = 5^5 = 20$ and $w = 7 \cdot 21^5 = 6$. The ElGamal ciphertext is $c = (20, 6)$.

Bob computes $6 \cdot 20^{-13} = 7$.

E *Exercise* 3.2. The above is an informal description of a public key encryption scheme. Implement the three algorithms $\mathcal{K}$, $\mathcal{E}$ and $\mathcal{D}$. Show that the triple $(\mathcal{K}, \mathcal{E}, \mathcal{D})$ is a public key encryption scheme.

Information Leakage For some groups, the ElGamal scheme may leak information about the message. This may or may not be problematic.

E *Exercise* 3.3. Suppose p is a prime such that $p - 1$ is divisible by a large prime and the discrete logarithm problem is hard in $\mathbb{F}_p^*$. Consider ElGamal based on $G = \mathbb{F}_p^*$. Let y be the encryption key, and let $c = (x, w)$ be an encryption of m under y. Show that

$$\left(\frac{m}{p}\right) = \begin{cases} \left(\frac{w}{p}\right) & \text{if } \left(\frac{x}{p}\right) \text{ or } \left(\frac{y}{p}\right) \text{ is } 1; \text{ and} \\ -\left(\frac{w}{p}\right) & \text{otherwise.} \end{cases}$$

E *Example* 3.5. Continuing Example 3.4, we find $\left(\frac{21}{23}\right) = -1$, $\left(\frac{20}{23}\right) = -1$ and $\left(\frac{6}{23}\right) = 1$. This means that $\left(\frac{m}{23}\right)$ should be -1, and indeed, $\left(\frac{7}{23}\right) = -1$.

E *Exercise* 3.4. Suppose p is a prime such that the discrete logarithm problem is hard in $\mathbb{F}_p^*$. Suppose also that $p - 1$ is divisible by a (small) known prime ℓ, and that $m \in \mathbb{F}_p^*$ has order ℓ. Show that m can be recovered from an encryption of m using essentially a small multiple of $\sqrt{\ell}$ group operations.

Malleability When the attacker sees the encryption key and some ciphertexts, we say that we have a *chosen plaintext* attack. The attacker may also be able to deduce some information about the decryption of other ciphertexts. We usually consider a situation where the adversary wants to learn something about the decryption of some ciphertexts, and may ask for the decryption of other ciphertexts. This is known as a *chosen ciphertext* attack.

We begin by showing that ElGamal is *malleable*, in that even if you do not know the decryption of a ciphertext, it is still possible to create new ciphertexts that decrypt to the same or related messages.

E *Exercise* 3.5. Suppose $c = (x, w)$ is an encryption of an unknown message m. Show how to create an encryption $c' = (x', w')$ of m where $x' \neq x$. Also show how to create an encryption of mm' for any $m' \in G$.

We get a chosen ciphertext attack against the confidentiality of ElGamal.

E *Exercise* 3.6. Suppose Alice sends Bob the ciphertext $c = (x, w)$ and Eve learns this ciphertext. Suppose also that Eve can trick Bob into decrypting a ciphertext distinct from c. Explain how Eve can learn the decryption of c.

These exercises show that ElGamal, in particular over $\mathbb{F}_p^*$, does not provide confidentiality. Neither does ElGamal provide non-malleability.

However, if the symmetric cryptosystem used provides confidentiality and integrity, it can be proven under reasonable assumptions that the public key encryption scheme from Exercise 3.1 is non-malleable and provides confidentiality. We return to this in Chapter 8.

3.3   RSA

In this section we shall develop the famous RSA (Rivest-Shamir-Adleman) public key encryption scheme. We begin with the Pohlig-Hellman exponentiation cipher, which can be interpreted as a public key encryption scheme, albeit an insecure one. The problem is that if the group order is known, it is easy to recover the decryption key from the encryption key.

The exponentiation cipher can be adapted to a different algebraic structure, where we are able to prove that recovering a decryption key is as hard as factoring certain integers. While this does not prove that the scheme is secure if it is hard to factor those integers, it turns out that factoring seems to be the best way to attack the cryptosystem. We show a number of flaws in this cryptosystem and discuss various ways of improving the system.

3.3.1   Preliminaries

The Pohlig-Hellman exponentiation cipher from Exercise 1.23 in Section 1.2.5 is based on a cyclic group G of order N. The key is an integer k relatively prime to N, and the two block cipher maps are

$$(k, x) \mapsto x^k \qquad \text{and} \qquad (k, y) \mapsto y^{k^{-1}},$$

where k^{-1} is any inverse of k modulo N.

Observe now that we can turn the Pohlig-Hellman cipher into a public key encryption scheme as follows. The key generation algorithm chooses an integer e from the integers between 0 and N that are invertible modulo N. It then finds some inverse d of e modulo the group order N. The encryption key ek is e, while the decryption key dk is d.

The encryption algorithm takes as input an encryption key $ek = e$ and a message $m \in G$ and outputs the ciphertext $c = m^e$.

The decryption algorithm takes as input a decryption key $dk = d$ and a ciphertext $c \in G$ out outputs the message $m = c^d$.

E | *Exercise* 3.7. The above is an informal description of a public key encryption scheme. Implement the three algorithms $\mathcal{K}$, $\mathcal{E}$ and $\mathcal{D}$. Show that the triple $(\mathcal{K}, \mathcal{E}, \mathcal{D})$ is a public key encryption scheme.

Unfortunately, this public key encryption scheme is trivially insecure. An adversary that sees the encryption key e can easily compute an inverse of e modulo the group order N, which is assumed known. (It may be impossible to compute the exact inverse found by the key generation algorithm, but any inverse can be used to compute the decryption map.)

However, if the group order was unknown, there would be no obvious way to compute a decryption key from the encryption key. The first thing we do is to note that Pohlig-Hellman works equally well in any finite group.

E | *Exercise* 3.8. Let G be a group of order N, and let e and d be inverses modulo (a multiple of) N. Show that the maps $x \mapsto x^e$ and $y \mapsto y^d$ are inverse maps.

The key generation algorithm needs to compute inverses modulo the group order, which usually requires that the key generation algorithm knows the group order. This means that we cannot fix a single group. Instead, the key generation algorithm must choose a group in such a way that it knows the group order. Then it must include in the encryption key a description of the group such that the group order is hard to compute from that description.

Before we continue, we observe that for the Pohlig-Hellman cipher, $ed \equiv 1$ (mod N), which means that $ed - 1$ is a multiple of N. Likewise, Exercise 3.8 says that if we know a multiple of the group order, we can find something that is effectively a decryption key.

3.3.2 The RSA Cryptosystem

Before we define the RSA cryptosystem, we extend the result of Exercise 3.8 to a specific *ring*.

E | *Exercise* 3.9. Let n be the product of two large primes p and q, and let e and d be inverses modulo (a multiple of) $\operatorname{lcm}(p - 1, q - 1)$. Show that the maps $x \mapsto x^e$ and $y \mapsto y^d$ on $\mathbb{Z}_n$ are inverses.

Hint: Prove that p divides the difference $x^{ed} - x$ for any integer x. Repeat for q. Then conclude that the product divides the difference.

E | *Example* 3.6. The *textbook RSA* public key encryption scheme $(\mathcal{K}, \mathcal{E}, \mathcal{D})$ works as follows.

- The *key generation* algorithm $\mathcal{K}$ chooses two large primes p and q. It computes the *RSA modulus* $n = pq$, chooses e and finds d such that

$ed \equiv 1 \pmod{\text{lcm}(p-1, q-1)}$. Finally, it outputs $ek = (n, e)$ and $dk = (n, d)$. The message set associated to ek is $\{0, 1, 2, \ldots, n-1\}$.

- The *encryption* algorithm $\mathcal{E}$ takes as input an encryption key (n, e) and a message $m \in \{0, 1, \ldots, n-1\}$. It computes $c = m^e \bmod n$ and outputs the ciphertext c.

- The *decryption* algorithm $\mathcal{D}$ takes as input a decryption key (n, d) and a ciphertext c. It computes $m = c^d \bmod n$ and outputs the message m.

Note that the encryption exponent e may be very small, even 3.

E *Exercise* 3.10. The above is an informal description of a public key encryption scheme. Implement (use some suitable method to choose p, q and e) the three algorithms $\mathcal{K}$, $\mathcal{E}$ and $\mathcal{D}$ are. Show that the triple $(\mathcal{K}, \mathcal{E}, \mathcal{D})$ is a public key encryption scheme.

Hint: Use Exercise 3.9.

E *Example* 3.7. Bob chooses the two primes $p = 41$ and $q = 53$ and computes $n = 41 \cdot 53 = 2173$. He chooses $e = 3$ and computes an inverse $d = 347$ of 3 modulo $(41 - 1)(53 - 1)$. The encryption key is $(2173, 3)$, the decryption key is $(2173, 347)$.

Alice wants to encrypt the message $m = 1234$. She knows Bob's encryption key. She computes $c = 1234^3 \bmod 2173 = 884$.

Bob computes $884^{347} \bmod 2173 = 1234$.

As we saw in Section 3.3.1, a public key encryption scheme is obviously useless if it is easy to deduce the decryption key from the encryption key. We shall now consider this problem for the RSA cryptosystem.

Let p and q be distinct, large primes, and let $n = pq$. Consider the group $\mathbb{Z}_n^*$, which is isomorphic to $\mathbb{Z}_p^* \times \mathbb{Z}_q^*$. It is clear that if we know p and q, we can find $N = (p-1)(q-1)$. Conversely, if we know n and N, then we can easily recover the factorisation.

T **Proposition 3.1.** *Let n be a product of two distinct primes p and q, and let N be the order of $\mathbb{Z}_n^*$. Then p and q are zeros of the polynomial $f(X) = X^2 + (N - n - 1)X + n$.*

Proof. First, note that $N = n - (p + q) + 1$, so

$$p + q = n + 1 - N.$$

Now consider the polynomial $f(X) = (X - p)(X - q) = X^2 - (p+q)X + n$. □

Note that if we know N, then we know the polynomial's coefficients, and if we know the coefficients, we can easily compute the zeros of the polynomial using the usual formula for the zeros of a quadratic polynomial.

If we only know n and a multiple of l of N, we cannot use the above result, but we can still recover the factorisation of n.

E| *Exercise 3.11.* Let n be a product of two distinct primes p and q. Let x and y be integers such that

$$x \equiv y \pmod{p} \qquad \text{and} \qquad x \not\equiv y \pmod{q}.$$

Show that $\gcd(x - y, n) = p$.

T| **Lemma 3.2.** *Let n be a product of two distinct large primes p and q, and let k be an odd multiple of $\mathrm{lcm}(p-1, q-1)/2$. Then for at least half of all integers z between 0 and n that are relatively prime to n,*

$$z^k \equiv \pm 1 \pmod{p} \qquad \text{and} \qquad z^k \equiv \mp 1 \pmod{q}. \qquad (3.1)$$

E| *Exercise 3.12.* Prove Lemma 3.2.

T| **Proposition 3.3.** *Let n be a product of two distinct primes p and q. For any $\kappa > 0$, given a multiple l of N, we can compute p and q with probability $1 - 2^{-\kappa}$ using at most $\lceil 4\kappa \log_2 l \rceil$ arithmetic operations and $\kappa \log_2 l$ gcd computations.*

Proof. Write $l = 2^t s$, $p - 1 = 2^{t_p} s_p$ and $q - 1 = 2^{t_q} s_q$ with s, s_p and s_q all odd. We may assume that $t_p \geq t_q$. It is then clear that $2^{t_p - 1} s$ is an odd multiple of $\mathrm{lcm}(p - 1, q - 1)/2$.

Lemma 3.2 then says that for half of all z between 0 and n that are relatively prime to n,

$$\gcd(z^{2^{t_p - 1} s} + 1, n) = p \text{ or } q.$$

For each value z chosen uniformly at random from $\{x \mid 0 < x < n, \gcd(x, n) = 1\}$ the probability that the above gcd computation does not produce a proper factor of n is $1/2$. The probability that a proper factor of n has not appeared after κ independently sampled z values is $2^{-\kappa}$.

Obviously, we do not know t_p. But for each value of z chosen, we can simply try all possible values for t_p. Since we know that $t_p \leq t < \log_2 l$, this requires at most $\log_2 l$ gcd computations per z value.

We can compute the greatest common divisor by computing $z^{2^i s}$ modulo n before computing the gcd. We can compute this using the algorithm from Exercise 2.4 using $4 \log_2 l$ arithmetic operations. By computing first $z^{2^0 s}$ first, we can compute $z^{2^1 s}, \ldots, z^{2^{t-1} s}$ successively, thereby computing all t values modulo n using just $4 \log_2 l$ arithmetic operations. Doing this for at most κ distinct z values then requires at most $4\kappa \log_2 l$ arithmetic operations. □

E| *Exercise 3.13.* The proof of Proposition 3.3 essentially describes an algorithm for factoring a product of large primes p and q given a multiple l of $\mathrm{lcm}(p - 1, q - 1)$. Implement this algorithm and restate Proposition 3.3 as a statement about the algorithm's time complexity (in terms of arithmetic operations and gcd computations, ignoring any other form of computation involved).

Example 3.8. Consider $n = 2173$. Suppose we know that 35360 is a multiple of the order of $\mathbb{Z}_{2173}^*$. We write $35360 = 2^5 \cdot 1105$.

We begin with $z = 2$ and compute $2^{1105} \bmod 2173 = 401$. We then compute $401^2 \bmod 2173 = 2172$ and $2172^2 \bmod 2173 = 1$. This did not give us the factors of 2173.

We try again with $z = 3$ and compute $3^{1105} \bmod 2173 = 454$. We then compute $454^2 \bmod 2173 = 1854$ and $1854^2 \bmod 2173 = 1803$. At this point, we compute $\gcd(1803 - 1, 2173) = 53$. We have found that $2173 = 53 \cdot 41$.

Proposition 3.3 says that if someone knows a multiple of N, they can factor. Therefore, if we believe that it is difficult to factor integers like n, then it is also hard to deduce the group order of $\mathbb{Z}_n^*$ or any multiple of it, and therefore it is hard to deduce the RSA decryption key from the RSA encryption key.

3.3.3 Attacks

We have seen that as long as the RSA modulus n chosen by the key generation algorithm is hard to factor, the above public key encryption scheme is not obviously useless. In fact, it turns out that it is useful.

The best strategy for recovering a completely unknown m from $c = m^e \bmod n$, computing eth *roots*, seems to be to factor n. However, as the security definitions in Section 3.1 make clear, the adversary does not need to recover a completely unknown m to break the system. In this section, we shall consider a few attacks on the public key encryption scheme from Exercise 3.10 that work essentially without computing eth roots.

Deterministic Encryption   The RSA public key encryption scheme is *deterministic*, that is, the encryption algorithm is deterministic and the encryption key and the message fully determine the ciphertext. This is a problem.

Exercise 3.14. Show that $\mathcal{D}(dk, c) = m$ if and only if $\mathcal{E}(ek, m) = c$.

Exercise 3.15. Let S be a fixed and known set of messages, and let c be an encryption of some message from S. Show that we can decide what the decryption of c is using at most $|S|$ encryptions.

Information Leakage   Part of the message encrypted will leak, which may or may not be a problem.

[E] *Exercise* 3.16. Show that both e and d will be odd. Show that

$$\left(\frac{m}{n}\right) = \left(\frac{c}{n}\right).$$

Malleability   Just like ElGamal, RSA is *malleable* and this can be used in a *chosen ciphertext* attack.

[E] *Exercise* 3.17. Suppose c is an encryption of an unknown message m. Show how to create an encryption c' of mm' for any m' in the message space.

[E] *Exercise* 3.18. Suppose Alice sends Bob the ciphertext c and Eve learns this ciphertext. Suppose also that Eve can trick Bob into decrypting a ciphertext distinct from c. Explain how Eve can learn the decryption of c.

Short Messages   We built a public key encryption scheme based on Diffie-Hellman and a symmetric encryption scheme. We can do the same based on RSA and a symmetric cryptosystem $(\mathfrak{K}_s, \mathfrak{P}, \mathfrak{C}, \mathcal{E}_s, \mathcal{D}_s)$. The natural idea is to think of the symmetric cryptosystem keys as integers and do as follows:

- The *key generation* algorithm is the RSA key generation algorithm.
- The *encryption* algorithm takes an encryption key $ek = (n, e)$ and a message $m \in \mathfrak{P}$ as input. It samples $k \xleftarrow{r} \mathfrak{K}_s$, and computes $x \leftarrow k^e \bmod n$ and $w \leftarrow \mathcal{E}_s(k, m)$. It outputs $c = (x, w)$.
- The *decryption* algorithm takes a decryption key $dk = (n, d)$ and a ciphertext $c = (x, w)$ as input. It computes $k \leftarrow x^d \bmod n$ and $m \leftarrow \mathcal{D}_s(k, w)$. If $k \notin \mathfrak{K}_s$ or decryption of w fails, it outputs $\perp$, otherwise m.

But symmetric keys are usually much smaller than an RSA modulus.

[E] *Exercise* 3.19. Suppose $n \approx 2^{2000}$, $e = 3$ and $\mathfrak{K}_s = \{0, 1, 2, \ldots, 2^{256} - 1\}$. Given an encryption c of an unknown key k, show how you can easily recover k.

Small Integer Roots of Polynomial Equations   A first solution to this problem is to add a large, fixed integer such as 2^{1999} to the key before raising it to the eth power, and subtracting it after raising the RSA ciphertext to the dth power. That is, $x \equiv (2^{1999} + k)^e \pmod{n}$. But for small e it is easy to find small integer roots of modular univariate equations such as $(2^{1999} + X)^e - x \equiv 0 \pmod{n}$, and $k < 2^{256}$ is a small integer root of this equation.

A better solution is to add a random multiple of a large, fixed integer. That is, $x \equiv (r2^{256} + k)^e \pmod{n}$. We recover k by computing the remainder when divided by 2^{256} after raising x to the dth power.

While this seems to work, there are some worries. First, even if it is hard to compute eth roots in $\mathbb{Z}_n$, computing parts of the root, such as the key k, may be easier. Second, it may be possible to modify x such that k does not change, only r. Since the decryption algorithm discards r, it will accept the modified ciphertext as valid, which may be a problem for some applications.

3.3.4 Secure Variants

There is a simple solution that uses a *random-looking* function to derive the key from a random number, which is raised to the eth power. The random-looking function is called a *hash function* or *key derivation function*.

Example 3.9. Suppose h is a function from $\{0, 1, \ldots, n-1\}$ to $\mathfrak{K}_s$. We get the following cryptosystem.

- The *key generation* algorithm is the RSA key generation algorithm.

- The *encryption* algorithm takes an encryption key $ek = (n, e)$ and a message $m \in \mathfrak{P}$ as input. It samples $r \xleftarrow{r} \{0, 1, \ldots, n-1\}$, and computes $k \leftarrow h(r)$, $x \leftarrow r^e \bmod n$ and $w \leftarrow \mathcal{E}_s(k, m)$. It outputs $c = (x, w)$.

- The *decryption* algorithm takes a decryption key $dk = (n, d)$ and a ciphertext $c = (x, w)$ as input. It computes $r \leftarrow x^d \bmod n$, $k \leftarrow h(r)$ and $m \leftarrow \mathcal{D}_s(k, w)$. If decryption of w fails, it outputs $\perp$, otherwise m.

Unless you know all of r, you know very little about the key k, since h is random-looking. Furthermore, any change in x will lead to a change in r, which will lead to an unpredictable change in k because h is random-looking. If k has changed unpredictably, it will be hard to modify w so that it is a valid ciphertext under the modified k, so it will be hard to get the decryption algorithm to say anything but $\perp$.

Exercise 3.20. The above is an informal description of a public key encryption scheme. Implement (reuse key generation from Exercise 3.10) the three algorithms $\mathcal{K}$, $\mathcal{E}$ and $\mathcal{D}$. Show that $(\mathcal{K}, \mathcal{E}, \mathcal{D})$ is a public key encryption scheme.

Exercise 3.21. Suppose you want to send the same message m to l recipients with public keys $(n_1, e_1), (n_2, e_2), \ldots, (n_l, e_l)$. One simple approach is to use the same r and compute $x_i \leftarrow r^{e_i} \bmod n_i$, $i = 1, 2, \ldots, l$, $k \leftarrow h(r)$ and $w \leftarrow \mathcal{E}_s(k, m)$. You send (x_i, w) to the ith recipient.
  Suppose $e_1 = e_2 = \cdots = e_l \leq l$. Show how you can easily recover r.

Remark. Exercise 3.21 does not show an attack on the public key encryption scheme from Exercise 3.20. Instead, it illustrates one way to misuse a cryptosystem and thereby introduce weaknesses. In this case, reusing the randomness r for more than one encryption introduced the weakness. In general, it is implicitly assumed that randomness used by key generation and encryption algorithms is never reused and does not leak out of the algorithm. If those assumptions are violated, weaknesses may result, as shown by the exercise.

Informally. *If it is hard to compute eth roots modulo n for (n, e) as output by the key generation algorithm $\mathcal{K}$, and $(\mathfrak{K}_s, \mathfrak{P}, \mathfrak{C}, \mathcal{E}_s, \mathcal{D}_s)$ provides confidentiality and integrity, then the above public key encryption scheme provides confidentiality and non-malleability.*

In situations with multiple recipients of the same message, other schemes may be more convenient. One solution is based on a Feistel-like permutation. Suppose we have three groups G_1, G_2 and G_3 such that $G_1 \times G_2 \times G_3$ can be considered a subset of $\{0, 1, \ldots, n - 1\}$. Suppose also that we have three random-looking functions $h_1 : G_2 \times G_3 \to G_1$, $h_2 : G_1 \to G_2$ and $h_3 : G_1 \to G_3$. Then we can construct two permutations on $G_1 \times G_2 \times G_3$ as

$$\pi_1(y_1, y_2, y_3) = (y_1, y_2 + h_2(y_1), y_3 + h_3(y_1)) \text{ and}$$
$$\pi_2(y_1', y_2', y_3') = (y_1' + h_1(y_2', y_3'), y_2', y_3').$$

The inverses of these two permutations are obvious. Our cryptosystem will use the composition $\pi = \pi_2 \circ \pi_1$.

Example 3.10. Let G_1, G_2, G_3 and π be as above, and suppose $\mathfrak{K}_s \subseteq G_2$. The cryptosystem works as follows.

- The *key generation* algorithm is the RSA key generation algorithm.

- The *encryption algorithm* takes an encryption key $ek = (n, e)$ and a message $m \in \mathfrak{P}$ as input. It samples $r \xleftarrow{r} G_1$ and $k \xleftarrow{r} \mathfrak{K}_s$, and computes $x \leftarrow \pi(r, k, 0)^e \bmod n$ and $w \leftarrow \mathcal{E}_s(k, m)$. It outputs $c = (x, w)$.

- The *decryption algorithm* takes a decryption key $dk = (n, d)$ and a ciphertext d as input. It computes $(r, k, y_3) \leftarrow \pi^{-1}(x^d \bmod n)$ and $m \leftarrow \mathcal{D}_s(k, w)$. If $k \notin \mathfrak{K}_s$ or $y_3 \neq 0$ or the decryption of w failed, it outputs $\perp$. Otherwise it outputs m.

Exercise 3.22. The above is an informal description of a public key encryption scheme. Implement (choose some sensible hash functions, and reuse key generation from Exercise 3.10) the three algorithms $\mathcal{K}$, $\mathcal{E}$ and $\mathcal{D}$ are. Show that the triple $(\mathcal{K}, \mathcal{E}, \mathcal{D})$ is a public key encryption scheme.

Suppose we know x, that $x^d \bmod n = (y_1', y_2', y_3')$ and that $\pi^{-1}(y_1', y_2', y_3') = (r, k, 0)$.

Since $h_2(r)$ is added to the key to get y_2', it is impossible to recover k without knowing r. But h_1 is used to hide r, so it is impossible to recover all of r without knowing both y_1', y_2' and y_3'.

Furthermore, any change in y_1' will lead to a change in r, which will lead to an unpredictable change in $h_3(r)$. Any change in y_2' or y_3' will lead to an unpredictable change in r. If r has changed, it is unlikely that π^{-1} will result in 0 in the third coordinate, which means that the decryption algorithm will reject the ciphertext.

It is possible to prove that under reasonable assumptions, variants of the above public key cryptosystem provide confidentiality and non-malleability.

3.4 FACTORING INTEGERS

The cryptosystems described in Section 3.3.4 seem to be secure if it is hard to compute eth roots modulo the RSA modulus. It seems like the best method to compute eth roots modulo the RSA modulus n is to factor n.

Informally. *It is conjectured that computing eth roots modulo a well-chosen RSA modulus is not (much) easier than factoring the RSA modulus.*

Under this conjecture, the study of the security of the RSA-based cryptosystems from Section 3.3.4 reduces to the study of factoring RSA moduluses.

Throughout this section, unless otherwise stated we shall consider an integer n that is the product of two large primes p and q, with $q < p$. We do this to simplify the exposition, but we note that most of the factoring algorithms we consider will work for other composite numbers, and often much better.

We shall usually estimate the work required by our algorithms in terms of arithmetic operations. By arithmetic operations, we mean additions, subtractions, multiplications or divisions of integers of about the same size as n. We ignore additions and subtractions of small constants, as well as multiplication and division by 2.

We begin with the simplest possible factoring algorithm, trial division.

T **Proposition 3.4.** *A factor of a composite integer n can be found using at most $\sqrt{n}$ arithmetic operations.*

E *Exercise 3.23.* Describe an algorithm for finding a factor of a composite integer that proves Proposition 3.4. Implement this algorithm and restate Proposition 3.4 as a statement about the algorithm's time complexity (in terms of arithmetic operations, ignoring any other form of computation involved).

Requirement 3.1. Let q be the smallest divisor of n. Then q arithmetic operations should be an infeasible computation.

Having sufficiently large prime divisors is easy to arrange.

E *Exercise 3.24.* Suppose n is a product of one or more prime powers, and that the biggest prime divisor is p. Show that the prime factorisation of n can be computed using at most $p + \log_2 n$ arithmetic operations. (Compare with Proposition 2.15 and Exercise 2.56.)

3.4.1 Fermat Factoring

If p and q are numbers of roughly the same size, then p and q should be close to $\sqrt{n}$, but their average should be closer. Searching for the average close to $\sqrt{n}$ may make sense.

Let $t = (p+q)/2$ and let $s = (p-q)/2$. Then

$$t^2 - s^2 = (t+s)(t-s) = pq = n.$$

This means that we can decide if some integer τ equals t by computing $\tau^2 - n$ and checking if it is an integer square.

T **Lemma 3.5.** *Given an integer $k > 1$, we can compute the integer square root of k if it exists using at most $3\log_2 k + 3$ arithmetic operations.*

E *Exercise 3.25.* Describe an algorithm for computing an integer square root if it exists that proves Lemma 3.5. Implement this algorithm and restate Lemma 3.5 as a statement about the algorithm's time complexity (in terms of arithmetic operations, ignoring any other form of computation involved).

E *Exercise 3.26.* Compute the square root of 16129 (by hand) using the algorithm from Exercise 3.25.

Remark. The algorithm implied by Lemma 3.5 is not the best. Variants of Newtons method will typically be much faster. Furthermore, if we only want to know if a given integer is square, we can detect most non-squares very quickly by checking if they are squares modulo small primes.

T **Lemma 3.6.** *Let $n = pq$ where p, q are large primes. Then $(p+q)/2 - \sqrt{n} \leq |p-q|/2$.*

Proof. The claim holds if $p = q$. We may therefore assume that $p > q$. With $t = (p+q)/2$ and $s = (p-q)/2$ as above, we have that

$$t - s = \sqrt{(t-s)^2} < \sqrt{(t-s)(t+s)} = \sqrt{n},$$

from which the claim follows. $\qquad\qquad\qquad\qquad\qquad\qquad\qquad\square$

T **Theorem 3.7** (Fermat factoring). *Let n be a product of two large primes p, q. Then a factor of n can be found using at most $3|p-q|(2+\log_2 n)/2 + 1$ arithmetic operations.*

Proof. We are looking for an integer τ such that $\tau = (p-q)/2$. We know that $\tau > \sqrt{n}$, so we may consider the distance $\tau - \lceil\sqrt{n}\rceil$. By Lemma 3.6, this distance is at most $|p-q|/2$. In other words, there is an $i < |p-q|/2$ and integer σ such that

$$(\lceil\sqrt{n}\rceil + i)^2 - n = \sigma^2.$$

We can find the first integer square among the $|p-q|/2$ integers $(\lceil\sqrt{n}\rceil+i)^2-n$ using at most $3\log_2 n+3+3$ arithmetic operations per integer, by Lemma 3.5.

Once we find $t = \lceil\sqrt{n}\rceil + i$ and the square root s, we find a proper factor $t - s$ of n using a single arithmetic operation. The claim follows. □

E *Exercise 3.27.* The proof of Theorem 3.7 essentially describes an algorithm for finding a factor of a composite integer. Implement this algorithm and restate Theorem 3.7 as a statement about the algorithm's time complexity (in terms of arithmetic operations, ignoring any other form of computation involved).

E *Example 3.11.* Consider $n = 1173$. We get that $\lceil\sqrt{n}\rceil = 35$.
   Then $35^2 - 1173 = 52$, which is not a square.
   Then $(35 + 1)^2 - 1173 = 123$, which is not a square.
   Then $(35 + 2)^2 - 1173 = 196 = 14^2$. We get that

$$1173 = (37 - 14)(37 + 14) = 23 \cdot 51.$$

E *Exercise 3.28.* Factor 1683557 using the algorithm from the above proof.

E *Exercise 3.29.* We can use a variant of the Sieve of Eratosthenes to find primes quickly. The idea is that we sieve over a small integer range to quickly exclude numbers divisible by small primes. The remaining numbers are then much more likely to be prime. This reduces the number of expensive primality tests we must do before we find a prime.

We can adapt this to RSA key generation by sieving over a range likely to contain two primes, which will be our p and q. Explain why this is a bad idea.

We arrive at the following requirement.

Requirement 3.2. Let $n = pq$. Then $|p - q|$ arithmetic operations should be an infeasible computation.

E *Exercise 3.30.* Suppose we choose two numbers x and y independently and uniformly at random from the range $\{2^{k-1}, 2^{k-1} + 1, \ldots, 2^k - 1\}$. What is the expected value of $|x - y|$?

3.4.2 Pollard's $p - 1$

We saw earlier that if we have a multiple of $\mathrm{lcm}(p - 1, q - 1)$, then we can factor n. However, it turns out that if we have a multiple of $p-1$ or $q-1$, we can usually also factor n.

T **Proposition 3.8.** *Let n be the product of two distinct large primes p and q, and let k be a multiple of $p - 1$, but not a multiple of $q - 1$. Then for at least* *half of all integers z between 0 and n that are relatively prime to n,*

$$z^k \equiv 1 \quad (\text{mod } p) \qquad and \qquad z^k \not\equiv 1 \quad (\text{mod } q). \tag{3.2}$$

E| *Exercise* 3.31. Prove Proposition 3.8

If we know z and k such that (3.2) holds, we can factor n with Exercise 3.11.

How do we find a multiple of $p - 1$? We could hope that $p - 1$ is only divisible by the l smallest primes $\ell_1, \ell_2, \ldots, \ell_l$. The largest prime power that could then divide the (unknown) value $p-1$ is $\ell_i^{\lfloor \log n / \log \ell_i \rfloor}$, we could construct a multiple using

$$k = \prod_{i=1}^{l} \ell_i^{\lfloor \log n / \log \ell_i \rfloor}.$$

In practice, $k = \ell_l!$ will work reasonably well.

E| *Example* 3.12. Consider $n = 1007$. We choose $z = 2$, and try the bound $\ell_l = 6$ and $k = 6! = 720$.

We compute $2^{720} \equiv 153$ (mod 1007), and then $\gcd(153 - 1, 1007) = 19$. We find that $1007 = 19 \cdot 53$.

E| *Exercise* 3.32. Using the above approach, factor 1829.

We arrive at the following requirement.

Requirement 3.3. Suppose $n = pq$. Let k_1 be the largest divisor of $p-1$, and let k_2 be the largest divisor of $q - 1$. Then $\min\{k_1, k_2\}$ arithmetic operations should be an infeasible computation.

Choosing safe primes p and q, such that $(p - 1)/2$ and $(q - 1)/2$ are also prime, satisfies this requirement. Choosing sufficiently large primes also works.

3.4.3 Pollard's ρ Method

We shall reuse the techniques involved in Pollard's ρ method for computing discrete logarithms from Section 2.2.5. Our goal this time is to construct three random-looking sequences of integers, one of which will collide and provide us with our factorisation.

Let s_1 be an integer between 0 and n. We construct a sequence of integers $s_1, s_2, \ldots$ using the rule

$$s_{i+1} = s_i^2 + 1 \bmod n. \tag{3.3}$$

We define a second sequence $s_{q,1}, s_{q,2}, \ldots$ by $s_{q,i} = s_i \bmod q$. Note that

$$s_{q,i+1} = s_{q,i}^2 + 1 \bmod q. \tag{3.4}$$

T **Lemma 3.9.** *Let the sequences $s_1, s_2, \ldots$ and $s_{q,1}, s_{q,2}, \ldots$ be as above. Suppose indexes i, j exist such that $s_{q,i} = s_{q,j}$. Then $\gcd(s_i - s_j, n) > 1$. If $s_i \neq s_j$, then $\gcd(s_i - s_j, n) = q$.*

Proof. If $s_i = s_j$, then $\gcd(s_i - s_j, n) = n$, so suppose $s_i \neq s_j$. Since

$$s_i \equiv s_{q,i} \equiv s_{q,j} \equiv s_j \pmod{q},$$

we see that q divides the difference $s_i - s_j$. But $s_i \neq s_j$, so we cannot also have that p divides $s_i - s_j$, and the claim follows. □

E *Exercise 3.33.* In this exercise, we will use our knowledge of the factors of n to better see what is happening. Let $n = 2573 = 31 \cdot 83$ and $s_1 = 2380$.

(a) Compute the first 15 terms of the sequences from (3.3) and (3.4).

(b) Find by inspection the first repetition in the sequence $s_{q,1}, s_{q,2}, \ldots$.

(c) Use Lemma 3.9 and the repetition found to factor $n = 2573$.

T **Proposition 3.10.** *Let $s_1, s_2, \ldots$ and $s_{q,1}, s_{q,2}, \ldots$ be defined as above. Suppose k is the smallest integer such that $s_{q,k} = s_{q,k'}$ for some $k' < k$. Then distinct indexes i, j can be found such that $\gcd(s_i - s_j, n) > 1$ using at most $9k$ arithmetic operations and k gcd computations.*

Proof. We consider the sequences $u_1, u_2, \ldots$ and $u_{q,1}, u_{q,2}, \ldots$ given by $u_j = s_{2j}$ and $u_{q,j} = s_{q,2j}$. For some i, which is at most k, we have $u_{q,i} = s_{q,i}$.

We can compute successively the pairs $(s_1, u_1), (s_2, u_2), \ldots$ using the rule

$$(s_{i+1}, u_{i+1}) = (s_i^2 + 1 \bmod n, (u_i^2 + 1)^2 + 1 \bmod n).$$

We compute $\gcd(s_i - u_i, n)$ for each iteration, which by Lemma 3.9 will be not 1 for some $i \leq k$. Computing each new pair requires 9 arithmetic operations. □

E *Exercise 3.34.* The proof of Proposition 3.10 implies an algorithm that eventually computes a greatest common divisor larger than 1. Implement this algorithm and restate Proposition 3.10 as a statement about the algorithm's time complexity (in terms of arithmetic operations and gcd computations).

Also, suppose the elements $s_1, s_2, \ldots, s_k$ are all distinct. Show that then the greatest common divisor eventually computed is a proper divisor of n.

Example 3.13. Consider $n = 1007$.
We begin with $s_1 = 2$, $u_1 = 2^2 + 1 = 5$, and compute:

$$s_2 = 5 \qquad u_2 = 677 \qquad \gcd(677 - 5, 1007) = 1$$
$$s_3 = 26 \qquad u_3 = 886 \qquad \gcd(886 - 26, 1007) = 1$$
$$s_4 = 677 \qquad u_4 = 886 \qquad \gcd(886 - 677, 1007) = 19$$

We get that $1007 = 19 \cdot 53$.

Exercise 3.35. Use the algorithm from Exercise 3.34 to factor $n = 2573$.

Let E be the event that the L first elements of $s_{q,1}, s_{q,2}, \dots$ are all distinct, and let E' be the event that the L first elements of $s_1, s_2, \dots$ are all distinct. Then we can define two functions

$$\theta(L, n) = \Pr[E] \qquad \text{and} \qquad \gamma(L, n) = 1 - \Pr[E'],$$

where the probability is taken over the choice of s_1.

Theorem 3.11. *Let n be a product of two distinct primes p and q. Then a proper factor of n can be computed using at most $9L$ arithmetic operations and L gcd computations, except with probability $\theta(L, n) + \gamma(L, n)$.*

Proof. Some integer will appear twice among the L first elements of the sequence $s_{q,1}, s_{q,2}, \dots$ except with probability $\theta(L, n)$.
There will be no repetitions among the L first elements of the sequence $s_1, s_2, \dots$ except with probability $\gamma(L, n)$.
By Exercise 3.34 there is an algorithm that computes a proper factor of n using at most $9L$ arithmetic operations and L gcd computations. $\square$

Now suppose the two sequences are "random-looking" with respect to repetitions. This means that since the elements of the second sequence come from a much smaller set, we expect a repetition in that sequence long before we have a repetition in the larger sequence. It seems plausible that the sequences we have defined above are "random-looking" with respect to repetitions. Therefore, we make the following conjecture.

Conjecture 3.12. *Roughly approximate, we have that*

$$\theta(L, n) \approx \exp\left(-\frac{L(L-1)}{2q}\right) \qquad \gamma(L, n) \approx \frac{L(L-1)}{2n}.$$

Conjecture 3.12 and Theorem 3.11 gives us the following requirement.

Requirement 3.4. Suppose $n = pq$, with $q < p$. Then $\sqrt{q}$ arithmetic operations should be an infeasible computation.

3.4.4   Index Calculus

Index calculus for factoring is very similar to index calculus for discrete logarithms. We begin with the observation that knowledge of more than two square roots modulo n of the same number allows us to factor n.

Proposition 3.13. *Let n be the product of two distinct, large primes p and q. Let z be a square modulo n with $\gcd(z, n) = 1$. Then z has four square roots modulo n.*

Suppose further that x and y are square roots of z modulo n satisfying

$$x \not\equiv \pm y \pmod{n}.$$

Then $\gcd(x - y, n)$ is a proper divisor of n.

Proof. If z is a square modulo n, then x exists such that $x^2 \equiv z \pmod{n}$, which means that x^2 is congruent to z modulo both p and q. It follows that x and $-x$ are square roots of z modulo both p and q, and they are distinct since z is relatively prime to n. The Chinese remainder theorem then says that these can be combined into four square roots of z modulo n.

Since $x^2 \equiv z \equiv y^2 \pmod{n}$, we know that n divides $x^2 - y^2 = (x-y)(x+y)$. But since $x \not\equiv \pm y \pmod{n}$, we know that n does not divide $x - y$ and $x + y$. It follows that $\gcd(x - y, n)$ is a proper divisor of n. □

Example 3.14. Consider $n = 314791$ and the two numbers 125823 and 17500. We see that

$$125823^2 \equiv 273148 \equiv 17500^2 \pmod{314791} \quad \text{and}$$
$$125823 \not\equiv \pm 17500 \pmod{314791}.$$

A quick computation gives us

$$\gcd(125823 - 17500, 314791) = 727.$$

The next idea is that if we have a sufficient number of relations between random integers and small primes modulo n, then linear algebra will allow us to construct a square root of a product of these random integers.

Proposition 3.14. *Let $t_1, t_2, \ldots, t_{l+1}, \ell_1, \ell_2, \ldots, \ell_l$ be integers satisfying*

$$t_i \equiv \prod_{j=1}^{l} \ell^{s_{ij}}. \tag{3.5}$$

Then using at most $(l + 1)^3$ arithmetic operations, we can compute $\alpha_1, \alpha_2, \ldots, \alpha_{l+1} \in \{0,1\}$ (not all zero) such that

$$\prod_{i=1}^{l+1} t_i^{\alpha_i} \equiv \left(\prod_{j=1}^{l} \ell_j^{\frac{1}{2} \sum_{i=1}^{l+1} \alpha_i s_{ij}} \right)^2 \pmod{n}.$$

E| *Exercise* 3.36. Prove Proposition 3.14.

E| *Example* 3.15. Consider $n = 314791$. We have six relations:

$$5000 = 2^3 \cdot 5^4 \qquad\qquad 118800 = 2^4 \cdot 3^3 \cdot 5^2 \cdot 11$$
$$7425 = 3^3 \cdot 5^2 \cdot 11 \qquad\qquad 882 = 2 \cdot 3^2 \cdot 7^2$$
$$25410 = 2 \cdot 3 \cdot 5 \cdot 7 \cdot 11^2 \qquad\qquad 61250 = 2 \cdot 5^4 \cdot 7^2$$

We get the matrix

$$\begin{pmatrix} 1 & 0 & 0 & 0 & 0 \\ 0 & 1 & 0 & 0 & 1 \\ 0 & 1 & 0 & 0 & 1 \\ 1 & 0 & 0 & 0 & 0 \\ 1 & 1 & 1 & 1 & 0 \\ 1 & 0 & 0 & 0 & 0 \end{pmatrix}$$

We could use Gaussian elimination to find vectors in the matrix kernel, but for such a small matrix, it is easier to find them by inspection. We quickly see that $(1, 0, 0, 1, 0, 0)$ is in the matrix kernel, and we get the square

$$5000 \cdot 882 \equiv 2926 \equiv (2^2 \cdot 3 \cdot 5^2 \cdot 7)^2 \pmod{314791}.$$

Another quick look at the matrix tells us that $(0, 1, 1, 0, 0, 0)$ is in the matrix kernel, and we get the square

$$118800 \cdot 7425 \equiv 45618 \equiv (2^2 \cdot 3^3 \cdot 5^2 \cdot 11)^2 \pmod{314791}.$$

Finally, we see that $(1, 0, 0, 0, 0, 1)$ is in the matrix kernel, which says that

$$5000 \cdot 61250 \equiv 273148 \equiv (2^2 \cdot 5^4 \cdot 7)^2 \pmod{314791}.$$

We have found three squares.

If we already know squares of the random integers from our relations, we can find a square of the product. This second square will be independent of the above one. Which means that we will be able to factor n half the time.

T| **Theorem 3.15.** *Suppose n is a product of two distinct large primes, and that a fraction σ of the squares modulo n are divisible only by the small primes $\ell_1, \ell_2, \ldots, \ell_l$.*

Then we can find a proper factor of n with probability $1/2$ using an expected

$$\sigma^{-1}(l+1)(l+\log_2 n+2)+(l+1)^3+2l^2+2(l+1)\log_2 n+2l$$

arithmetic operations and one gcd computation.

Proof. We shall find a number and two independent square roots modulo n.

We begin by choosing random numbers r between 0 and n for each number checking if $t = r^2 \bmod n$ factors as a product of the small primes. In this way, we eventually find $l+1$ relations of the form (3.5).

By Proposition 3.14, we can find $\alpha_1, \alpha_2, \ldots, \alpha_{l+1} \in \{0, 1\}$ such that

$$\left(\prod_{i=1}^{l+1} r_i\right)^2 \equiv \left(\prod_{j=1}^{l} \ell_j^{\frac{1}{2}\sum_{i=1}^{l+1}\alpha_i s_{ij}}\right)^2 \quad (\bmod\ n).$$

In other words, we have two square roots modulo n of the same number. Since the coefficients s_{ij} depend only on the square modulo n of the random numbers $r_1, \ldots, r_{l+1}$, the two square roots are also independent.

By Proposition 3.13, we can then factor n with probability $1/2$ using one gcd computation.

For each random number r, squaring modulo n requires 2 arithmetic operations. Checking if the square factors as a product of $\ell_1, \ldots, \ell_l$ requires at most $\lfloor l+\log_2 n\rfloor$ arithmetic operations.

We expect to find $l+1$ relations after trying $\sigma^{-1}(l+1)$ random numbers, which means that we need at most

$$\sigma^{-1}(l+1)(l+\log_2 n+2)$$

arithmetic operations to generate the relations. We then need at most $(l+1)^3$ arithmetic operations to find $\alpha_1, \ldots, \alpha_{l+1}$. (Strictly speaking, we work with a binary matrix, so these arithmetic operations are much faster.)

Finally, we need at most $2l^2$ arithmetic operations to compute the sums $\sum_{i=1}^{l+1}\alpha_i s_{ij}$, then at most $2(l+1)\log_2 n$ arithmetic operations to compute the small primes raise to these sums. Finally, we require at most $2l$ multiplications to compute the two independent square roots. $\square$

Example 3.16. We choose random numbers, eventually finding these relations:

$$237625^2 \equiv 5000 \equiv 2^3 \cdot 5^4 \quad (\bmod\ 314791)$$
$$90962^2 \equiv 118800 \equiv 2^4 \cdot 3^3 \cdot 5^2 \cdot 11 \quad (\bmod\ 314791)$$
$$180136^2 \equiv 7425 \equiv 3^3 \cdot 5^2 \cdot 11 \quad (\bmod\ 314791)$$
$$57593^2 \equiv 882 \equiv 2 \cdot 3^2 \cdot 7^2 \quad (\bmod\ 314791)$$
$$228218^2 \equiv 25410 \equiv 2 \cdot 3 \cdot 5 \cdot 7 \cdot 11^2 \quad (\bmod\ 314791)$$
$$211193^2 \equiv 61250 \equiv 2 \cdot 5^4 \cdot 7^2 \quad (\bmod\ 314791)$$

As we saw in Example 3.15, we can use these relations to attempt to find two non-trivial square roots of the same number. And as we see in Example 3.14, once we have these non-trivial square roots, an easy gcd computation gives us a factor of 314791.

Note that if we do all these calculations and fail to factor n, then we do not have to all the work over again. In practice, discarding a few of our old relations and replacing them by a few new ones will give us another go. In this way, we will quickly factor n without doing significantly more work.

We also expect l to be much larger than $\log_2 n$, so we expect to be able to factor n using approximately

$$\sigma^{-1} l^2 + l^3$$

arithmetic operations. It is reasonable to assume that random squares modulo n are as likely to be products of small primes as random integers. As in the discussion in Section 2.4.3, we can choose l to factor using approximately

$$\exp\left(\sqrt{8}\sqrt{\log n \log \log n}\right)$$

arithmetic operations.

We have arrived at the following requirement.

Requirement 3.5. $\exp(\sqrt{8}\sqrt{\log n \log \log n})$ arithmetic operations should be an infeasible computation.

There are much better factoring algorithms. We shall not study these algorithms, but we note that the above requirement is not the final requirement.

3.5 LATTICES

Many different mathematical concepts share the name *lattice*, but the one we are interested in is essentially integer linear combinations of linearly independent vectors in a real vector space. This notion of lattice arises naturally in many areas of mathematics, and also in many applications, leading to a rich and varied theory that was well-developed long before its use in cryptography.

Lattices have many uses in cryptography, and have for instance been used to attack many cryptographic constructions. There have been many attempts to construct cryptographic systems based on lattices, and some of these have failed. Recently, the theory of lattice-based cryptography has grown significantly, especially related to so-called *homomorphic* cryptosystems.

However, at the moment, we are not interested in using lattices to attack cryptosystems or these recent constructive developments, but rather the fact that there does not seem to be any fast quantum algorithms for solving difficult lattice-related problems. This makes lattice-based cryptography into a candidate for *quantum-safe* cryptography.

3.5.1 Lattice Basics

There are several common, equivalent ways to define lattices. We have chosen a variant that is convenient for our purposes.

Definition 3.2. A *lattice* is a subgroup of $\mathbb{R}^n$ of the form

$$\Lambda = \left\{ \sum_{i=1}^{r} a_i \mathbf{b}_i \,\middle|\, a_1, \ldots, a_r \in \mathbb{Z} \right\},$$

where $\mathbf{b}_1, \ldots, \mathbf{b}_r$ are linearly independent vectors in $\mathbb{R}^n$.

A *basis* for a lattice Λ is any set of linearly independent vectors $\mathbf{c}_1, \mathbf{c}_2, \ldots, \mathbf{c}_{r'}$ such that $\Lambda = \{ \sum_i a_i \mathbf{c}_i \mid a_i \in \mathbb{Z} \}$. We can let a basis $\mathbf{b}_1, \ldots, \mathbf{b}_r$ be the rows of a $r \times n$ matrix $\mathbf{B}$. Then

$$\Lambda = \{ \mathbf{aB} \mid \mathbf{a} \in \mathbb{Z}^n \}.$$

The matrix $\mathbf{B}$ is called a basis matrix or just a basis. Since the rows of $\mathbf{B}$ are linearly independent, the rank of the matrix is equal to the number of rows.

Example 3.17. The vectors $(1, 0, 1, 0)$, $(1, 1, 1, 0)$, $(1, -1, -1, 1)$ and $(0, 1, -1, -1)$ are linearly independent and generate a lattice. We get the basis matrix

$$\mathbf{B} = \begin{pmatrix} 1 & 0 & 1 & 0 \\ 1 & 1 & 1 & 0 \\ 1 & -1 & -1 & 1 \\ 0 & 1 & -1 & -1 \end{pmatrix}.$$

Exercise 3.37. Let Λ be the lattice with the basis from Example 3.17. Which of $(6, 3, -4, -1)$, $(3, 6, 9, 12)$, $(1, 2, 3, 5)$ and $(1, 2, 1, 2)$ are lattice vectors?

Proposition 3.16. *Let Λ be a lattice, and let $\mathbf{B}$ and $\mathbf{C}$ be two bases for Λ. Then $\mathbf{B}$ and $\mathbf{C}$ have the same rank r, and there exists an $r \times r$ invertible integer matrix $\mathbf{U}$ such that $\mathbf{UB} = \mathbf{C}$ and $\mathbf{U}^{-1}$ is an integer matrix.*

Proof. Every row of $\mathbf{B}$ is in Λ, which is a subset of the row space of $\mathbf{C}$, so the row space of $\mathbf{B}$ is a subspace of the row space of $\mathbf{C}$. The converse also holds, so the matrices have the same row space and therefore also the same rank.

For every row $\mathbf{b}_i$ in $\mathbf{B}$, there exists an integer vector $\mathbf{a}_i \in \mathbb{Z}^r$ such that $\mathbf{b}_i = \mathbf{a}_i \mathbf{C}$. Combining these expression, we get an $r \times r$ integer matrix $\mathbf{U}$ such that $\mathbf{B} = \mathbf{UC}$. In the same way, we get an $r \times r$ integer matrix $\mathbf{V}$ such that $\mathbf{C} = \mathbf{VB}$, and $\mathbf{B} = \mathbf{UVB}$. Since $\mathbf{B}$ and $\mathbf{C}$ have maximal rank (and therefore have right-inverses) $\mathbf{U}$ and $\mathbf{V}$ must be inverses. □

E *Exercise 3.38.* The matrix

$$
\mathbf{C} = \begin{pmatrix} -1 & -3 & -4 & 0 \\ 2 & -6 & -3 & 4 \\ 9 & -1 & 9 & 9 \\ 6 & 7 & 12 & 3 \end{pmatrix}
$$

is another basis matrix for the lattice Λ from Example 3.17. Find the integer matrices taking $\mathbf{B}$ to $\mathbf{C}$ and back again, and check that they are inverses.

D **Definition 3.3.** The *rank* of a lattice is the number of vectors in a basis. If Λ in $\mathbb{R}^n$ has rank n, we say that Λ is a *full rank* lattice.

3.5.2 Gram-Schmidt

The Gram-Schmidt algorithm is a general inner product space algorithm, which should be well-known from linear algebra. We will need it later on, so it makes sense to repeat its properties here. In the following, $\langle \cdot, \cdot \rangle$ denotes the usual inner product on $\mathbb{R}^n$. The Gram-Schmidt algorithm takes a vector space basis $\mathbf{b}_1, \ldots, \mathbf{b}_r$ as input and constructs an orthogonal basis $\mathbf{b}_1^*, \ldots, \mathbf{b}_r^*$ recursively, beginning with $\mathbf{b}_1^* = \mathbf{b}_1$ and then computing

$$
\mu_{ij} = \frac{\langle \mathbf{b}_i, \mathbf{b}_j^* \rangle}{\langle \mathbf{b}_j^*, \mathbf{b}_j^* \rangle}, \quad 1 < i \le r, 1 \le j < i, \quad \text{and} \quad \mathbf{b}_i^* = \mathbf{b}_i - \sum_{j=1}^{i-1} \mu_{ij} \mathbf{b}_j^*. \tag{3.6}
$$

If we rearrange the equation defining $\mathbf{b}_i^*$ as $\mathbf{b}_i = \mathbf{b}_i^* + \sum_{j=1}^{i-1} \mu_{ij} \mathbf{b}_j^*$, we get a lower-triangular matrix and the matrix equation

$$
\mathbf{B} = \begin{pmatrix} \mathbf{b}_1 \\ \vdots \\ \mathbf{b}_r \end{pmatrix} = \begin{pmatrix} 1 & & 0 \\ & \ddots & \\ \mu_{ij} & & 1 \end{pmatrix} \begin{pmatrix} \mathbf{b}_1^* \\ \vdots \\ \mathbf{b}_r^* \end{pmatrix}. \tag{3.7}
$$

E *Exercise 3.39.* Let $\mathbf{b}_1, \mathbf{b}_2, \ldots, \mathbf{b}_r \in \mathbb{R}^n$ be a lattice basis. Show that we can compute the Gram-Schmidt basis $\mathbf{b}_1^*, \mathbf{b}_2^*, \ldots, \mathbf{b}_r^*$ and coefficients μ_{ij}, $1 \le j < i \le r$ using $2nr^2 - 1$ arithmetic operations.

E *Exercise 3.40.* The answer to Exercise 3.39 implies the existence of an algorithm to compute a Gram-Schmidt basis. Implement this algorithm.

E *Exercise 3.41.* Let Λ be the lattice from Example 3.17 with the basis matrix $\mathbf{B}$. Compute the Gram-Schmidt basis corresponding to this basis.

E *Exercise* 3.42. Let $\mathbf{b}_1, \mathbf{b}_2, \ldots, \mathbf{b}_r$ be a basis for a lattice Λ, let $\mathbf{B}$ be the corresponding matrix, and let $\mathbf{b}_1^*, \ldots, \mathbf{b}_r^*$ be the corresponding Gram-Schmidt orthogonal basis.

(a) Show that $\mathbf{b}_1^*, \ldots, \mathbf{b}_r^*$ can be extended to an orthogonal basis $\mathbf{b}_1^*, \ldots, \mathbf{b}_r^*, \mathbf{c}_1^*, \ldots, \mathbf{c}_{n-r}^*$ for $\mathbb{R}^n$.

(b) Let $\mathbf{e}_1, \mathbf{e}_2, \ldots, \mathbf{e}_r$ be an orthonormal basis for $\mathbb{R}^r$. Let $\mathbf{P}$ be the linear map from $\mathbb{R}^n$ to $\mathbb{R}^r$ that takes $\mathbf{b}_i^*$ to $\|\mathbf{b}_i^*\|\mathbf{e}_i$, $1 \leq i \leq r$, and $\mathbf{c}_i^*$ to 0, $1 \leq i \leq n - r$. Show that for any vector $\mathbf{v} \in \mathrm{span}\{\mathbf{b}_1^*, \ldots, \mathbf{b}_r^*\}$, $\|\mathbf{v}\| = \|\mathbf{v}\mathbf{P}\|$.

(c) Show that the image $\Lambda\mathbf{P}$ of Λ under $\mathbf{P}$ is a full-rank lattice, and $\mathbf{b}_1\mathbf{P}, \mathbf{b}_2\mathbf{P}, \ldots, \mathbf{b}_r\mathbf{P}$ is a basis for $\Lambda\mathbf{P}$.

3.5.3 The Fundamental Domain

Let $\mathbf{b}_1, \ldots, \mathbf{b}_r$ be a basis for a lattice Λ, $\mathbf{B}$ in matrix form. The *fundamental domain* of the lattice is the parallelepiped

$$\left\{ \sum_{i=1}^{r} a_i \mathbf{b}_i \mid 0 \leq a_i < 1 \right\}.$$

When the lattice has full rank, it can be shown that the volume of the fundamental domain is $|\det(\mathbf{B})|$. The volume of the fundamental domain is independent of the basis, since any two bases are related by an invertible integer matrix, which has determinant ± 1. We call this volume the *determinant* or the *volume* of the lattice, denoted by $\det(\Lambda)$, or sometimes by $\det(\mathbf{b}_1, \mathbf{b}_2, \ldots, \mathbf{b}_r)$.

Since a determinant is unchanged by elementary row operations, if $\mathbf{b}_1, \ldots, \mathbf{b}_n$ is a basis for a full-rank lattice Λ, then

$$\det(\Lambda) = |\det(\mathbf{B})| = \prod_i \|\mathbf{b}_i^*\| \leq \prod_i \|\mathbf{b}_i\|.$$

The value

$$\frac{\prod \|\mathbf{b}_i\|}{\det(\Lambda)}$$

is called the *orthogonality defect* of the basis.

E *Exercise* 3.43. We continue Exercise 3.42. Show that $\det(\Lambda\mathbf{P}) = \det(\mathbf{B}\mathbf{P}) = \prod_{i=1}^{r} \|\mathbf{b}_i^*\|$.

When $r < n$, we define the volume of the fundamental domain to be

$$\det(\Lambda) = \sqrt{|\det(\mathbf{B}\mathbf{B}^T)|} = \prod_{i=1}^{r} \|\mathbf{b}_i^*\|.$$

Another interesting property of the fundamental domain of a full-rank lattice is that any point in space can be expressed uniquely as the sum of a lattice point and a point in the fundamental domain.

E *Exercise 3.44.* Let Λ be a lattice with basis $\mathbf{b}_1, \mathbf{b}_2, \dots, \mathbf{b}_n$ and let $\mathbf{x} \in \mathbb{R}^n$. Show that unique integers $a_1, a_2, \dots, a_n$ and real numbers $\alpha_1, \alpha_2, \dots, \alpha_n \in [0, 1)$ exist such that

$$\mathbf{x} = \sum_i a_i \mathbf{b}_i + \sum_i \alpha_i \mathbf{b}_i.$$

Note that the fundamental domain depends on the basis for the lattice. We shall use the notation $\mathbf{x} \bmod \mathbf{B}$ to denote the unique point in the fundamental domain. If two vectors $\mathbf{x}_1, \mathbf{x}_2$ satisfy $\mathbf{x}_1 \bmod \mathbf{B} = \mathbf{x}_2 \bmod \mathbf{B}$, then $\mathbf{x}_2 - \mathbf{x}_1 \in \Lambda$. We shall write both $\mathbf{x}_1 \equiv \mathbf{x}_2 \pmod{\mathbf{B}}$ and $\mathbf{x}_1 \equiv \mathbf{x}_2 \pmod{\Lambda}$.

For a full-rank lattice, the vector $\boldsymbol{\alpha}$ is sometimes denoted by $(\mathbf{x}\mathbf{B}^{-1}) \bmod 1$, but we shall not use this notation. Instead, we shall write $(\mathbf{x} \bmod \mathbf{B})\mathbf{B}^{-1}$.

3.5.4 Dual Lattice

Corresponding to dual vector spaces, we shall define the dual lattice as the set of linear integer-valued functions (functionals) on the lattice, and then identify this set of functions with a particular lattice.

D **Definition 3.4.** Let $\Lambda \subseteq \mathbb{R}^n$ be a lattice. The *dual lattice* is $\Lambda^* = \{f : \Lambda \to \mathbb{Z} \mid f \text{ linear}\}$.

We would like to study the structure of Λ^*. It is clearly closed under addition and multiplication by integers. Recall that for any functional f on $\mathbb{R}^n$, there is a vector $\mathbf{w}_f$ such that for any $\mathbf{v} \in \mathbb{R}^n$ we have that $f(\mathbf{v}) = \mathbf{v}\mathbf{w}_f^T$. If f is a functional on a subspace, we can we choose $\mathbf{w}_f$ to be in the subspace. Since any $f \in \Lambda^*$ is also a functional on $\mathbb{R}^n$ (or span Λ if the lattice is not a full-rank lattice), we can associate the set

$$S_\Lambda = \{\mathbf{w}_f \in \text{span } \Lambda \mid f \in \Lambda^*\}$$

with Λ^*.

E *Exercise 3.45.* Show that the set S_Λ is closed under addition and multiplication by integers, and that the natural map $f \mapsto \mathbf{w}_f$ is a bijection and respects addition and multiplication by integers.

T **Proposition 3.17.** *Let $\mathbf{B}$ be a basis matrix for Λ. The set S_Λ is a lattice with basis matrix* $(\mathbf{B}\mathbf{B}^T)^{-T}\mathbf{B}$.

Proof. If $\mathbf{u}, \mathbf{v}$ are integer tuples, $\mathbf{w} = \mathbf{u}(\mathbf{B}\mathbf{B}^T)^{-T}\mathbf{B}$ and $\mathbf{v} = \mathbf{v}\mathbf{B}$, then

$$\mathbf{v}\mathbf{w}^T = \mathbf{v}\mathbf{B}\mathbf{B}^T(\mathbf{B}\mathbf{B}^T)^{-1}\mathbf{u}^T = \mathbf{v}\mathbf{u}^T \in \mathbb{Z},$$

so the lattice generated by $(\mathbf{B}\mathbf{B}^T)^{-T}\mathbf{B}$ is a subset of S_Λ.

Now suppose $\mathbf{w}_f \in S_\Lambda$. Then $\mathbf{w}_f = \mathbf{u}\mathbf{B}$ for some $\mathbf{u} \in \mathbb{R}^r$. Also, since the rows of $\mathbf{B}$ are in Λ, we must have that $\mathbf{B}\mathbf{w}_f^T$ is an integer tuple. Then

$$\mathbf{w}_f = \mathbf{u}\mathbf{B} = \mathbf{u}(\mathbf{B}\mathbf{B}^T)^T(\mathbf{B}\mathbf{B}^T)^{-T}\mathbf{B} = \mathbf{u}\mathbf{B}\mathbf{B}^T(\mathbf{B}\mathbf{B}^T)^{-T}\mathbf{B} = \mathbf{w}_f\mathbf{B}^T(\mathbf{B}\mathbf{B}^T)^{-T}\mathbf{B}$$

which proves that $\mathbf{w}_f$ lies in the lattice generated by $(\mathbf{B}\mathbf{B}^T)^{-T}\mathbf{B}$. □

We identify Λ^* with the set S_Λ, so that Λ^* is obviously a lattice.

E | *Exercise* 3.46. Show that the dual lattice of Λ^* is Λ.

T | **Proposition 3.18.** *Let Λ be a lattice with dual Λ^*. Let $\mathbf{B}$ be a basis for Λ and let $\mathbf{C}$ be a basis for Λ^*. Then $\mathbf{BC}^T$ is an integer matrix with integral inverse.*

Proof. First, note that $(\mathbf{BB}^T)^{-T}\mathbf{B}$ is a basis for Λ^*, so there is an integer matrix with integral inverse $\mathbf{U}$ such that $\mathbf{C} = \mathbf{U}(\mathbf{BB}^T)^{-T}\mathbf{B}$. But then

$$\mathbf{BC}^T = \mathbf{BB}^T(\mathbf{BB}^T)^{-1}\mathbf{U}^T = \mathbf{U}^T,$$

which proves the claim. □

3.5.5 p-ary Lattices

Several lattices we shall encounter originate with problems over prime finite fields. In these cases, it makes sense to study certain special lattices.

D | **Definition 3.5.** Let $p\mathbb{Z}^n = \{p\mathbf{u} \mid \mathbf{u} \in \mathbb{Z}^n\}$. A lattice is a *$p$-ary lattice* if $p\mathbb{Z}^n \subseteq \Lambda \subseteq \mathbb{Z}^n$.

Let $\mathbf{M}$ be any $n \times r$ integer matrix. Then let

$$\Lambda_p(\mathbf{M}) = \{\mathbf{u} \in \mathbb{Z}^n \mid \exists \mathbf{a} \in \mathbb{Z}^r : \mathbf{aM} \equiv \mathbf{u} \pmod{p}\} \text{ and}$$
$$\Lambda_p^\perp(\mathbf{M}) = \{\mathbf{u} \in \mathbb{Z}^n \mid \mathbf{Mu}^T \equiv \mathbf{0} \pmod{p}\}.$$

Remark. $\mathbf{M}$ can be thought of as a generating matrix for a block code over $\mathbb{F}_p$, in which case $\mathbf{M}^T$ would be a parity check matrix for a dual code.

T | **Proposition 3.19.** *Let Λ be a lattice. The following are equivalent:*

(i) Λ is p-ary;
(ii) there exists a matrix $\mathbf{M}$ such that $\Lambda = \Lambda_p(\mathbf{M})$; and
(iii) there exists a matrix $\mathbf{M}'$ such that $\Lambda = \Lambda_p^\perp(\mathbf{M}')$.

Before we prove the proposition, we shall need a small result about the existence of certain matrices related to subspaces of $\mathbb{F}_p^n$. (Again, this is related to linear codes and their generator and parity check matrices.)

E | *Exercise* 3.47. Let V be any non-trivial proper subspace of $\mathbb{F}_p^n$ of dimension r. Show that there exists matrices $\mathbf{M} \in \mathbb{F}_p^{r \times n}$ and $\mathbf{M}' \in \mathbb{F}_p^{n \times (n-r)}$ such that V is the row space of $\mathbf{M}$ and the left kernel of $(\mathbf{M}')^T$.

E | *Exercise* 3.48. Prove of Proposition 3.19.

When we consider the dual lattice $\Lambda_p^*(\mathbf{M})$ as an integer lattice, it is clear that $\Lambda_p^\perp(\mathbf{M}) \subseteq \Lambda_p^*(\mathbf{M})$. Likewise, since for any vector $\mathbf{u} \in \Lambda_p(\mathbf{M})$ and any vector $\mathbf{v} \in \Lambda_p^\perp(\mathbf{M})$ we have that

$$\mathbf{u}\mathbf{v}^T = \mathbf{a}\mathbf{M}\mathbf{v}^T \equiv 0 \pmod{p},$$

it follows that $\frac{1}{p}\mathbf{v} \in \Lambda_p^*(\mathbf{M})$ and that

$$\frac{1}{p}\Lambda_p^\perp(\mathbf{M}) \subseteq \Lambda_p^*(\mathbf{M}).$$

In fact, we have equality, since for any $\mathbf{u} \in \Lambda_p(\mathbf{M})$ and $\mathbf{v} \in \Lambda_p^*(\mathbf{M})$ we have that $\mathbf{u}\mathbf{v}^T \in \mathbb{Z}$, which means that $\mathbf{u}(p\mathbf{v})^T \equiv 0 \pmod{p}$.

T | **Proposition 3.20.** *Let Λ be a p-ary lattice and let $\mathbf{M}$ be such that $\Lambda = \Lambda_p(\mathbf{M})$. Then*

$$\frac{1}{p}\Lambda_p^\perp(\mathbf{M}) = \Lambda_p^*(\mathbf{M}).$$

3.5.6 Short Vectors

The problems where the theory of lattices originated often involved finding short integer linear combinations or deciding if such existed. It will turn out that short non-zero vectors in lattices are important also for cryptography.

D | **Definition 3.6.** Let Λ be a lattice. The *i-th successive minimum* $\lambda_i(\Lambda)$ is the minimal real number such that there are i linearly independent vectors of length at most $\lambda_i(\Lambda)$ in Λ.

For a lattice of rank r, there are r successive minima, with $0 < \lambda_1(\Lambda) \leq \lambda_2(\Lambda) \leq \cdots \leq \lambda_n(\Lambda)$, and $\lambda_1(\Lambda)$ is the shortest length of a non-zero vector in Λ. Note that $\lambda_2(\Lambda)$ does not have to be the second shortest length of a non-zero vector, since we require independent vectors.

T | **Fact 3.21.** *There exists a constant γ_n, depending only on n, such that for any lattice Λ*

$$\lambda_1(\Lambda)^2 < \gamma_n \det(\Lambda)^{2/n}.$$

A natural question related to lattices is to ask what $\lambda_1(\Lambda)$ is. For large n, it can be shown that $n/(2\pi e) \leq \gamma_n$, and heuristic arguments suggests that an "average" lattice will not contain non-zero vectors much shorter than $\sqrt{n/(2\pi e)} \det(\Lambda)^{1/n}$.

More practically important is finding a lattice vector of length $\lambda_1(\Lambda)$. There will be more than one non-zero vector of minimal length, so the shortest vector problem will not have a unique answer. For some applications, merely finding a short vector is sufficient.

Definition 3.7. The *shortest vector problem* for a lattice Λ is to find a vector $\mathbf{u} \in \Lambda$ such that $\|\mathbf{u}\| = \lambda_1(\Lambda)$. The γ-*approximate shortest vector problem* for a lattice Λ is to find a vector $\mathbf{u} \in \Lambda$ such that $\|\mathbf{u}\| \leq \gamma\lambda_1(\Lambda)$.

A slightly different problem is to find a lattice vector close to some given point. It also has an approximate version.

Definition 3.8. The *closest vector problem* for a lattice $\Lambda \subseteq \mathbb{R}^n$ and $\mathbf{x} \in \mathbb{R}^n$ is to find $\mathbf{u} \in \Lambda$ such that for any $\mathbf{v} \in \Lambda$, $\|\mathbf{u} - \mathbf{x}\| \leq \|\mathbf{v} - \mathbf{x}\|$.

The γ-*approximate closest vector problem* for a lattice $\Lambda \subseteq \mathbb{R}^n$ and $\mathbf{x} \in \mathbb{R}^n$ is to find $\mathbf{u} \in \Lambda$ such that for any $\mathbf{v} \in \Lambda$, $\|\mathbf{u} - \mathbf{x}\| \leq \gamma\|\mathbf{v} - \mathbf{x}\|$.

An exhaustive search for short or close vectors will quickly become infeasible as the lattice dimension grows. There is evidence that these lattice problems are hard in a very fundamental way. However, the hardness depends on the lattice and on how the lattice is described.

Exercise 3.49. Suppose Λ has a given orthogonal basis $\mathbf{b}_1, \mathbf{b}_2, \ldots, \mathbf{b}_n$. Show how to find a shortest vector in Λ and all the successive minima, essentially without any arithmetic.

If the basis is nearly orthogonal, the same approach as in the exercise will find short vectors and make it easier to find a shortest vector.

Exercise 3.50. Let $\mathbf{B}$ be an invertible $n \times n$ matrix and let $\mathbf{b}_1, \mathbf{b}_2, \ldots, \mathbf{b}_n$ correspond to the n rows of $\mathbf{B}$. Let $\mathbf{x} \in \mathbb{R}^n$ and let $(\alpha_1, \alpha_2, \ldots, \alpha_n) = \mathbf{x}\mathbf{B}^{-1}$. Show that $\mathbf{x} = \alpha_1\mathbf{b}_1 + \alpha_2\mathbf{b}_2 + \cdots + \alpha_n\mathbf{b}_n$.

For the following exercise, it is convenient to introduce the following notation: $\lfloor\alpha\rceil$ is the nearest integer to α (rounding halves towards even numbers), and for a vector $\boldsymbol{\alpha} = (\alpha_1, \alpha_2, \ldots, \alpha_n)$,

$$\lfloor\boldsymbol{\alpha}\rceil = (\lfloor\alpha_1\rceil, \lfloor\alpha_2\rceil, \ldots, \lfloor\alpha_n\rceil).$$

Exercise 3.51. Suppose Λ has a given orthogonal basis $\mathbf{B}$, and let $\mathbf{x} \in \mathbb{R}^n$. Suppose $\mathbf{a} = \lfloor\mathbf{x}\mathbf{B}^{-1}\rceil$. Explain why $\mathbf{a}\mathbf{B}$ is the closest vector in Λ to $\mathbf{x}$.

If the basis is nearly orthogonal, the same approach as in the exercise will tend to find the closest vector if $\mathbf{x}$ was reasonably close to a lattice point.

Exercise 3.52. Let Λ be the lattice from Example 3.17 with the basis matrix $\mathbf{B}$ from that exercise and the basis matrix $\mathbf{C}$ given in Exercise 3.38. Consider the vector $(2, 2, 2, 2) \in \mathbb{R}^4$. Use the rounding strategy from Exercise 3.51 to find two lattice points. Are both of them "close" to $(2, 2, 2, 2)$. Can you decide if there are lattice points closer to $(2, 2, 2, 2)$ than you have already found?

Finding the closest lattice point in a lattice Λ has a related problem on the dual lattice Λ^*. Note that if $\mathbf{u}$ is a lattice point such that $\|\mathbf{x} - \mathbf{u}\|$ is minimal, then $\mathbf{f} = \mathbf{x} - \mathbf{u}$ has minimal length among the vectors satisfying $\mathbf{f} \equiv \mathbf{x} \pmod{\Lambda}$.

Let $\mathbf{C}$ be a basis for Λ^*. Then $\mathbf{f}\mathbf{C}^T = \mathbf{x}\mathbf{C}^T - \mathbf{u}\mathbf{C}^T$, which means that the difference between $\mathbf{f}\mathbf{C}^T$ and $\mathbf{x}\mathbf{C}^T$ is an integral vector.

3.6 LATTICE-BASED CRYPTOSYSTEMS

While the first cryptosystem we shall look at, the GGH system in Section 3.6.1, is phrased directly in terms of lattices, many cryptosystems do not directly use lattices, yet their security is essentially based on certain lattice problems being difficult to solve.

We now discuss three systems, GGH, Regev's Learning-with-errors cryptosystem and NTRU. GGH is directly based on lattices. Regev's cryptosystem is based on the difficulty of solving an overdefined linear system of equations that contains errors. NTRU is based on the difficulty of expressing an element in a polynomial ring as a fraction where the numerator and denominator are both polynomials with short coefficients.

Remark. When we discussed public key encryption schemes based on Diffie-Hellman and RSA, we spent some time discussing attacks against simple variants of these schemes. Similar attacks exist against the cryptosystems in this section, or at least against simplified versions. We do not repeat the discussion.

3.6.1 The GGH Cryptosystem

One idea for a symmetric encryption scheme based on lattices is to have a lattice Λ with a nearly orthogonal basis $\mathbf{B}$ as a secret key. To encrypt we somehow encode the message as a lattice vector $\mathbf{u} \in \Lambda$ and then add *random noise* $\mathbf{f}$ to that vector to get a ciphertext $\mathbf{x} = \mathbf{u} + \mathbf{f}$. To decrypt, we can use our nearly orthogonal basis to find the closest vector $\mathbf{u}$ to $\mathbf{x}$ (using the technique from Exercise 3.51), and then decode the lattice point to recover the message.

It is clear that we need to limit the magnitude of the random noise, since if it is too big, we will no longer be able to recover $\mathbf{u}$ as the closest vector. This could happen because $\mathbf{u}$ is no longer the closest vector to $\mathbf{x}$, or because our basis is not orthogonal and so does not perfectly solve the closest vector problem.

E | *Exercise* 3.53. Let $\mathbf{B}$ be a basis for a lattice Λ. Show that with $\mathbf{u} \in \Lambda$ and $\mathbf{x} = \mathbf{u} + \mathbf{f}$, then $\lfloor \mathbf{x}\mathbf{B}^{-1} \rceil \mathbf{B} = \mathbf{u}$ if and only if $\lfloor \mathbf{f}\mathbf{B}^{-1} \rceil = \mathbf{0}$.

E | *Exercise* 3.54. Recall that for a real vector $\boldsymbol{\alpha} = (\alpha_1, \alpha_2, \ldots, \alpha_n)$, we have the norms $\|\boldsymbol{\alpha}\|_1 = \sum_i |\alpha_i|$ and $\|\boldsymbol{\alpha}\|_\infty = \max_i |\alpha_i|$.

Let $\mathbf{B}$ be a basis for a lattice Λ, let ρ be a bound on the $\|\cdot\|_1$ norm of the columns of $\mathbf{B}^{-1}$. Show that for any vector $\mathbf{f}$, we have that

$$\|\mathbf{f}\mathbf{B}^{-1}\|_\infty \le \rho\|\mathbf{f}\|_\infty.$$

Explain how this can be used to find a bound on the random noise when encrypting, to ensure decryption still works.

It is tempting to turn this idea into a public key encryption scheme by publishing a basis for the lattice. Obviously, we cannot publish our nearly orthogonal basis $\mathbf{B}$, since this is essentially the decryption key.

Recall that any lattice Λ with basis matrix $\mathbf{B}$, if $\mathbf{U}$ is an integer matrix with determinant ± 1, then $\mathbf{C} = \mathbf{U}\mathbf{B}$ is another basis matrix for Λ.

One idea is then to create and publish a different basis for the lattice, one that is not nearly orthogonal, and therefore cannot be used to find the closest vector using the approach from Exercise 3.51. One possible choice is to use the Hermite normal form for $\mathbf{B}$ as $\mathbf{C}$. (The Hermite normal form is in some sense the worst possible form we can give the public basis, since any adversary could compute the Hermite normal form on his own.)

As usual, while we could attempt to embed the message in the vector $\mathbf{u}$, it makes more sense to use $\mathbf{u}$ and $\mathbf{f}$ as keys for a symmetric cryptosystem. As we did in Example 3.9, we shall use a function to derive the key from the lattice point and the noise, a so-called *key derivation function*.

Example 3.18. The GGH (Goldreich-Goldwasser-Halevi) public key encryption scheme $(\mathcal{K}, \mathcal{E}, \mathcal{D})$ is based on lattices in $\mathbb{R}^n$, a key derivation function $kdf : \mathbb{R}^n \times \mathbb{R}^n \to \mathfrak{K}_s$ and a symmetric cryptosystem $(\mathfrak{K}_s, \mathfrak{P}, \mathfrak{C}, \mathcal{E}_s, \mathcal{D}_s)$ and works as following.

- The *key generation* algorithm $\mathcal{K}$ chooses a lattice Λ by choosing a basis matrix $\mathbf{B}$ that is nearly orthogonal. It chooses an integer matrix $\mathbf{U}$ with determinant ± 1 and computes $\mathbf{Y} = \mathbf{U}\mathbf{B}$. It outputs $ek = \mathbf{Y}$ and $dk = \mathbf{B}$.

- The *encryption* algorithm $\mathcal{E}$ takes as input an encryption key $ek = \mathbf{Y}$ and a message $m \in \mathfrak{P}$. It chooses a random vector $\mathbf{r} \in \mathbb{Z}^n$ and random noise $\mathbf{f}$. Then it computes $\mathbf{u} = \mathbf{r}\mathbf{Y}$, $\mathbf{x} = \mathbf{u} + \mathbf{f}$, and encrypts the message as $w = \mathcal{E}_s(kdf(\mathbf{u}, \mathbf{f}), m)$. It outputs the ciphertext $c = (\mathbf{x}, w)$.

- The *decryption* algorithm $\mathcal{D}$ takes as input a decryption key $dk = \mathbf{B}$ and a ciphertext $c = (\mathbf{x}, w)$. It computes $\mathbf{u} = \lfloor \mathbf{x}\mathbf{B}^{-1} \rceil \mathbf{B}$, $\mathbf{f} = \mathbf{x} - \mathbf{u}$, and $m = \mathcal{D}_s(kdf(\mathbf{u}, \mathbf{f}), w)$. If $\mathcal{D}_s$ outputs the special symbol $\perp$, then $\mathcal{D}$ outputs $\perp$, otherwise m.

Remark. We have not said how to choose $\mathbf{B}$ or $\mathbf{U}$. Nor will we.

E *Exercise 3.55.* The above is an informal description of a public key encryption scheme. Implement (use some simple method to choose $\mathbf{B}$, $\mathbf{U}$, $\mathbf{r}$ and $\mathbf{f}$) the three algorithms $\mathcal{K}$, $\mathcal{E}$ and $\mathcal{D}$. Show that the triple $(\mathcal{K}, \mathcal{E}, \mathcal{D})$ is a public key encryption scheme.

3.6.2 Regev's Cryptosystem

Let p be a prime. Let $\mathbf{g}_1, \mathbf{g}_2, \ldots, \mathbf{g}_l \in \mathbb{F}_p^n$ be a set of randomly chosen vectors that contains n linearly independent vectors. Let $\mathbf{b} \in \mathbb{F}_p^n$, and let

$$\beta_i = \mathbf{g}_i \cdot \mathbf{b}, \qquad i = 1, 2, \ldots, l.$$

If we learn the value of these dot products, we can recover the vector $\mathbf{b}$.

E *Exercise 3.56.* Given $\mathbf{g}_1, \ldots, \mathbf{g}_l$ and $\beta_1, \ldots, \beta_l$, show how to compute $\mathbf{b}$.

If we add a bit of randomness to the dot product, it turns out that finding the value $\mathbf{b}$ becomes much more difficult. Let χ be a probability distribution on $\mathbb{F}_p$ and let $f_1, f_2, \ldots, f_l$ be sampled independently according to χ. Let

$$y_i = \mathbf{g}_i \cdot \mathbf{b} + f_i, \qquad i = 1, 2, \ldots, l.$$

Finding $\mathbf{b}$ given $y_1, y_2, \ldots, y_l$ is known as the *learning with errors* (LWE) problem (we want to learn $\mathbf{b}$ from a set of equations with errors in them).

E *Exercise 3.57.* Show that if $l = n$, then the learning with errors problem is impossible to answer with (close to) certainty. (That is, any algorithm trying to solve the learning with errors problem must fail sometimes.)

We shall need one particular property for the probability distribution χ. Except with small probability, when $f_1, f_2, \ldots, f_l$ have been sampled independently from χ, then for almost all subsets $S \subseteq \{1, 2, \ldots, l\}$ we have that

$$\sum_{i \in S} f_i = k + \langle p \rangle,$$

for some integer k with $|k| < p/4$. It can be shown that χ can be chosen such that this requirement is satisfied, and it still seems hard to solve the learning with errors problem.

E *Example 3.19. Regev's learning with errors cryptosystem* works as follows:

- The *key generation* algorithm $\mathcal{K}$ chooses random vectors $\mathbf{g}_1, \mathbf{g}_2, \ldots, \mathbf{g}_l$ from $\mathbb{F}_p^n$ such that there are n linearly independent vectors. It chooses a random vector $\mathbf{b} \in \mathbb{F}_p^n$, samples errors $f_1, f_2, \ldots, f_l$ independently according to χ and computes $y_i = \mathbf{g}_i \cdot \mathbf{b} + f_i$ for $i = 1, 2, \ldots, l$. It then outputs $ek = (\mathbf{g}_1, \mathbf{g}_2, \ldots, \mathbf{g}_l, y_1, y_2, \ldots, y_l)$ and $dk = \mathbf{b}$.

- The *encryption* algorithm $\mathcal{E}$ takes as input an encryption key $ek = (\mathbf{g}_1, \mathbf{g}_2, \ldots, \mathbf{g}_l, y_1, y_2, \ldots, y_l)$ and a message $m \in \{0, 1\}$. It chooses a random subset $S \subseteq \{1, 2, \ldots, l\}$ and computes

$$\mathbf{x} = \sum_{i \in S} \mathbf{g}_i \qquad \text{and} \qquad w = m\frac{p-1}{2} + \sum_{i \in S} y.$$

It outputs the ciphertext $c = (\mathbf{x}, w)$.

- The *decryption* algorithm $\mathcal{D}$ takes as input a ciphertext $c = (\mathbf{x}, w)$. It computes $w - \mathbf{x} \cdot \mathbf{b} = t + \langle p \rangle$, where $|t| < p/2$ If $|t| < p/4$, it outputs $m = 0$, otherwise it outputs $m = 1$.

E *Exercise* 3.58. The above is an informal description of a public key encryption scheme. Implement (choose some χ) the three algorithms $\mathcal{K}$, $\mathcal{E}$ and $\mathcal{D}$. Show that the triple $(\mathcal{K}, \mathcal{E}, \mathcal{D})$ is a public key encryption scheme.

Remark. Note that this cryptosystem can only encrypt a single bit, so it is not at all practical. It is, however, of great theoretical interest, in particular because of the learning with errors problem.

Attacks   We can attack the system by solving a lattice problem.

First, we reformulate the scheme in terms of matrices. Let $\mathbf{G}$ be the matrix with rows $g_1, g_2, \ldots, g_l$, let $\mathbf{b} = (b_1, b_2, \ldots, b_l)$, $\mathbf{f} = (f_1, f_2, \ldots, f_l)$ and $\mathbf{y} = (y_1, y_2, \ldots, y_n)$. Then for some very short vector $\mathbf{r}$ we have

$$\mathbf{y} = \mathbf{bG} + \mathbf{f}, \qquad \mathbf{x} = \mathbf{rG}, \qquad w = \mathbf{ry}^T + m\frac{p-1}{2}.$$

Let $\mathbf{x}$ be any vector in $\mathbb{Z}^n$ such that $\mathbf{x} \equiv \mathbf{y} \pmod{p}$. The matrix $\mathbf{G}$ defines a p-ary lattice $\Lambda = \Lambda_p(\mathbf{G})$. We may assume that the error distribution for the learning with errors problem is such that the length of $\mathbf{f}$ is smaller than half the length of the shortest vector in Λ. In other words, if $\mathbf{u}$ is a vector in Λ that is closest to $\mathbf{x}$, then $\mathbf{x} - \mathbf{u} \equiv \pm\mathbf{f} \pmod{p}$. Once we know the error vector, we can compute $\mathbf{b}$ using linear algebra. In other words, we can recover the secret key if we can find the closest vector.

3.6.3   NTRU Encrypt

Superficially, the NTRU cryptosystem is based on polynomial arithmetic. However, the best attacks on the cryptosystem comes from using lattices.

We begin by considering three rings, $R = \mathbb{Z}[X]/\langle X^n - 1 \rangle$, $R_p = \mathbb{Z}_p[X]/\langle X^n - 1 \rangle$ and $R_q = \mathbb{Z}_q[X]/\langle X^n - 1 \rangle$. Note that there are natural homomorphisms from R to the other two rings (but not between the rings), so any element in R may be considered also to be an element of R_p and R_q.

Strictly speaking, the elements in these rings are not polynomials, but cosets. As usual, we take the lowest-degree representative whenever we need a representative. Furthermore, when we consider elements in R_q as polynomials, the coefficients are integers modulo q. We shall pull them back to R by choosing the coefficient representatives with minimal absolute value, which means that the coefficients will be integers between $-q/2$ and $q/2$.

Consider a polynomial $a \in R$ such that a is invertible in R_p and R_q, and let b be any other polynomial in R. Let y be a polynomial such that $y = b/a$ in R_q. Suppose r, f are polynomials and that $c = pry + f$ has been computed in R_q. Then we have that

$$ac = prb + af$$

in R_q. Furthermore, if the polynomials rb and af have only "small" coefficients, then this equality also holds in R, which means that if $u \in R$ is equal to ac computed in R_q, then $u = af$ in R_p, or $f = u/a$ in R_p.

Example 3.20. The *NTRU* public key encryption scheme $(\mathcal{K}, \mathcal{E}, \mathcal{D})$ based on the above equations with parameters n, p and q, a symmetric cryptosystem $(\mathfrak{K}_s, \mathfrak{P}, \mathfrak{C}, \mathcal{E}_s, \mathcal{D}_s)$ and a key derivation function $h : R_q \times R_p \to \mathfrak{K}_s$ works as follows:

- The *key generation* algorithm $\mathcal{K}$ chooses polynomials b and a with "small" coefficients such that a is invertible in R_p and R_q. It computes $y \leftarrow b/a$ in R_q. It then outputs $ek = y$ and $dk = a$.

- The *encryption* algorithm $\mathcal{E}$ takes as input an encryption key $ek = y$ and a message $m \in \mathfrak{P}$. It chooses polynomials r and f with "small" coefficients and computes $c \leftarrow pry + f$ in R_q, $k \leftarrow h(c, f)$ and $w \leftarrow \mathcal{E}_s(k, m)$. It outputs the ciphertext $c = (c, w)$.

- The *decryption* algorithm $\mathcal{D}$ takes as input a decryption key $dk = a$ and a ciphertext $c = (c, w)$. It computes $u = ac$ in R_q and $f = u/a$ in R_p. Then it computes $k \leftarrow h(c, f)$ and $m \leftarrow \mathcal{D}_s(k, w)$. If $\mathcal{D}_s$ outputs the special symbol $\bot$ indicating decryption failure, then $\mathcal{D}$ outputs $\bot$, otherwise it outputs m.

Exercise 3.59. The above is an informal description of a public key encryption scheme. Implement (use some simple method to choose polynomials) the three algorithms $\mathcal{K}$, $\mathcal{E}$ and $\mathcal{D}$. Show that the triple $(\mathcal{K}, \mathcal{E}, \mathcal{D})$ is a public key encryption scheme (under reasonable assumptions on what "small" means).

Attacks Let y be represented by the polynomial $y_0 + y_1 X + \cdots + y_{n-1} X^{n-1} \in \mathbb{Z}[X]$. We can construct a matrix

$$\mathbf{M} = \begin{pmatrix} \mathbf{I} & \mathbf{Y} \\ \mathbf{0} & q\mathbf{I} \end{pmatrix}$$

where $\mathbf{I}$ is a $n \times n$ identity matrix, and the matrix $\mathbf{Y}$ consists of the cyclic shifts of the coefficients of above representation of the y polynomial. That is, the ith row (starting at 0) of $\mathbf{Y}$ is $(y_{n-i}, y_{n-1}, y_0, y_1, \ldots, y_{n-i-1})$. This $2n \times 2n$ matrix has full rank, so it generates a full rank lattice Λ.

If a and b are represented by $a_0 + a_1 X + \cdots + a_{n-1} X^{n-1} \in \mathbb{Z}[X]$ and $b_0 + b_1 X + \ldots b_{n-1} X^{n-1} \in \mathbb{Z}[X]$, we have that in R

$$\left(\sum_{i=0}^{n-1} a_i X_i \right) \left(\sum_{j=0}^{n-1} y_j X_j \right) = \sum_{i=0}^{n-1} X_i \sum_{j=0}^{n-1} a_{i+j} y_{n-j} = \sum_{i=0}^{n-1} b_i X^i + \sum_{i=0}^{n-1} q w_i X^i$$

for some polynomial $w = w_0 + w_1 X + \cdots + w_{n-1} X^{n-1} \in \mathbb{Z}[X]$. It follows that with $\mathbf{a} = (a_0, a_1, \ldots, a_{n-1}, w_0, w_1, \ldots, w_{n-1})$, we have

$$\mathbf{aM} = (a_0, a_1, \ldots, a_{n-1}, b_0, b_1, \ldots, b_{n-1}).$$

Since a_i and b_j are small, this vector's length is a small multiple of $\sqrt{2n}$.

We have already noted that an "average" lattice of dimension $2n$ will not have vectors much shorter than $\sqrt{2n/(2\pi e)} \det(\Lambda)^{1/(2n)}$. The determinant of our lattice is q^n, so we compare a small multiple of $\sqrt{2n}$ with $\sqrt{2nq/(2\pi e)}$. If n is not too small, our lattice will have a very short vector. It is then not too unreasonable to hope that a short vector in the lattice will be this short vector (or one closely related to it).

In other words, if we can find short vectors in the lattice Λ, we have a reasonable hope of finding the NTRU decryption key.

3.7 LATTICE ALGORITHMS

We have now seen how we can attack several cryptosystems by either finding a closest vector in a lattice or finding a short vector in a lattice. It follows that in order to understand the security of these cryptosystems, we must understand how to find short vectors and closest vectors.

We begin by developing an algorithm for enumerating all the lattice vectors in a ball around the origin. It will turn out that this algorithm is sensitive to certain properties of the basis. We shall then study the famous LLL (Lenstra–Lenstra–Lovász) algorithm which will produce a basis that is (among other things) a much better starting point for the enumeration algorithm.

3.7.1 Enumerating Short Vectors

We would like to find a short vector in a lattice. One idea would simply be to enumerate all linear combinations of the basis vectors with some bound on the coefficients. Unfortunately, short vectors could in principle come from linear combinations with large coefficients. Instead, we shall use the Gram-Schmidt basis to bound the size of the coefficients.

We shall find all points $\mathbf{u}$ in a lattice with $\|\mathbf{u}\|^2 \leq A^2$ for some bound A^2. Let $\mathbf{b}_1, \mathbf{b}_2, \ldots, \mathbf{b}_n$ be a basis for Λ, and let $\mathbf{b}_1^*, \ldots, \mathbf{b}_n^*$ be the corresponding

Gram-Schmidt basis. Note that for any vector $\mathbf{u} \in \Lambda$, we can write it as a linear combination of the Gram-Schmidt basis vectors $\mathbf{u} = \sum_i \alpha_i \mathbf{b}_i^*$ and then

$$\|\mathbf{u}\|^2 = \sum_{i=1}^n \alpha_i^2 \|\mathbf{b}_i^*\|^2.$$

Recall that $\mathbf{b}_n^*$ is the part of $\mathbf{b}_n$ that is orthogonal to all the earlier basis vectors. This means when $\mathbf{u} = \sum_i a_i \mathbf{b}_i = \sum_i \alpha_i \mathbf{b}_i^*$, then $a_n = \alpha_n$. Therefore, if $\alpha \|\mathbf{b}_n^*\| > A$ then $\|\mathbf{u}\| > A$.

We therefore begin by enumerating vectors with n-th coordinate a_n between $-\lfloor A/\|\mathbf{b}_n^*\| \rfloor$ and $+\lfloor A/\|\mathbf{b}_n^*\| \rfloor$.

Given a_n, we can now consider the possibilities for a_{n-1}. Of course, this time, the contribution in the direction of $\mathbf{b}_{n-1}^*$ is that given by $a_{n-1}\mathbf{b}_{n-1}$ and $a_n\mathbf{b}_n$, where the latter's contribution is $a_n\mu_{n,n-1}\|\mathbf{b}_{n-1}^*\|$. So given a_n, we want to enumerate all the $(n-1)$-th coordinates a_{n-1} such that

$$(a_{n-1} + a_n\mu_{n,n-1})^2 \|\mathbf{b}_{n-1}^*\|^2 + a_n^2\|\mathbf{b}_n^*\|^2 \leq A^2.$$

In general, given $a_{i+1}, \ldots, a_n$, we consider the possibilities for a_i. Again, we want to enumerate all a_i such that

$$\left(a_i + \sum_{j=i+1}^n a_j\mu_{ji}\right)^2 \|\mathbf{b}_i^*\|^2 + \sum_{j=i+1}^n \left(a_j + \sum_{k=j+1}^n a_k\mu_{k,j}\right)^2 \|\mathbf{b}_j^*\|^2 \leq A^2.$$

For some choices of $a_{i+1}, \ldots, a_n$ there may be no possible choices for a_i, in which case we stop and continue with other choices for $a_{i+1}, \ldots, a_n$.

Whenever we find a non-empty region for a_1 and enumerate those values, we enumerate lattice vectors of length less than A. It is clear that for any lattice point $\mathbf{u}$ of length less than A, this point must be among the lattice point eventually enumerated.

This enumeration algorithm must enumerate a large number of possible coordinates. For the ith coordinate, we can upper-bound the number of possibilities for a_i by $A/\|\mathbf{b}_i^*\|$, so the total can be bounded by

$$\prod_{i=1}^n \frac{A}{\|\mathbf{b}_i^*\|} = \frac{A^n}{\det(\Lambda)}.$$

Exercise 3.60. The above discussion describes an algorithm for enumerating all the vectors in a lattice shorter than some bound. Implement this algorithm.

Exercise 3.61. Consider the lattice given by the basis matrix $\mathbf{C}$ given in Exercise 3.38. Enumerate all the vectors in the lattice of length at most 3.

We can solve the closest vector problem if we can find an estimate for the closest vector and a useful upper bound on how far the estimate is from the closest point. Given this, we can enumerate all the short vectors and thereby all the lattice points close to the estimate, one of which must be closest. There are faster algorithms for finding a shortest or closest vector.

3.7.2 LLL Algorithm

As we saw, if we have an orthogonal basis, we can solve the closest vector problem, and if we have a nearly orthogonal basis, we can solve closest vector problem if the closest vector is close enough to a lattice point.

The natural question is how to find a reasonably good basis that will allow us to solve the closest vector problem. The first goal should be to be precise about what we mean by "reasonably good".

Definition 3.9. Let $\mathbf{b}_1, \mathbf{b}_2, \ldots, \mathbf{b}_n$ be a lattice basis, with Gram-Schmidt basis $\mathbf{b}_1^*, \ldots, \mathbf{b}_n^*$ and Gram-Schmidt coefficients μ_{ij}, as defined in (3.6). Let $\frac{1}{4} < \delta < 1$ be a real number. We say that the basis is δ-*LLL-reduced* if

$$|\mu_{ij}| \leq \frac{1}{2} \qquad\qquad \text{for all } 1 \leq j < i \leq n, \text{ and} \qquad (3.8)$$
$$\delta \|\mathbf{b}_{i-1}^*\|^2 \leq \|\mathbf{b}_i^*\|^2 + \mu_{i,i-1}^2 \|\mathbf{b}_{i-1}^*\|^2 \qquad \text{for all } 2 < i \leq n. \qquad (3.9)$$

When a basis satisfies (3.8) we cannot easily make the basis vectors more orthogonal. When the basis satisfies (3.9), the basis vectors of the Gram-Schmidt orthogonalisation will be ordered roughly according to length.

Exercise 3.62. A common choice for δ is $3/4$. Show that (3.9) then implies

$$\|\mathbf{b}_{i-1}^*\|^2 \leq 2\|\mathbf{b}_i^*\|^2 \text{ for all } 2 \leq i \leq n.$$

Hint: You may use the fact that (3.8) also must hold.

That an LLL-reduced basis is somehow a good basis can be seen from the following fact, which we state without proof.

Fact 3.22. *Suppose* $\mathbf{b}_1, \mathbf{b}_2, \ldots, \mathbf{b}_n$ *with corresponding basis matrix* $\mathbf{B}$ *is a $3/4$-LLL-reduced basis for a lattice* Λ. *Then* $\|\mathbf{b}_1\| \leq 2^{(n-1)/2}\lambda_1(\Lambda)$. *Also, if* $\mathbf{x} \in \mathbb{R}^n$, *then*

$$\|\mathbf{x} - \lfloor \mathbf{x}\mathbf{B}^{-1} \rceil \mathbf{B}\| \leq (1 + 2n(9/2)^{n/2})\|\mathbf{x} - \mathbf{u}\| \text{ for any } \mathbf{u} \in \Lambda.$$

We know that $\lambda_1(\Lambda) \leq \sqrt{\gamma} \det(\Lambda)^{1/n}$, which means that if we have an LLL-reduced basis and use $\|\mathbf{b}_1\|$ as our search bound, the enumeration approach from the previous section will have to enumerate at most

$$\frac{(2^{(n-1)/2}\gamma^{1/2} \det(\Lambda)^{1/n})^n}{\det(\Lambda)} = 2^{n(n-1)/2}\gamma^{n/2}$$

points. While it does not affect the upper bound we deduced, having the Gram-Schmidt vectors not too small will decrease the total number of points the algorithm will iterate over.

We also note that the LLL-reduced basis will give us an estimate for the closest vector problem. (There are better ways to use the LLL-reduced basis.)

Having an LLL-reduced basis is therefore very useful. The question is how to find an LLL-reduced basis.

We will now discuss two ways to modify a lattice basis. The first is to use the initial basis vectors to modify the later basis vectors. This clearly results in a new basis for the same lattice, but the new basis will have the same Gram-Schmidt orthogonalisation.

E | *Exercise* 3.63. Let $\mathbf{b}_1, \mathbf{b}_2, \dots, \mathbf{b}_n$ be a lattice basis, and let $\mathbf{c}_1, \dots, \mathbf{c}_n$ be another basis for the lattice defined by

$$\mathbf{c}_i = \mathbf{b}_i - \sum_{j=1}^{i-1} \alpha_{ij} \mathbf{b}_j, \quad \alpha_{ij} \in \mathbb{Z}. \tag{3.10}$$

Let $\mathbf{b}_1^*, \dots, \mathbf{b}_n^*$ and $\mathbf{c}_1^*, \dots, \mathbf{c}_n^*$ be the corresponding Gram-Schmidt orthogonalisation. Show that $\mathbf{b}_i^* = \mathbf{c}_i^*$ for $i = 1, 2, \dots, n$.

With this result in mind, we now turn to (3.7) which describes the relationship between a basis and its Gram-Schmidt orthogonalisation,

$$\mathbf{B} = \begin{pmatrix} 1 & & 0 \\ & \ddots & \\ \mu_{ij} & & 1 \end{pmatrix} \mathbf{B}^*.$$

By adding or subtracting a suitable integer multiple of the rightmost column of the Gram-Schmidt coefficient matrix, we can ensure that $|\mu_{n,n-1}| \leq 1/2$. Then, by subtracting suitable integer multiples of the two rightmost columns, we can ensure that $|\mu_{n-1,n-2}| \leq 1/2$ and $|\mu_{n,n-2}| \leq 1/2$. In this way we can ensure that any element below the diagonal in the coefficient matrix has absolute value at most $1/2$.

Since column arithmetic in the coefficient matrix on the right-hand side of the equation corresponds to row arithmetic in the basis matrix on the left hand side, the above procedure essentially tells us how to modify a lattice basis so that its Gram-Schmidt coefficients satisfy (3.8). The next exercise details a more algorithmic way to do this computation.

E | *Exercise* 3.64. Let $\mathbf{b}_1, \mathbf{b}_2, \dots, \mathbf{b}_n$ be a lattice basis, let $\mathbf{b}_1^*, \mathbf{b}_2^*, \dots, \mathbf{b}_n^*$ be the Gram-Schmidt orthogonalisation with coefficients μ_{ij}, $1 < i \leq n$.

Suppose $|\mu_{ij}| \leq 1/2$ for $1 \leq j < i < k$ for some $k \leq n$. Define $\mathbf{b}_k^{(i)}$ for $1 \leq i \leq k$ by $\mathbf{b}_k^{(k)} = \mathbf{b}_k$ and

$$\mathbf{b}_k^{(i)} = \mathbf{b}_k^{(i+1)} - \left\lceil \frac{\langle \mathbf{b}_k^{(i+1)}, \mathbf{b}_i^* \rangle}{\langle \mathbf{b}_i^*, \mathbf{b}_i^* \rangle} \right\rfloor \mathbf{b}_i.$$

Consider now the new basis $\mathbf{b}_1, \dots, \mathbf{b}_{k-1}, \mathbf{b}_k^{(1)}, \mathbf{b}_{k+1}, \dots, \mathbf{b}_n$.

(a) Show that the new basis has the same Gram-Schmidt orthogonalisation.

(b) Let μ'_{ij} be the Gram-Schmidt coefficients of the new basis. Show that $|\mu'_{ij}| \le 1/2$ for $1 \le j < i \le k$.

(c) Show that the above procedure essentially gives us an algorithm that for any lattice basis computes a new, equivalent basis with the same Gram-Schmidt orthogonalisation that also satisfies (3.8).

(d) Show that the resulting algorithm will compute the new basis using at most $2n^3$ arithmetic operations.

This result tells us that we can not only modify a lattice basis such that (3.8) holds without changing the Gram-Schmidt orthogonalisation, but we can also do this relatively quickly.

Next, we want to modify our basis by changing the order of the basis vectors. Unlike the previous modification, this will change the resulting Gram-Schmidt orthogonal basis. Before we describe and analyse this change, we need to define a new kind of volume for lattices that weights the basis vectors differently. For a basis $\mathbf{b}_1, \mathbf{b}_2, \ldots, \mathbf{b}_n$ with corresponding Gram-Schmidt orthogonalisation $\mathbf{b}_1^*, \mathbf{b}_2^*, \ldots, \mathbf{b}_n^*$, define

$$d(\mathbf{b}_1, \mathbf{b}_2, \ldots, \mathbf{b}_n) = \prod_{i=1}^{n} \prod_{j=1}^{i-1} \|\mathbf{b}_j^*\| = \prod_{i=1}^{n} \|\mathbf{b}_i^*\|^{n-i+1}.$$

Suppose $\mathbf{b}_1, \ldots, \mathbf{b}_n$ satisfies (3.8), but not (3.9), and i is the first index where the latter condition fails. Now suppose we change the order of the ith and $i+1$th basis vectors, so instead of considering the basis $\mathbf{b}_1, \mathbf{b}_2, \ldots, \mathbf{b}_i, \mathbf{b}_{i+1}, \ldots, \mathbf{b}_n$, we instead consider the basis $\mathbf{c}_1, \ldots, \mathbf{c}_n$ given by

$$\mathbf{c}_j = \begin{cases} \mathbf{b}_{i+1} & j = i \\ \mathbf{b}_i & j = i+1 \\ \mathbf{b}_j & \text{otherwise.} \end{cases}$$

E *Exercise* 3.65. Let $\mathbf{b}_1, \ldots, \mathbf{b}_n$ and $\mathbf{c}_1, \ldots, \mathbf{c}_n$ be as above, and let $\mathbf{b}_1^*, \ldots, \mathbf{b}_n^*$ and $\mathbf{c}_1^*, \ldots, \mathbf{c}_n^*$ be the corresponding Gram-Schmidt orthogonalisations. Let μ_{ij} be the Gram-Schmidt coefficients for $\mathbf{b}_1^*, \ldots, \mathbf{b}_n^*$.

(a) Show that $\mathbf{b}_j^* = \mathbf{c}_j^*$ when $j \ne i, i+1$.

(b) Show that $\mathbf{c}_i^* = \mathbf{b}_{i+1}^* + \mu_{i+1,i}\mathbf{b}_i^*$.

(c) Show that

$$\prod_{j=1}^{k} \|\mathbf{b}_j^*\| = \prod_{j=1}^{k} \|\mathbf{c}_j^*\|$$

when $i \ne k$.

(d) Show that
$$\frac{d(\mathbf{c}_1, \ldots, \mathbf{c}_n)}{d(\mathbf{b}_1, \ldots, \mathbf{b}_n)} \leq \sqrt{\delta}.$$

E *Exercise* 3.66. Show that for any lattice $\Lambda \in \mathbb{Z}^n$, then regardless of basis, there is a lower bound to $d(\mathbf{b}_1, \ldots, \mathbf{b}_n)$ that is strictly larger than zero.

We are now ready to prove that a basis satisfying (3.8) and (3.9) can be computed relatively quickly.

T **Theorem 3.23.** *Let* **B** *be a basis for a lattice* $\Lambda \in \mathbb{Z}^n$. *Then a* δ-*LLL-reduced basis* **C** *can be computed using less than* $10n^3 \log d(\mathbf{B})/ \log \delta^{-1}$ *arithmetic operations.*

Proof. We begin by constructing a sequence of lattice bases $\mathbf{B}_1, \mathbf{B}_2, \ldots$ as follows. The initial basis is $\mathbf{B}_1 = \mathbf{B}$. We construct the $(k+1)$th basis from the kth basis $\mathbf{B}_k$ using the following two steps.

1. Use the algorithm from Exercise 3.64 to create a basis $\mathbf{B}'_k$.

2. If the basis $\mathbf{B}'_k$ does not satisfy (3.9), and i is the first index where this condition fails we swap the ith and the $(i+1)$th basis vectors.

If $\mathbf{B}_k$ satisfies (3.8) and (3.9), then $\mathbf{B}_{k+1} = \mathbf{B}_k$. Furthermore, if we did not swap two basis vectors when creating $\mathbf{B}_{k+1}$, it satisfies (3.8) and (3.9). When we change the order of two basis vectors, Exercise 3.65 says that

$$\frac{d(\mathbf{B}_{k+1})}{d(\mathbf{B}_k)} \leq \sqrt{\delta} \qquad \Rightarrow \qquad d(\mathbf{B}_{k+1}) \leq \delta^{k/2} d(\mathbf{B}_1).$$

Since $d(\mathbf{B}_{k+1}) \geq 1$, it follows that when $k > 2 \log d(\mathbf{B}_1)/ \log \delta^{-1}$ there can be no more changes of order, and we have computed a δ-LLL-reduced basis.

By Exercise 3.64 the first step requires at most $2n^3$ arithmetic operations. To do the second step, we need the Gram-Schmidt coefficients which we can compute using at most $2n^3$ arithmetic operations by Exercise 3.39. Finally, we can find the first index for which (3.9) does not hold using less than $3n^2$ arithmetic operations. This means that we can compute $\mathbf{B}_{k+1}$ from $\mathbf{B}_k$ using less than $5n^3$ arithmetic operations. The claim follows. □

E *Exercise* 3.67. The proof of Theorem 3.23 essentially describes an algorithm for computing an LLL-reduced basis. Implement this algorithm and restate Theorem 3.23 as a statement about the algorithm's time complexity (in terms of arithmetic operations, ignoring any other form of computation involved).

It has been sensible to measure runtime in terms of arithmetic operations, but the LLL algorithm does arithmetic with rational numbers, so we need to consider the size of the numbers involved, which sometimes grows quickly. It can be proved that the size of the rational numbers involved do not grow too quickly, although it significantly affects the time required to run the algorithm.

The algorithm implied by the proof is inefficient. In practice, there is no need to constantly recompute all the Gram-Schmidt coefficients.

E *Exercise* 3.68. Consider the lattice Λ given in Example 3.17 and the basis matrix **B** given in the example and the basis matrix **C** given in Exercise 3.38. For each basis, compute a 3/4-LLL-reduced basis.

3.8 THE PUBLIC KEY INFRASTRUCTURE PROBLEM

As we have seen, we can construct plausible cryptosystems where anyone who knows the encryption key can encrypt messages, but only those who know the decryption key can decrypt messages.

One problem remains. Alice wants to send a message to Bob. How does she get Bob's encryption key? Suppose Alice finds a key that she thinks belong's to Bob, but which in reality belongs to Eve who has the corresponding decryption key. Eve can then easily decrypt Alice's ciphertext.

One possible solution is a trusted public key directory, modelled on a telephone directory, listing people and their public keys. It seems impractical to have printed copies of this directory, so it would have to be an online service, which begs the question: How can Alice be sure that the encryption key she just fetched came from the directory, and not from Eve?

The *public key infrastructure* problem is Alice's problem of getting hold of Bob's public key.

Another interesting problem is what happens when Bob receives Alice's ciphertext. How does he know that it comes from Alice, and not from Eve?

It turns out that both of these problems have plausible solutions involving digital signatures.

CHAPTER 4

Digital Signatures

We shall now consider a variant of the original cryptographic problem. Alice wants to send messages to Bob via some channel. Eve has access to the channel, and she may tamper with anything sent over the channel, and even introduce her own messages. Alice wants her messages to Bob to arrive without modification, or if it they have been tampered with, Bob should notice.

The obvious solution is for Alice and Bob to have a shared secret and use a message authentication code. But Bob may have many correspondents, and may not want to manage shared secrets with each of them.

What if Alice wants to make her message public and let anyone, even people Alice has never met, convince themselves that Alice sent the message?

The solution is digital signatures, which resemble message authentication codes. As for public key encryption, we have two keys, one key for creating tags and another key for verifying them. The tags are called signatures and creating them is called signing. Alice makes the verification key public key, allowing anyone to verify, and keeps the signing key secret.

Before we begin the design of signature schemes, we shall discuss hash functions, a tool that we can use to extend the plaintext space for signatures, simplifying the design of signature schemes.

The first class of signature schemes we study are based on the famous RSA cryptosystem. At first sight, this system looks like a "dual" of the textbook RSA public key encryption scheme, but this similarity is superficial.

The second class of signature schemes we study is based on a much deeper theory, namely how to argue convincingly that something is true without revealing the evidence for *why* it is true.

We shall discuss few new computational algorithms in this chapter, since we are in effect reusing analysis we did in the previous two chapters.

Before we end, we briefly discuss signatures that are not based on number-theoretic problems, as well as how to use signatures to protect key exchange.

4.1   DEFINITIONS

Alice, Bob and a number of other people want to be able to send messages to each other, and they want to notice any tampering with those messages. Alice does not want to manage a long-term secret for each correspondent, so symmetric key techniques such as message authentication codes cannot be used. Alice is willing to manage public information for each correspondent.

In this situation, what is needed is digital signatures.

Definition 4.1. A *(digital) signature* scheme consists of:

- The *key generation* algorithm $\mathcal{K}$ takes no input and outputs a *signing key sk* and a *verification key vk*. To each key pair there is an associated message set denoted by $\mathfrak{M}_{sk}$ or $\mathfrak{M}_{vk}$.

- The *signing* algorithm $\mathcal{S}$ takes as input a signing key *sk* and a message $m \in \mathfrak{M}_{sk}$ and outputs a *signature* σ.

- The *verification* algorithm $\mathcal{V}$ takes as input a verification key *vk*, a message $m \in \mathfrak{M}_{vk}$ and a signature σ, and outputs either 0 or 1.

For any key pair (vk, sk) output by $\mathcal{K}$ and any message $m \in \mathfrak{M}_{vk}$, we require

$$\mathcal{V}(vk, m, \mathcal{S}(sk, m)) = 1.$$

If the verification algorithm outputs 1, the signature is *valid*, otherwise it is *invalid*. A valid signature created without the signing key is a *forgery*.

Informally. A signature scheme is *secure* if it is hard to create a forgery, even when you can see valid signatures on many different messages.

4.2   HASH FUNCTIONS

A vital component for designing practical digital signature schemes is the *hash function*, which is typically a function with a large domain and a small co-domain. The idea is that we can easily build signature schemes, but often we get a scheme with a very small message space. Moreover, many of the schemes we build suffer from a weakness where it is very easy to come up with signatures on random messages, even without the signing key. If we combine our primitive signature schemes with a suitable hash function, we can extend the message space and protect against these designed-in weaknesses.

The idea for the construction is that instead of signing the message itself, we shall sign a hash of the message. Let $(\mathcal{K}, \mathcal{S}', \mathcal{V}')$ be a signature scheme and let $h : S \to T$ be a hash function such that T is a subset of the signature scheme's message space. We construct a new signature scheme $(\mathcal{K}, \mathcal{S}, \mathcal{V})$ with message space S as follows. The key generation algorithm is unchanged. The signing algorithm creates a signature of a message m under the signing key *sk* by computing $\mathcal{S}'(sk, h(m))$. On input of vk, m and σ, the verification algorithm outputs $\mathcal{V}'(vk, h(m), \sigma)$.

Signing a hash of the message could be a security problem if we could find two messages that have the same hash. A signature on one of the messages would also be a signature on the other message, which would be bad. Ideally, we would like the hash function to be injective, but that would not allow us to expand the message space. Instead, we shall settle for a hash function that merely "looks" injective.

Definition 4.2. Let $h : S \to T$ be a function. A *preimage* of $u \in T$ is an element $s \in S$ such that $h(s) = u$. A *second preimage* for $s_1 \in S$ is an element $s_2 \in S$ such that $s_1 \neq s_2$ while $h(s_1) = h(s_2)$. A *collision* for h is a pair of distinct elements $s_1, s_2 \in S$ such that $h(s_1) = h(s_2)$.

The following two definitions are informal. We return to these concepts in Section 7.5.

Informally. Let $h : S \to T$ be a function. We say that it is *one-way* if it is an infeasible computation to find a preimage of a random $u \in T$ and to find a second preimage for a random $s \in S$.

Informally. Let $h : S \to T$ be a function. We say that it is *collision-resistant* if it is an infeasible computation to find collisions for h and to find a second preimage for a random $s \in S$.

We quickly note that if you can find second preimages, you can also find collisions. The converse does not have to be true. It follows that if finding a collision is an infeasible computation, the hash function will be collision-resistant, and it will behave like an injective function in practice.

It would be natural if the ability to find preimages implied the ability to find second preimages. This is not true, as implied by the following exercise.

Exercise 4.1. Let $h : S \to T$ be a hash function, and suppose that $T \subseteq S$. Let $h' : S \to \{0, 1\} \times T$ be the hash function defined by

$$h'(s) = \begin{cases} (0, s) & s \in T, \\ (1, h(s)) & \text{otherwise.} \end{cases}$$

Show that for half of all elements of $\{0, 1\} \times T$, it is easy to find preimages, but for none of those preimages there exists a second preimage.

We see that if our hash function is collision-resistant, the hash function behaves as an injective function and the above signature scheme $(\mathcal{K}, \mathcal{S}, \mathcal{V})$ is no less secure than the original scheme $(\mathcal{K}, \mathcal{S}', \mathcal{V}')$.

If our hash function is one-way, the above scheme may actually be more secure than the original scheme.

Another security idea for hash functions is *random-looking*, which is that the hash function should in some sense look like an "average" function. We have used this for key derivation functions.

4.2.1 Attacks

A generic collision finder is to hash random messages until a collision is found. Let $h : S \to T$ be a hash function. Choose a large number of random messages $s_1, s_2, \ldots, s_l$ and compute their hashes. We store the messages and their hashes in a list sorted by the hashes. By the birthday paradox, if l is roughly $\sqrt{|T|}$, we should have a reasonable likelyhood of finding collisions.

This result gives us a minimal size for the set T, namely that $\sqrt{|T|}$ should be an infeasible computation. However, the attack described above requires a lot of memory. The following algorithm uses ideas similar to Pollard's rho method. It and similar algorithms are often called Pollard's lambda method. (Illustrations of the single self-colliding sequence in the rho method on a blackboard often resemble the Greek letter ρ. Illustrations of the two colliding sequences in the lambda method often resemble the Greek letter λ.)

E | *Exercise* 4.2. Let $l = 10\lfloor \sqrt{|T|} \rfloor$. Let $g : T \to S$ be an injective function, and let $f : T \to T$ be the function defined by $f(u) = h(g(u))$. Let u_0 and u_0' be two distinct elements of T. Define two sequences by the equations

$$u_i = f(u_{i-1}) \qquad \text{and} \qquad u_i' = f(u_{i-1}').$$

(a) Suppose $u_j' = u_l$ for some $j < 2l$, and $u_j' \neq u_0$ for any $j < l$. Show how to find, given j, a collision in h using at most $3l$ evaluations of g and h.

(b) Consider two sequences of random elements of length l. Argue that there will be a common element with reasonable probability.

(c) Argue informally that the above result should apply to our sequences if h is random-looking.

(d) Suppose h is random-looking. Argue that with reasonable probablity, you can find a collision in h using about $6l$ hash evaluations.

4.2.2 Compression Functions

Typically, the domain of a hash function is much larger than the range. For example, $\log_2 |T|$ will typically be between one hundred and a few thousand, while $\log_2 |S|$ is 2^{64} or higher. A hash function where the domain is larger than the range, but not by much, is often called a *compression function*. (Note that naming often depends on intended use, rather than exact properties).

We are interested in compression functions for two reasons. First of all, it is probably easier to construct compression functions than large-domain hash functions. And second, we have efficient constructions that turn secure compression functions into secure hash functions.

We begin by discussing the *Merkle-Damgård* construction. So let $f : S' \to T$ be a compression function. Suppose further that there is a set $\mathfrak{A}$ such that $\{0,1\} \times \mathfrak{A} \times T$ is either a subset of S' or trivially injects into S'. Now consider the restriction of f to $\{0,1\} \times \mathfrak{A} \times T$.

We shall now construct a hash function. The domain is the set of all finite sequences of elements from $\mathfrak{A}$, denoted by $\mathfrak{A}^*$. Let $u_0 \in T$ be a fixed element of T. The value s we want to hash is a sequence $s_1 s_2 \ldots s_l$ of elements from $\mathfrak{A}$. The function h is computed using the rule

$$u_1 = f(1, s_1, u_0), \qquad u_i = f(0, s_i, u_{i-1}), \quad 2 \leq i \leq l, \qquad h(s) = u_l.$$

The cost (in terms of compression function evaluations) of computing h is linear in the length of the message to be hashed. If it is easy to compute f, then computing h is quite efficient.

If the compression function is one-way and collision-resistant, we would like the above defined hash function to be both one-way and collision-resistant.

Theorem 4.1. *Given a collision (s, s') in the above constructed hash function, we can find a collision in the compression function f using at most $2l$ evaluations of f, where the length of s and s' is at most l.*

Proof. Let $s = s_1 s_2 \ldots s_l$ and $s' = s'_1 s'_2 \ldots s'_{l'}$, with $l' \leq l$. We know that

$$f(0, s_l, u_{l-1}) = f(0, s'_{l'}, u'_{l'-1}).$$

Either we have our collision, or $s_l = s_{l'}$ and $u_{l-1} = u'_{l'-1}$. If the latter, then

$$f(0, s_{l-1}, u_{l-2}) = f(0, s'_{l'-1}, u'_{l'-2}).$$

Continuing in this way, either we find a collision or the beginning of s.

If $l = l'$, we must have $s_1 \neq s'_1$ since $s \neq s'$, which gives us a collision since

$$f(1, s_1, u_0) = f(1, s'_1, u_0).$$

If $l' < l$, we get a collision because

$$f(0, s_{l-l'+1}, u_{l-l'}) = f(1, s'_1, u_0).$$

We find this collision by first computing $u_1, u_2, \ldots, u_{l-l'}$, then computing the pairs $(u_{l-l'+1}, u'_1), (u_{l-l'+2}, u'_2), \ldots, (u_l, u'_{l'})$, one which is our collision. $\square$

The theorem says that from any collision in the constructed hash function h, it is easy to find a collision in the compression function f. Which means that if our compression function is collision-resistant, the constructed hash function is also collision-resistant.

E *Exercise* 4.3. Consider the above construction. Suppose you have a "magic box" that for any $u \in T$ will provide you with a reasonable-length preimage of u under h. Show that you can use this oracle to find preimages for any $u \in T$ under f.

As for collision-resistance, the consequence of the above exercise is that if we can construct a compression function where finding preimages is an infeasible computation, we can construct a hash function where finding preimages is an infeasible computation.

To conclude that the hash function is one-way, we must also consider second preimages. Unlike for preimages and collision, being able to find second preimages for h does not seem to imply the ability to find second preimages of f. But the ability to find second preimages for h implies the ability to find collisions for h, which implies the ability to find collisions for f. That is, if we can find second preimages for h, then we can find collisions in f.

This means that if f is one-way and collision-resistant, then h is one-way and collision-resistant hash function.

Note that the construction uses the 0 and the 1 to differentiate the start of the iteration. This is important in the proof, since without this differentiation, we could have run into problems when messages of different length collided.

E *Exercise* 4.4. Suppose we have a compression function $f : \mathfrak{A} \times T \to T$, and that $\{0, 1, \ldots, 2^{64} - 1\}$ is a subset of $\mathfrak{A}$. Let u_0 be a fixed element of T.

We define two hash functions for messages that are sequences of elements from $\mathfrak{A}$ of length less than 2^{64}, using the two recursive formulas

$$u_1 = f(l, u_0), \qquad\qquad u_{i+1} = f(s_i, u_i), \qquad 1 \le i \le l,$$

and

$$u_i = f(s_i, u_{i-1}), \qquad 1 \le i \le l, \qquad u_{l+1} = f(l, u_l).$$

In either case, the hash of the message is u_{l+1}.

For each hash function, state and prove a result similar to that of Theorem 4.1 for that hash function.

Note that to use the first construction in Exercise 4.4, you must know the length of the message before you begin hashing it. There are reasonable cases where you want to begin hashing a message before you know the entire message, and in particular before you know the length of the entire message.

Hash functions are used for many things in cryptography, and the security requirements vary greatly with the application. One example is to use a hash function as a message authentication code, simply by computing $\mu(k, m) = h(k\|m)$, where $k\|m$ denotes the concatenation of the key and the message. As the following exercise shows, our construction cannot be used like this.

E *Exercise* 4.5. Let $h : \mathfrak{A}^* \to T$ be the hash function constructed at the start of the section. Suppose you are given a value u such that $u = h(m)$. Show that you can easily compute $h(m||m')$ for any m', even when you do not know m, only u. (Here, $m||m'$ denotes the concatenation of m and m'.)

4.2.3 Constructing a Compression Function

Let G be a cyclic group of prime order n, and let x and y be non-zero elements. Then we can construct a compression function $f_{x,y} : \{0, 1, 2, \ldots, n-1\} \times \{0, 1, 2, \ldots, n-1\} \to G$ as

$$f_{x,y}(u, v) = x^u y^v.$$

When x and y are clear from context, we shall write simply f for $f_{x,y}$.

If it is hard to compute discrete logarithms in G, then it is hard to find both preimages and collisions for this compression function, provided x and y have been chosen at random from G.

T **Theorem 4.2.** *Suppose we know a collision $((u, v), (u', v'))$ for f. Then we can compute $\log_x y$ using 3 arithmetic operations.*

Proof. If we have a collision, we know that $x^u y^v = x^{u'} y^{v'}$. Since $(u, v) \neq (u', v')$ and the above equation holds, we have that $u \neq u'$ and $v \neq v'$. This means that

$$y = x^{-(u-u')(v-v')^{-1}}.$$
□

E *Exercise* 4.6. Suppose you have a "magic box" that for any x, y, z of your choice will find one preimage of z under the hash function $f_{x,y}$. Explain how you can use this "magic box" to compute discrete logarithms in G.

Using this compression function and the construction from the previous section, we have a one-way and collision-resistant hash function. However, this hash function is of theoretical interest only, since we have much faster constructions that also have other interesting properties.

4.3 RSA SIGNATURES

We briefly recall the textbook RSA public key cryptosystem. The key generation algorithm chooses two primes p and q and finds e and d such that $ed \equiv 1 \pmod{\operatorname{lcm}(p-1, q-1)}$. The encryption key is (n, e), where $n = pq$ and the decryption key is (n, d). To encrypt a message $m \in \{0, 1, \ldots, n-1\}$, we compute $c = m^e \bmod n$. To decrypt a ciphertext c, we compute $m = c^d \bmod n$.

We can construct a very simple signature scheme based on this.

E *Example* 4.1. The *textbook RSA* signature scheme $(\mathcal{K}, \mathcal{S}, \mathcal{V})$ works as follows.

- The *key generation* algorithm $\mathcal{K}$ is identical to the one from Example 3.6 and outputs $vk = (n, e)$ and $sk = (n, d)$. The message set is identical.

- The *signing* algorithm $\mathcal{S}$ takes as input a signing key (n, d) and a message $m \in \{0, 1, 2, \ldots, n - 1\}$. It computes $\sigma = m^d \bmod n$ and outputs the signature σ.

- The *verification* algorithm $\mathcal{V}$ takes as input a verification key (n, e), a message $m \in \{0, 1, 2, \ldots, n - 1\}$ and a signature $\sigma \in \{0, 1, 2, \ldots, n - 1\}$. It outputs 1 if $\sigma^e \equiv m \pmod{n}$, otherwise 0.

E *Exercise* 4.7. The above is an informal description of a signature scheme. Implement the three algorithms $\mathcal{K}$, $\mathcal{S}$ and $\mathcal{V}$. Show that the triple $(\mathcal{K}, \mathcal{S}, \mathcal{V})$ is a signature scheme.

4.3.1 Attacks

As for textbook RSA public key encryption, as long as the RSA modulus n chosen by the key generation algorithm is hard to factor, the above digital signature scheme is useful. But it is not entirely secure.

E *Exercise* 4.8. Suppose $(n, 3)$ is a verification key. Forge a signature on $m = 8$.

E *Exercise* 4.9. Suppose (n, e) is a verification key. Explain how to create a random message with a forged signature.

Malleability Just like the textbook RSA encryption scheme, the RSA signature scheme is malleable, and this can be used to create forgeries.

E *Exercise* 4.10. Suppose you have two messages m, m' and signatures σ, σ' on those messages under the verification key (n, e). Show how to construct a signature on the product $mm' \bmod n$.

E *Exercise* 4.11. Suppose Eve wants to have Alice' signature on a message m. Suppose also that she is capable of getting Alice to sign any other message. Show how Eve can use this to forge a signature on m.

Short Messages We want to use the idea from Section 4.2 with a hash function $h : S \to T$ that is both collision-resistant and one-way, and where T is a set of integers all smaller than any RSA modulus we choose.

E *Example* 4.2. Consider a hash function $h : S \to T$ where T is a set of non-negative integers smaller than any RSA modulus we choose. The *hashed RSA* signature scheme works as follows.

- The *key generation* algorithm is the usual RSA key generation algorithm.

- The *signing* algorithm $\mathcal{S}$ takes a signing key $sk = (n, d)$ and a message $m \in S$ as input. It computes $\sigma = (h(m))^d \bmod n$ and outputs the signature σ.

- The *verification* algorithm $\mathcal{V}$ takes as input a verification key (n, e), a message $m \in S$ and a signature $\sigma \in \{0, 1, 2, \ldots, n-1\}$. It outputs 1 if $\sigma^e \equiv h(m) \pmod{n}$, otherwise 0.

E *Exercise* 4.12. The attacks from Exercises 4.8 and 4.9 now fail. Why?
Hint: The hash function is one-way.

E *Exercise* 4.13. Suppose that the hash function used is also random-looking. Explain why the attacks from Exercises 4.10 and 4.11 become more difficult.

However, most practical hash functions have outputs that are very short relative to an RSA modulus, and this allows us to develop an attack using ideas from index calculus.

E *Exercise* 4.14. Let (n, e) be a verification key with corresponding signing key (n, d). Suppose you have messages $m_1, m_2, \ldots, m_l$, and integers $\ell_1, \ell_2, \ldots, \ell_l$, $\sigma_1, \sigma_2, \ldots, \sigma_l$ and s_{ij}, $1 \le i, j \le l$, such that

$$h(m_i) = \prod_{j=1}^{l} \ell_j^{s_{ij}}, \quad 1 \le i \le l \quad \text{and} \quad h(m_i) \equiv \sigma_i^e \pmod{n}.$$

We shall assume that the matrix $S = (s_{ij})$ is invertible modulo e, and that $R = (r_{ki})$ is an inverse modulo e.

(a) Show that

$$\sum_{i=1}^{l} r_{ki} s_{ij} = \delta_{kj} + t_{kj} e$$

where $\delta_{kj} = 1$ if $k = j$, otherwise $\delta_{kj} = 0$.

(b) Show that

$$\prod_{i=1}^{l} \sigma_i^{r_{ki}} \equiv \ell_k^d \prod_{j=1}^{l} \ell_j^{t_{kj}} \pmod{n}.$$

(c) Explain how we can easily compute $\ell_k^d \bmod n$ from the above.

(d) Suppose you are given a message m such that

$$h(m) = \prod_{i=1}^{l} \ell_i^{u_i}.$$

Explain how you, given the above, can easily compute $h(m)^d \bmod n$.

If we let $\ell_1, \ell_2, \ldots, \ell_l$ be small primes, then if our hash function h has small integers as output, we can quickly find messages $m, m_1, m_2, \ldots, m_l$ such that their hashes are products of powers of our small primes. Which means that if we get signatures on $m_1, m_2, \ldots, m_l$, we can construct a forgery on m.

4.3.2 Secure Variants

It turns out that it is quite easy to fix the hashed RSA signature scheme discussed above. All we need is a random-looking, one-way, collision-resistant hash function whose domain is almost all of $\{0, 1, 2, \ldots, n - 1\}$. In which case the hashed RSA scheme is known as the *full domain hash RSA* signature scheme, or *RSA-FDH*.

E | *Exercise* 4.15. The attack in Exercise 4.14 fails against RSA-FDH. Why?

4.4 SCHNORR SIGNATURES

We shall now develop the well-known Schnorr signature scheme. Variants of the Schnorr scheme are widely used, but Schnorr signatures are more interesting for our purposes. For the remainder of this section, let G be a group of prime order n, and let g be a generator.

4.4.1 How to Prove That You Know a Secret

We begin with a very different question. Suppose Alice knows a secret number $a \in \{0, 1, 2, \ldots, n - 1\}$. Bob does not know the secret number, but he knows $x \in G$ such that $x = g^a$. Alice wants to convince Bob that she really knows a. From Bob's point of view, he will have a conversation with someone and either accept or reject the claim that someone is Alice.

Of course there is an adversary. Eve wants to cheat Bob by having a conversation with Bob that ends with Bob accepting, believing Eve is Alice. To this end, Eve may have conversations with Alice in order to somehow learn a, so that she can trivially pretend to be Alice.

The latter point explains why Alice cannot convince Bob simply by revealing a to Bob. While Bob is honest, Alice will also want to convince Eve that she knows a, after which Eve could impersonate Alice.

One thing Alice could do was to choose a random number r, compute $\alpha = g^r$ and $\gamma = r + a \bmod n$, and then show Bob α and γ. Bob accepts if

$$g^\gamma \overset{?}{=} \alpha x.$$

The idea is that the above protocol does not reveal a to Bob, because Alice just as well could choose a random γ and compute α as $g^\gamma x^{-1}$.

E *Exercise* 4.16. Explain how Eve can choose α and γ to make Bob accept, even though she only knows x.

We can improve on this procedure as follows. Instead of showing Bob both α and γ at once, Alice first shows Bob α. Then Bob is allowed to choose if he wants to see $\gamma = r$ or $\gamma = r + a$ mod n. He accepts if

$$g^\gamma \stackrel{?}{=} \alpha \qquad \text{or} \qquad g^\gamma \stackrel{?}{=} \alpha x, \text{ respectively.}$$

Note that we can encode Bob's choice β as a 0 or a 1, in which case the above formulas can be reduced to

$$\gamma = r + \beta a \bmod n \qquad \text{and} \qquad g^\gamma \stackrel{?}{=} \alpha x^\beta. \tag{4.1}$$

E *Exercise* 4.17. (a) Suppose Bob tells Eve what his choice β will be before Eve chooses α. Explain how Eve can choose α and γ to convince Bob that she knows a, even though she only knows x.

If Bob does not tell Eve his choice early, explain why Eve can guess his choice, proceed as above and cheat Bob with probability $1/2$.

(b) Suppose that Eve has chosen α and that she knows responses that Bob will accept, for each of Bob's choices. That is, Eve knows γ_0 and γ_1 such that $g^{\gamma_\beta} = \alpha x^\beta$. Show that Eve can easily compute a from γ_0 and γ_1.

We wanted to ensure that if Alice runs this protocol with Eve, then Eve learns nothing about Alice's secret. We shall argue that if Eve can learn something from Alice, she can learn the same thing without Alice.

So suppose Eve gets α from Alice, chooses β, receives γ and from that exchange learns something about a. Now Eve decides to do without Alice. Instead, she makes a guess β' at what she will choose upon seeing α. Then she proceeds according to the first part of Exercise 4.17. With probability $1/2$, she will guess her choice correctly, in which case her conversation would proceed exactly as if she were talking to Alice, which means that she would learn something about a. Of course, with probability $1/2$, Eve will not guess correctly, so she may not learn anything, but in this case she can just try again.

We also wanted to ensure that Eve cannot cheat Bob. From the above exercise we know that Eve can successfully pretend to know a with probability $1/2$, but unless she knows a, she cannot succeed with any greater probability.

It follows that Alice can convince Bob that she almost certainly knows a by repeating the above protocol many times. For each repetition, Eve would have probability $1/2$ of cheating successfully, but for k repetitions her success probability sinks to 2^{-k}.

Doing k repetitions is quite inefficient, of course. Instead, we can do something slightly different. The problem is that Eve can guess Bob's choice and choose α based on that. But note that in (4.1), there is nothing that forces β to be just 0 or 1.

The protocol can then work as follows.

1. Alice chooses r uniformly at random from $\{0, 1, 2, \ldots, n-1\}$, computes $\alpha = g^r$, and sends α to Bob.

2. Bob chooses β uniformly at random from $\{0, 1, 2, \ldots, 2^k - 1\}$ and sends β to Alice.

3. Alice computes $\gamma = r + \beta a \bmod n$ and sends γ to Bob.

Bob accepts that Alice knows a if

$$g^\gamma \overset{?}{=} \alpha x^\beta.$$

By the above arguments, the probability that Eve cheats Bob should not be much larger than 2^{-k}. One problem is that our argument for why Eve does not learn anything from Alice was exactly the argument that proved that Eve could cheat Bob. Which means that strictly speaking, we no longer have an argument for why Eve will not learn anything by talking to Alice.

However, if Eve chooses without looking at Alice's α, then the argument from the first part of Exercise 4.17 applies, and we can use it to show that in this case Eve cannot learn anything about Alice's message.

Unfortunately, for the same reason, Bob cannot reveal his choice before Alice reveals her α. The question is, how can we force Bob to choose before Alice reveals her α, but without Bob revealing his choice?

Ⓔ *Exercise* 4.18. Let $h : S \to T$ be a hash function such that $\{0, 1, 2, \ldots, 2^{2k} - 1\} \times \{0, 1, 2, \ldots, 2^k - 1\}$ is a subset of the domain.

Suppose Bob chooses o and β and computes $c = h(o, \beta)$. He sends c to Alice. The protocol then proceeds by Alice sending α to Bob, who responds with his already chosen β and the random number o. Alice verifies that c equals $h(o, \beta)$ and responds with the correct γ, which Bob verifies as usual.

Under reasonable assumptions on the hash function h, it can be shown that c does not reveal anything about β, and that Bob cannot find different o', β' such that $h(o', \beta') = c$.

Argue why Eve cannot cheat Alice (to learn something about a) or Bob (to convince him that Eve knows a) using the above protocol.

A different approach can be used if we have a random-looking hash function $h : S \to T$ where $G \times G$ is a subset of S and $T = \{0, 1, \ldots, 2^k - 1\}$. Instead of choosing a random challenge, Bob can instead compute the challenge as $\beta = h(x, \alpha)$. This is the *Fiat-Shamir heuristic*.

Since Eve no longer chooses her challenge when trying to cheat Alice, Eve will not be looking at α before deciding on her challenge, so as we have argued above, she should not learn anything new.

When Eve is trying to cheat Bob, however, she can know what challenge Bob will choose without actually sending α to Bob. She cannot know β until after she has chosen α, but she will still have the ability to look at many αs with corresponding βs, before she sends one α to Bob. While this does give her increased power, it can easily be neutralised by increasing k.

If Eve can compute β before sending α to Bob, so can Alice. We can therefore simplify the process. Alice chooses r, computes $\alpha = g^r$, $\beta = h(x, \alpha)$, and $\gamma = r + \beta a \bmod n$. She then sends α, β and γ to Bob. Bob verifies that

$$\beta \stackrel{?}{=} h(x, \alpha) \qquad \text{and} \qquad g^\gamma \stackrel{?}{=} \alpha x^\beta.$$

An equivalent verification equation can be

$$\alpha \stackrel{?}{=} g^\gamma x^{-\beta}.$$

If the hash function is collision-resistant (and a "random-looking" hash function should be), this equation will hold in practice if and only if

$$h(x, g^\gamma x^{-\beta}) \stackrel{?}{=} \beta.$$

This opens for further simplification. The process now works as follows. Alice chooses r, computes $\alpha = g^r$, $\beta = h(x, \alpha)$ and $\gamma = r - \beta a \bmod n$. She then sends β and γ to Bob. Bob verifies that

$$h(x, g^\gamma x^\beta) \stackrel{?}{=} \beta.$$

4.4.2 Schnorr Signatures

The Schnorr signature scheme is based on the ideas on how to prove that you know something, but the proofs are augmented by including something extra in the hash that generates the challenge.

Example 4.3. Suppose $\mathfrak{P}$ is a set of messages, and we have a random-looking hash function $h : S \to T$, where $G \times G \times \mathfrak{P}$ is a subset of S and $T = \{0, 1, \ldots, 2^k - 1\}$. The *Schnorr* signature scheme $(\mathcal{K}, \mathcal{S}, \mathcal{V})$ works as follows.

- The *key generation* algorithm $\mathcal{K}$ samples a number a uniformly at random from the set $\{0, 1, 2, \ldots, n - 1\}$. It computes $x = g^a$ and outputs $vk = x$ and $sk = a$. The message set associated to vk is $\mathfrak{P}$.

- The *signing* algorithm $\mathcal{S}$ takes as input a signing key a and a message $m \in \mathfrak{P}$. It samples $r \xleftarrow{r} \{0, 1, 2, \ldots, n - 1\}$, and computes $\alpha = g^r$, $\beta = h(x, \alpha, m)$ and $\gamma = r - \beta a \bmod n$. It outputs the signature (β, γ).

- The *verification* algorithm $\mathcal{V}$ takes as input a verification key x, a message $m \in \mathfrak{P}$ and a signature (β, γ). It outputs 1 if $h(x, g^\gamma x^\beta, m) \stackrel{?}{=} \beta$, otherwise 0.

E *Exercise* 4.19. The above is an informal description of a signature scheme. Implement the three algorithms $\mathcal{K}$, $\mathcal{S}$ and $\mathcal{V}$. Show that the triple $(\mathcal{K}, \mathcal{S}, \mathcal{V})$ is a signature scheme.

The Schnorr signatures and related schemes are famously sensitive to faulty random number generation, as the following exercise shows.

E *Exercise* 4.20. Let x be a verification key for the Schnorr signature scheme. Suppose that under this verification key, (β, γ) is a valid signature on m, and (β', γ') is a valid signature on m', $m \neq m'$. Suppose further that

$$g^\gamma x^\beta = g^{\gamma'} x^{\beta'}.$$

Explain why the existence of these two signatures suggest a malfunction in random number generation, and show how to recover the signing key corresponding to x using only a handful of arithmetic operations.

4.5   HASH-BASED SIGNATURES

Signature schemes based on factoring (Section 4.3) or discrete logarithms (Section 4.4) are all vulnerable to Shor's algorithm (which we shall study in Chapter 5) if someone ever builds a suitable quantum computer. For encryption or key exchange algorithms, potential future quantum computers will often be a problem, since today's adversaries may be storing today's intercepted ciphertexts so that they can read the corresponding messages after they have built a sufficiently large quantum computer. The quantum computer threat is less acute for many applications of digital signatures, since being able to forge a signature in a decade or two will not compromise today's use of the system. However, sufficiently large quantum computers may be built, so it makes sense to prepare by designing quantum-safe signature schemes.

4.5.1   Lamport's One-time Signatures

Lamport's *one-time* signature scheme, is based on a one-way hash function.

E *Example* 4.4. Let $h : S \to T$ be a hash function. *Lamport's one-time signature scheme* $(\mathcal{K}_1, \mathcal{S}_1, \mathcal{V}_1)$ works as follows.

- The *key generation* algorithm $\mathcal{K}_1$ samples l pairs $(s_{i0}, s_{i1}) \xleftarrow{r} S^2$, $i = 1, 2, \ldots, l$, computes $u_{ij} \leftarrow h(s_{ij})$ for $i \in \{1, 2, \ldots, l\}$, $j \in \{0, 1\}$, and outputs $vk = ((u_{10}, u_{11}), \ldots, (u_{l0}, u_{l1}))$ and $sk = ((s_{10}, s_{11}), \ldots, (s_{l0}, s_{l1}))$.

- The *signing* algorithm $\mathcal{S}_1$ takes as input a signing key $sk = ((s_{10}, s_{11}), \ldots, (s_{l0}, s_{l1}))$ and a message $m = m_1 m_2 \ldots m_l \in \{0, 1\}^l$. The signature is $(s_{1,m_1}, s_{2,m_2}, \ldots, s_{l,m_l})$.

- The verification algorithm $\mathcal{V}_1$ takes as input a verification key $vk = ((u_{10}, u_{11}), \ldots, (u_{l0}, u_{l1}))$, a message $m = m_1 m_2 \ldots m_l \in \{0, 1\}^l$ and a signature $(v_1, v_2, \ldots, v_l)$. It outputs 1 if $u_{i,m_i} = h(v_i)$ for $i = 1, 2, \ldots, l$, otherwise 0.

As usual, scheme can be augmented with another collision resistant hash function $h' : \mathfrak{P} \to \{0, 1\}^l$ to increase the size of the plaintext set.

E | *Exercise 4.21.* The above is an informal description of a signature scheme. Implement the three algorithms $\mathcal{K}_1$, $\mathcal{S}_1$ and $\mathcal{V}_1$. Show that the triple $(\mathcal{K}_1, \mathcal{S}_1, \mathcal{V}_1)$ is a signature scheme.

E | *Exercise 4.22.* Lamport's scheme is a one-time scheme. Show that you can create a forgery given signatures on two messages that differ in two positions.

E | *Exercise 4.23.* Consider a hash function with $T = \{0, 1\}^{256}$. If we use this hash function with Lamport's scheme, how long is a signature (in bits)? And how long is the secret key (in bits)? Compare with the length of the signatures of RSA signatures (with $n \approx 2^{3072}$) and Schnorr signatures (with $p \approx 2^{256}$).

Remark. The verification key is large since two hash values are needed to verify each message bit. One way to reduce its size is to use a hash function $h : T \to T$ and a pair of *hash chains* of length k to sign an integer in $\{0, 1, \ldots, k\}$.

An ith preimage of u is a value s such that

$$(\underbrace{h \circ h \circ \cdots \circ h}_{i \text{ times}})(s) = u.$$

Given two hash values u and u', we can encode an integer i as an ith preimage of u and an $(k - i)$th preimage of u'.

In this way, at the cost of $2k$ hash computations during key generation and k hash computations during signing and verification, we can sign k distinct values which makes signatures much shorter.

Remark. A one-time scheme $(\mathcal{K}_1, \mathcal{S}_1, \mathcal{V}_1)$ is of limited use. One way to make this scheme more useful is to use a tree of signatures. Each leaf node contains a key pair. Each internal node in the tree contains a key pair and a signature on the verification keys of the child nodes.

The verification key would be the root node's verification key. To sign a message, we sign it using one of the leaf node signing keys. The total signature consists of the one-time signature along with all the verification keys and one-time signatures needed to connect the message to the root verification key.

There are a number of problems with this approach. Even though the depth of the tree is logarithmic in the total number of messages we want to sign, signatures are very long. Also, we need to keep track of which keys have been used, which means that the system is stateful. We also need to keep track of many signing keys, which means that the system state is large.

4.5.2 Merkle Signatures

A one-time scheme $(\mathcal{K}_1, \mathcal{S}_1, \mathcal{V}_1)$ is of limited use. However, we can use a *Merkle tree* to turn the one-time scheme into a somewhat more practical scheme that can sign many messages. The idea is to create a hash tree (*Merkle tree*) where each internal node contains the hash of its children, while each leaf node contains the hash of a verification key. An example is shown in Figure 4.1.

We first define an indexing scheme for the nodes in a binary tree. We define the indexing recursively. Let $(0,0)$ be the index of the root node. If a node has index (i, j), its left child will have index $(i + 1, 2j)$ and its right child will have index $(i + 1, 2j + 1)$. An index (i, j) will therefore satisfy $0 \leq j < 2^i$.

An index (i, j) contains the path from the root node to the node: write j as an i-digit binary number, begin at the most significant digit and interpret a 0 as "left" and a 1 as "right". The indexes of the nodes in the path to (k, j) will be $(0, 0), (1, \lfloor j/2^{k-1} \rfloor), (2, \lfloor j/2^{k-2} \rfloor), \ldots, (k - 1, \lfloor j/2 \rfloor), (k, j)$.

It is convenient to define some notation for the tree. We denote the node on the ith level in the path to (k, j) as $j_i = \lfloor j/2^{k-i} \rfloor$. Since the left child always has an even index, while the right child has an odd index, the index of a nodes' sibling, as well as the left and right node in a sibling pair, is given by

$$sib(j) = \begin{cases} j - 1 & j \text{ odd,} \\ j + 1 & j \text{ even,} \end{cases} \quad left(j) = \begin{cases} j & j \text{ even,} \\ j - 1 & j \text{ odd,} \end{cases} \quad right(j) = left(j) + 1.$$

If the hash function is collision-resistant, then the root node in practice defines which verification keys are attached to the leaf nodes. In order to verify that the verification key vk_j is attached to the leaf node (k, j), we need only compute the hashes on the path from the root to (k, j) and verify that this hash chain is consistent with the hash on the root node. To do this, we also need the hashes stored on the siblings of the nodes in the path.

Example 4.5. *Merkle's signature scheme* $(\mathcal{K}_2, \mathcal{S}_2, \mathcal{V}_2)$ *based on a hash function* h *and a one-time signature scheme* $(\mathcal{K}_1, \mathcal{S}_1, \mathcal{V}_1)$ *works as follows.*

- The *key generation* algorithm $\mathcal{K}_2$ uses $\mathcal{K}_1$ to generate 2^k key pairs $(vk_0, sk_0), \ldots, (vk_{2^k-1}, sk_{2^k-1})$. It then computes the hashes on the internal nodes in the Merkle tree as

$$u_{i,j} = \begin{cases} h(u_{i+1,2j}, u_{i+1,2j+1}) & i = 0, 1, \ldots, k - 2, \\ h(vk_{2j}, vk_{2j+1}) & i = k - 1. \end{cases} \tag{4.2}$$

It then outputs $vk = u_{0,0}$ and $sk = ((vk_0, sk_0), \ldots, (vk_{2^k-1}, sk_{2^k-1}))$.

- The *signing* algorithm $\mathcal{S}_2$ takes as input a signing key $sk = ((vk_0, sk_0), \ldots, (vk_{2^k-1}, sk_{2^k-1}))$ and a message m. It chooses an index j and signs the message m as

$$\sigma_1 = \mathcal{S}_1(sk_j, m).$$

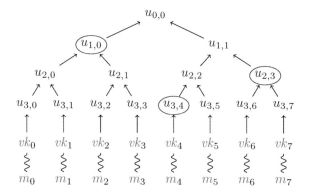

FIGURE 4.1 A Merkle tree with depth 3, allowing for 8 messages to be signed with one-time signature keys in the leaf nodes. Here $u_{3,j} = h(vk_j)$ and $u_{i,j} = h(u_{i+1,2j}, u_{i+1,2j+1})$. To verify that the one-time verification key vk_5 is part of a tree with root $u_{0,0}$, the hash values on the path to the root can be computed given the circled hash values.

Then it recomputes the hash values in the Merkle tree using (4.2). The signature is $(j, \sigma_1, vk_j, u_{1,sib(j_1)}, u_{2,sib(j_2)}, \ldots, u_{k-1,sib(j_{k-1})}, u_{k,sib(j)})$.

- The *verification* algorithm $\mathcal{V}_2$ takes as input a verification key $vk = u_{0,0}$, a message m and a signature $\sigma = (j, \sigma_1, vk, s_1, s_2, \ldots, s_k)$. It verifies that $\mathcal{V}_1(vk_j, m, \sigma_1) = 1$, and then computes a subset of the Merkle tree hashes as $u'_{k,j} = h(vk)$, $u'_{i,sib(j)} = s_i$ and $u'_{i-1,j_{i-1}} = h(u'_{i,left(j_i)}, u'_{i,right(j_i)})$. Finally, it outputs 1 if $u'_{0,0} = u_{0,0}$.

E *Exercise* 4.24. The above is an informal description of a signature scheme. Implement the three algorithms $(\mathcal{K}_2, \mathcal{S}_2, \mathcal{V}_2)$, up to the underlying one-time signature scheme. Show that the triple $(\mathcal{K}_2, \mathcal{S}_2, \mathcal{V}_2)$ is a signature scheme.

E *Exercise* 4.25. Consider a hash function with $T = \{0,1\}^{256}$ used with Lamport's one-time signatures (with the same hash function) and the above Merkle signatures. How long would a signature be (in bits), as a function of the number of messages that can be signed? How long would the secret key be? How many hash function computations would be done during key generation.

E *Exercise* 4.26. The Merkle signature scheme is not a one-time scheme. Show that you can create a forgery if you see two signatures for sufficiently distinct messages, but where both signatures have the same j.

If there are sufficiently many potential messages to be signed in this system, we could just choose leaf nodes at random and expect to avoid collisions.

However, this is impractical because of the effort involved in generating the key pair, which is essentially proportional to the number of leaf nodes.

The effort required to generate a key pair is proportional to the number of messages to be signed over the life of the key. This is impractical. However, there are a number of options for improving the Merkle tree, including the ideas from the final two remarks in Section 4.5.1.

- We can create a tree of Merkle trees, attaching the root of one tree to the leaf nodes of the parent tree. This would allow a trade-off between reducing key generation time and increasing the length of the signatures.

- We can use n-ary trees, to shorten signatures, allowing a trade-off between computational requirements and signature length.

- We can generate signing key material with a pseudo-random function, which would make the signing key small, at little computational cost.

- We could derive which leaf node to use from a hash of the message. If the hash function was collision-resistant, this would be secure. This would make the signature scheme stateless.

There are a number of interesting constructions for such hash functions.

4.6   SECURING DIFFIE-HELLMAN

We saw in Section 2.6 that the Diffie-Hellman protocol is subject to a man-in-the-middle attack, where Eve can run the Diffie-Hellman protocol separately with Alice and Bob, since Alice cannot distinguish Bob's bits from Eve's bits.

Alice and Bob can use signatures to protect their Diffie-Hellman key exchange. In this case, Alice has a signing key pair (sk_A, vk_A) and Bob has a signing key pair (sk_B, vk_B), and they both know the other's verification key.

1. Alice samples $a \xleftarrow{r} \{0, 1, 2, \ldots, n-1\}$, computes $x = g^a$ and sends x.

2. Bob receives x. He samples $b \xleftarrow{r} \{0, 1, 2, \ldots, n-1\}$, computes $y = g^b$, $z_B = x^b$ and $\sigma_B = \mathcal{S}(sk_B, m)$, where m is a message containing a key exchange *context*, vk_A, vk_B, x and y, and sends (y, σ_B).

3. Alice receives (y, σ_B). Alice verifies σ_B, computes $z_A = y^a$ and $\sigma_A = \mathcal{S}(sk_A, m')$, where m is a message containing a key exchange *context*, vk_A, vk_B, x, y and σ_B, and sends σ_A.

4. Bob receives σ_A from Alice. Bob verifies σ_A.

If either party's signature verification fails, that party stops immediately.

We shall also see that digital signatures can be used with public key encryption to solve the problem of who sent a given ciphertext.

4.7 THE PUBLIC KEY INFRASTRUCTURE PROBLEM

Before asymmetric encryption was invented, a shared secret was required for secure communication over insecure channels. As we have seen, the Diffie-Hellman key exchange, public key encryption and digital signatures have removed the need for a preexisting shared secret, but public keys (for encryption or signature verification) still need to be exchanged before communicating.

A *public key infrastructure* (PKI) is an infrastructure set up to move public keys from Alice to Bob in such a way that Bob can be sure that the public key he receives really belongs to Alice and is her current key, even if Alice and Bob have never communicated before.

As is often said, nothing will come of nothing, so Alice and Bob cannot hope to solve this problem on their own. One possible solution is the so-called *web of trust*. In this scheme, Alice and all her friends sign each other's public keys together with their unique names. Alice' public key, her name and her friend's signature is called a *certificate*. The certificate is interpreted as Alice' friend saying that the public key belongs to the named person, namely Alice.

Alice' friends in turn sign their friends' public keys, and so forth. If we consider people as vertices in a graph, with edges between friends who have signed each other's public keys, Alice and Bob need to find a path between themselves in this graph.

A more practical system relies on a trusted third party, usually called a *certificate authority*. Again, the trusted third party signs Alice' public key along with her unique name. If Alice and Bob both trust each other's certificate authorities, they can simply send their certificate to the other party, and then use an appropriate public key protocol.

Often, we cannot agree on one trusted third party. This means that Alice and Bob must deal with many trusted third parties, which is inconvenient. A variant of the web of trust can be used to reduce the number of trusted third parties Alice and Bob have to deal with, by having the many trusted third parties declare that they trust the other third parties.

Private keys are sometimes compromised, which means that Eve learns the key. When Alice discovers that someone knows her private key (and can thus impersonate her), Alice wants the certificate for that key pair to stop working. She notifies her certificate authority that the key pair has been compromised.

The certificate authority was not involved when Alice and Bob communicated, so somehow Bob must be told that Alice' certificate has been *revoked*. The traditional approach is for the certificate authority to maintain a list of revoked certificates (*certificate revocation list*). Anyone who relies on the certificate authority will periodically fetch an updated list of revoked certificates.

Since certificate revocation lists are fetched only periodically, there will typically be some time between Alice notifies her certificate authority until Bob stops accepting the certificate. Another problem with certificate revocation lists is that if there are many certificate authorities, managing the revocation lists becomes impractical.

One popular solution is for a certificate authority to provide a *certificate status service*. Any user may ask for the status of a given certificate. The certificate authority will reply with a signed message. If the certificate is valid (that is, not revoked), the message contains a statement to that effect and the current time. If the certificate has been revoked, the message contains a statement to that effect and the time of revocation.

While there are some irreducible problems, it is quite easy to design beautiful trust architectures for public key infrastructures in theory. In practice, deploying and operating public key infrastructures has turned out to be a difficult problem, typically involving complex legal, commercial, economic and social issues. Deployed infrastructures have often evolved through a series of compromises. The word beautiful is seldom used to describe them.

Today, vast networks of interoperable public key infrastructures have fallen somewhat out of fashion, often replaced by simpler systems with less functionality. In this book, we deal with this problem by observing that the design of cryptography and the design of public key infrastructures are largely orthogonal problems. Therefore, we shall design cryptography that does not deal with identities, but instead deals purely with keys. Then we design the cryptography to support *associated data* which can be used to tie keys and their usage to some *context*, which may involve identities and public key infrastructures.

CHAPTER 5

Factoring Using Quantum Computers

We shall now sketch a machine for quickly factoring an RSA modulus. This does not contradict the analysis in Section 3.4, because the machine we shall build is a *quantum computer*, exploiting quantum effects for computation.

Small quantum computers that can factor numbers with a few digits have been demonstrated. It is unclear[1] if we will ever be able to build sufficiently large and reliable quantum computers to factor a real-world RSA modulus, but for applications that require security decades into the future, quantum computers will certainly be a source of uncertainty for a long time.

An RSA modulus n is the product of two large primes p and q. Let $g \in \mathbb{Z}$ be relatively prime to n. Consider the function $f : \mathbb{Z} \to \mathbb{Z}$ given by $a \mapsto g^a \bmod n$. This function is *periodic*, and the period is the same as the multiplicative order of $g + \langle n \rangle$ in $\mathbb{Z}_n^*$. The period of this function therefore divides $(p-1)(q-1)$, and for most n and g the period will be close to $(p-1)(q-1)$.

We have previously seen that we can easily factor n if we know a multiple of $(p-1)(q-1)$. This means that if we can find the a multiple of the period of such a function, then by adapting our previous results, we can factor n.

So to meet our goal of factoring, it suffices to construct a machine that is capable of finding (a multiple of) the period of a periodic function $f : \mathbb{Z} \to \mathbb{Z}$. In the following, r will be the period of f.

5.1 BACKGROUND

Before we discuss quantum computing and using it for factoring, we must first discuss some very classical mathematics.

---

[1]I first heard the claim that we would certainly have large quantum computers in 20 years about 20 years ago. Sarcasm is easy, but should not be mistaken for argument.

DOI: 10.1201/9781003149422-5

5.1.1 Rational Approximations

A *rational approximation* of a real number is a fraction that is close to the real number. Continued fractions are a useful tool to find rational approximations.

Fact 5.1. *Let a, b, c, d be integers such that $|a/b - c/d| \leq 1/(2d^2)$. Then c/d is a convergent for the continued fraction expansion of a/b, and this convergent can be found using the Euclidian algorithm.*

Let N be an integer satisfying $N > n^2 > r^2$. Suppose we also have an integer k that is close to a multiple of the fraction N/r, that is, satisfying

$$k = l\frac{N}{r} + \delta_k \qquad \text{where } l, N, r \in \mathbb{Z} \text{ and } |\delta_k| \leq 1/2. \qquad (5.1)$$

Then

$$\left| \frac{k}{N} - \frac{l}{r} \right| \leq \frac{1}{2N}.$$

This means that l/r is a rational approximation of the number k/N. This approximation can be found quickly using the Euclidian algorithm. When we want a particular r, common factors between l and r may be lost.

Therefore, if we know N and find k satisfying (5.1), then we can often find a rational approximation l/r, which will give us r (up to common factors).

Example 5.1. Consider the rational number $k/N = 1365/2048$. In a procedure which is essentially the Euclidian algorithm, we get

$$\frac{1365}{2048} = \frac{1}{1 + \frac{683}{1365}} \qquad \frac{683}{1365} = \frac{1}{1 + \frac{682}{683}} \qquad \frac{682}{683} = \frac{1}{1 + \frac{1}{682}}$$

from which we get the continued fraction

$$\frac{1365}{2048} = \cfrac{1}{1 + \cfrac{1}{1 + \cfrac{1}{1 + \frac{1}{682}}}} \qquad \text{with convergents } 0, \frac{1}{1}, \frac{1}{2}, \frac{2}{3}, \frac{1365}{2048}.$$

Here we see that $2/3$ is a good rational approximation. Indeed, so would $4/6$, $8/12$, etc. also be.

Example 5.2. Consider the real number $k/N = 1195/2048$. Following the procedure from the previous example we get

$$\frac{1195}{2048} = \frac{1}{1 + \frac{853}{1195}} \qquad \frac{853}{1195} = \frac{1}{1 + \frac{342}{853}} \qquad \frac{342}{853} = \frac{1}{2 + \frac{169}{342}}$$

$$\frac{169}{342} = \frac{1}{2 + \frac{4}{169}} \qquad \frac{4}{169} = \frac{1}{42 + \frac{1}{4}}$$

from which we get the continued fraction

$$\frac{1195}{2048} = \cfrac{1}{1 + \cfrac{1}{1 + \cfrac{1}{2 + \cfrac{1}{2 + \frac{1}{42 + \frac{1}{4}}}}}}. \quad \text{with convergents } 0, \frac{1}{1}, \frac{1}{2}, \frac{3}{5}, \frac{7}{12}, \frac{297}{509}, \frac{1195}{2048}$$

Here we see that $7/12$ and $297/509$ are good rational approximations.

5.1.2 Discrete Fourier Transform

Let $\alpha = (\alpha_0, \alpha_1, \ldots, \alpha_{N-1}) \in \mathbb{C}^N$. Then a normalised *discrete Fourier transform* of α is $\beta \in \mathbb{C}^N$ given by

$$\beta_k = \frac{1}{\sqrt{N}} \sum_j \alpha_j \exp\left(2\pi i \frac{kj}{N}\right), \qquad 0 \le k < N.$$

It is normalised to preserve the norm of vectors, so $\|\alpha\| = \|\beta\|$. In fact, it is a linear map given by a unitary matrix (a normalised Vandermonde matrix).

E *Exercise 5.1.* Let $\omega = \exp(2\pi i/N) \in \mathbb{C}$, which is a primitive Nth root of unity. Let $\mathbf{V} = (v_{kj})$ be the Vandermonde matrix with $v_{kj} = \omega^{kj}$, $0 \le k, j < N$.

(a) Show that the $\overline{\omega^j} = \omega^{N-j}$, where $\bar{z}$ denotes complex conjugation.

(b) Show that $\sum_{j=0}^{N-1} (\omega^k)^j = 0$ for any $k \not\equiv 0 \pmod{N}$.

(c) Show that $\mathbf{V}\mathbf{V}^* = N\mathbf{I}$, where $\mathbf{V}^*$ denotes the conjugate transpose of $\mathbf{V}$.

(d) Show that when β is the discrete Fourier transform of α, then $\beta = \frac{1}{\sqrt{N}}\mathbf{V}\alpha$.

E *Exercise 5.2.* Let $\mathbf{U}$ be an $N \times N$ unitary matrix. Show that for any vector $\alpha \in \mathbb{C}^N$, $\|\mathbf{U}\alpha\| = \|\alpha\|$.

We want to study the Fourier coefficients of a very special complex vector. Let t_0 be an integer such that $0 \le t_0 < r$, and let m be minimal such that $mr + t_0 \ge N$. Let $\alpha \in \mathbb{C}^N$ be given by

$$\alpha_k = \begin{cases} \frac{1}{\sqrt{m}} & k = t_0 + jr, \text{ and} \\ 0 & \text{otherwise.} \end{cases} \tag{5.2}$$

Note that $\|\alpha\| = 1$ and that $1 - mr/N = 1 - (N - t_0)/N \le r/N \le 1/r$.

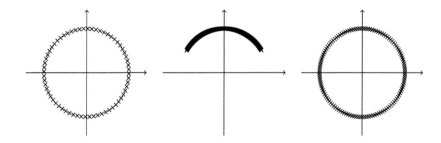

FIGURE 5.1 Plot of the terms in the final sum in (5.3) for $N = 2048$, $r = 12$ and $t_0 = 7$, and for $k = 1192, 1195, 1196$.

Applying the discrete Fourier transform gives us

$$\beta_k = \frac{1}{\sqrt{N}} \sum_{j=0}^{m-1} \frac{1}{\sqrt{m}} \exp\left(2\pi i \frac{k(t_0 + jr)}{N}\right)$$

$$= \frac{1}{\sqrt{N}} \frac{1}{\sqrt{m}} \exp\left(2\pi i \frac{k t_0}{N}\right) \sum_{j=0}^{m-1} \exp\left(2\pi i \frac{k j r}{N}\right).$$

(5.3)

E *Example* 5.3. Consider the vector $\boldsymbol{\alpha}$ from (5.2) for $N = 2048$, $r = 12$ and $t_0 = 7$. The terms of the sum in the expression for β_k from (5.3) has been plotted in the complex plane for $k = 1192, 1195, 1196$ in Figure 5.1.

Since the Fourier coefficient β_k is the sum of all the plotted complex numbers, it is easy to imagine that the sum of these complex numbers will tend to cancel out in the left and right cases, while for the middle case with $k = 1195$ there will be no cancellation. In other words, the amplitude of β_{1192} and β_{1196} will be small, while β_{1195} will be large.

E *Exercise* 5.3. Let $z \in \mathbb{R}$.

(a) Prove that $|1 - \exp(2\pi i z)|^2 = 4\sin^2(\pi z)$.

(b) If $0 \leq |z| \leq \pi/2$, prove that $(\sin z)/z \geq 2/\pi$.

We shall only care about the absolute value, which means that we shall only care about the sum factor. Furthermore, we shall only care about those k that are close to a multiple of the fraction N/r, that is those k satisfying (5.1).

For such a k, our sum factor becomes

$$\sum_{j=0}^{m-1} \exp\left(2\pi i \frac{kjr}{N}\right) = \sum_{j=0}^{m-1} \exp\left(2\pi i \frac{jr}{N}\left(l\frac{N}{r} + \delta_k\right)\right)$$

$$= \sum_{j=0}^{m-1} \exp(2\pi i\ jl\ )\exp\left(2\pi i \frac{jr}{N}\delta_k\right)$$

$$= \sum_{j=0}^{m-1} \exp\left(2\pi i \frac{jr}{N}\delta_k\right) = \sum_{j=0}^{m-1}\left(\exp\left(2\pi i \frac{r}{N}\delta_k\right)\right)^j$$

$$= \frac{1 - \exp\left(2\pi i \frac{r}{N}\delta_k m\right)}{1 - \exp\left(2\pi i \frac{r}{N}\delta_k\right)} = S.$$

Recall that when z is a very small real number, $|\sin z| \leq |z|$, while for $|z| < \pi/2$ we know that $|\sin z|$ is increasing. Since N is large relative to r, r/N will be small. Furthermore, $(m-1)r + t_0 \leq N < mr + t_0$, which means that mr/N is slightly smaller than 1. Using Exercise 5.3 we get

$$|S|^2 = \frac{\sin^2\left(\pi\frac{r}{N}\delta_k m\right)}{\sin^2\left(\pi\frac{r}{N}\delta_k\right)} \geq \frac{\sin^2\left(\pi\delta_k\right)}{\sin^2\left(\pi\frac{r}{N}\delta_k\right)} \geq \frac{\sin^2(\pi\delta_k)}{(\pi\delta_k r/N)^2} \geq \frac{N^2}{r^2}\left(\frac{2}{\pi}\right)^2.$$

For an index k satisfying (5.1), the discrete Fourier coefficient satisfies

$$|\beta_k|^2 \geq \frac{1}{N}\frac{1}{m}\frac{N^2}{r^2}\frac{4}{\pi^2} = \frac{1}{mr}\frac{N}{r}\frac{4}{\pi^2} \geq \frac{1}{mr}\frac{mr}{N}\frac{N}{r}\frac{4}{\pi^2} = \frac{1}{r}\frac{4}{\pi^2}.$$

Next, we want to sum these values for all k satisfying (5.1), and since there are at least $r - 1$ such k, we get an approximate lower bound

$$(r - 1)\frac{1}{r}\frac{4}{\pi^2} \approx \frac{4}{\pi^2}. \tag{5.4}$$

This result essentially says that when we sum the squared absolute values of all the Fourier coefficients, those corresponding to values of k satisfying (5.1) constitute a relatively large fraction of the total sum.

Example 5.4. Figure 5.2 shows a plot of the amplitude of the discrete Fourier transform of a function of the form (5.2). The locations of the amplitude peaks for the discrete fourier transform are given in Table 5.1.

The cumulative probability of these values is about 0.79. For only a few of these values does the rational approximation of k/N give us the period $r = 12$, but for all except $k = 0$ it is within a small multiple. The explanation is that the multiple l and the period r have common factors, which cancel.

Note, however, that we cannot compute the discrete Fourier transform, since N is too large, and even if we could, we could not easily find out which

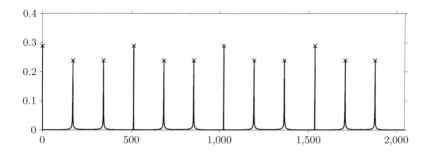

FIGURE 5.2 The amplitude of the discrete Fourier transform of a function of the form (5.2) with period $r = 12$ and $N = 2048$.

TABLE 5.1 The locations of the maximal amplitudes from Figure 5.2, with reference to (5.1).

k	$\lfloor k/(N/r) \rceil$	$|\delta|$	appr.	k	$\lfloor k/(N/r) \rceil$	$|\delta|$	appr.
0	0.0	0.0	-	171	1.0	0.002	1/12
341	2.0	0.002	1/6	512	3.0	0.0	1/4
683	4.0	0.002	1/3	853	5.0	0.002	5/12
1024	6.0	0.0	1/2	1195	7.0	0.002	7/12
1365	8.0	0.002	2/3	1536	9.0	0.0	3/4
1707	10.0	0.002	5/6	1877	11.0	0.002	11/12

values of k are included in the sum. So the discrete Fourier transform does not immediately help us find a k satisfying (5.1).

However, if we could arrange some probabilistic process in such a way that it will sample an integer from $\{0, 1, \ldots, N-1\}$ with probability according to the squared absolute value of the Fourier coefficients, then the above results says that it is reasonably likely that we will sample a k satisfying (5.1).

5.2 QUANTUM COMPUTATION

A *quantum bit* (often called *qubit*) is a physical system with two states. We interpret the two states as 0 and 1. A quantum system can be in a *superposition* of its two states, which is described by two complex numbers α and β such that $|\alpha|^2 + |\beta|^2 = 1$. The two values $|\alpha|^2$ and $|\beta|^2$ are the probabilities of measuring 0 and 1, respectively. We write the state as

$$\alpha|0\rangle + \beta|1\rangle.$$

A physical system can have more than two states, of course. If we number the states from 0 and to $N-1$, a superposition of states is described a complex vector $\boldsymbol{\alpha}$ with norm 1, and we write the state as

$$\sum_{j=0}^{N-1} \alpha_j |j\rangle.$$

Two systems (with N and N' states, respectively) can be *entangled*, which means that the outcome of measuring one system will influence the outcome of measuring the other system. We write the joint system state as

$$\sum_{k=0}^{N-1} \sum_{j=0}^{N'-1} \alpha_{kj} |k\rangle |j\rangle.$$

If we measure the second system, we will get the outcome j_0 with probability

$$\sum_{k=0}^{N-1} |\alpha_{kj_0}|^2$$

and if j_0 was the outcome, we will get the quantum state

$$\sum_{k=0}^{N-1} \frac{\alpha_{kj_0}}{\sqrt{\sum_{k=0}^{N-1} |\alpha_{kj_0}|^2}} |k\rangle |j_0\rangle.$$

When later measuring the first system, the probability of outcome k_0 will be

$$\frac{|\alpha_{k_0 j_0}|^2}{\sum_{k=0}^{N-1} |\alpha_{kj_0}|^2}.$$

Multiple quantum bits can be entangled, which means that the outcome of measuring one bit may influence the outcome of measuring the other bits. In this case, a system of l qubits will have $N = 2^l$ states. A system of multiple entangled qubits is called a *quantum register*.

5.2.1 Computing on Quantum Registers

Computations on quantum bits can be done using *quantum gates*. Any such computation can be described by an invertible linear map that preserves the norm, that is, a unitary matrix $\mathbf{U}$. That is, the transition

$$\sum_{j=0}^{N-1} \alpha_j |j\rangle \quad \overset{\mathbf{U}}{\longmapsto} \quad \sum_{j=0}^{N-1} \beta_j |j\rangle$$

is given by $\boldsymbol{\beta} = \mathbf{U}\boldsymbol{\alpha}$.

E *Example* 5.5. Consider a system with two states. We want to compute the function $f(z) = 1 - z$, the negation function, on this state. The unitary matrix

$$\mathbf{U} = \begin{pmatrix} 0 & 1 \\ 1 & 0 \end{pmatrix}$$

will compute this function, mapping the quantum state (α_0, α_1) to (α_1, α_0).

E *Example* 5.6. Consider a system of three qubits and the function $f(z_1, z_2, z_3) = (z_1, z_2, z_1 z_2 + z_3 \bmod 2)$, which negates z_3 if and only if $z_1 = 1 = z_2$. If we consider the natural basis $|000\rangle$, $|001\rangle$, $\ldots |110\rangle$, $|111\rangle$, the matrix

$$\mathbf{U} = \begin{pmatrix} 1 & 0 & 0 & 0 & 0 & 0 & 0 & 0 \\ 0 & 1 & 0 & 0 & 0 & 0 & 0 & 0 \\ 0 & 0 & 1 & 0 & 0 & 0 & 0 & 0 \\ 0 & 0 & 0 & 1 & 0 & 0 & 0 & 0 \\ 0 & 0 & 0 & 0 & 1 & 0 & 0 & 0 \\ 0 & 0 & 0 & 0 & 0 & 1 & 0 & 0 \\ 0 & 0 & 0 & 0 & 0 & 0 & 0 & 1 \\ 0 & 0 & 0 & 0 & 0 & 0 & 1 & 0 \end{pmatrix}$$

will compute exactly this function, swapping the quantum states 110 and 111, while leaving all other states unchanged. (It can be shown that this gate, the *Toffoli gate*, is universal for classical computations. Which means that this gate is sufficient in order to do any classical computation.)

E *Example* 5.7. Consider a two-state system. The gate given by the matrix

$$\mathbf{U} = \frac{1}{\sqrt{2}} \begin{pmatrix} 1 & 1 \\ 1 & -1 \end{pmatrix}$$

takes the state $|0\rangle$ to a superposition of $|0\rangle$ and $|1\rangle$, where each outcome will be equally likely. (We create superpositions with such gates.)

E *Exercise* 5.4. Show that the above three matrices are unitary.

Since unitary matrices are invertible, computations on quantum bits must also be invertible. But with a bit of care and effort any classical computation can be repeated on quantum bits. In particular, we can compute the function f discussed above, even though it is not invertible. Typically, we will have two quantum registers and map $|k\rangle|j\rangle$ to $|k\rangle|j + f(k)\rangle$, which is an invertible computation, at least if we use modular addition.

If we compute on a superposition, our result will be a superposition of the function values. In the example above

$$\sum_{k=0}^{N-1}\sum_{j=0}^{N'-1}\alpha_{kj}|k\rangle|j\rangle \quad\longmapsto\quad \sum_{k=0}^{N-1}\sum_{j=0}^{N'-1}\alpha_{kj}|k\rangle|j+f(k)\rangle.$$

5.2.2 Quantum Fourier Transform

Example 5.6 and the discussion in the previous section show that any classical computation can be replicated on a quantum computer. However, Example 5.7 shows that there are operations on quantum bits that do not correspond directly to classical computations. One such operation that will be important for us is the *quantum Fourier transform*.

The quantum Fourier transform on a system with N states is defined by

$$\sum_{j=0}^{N-1}\alpha_j|j\rangle \quad\overset{QFT}{\longmapsto}\quad \sum_{k=0}^{N-1}\beta_k|k\rangle \quad\text{with } \beta_k = \frac{1}{\sqrt{N}}\sum_{j=0}^{N-1}\alpha_j\exp(2\pi ijk/N).$$

This is of course exactly the same as our normalised discrete Fourier transform.

Unlike the discrete Fourier transform, it turns out that a quantum Fourier transform of a quantum system made up of l quantum bits (with $N = 2^l$) can be implemented using about l^2 elementary quantum gates. The reason for this is that the classical discrete Fourier transform must be computed using operations that involve all the complex numbers $\alpha_0, \alpha_1, \ldots$. The quantum Fourier transform is computed by manupulating the l quantum bits, which only indirectly manipulates the quantum probabilities $\alpha_0, \alpha_1, \ldots$.

We do not discuss how to compute the quantum Fourier transform.

5.3   FACTORING USING A QUANTUM COMPUTER

Our goal is to set up a quantum system with an amplitude distribution like (5.2) and apply the quantum Fourier transform, which will result in a quantum system where we know that the likelihood of measuring a k satisfying (5.1) is at least $4/\pi^2$. We then measure such a k and compute a rational approximation which gives us the period of the function f, which allows us to factor n.

We begin with a two entangled quantum registers, one with $\log_2 N$ qubits and the other with $\lceil\log_2 n\rceil$ qubits. The first register should contain a superposition of the integers $0, 1, 2, \ldots, N-1$, all with the same amplitude, while the second register should contain 0. We have the state

$$\sum_{j=0}^{N-1}\frac{1}{\sqrt{N}}|j\rangle|0\rangle.$$

Using a quantum circuit, we compute f on the first register, storing the result in the second register. Since any classical computation can be done on a quantum computer, computing f is easy. Our two quantum registers now contain a superposition of $(k, f(k))$ for $k = 0, 1, \ldots, N-1$, all being equally likely, and we have the new state

$$\sum_{j=0}^{N-1} \frac{1}{\sqrt{N}} |j\rangle |f(j)\rangle.$$

Next, we measure the second register. Suppose we measure the value s. Since we have not yet measured the first register, we do not know what it contains, but since it was entangled with the second register, every value we could possible measure must be consistent with s. Suppose t_0 is the smallest non-negative integer such that $f(t_0) = s$. Then the only values we can measure in the first register are the integers $t_0 + jr$.

Since we had a uniform probability before we measured the second register, we must have a uniform probability after measuring. We now have m possible states, so we have the state

$$\sum_{j=0}^{m-1} \frac{1}{\sqrt{m}} |t_0 + jr\rangle |s\rangle.$$

In other words, if we ignore the second register which is constant, our first register now has exactly the amplitudes given by (5.2).

We then use a second quantum circuit to compute the quantum Fourier transform on the first register. And then we measure the first register. As discussed, we will with significant probability measure a k satisfying (5.1), which will allow us to factor n.

Example 5.8. Consider $n = 35$. We choose $g = 2$ and want to compute (something close to) the period of $a \mapsto 2^a \bmod 35$. To this end, we have a quantum computer with two registers, the first of which has 2048 states.

We initialise the quantum registers with a superposition of all possible values in the first register and zero in the second register. We apply the exponentiation function and measure the function value in the second register. Suppose we get the answer 23, which implies that the amplitude of the collapsed first register corresponds to that from Example 5.3.

We then compute the quantum Fourier transform of the first register, and the end result is the amplitude shown in Example 5.4.

Then we measure the first register. Suppose we get the answer 1195. Rational approximation as in Example 5.2 gives us a period of 12. We compute $2^3 \equiv 8 \pmod{35}$ and immediately get that $\gcd(8 - 1, 35) = 7$.

(Even if we had measured 1365 and gotten the wrong period from the rational approximation in Example 5.1, we would still have factored 35. But that would have been just lucky.)

CHAPTER 6

Computational Problems

So far we have studied how to design cryptosystems and how to attack them, either by solving underlying computational problems or exploiting flaws in the design. Our goal is to design cryptosystems so that the only way to attack them is to solve some underlying computational problem.

Our first task is to define the computational problems precisely. The intuitive notion is to have a set of instances, a set of answers, and a function that defines the correct answers. We also need a probability space on the instances, typically defined by a sampling algorithm.

Sometimes, it is convenient to define problems without unique answers, in which case we replace the function with a relation. There are two interesting cases. One is where any answer will do. The other is where we look for a particular answer. Decision problems are examples of the latter case.

One powerful technique for studying the complexity of problems are so-called reductions, algorithms that use a solver for one problem in order to solve another problem. We sketch one example relating the difficulty of two important computational problems.

6.1 DEFINITIONS

A *problem* Prob consists of a set of *instances* $\mathfrak{I}$, a set of *answers* $\mathfrak{A}$, a relation R and a probability space, either on $\mathfrak{I}$ or on $\mathfrak{I} \times \mathfrak{A}$. We allow instances that are unrelated to any answers.

A τ-*solver* $\mathcal{A}$ for a problem Prob is an algorithm that takes an instance as input and outputs an answer. The runtime bound τ measures both the time taken by the algorithm and the time required to sample the instance (which is typically easy). If the sampling algorithm just samples an instance, we define E be the event that after we sample an instance and run the solver on the instance, the output is related to the instance. If the sampling algorithm also samples an answer, we define the E to be that the output equals the sampled answer. We define the *advantage* of the solver to be $\mathbf{Adv}^{\mathsf{Prob}}(\mathcal{A}) = \Pr[E]$. This is sometimes called the *success rate*.

DOI: 10.1201/9781003149422-6

The discrete logarithm problem from Definition 2.2 is another natural mathematical problem. As we saw in Chapter 2, which group we use is extremely important.

Example 6.1. Let G be a cyclic group of order n with generator G.

The *discrete logarithm problem* $\mathsf{DLog}_{G,g}$ has instance set G, answer set $\{0, 1, \ldots, n-1\}$ and the relation is the function $\log_g$. Instances are sampled from the uniform distribution on G.

In cryptography the discrete logarithm problem has its origins in the Diffie-Hellman key exchange, which gives us the Diffie-Hellman problem from Definition 2.1.

Example 6.2. Let G be a cyclic group of order n with generator G.

The (Computational) *Diffie-Hellman problem* $\mathsf{CDH}_{G,g}$ has instance set $G \times G$, answer set G and the relation (x, y) R z if and only if $\log_g x \cdot \log_g y \equiv \log_g z$ (mod n). Instances are sampled from the uniform distribution on $G \times G$.

Reductions We already observed that if you can compute discrete logarithms, you can also solve the Diffie-Hellman problem. The way this happens is via a *reduction*, which is an *interactive* algorithm $\mathcal{PR}$ that solves a problem Prob_1 and may ask for answers to some other problem Prob_2. We say that Prob_1 *reduces* to Prob_2, if the reduction interacting with a good solver for Prob_2 becomes a good solver for Prob_1. (Here, "good" informally means that the solver does not use too much runtime and has a useful advantage.)

Example 6.3. Consider $\mathsf{DLog}_{G,g}$ and $\mathsf{CDH}_{G,g}$. The reduction $\mathcal{PR}$ takes as input a pair (x, y), asks for the discrete logarithm r of x and outputs the answer y^r.

If a solver $\mathcal{A}$ for $\mathsf{DLog}_{G,g}$ has advantage ϵ, then $\mathcal{PR}$ interacting with $\mathcal{A}$ has advantage ϵ.

If Prob_1 reduces to Prob_2, this implies that Prob_2 cannot be easy unless Prob_1 is easy. One way to achieve our goal, as stated at the start of the chapter, is to design cryptosystems so that solving the underlying problems reduce to breaking the cryptosystems.

In some sense, we have already used reductions extensively. The Pohlig-Hellman algorithm for computing discrete logarithms can be considered a reduction of the discrete logarithm problem in a group to the discrete logarithm problems in the prime-ordered subgroups.

If Prob_1 reduces to Prob_2 and Prob_2 reduces to Prob_1, we often say that the problems are *equivalent*. In particular, this implies that one problem cannot be easy unless the other problem is also easy.

Remark. The words "reduction", "reduces" and "equivalent" are not precisely defined, and we shall not do so. Instead, we shall use the words informally.

The discrete logarithm problem has one very interesting property, namely that if $\log_g(xg^b) = a$, then $\log_g x \equiv a - b \pmod{n}$. This allows us to reduce the discrete logarithm problem to itself: on input of x, sample $b \xleftarrow{r} \{0, 1, \ldots, n-1\}$, ask for the logarithm a of xg^b and output $a - b \bmod n$. This is called *random self-reduction*. Why should we care? The result is that no instance of the discrete logarithm problem $\mathsf{DLog}_{G,g}$ is harder than any other.

E | *Exercise 6.1.* Show that $\mathsf{CDH}_{G,g}$ is random self-reducible.

E | *Exercise 6.2.* Use Proposition 2.2 to show $\mathsf{DLog}_{G,g}$ and $\mathsf{DLog}_{G,h}$ equivalent.

E | *Exercise 6.3.* Show that $\mathsf{CDH}_{G,g}$ and $\mathsf{CDH}_{G,h}$ are more or less equivalent.

Decision Problems   If $\mathfrak{A} = \{0,1\}$ and the probability distribution on $\mathfrak{A}$ induced by $\mathcal{PS}$ is uniform, we have a *decision problem*. A solver for a decision problem is called a *distinguisher*. With the usual definition of advantage, having advantage 0 is essentially the same as having advantage 1, so we define the advantage as $\mathbf{Adv}^{\mathsf{Prob}}(\mathcal{A}) = 2|\Pr[E] - 1/2|$.

E | *Example 6.4.* Let G be a cyclic group of order n with generator g.
   The *Decision Diffie-Hellman* (DDH) problem $\mathsf{DDH}_{G,g}$ has instance set $G \times G \times G$ and answer set $\{0,1\}$. Any tuple will be related to 1, and $(x, y, z) \mathrel{R} 0$ if and only if $\log_g x \cdot \log_g y \equiv \log_g z \pmod{n}$.
   The sampling algorithm samples $b \xleftarrow{r} \{0,1\}$, $r, t \xleftarrow{r} \{0, 1, \ldots, n-1\}$ and $z_1 \xleftarrow{r} G$, computes $x \leftarrow g^r$, $y \leftarrow g^t$ and $z_0 \leftarrow g^{rt}$, and outputs (x, y, z_b, b).
   Tuples of the form (g^r, g^t, g^{rt}) are called *DDH tuples*.

E | *Exercise 6.4.* Show that the Diffie-Hellman problem $\mathsf{DDH}_{G,g}$ is random self-reducible.

Remark. We consider the generator g to be fixed. Sometimes it cannot be fixed and must be included explicitly in the DDH tuple. As we have seen, this does not change DLog or CDH, but may change DDH.

E | *Exercise 6.5.* Show that $\mathsf{DDH}_{G,g}$ reduces to $\mathsf{CDH}_{G,g}$.

Decision problems are often phrased in terms of two probability spaces $\mathcal{X}_0$ and $\mathcal{X}_1$ on the set of instances, where the sampling algorithm samples $b \xleftarrow{r} \{0,1\}$, $z_0 \xleftarrow{r} \mathcal{X}_0$ and $z_1 \xleftarrow{r} \mathcal{X}_1$, and outputs (z_b, b). We talk about a distinguisher for $\mathcal{X}_0$ and $\mathcal{X}_1$. When the sampling algorithms for the probability spaces are obvious, we often talk about the probability distributions instead.

E *Example* 6.5. The Decision Diffie-Hellman problem is to distinguish between the uniform distribution on DDH tuples and the uniform distribution on G^3.

Oracle Problems   Reductions allow us to define (at least morally) a partial order on problems. Sometimes, it is an open question if there exists a good reduction from one problem to another. And sometimes, being able to solve one problem does not seem to help with solving a different problem. To formalise this notion, we define *oracle* problems, where the solver is an interactive algorithm that has access to one or more oracles that performs particular computations. This is of special interest for the Diffie-Hellman problem.

E *Example* 6.6. Let G be a cyclic group of order n with generator g.

Let $f : G^3 \to \{0, 1\}$ be such that $f(x, y, z) = 0$ if and only if (x, y, z) is a DDH tuple. Let $f_x : G^3 \to \{0, 1\}$ be such that $f_x(x', y, z) = 0$ if and only if $x' = x$ and $f(x, y, z) = 0$.

The *Gap Diffie-Hellman* (GDH) problem is the Diffie-Hellman problem where the solver has access to an oracle that computes f.

The *Strong Diffie-Hellman* (SDH) problem $\mathsf{SDH}_{G,g}$ is the Diffie-Hellman problem where the solver with input (x, y) has access to an oracle that computes f_x.

We discussed the runtime requirements for solvers earlier. For oracle problems, the number of questions to the oracle should also be accounted for. A (τ, l)-*solver* for an oracle problem is a solver that asks at most l questions to its oracle. If there are more than one oracle, we will bound the number of questions to each oracle.

Remark. The Gap Diffie-Hellman problem in some sense asks if the Computational Diffie-Hellman problem reduces to the Decision Diffie-Hellman problem. This means that a GDH solver is quite useless unless we know a DDH solver. In fact, we cannot even test if a claimed GDH solver works. However, the GDH problem is still a well-defined oracle problem and a legitimate object of study.

Reductions involving oracle problems are more tricky, since we not only have to produce suitable input for the solver, we also have to answer its questions to its oracles. But in some cases it is easy.

E *Exercise* 6.6. Show that GDH reduces to SDH, and that SDH reduces to CDH.

Remark. It is sometimes convenient to think of reductions as algorithms that have access to an oracle that solves the underlying problem. But we must remember that a reduction may not have access to a perfect solver for the underlying problem, so we have to consider an oracle that may be *unreliable*, so that it may not always return the correct answer.

E *Exercise* 6.7. Use random self-reducibility for CDH to show that an unreliable CDH solver may be made reliable by running it many times and comparing the answers.

The Decision Diffie-Hellman problem is random self-reducible. In some cases, it is convenient to reformulate the problem into an interactive problem where one may ask an oracle for *samples*, to which the oracle responds with a tuple of group elements. These tuples are either all DDH tuples, or they are all random tuples. The problem then is no longer to decide how some concrete instance was sampled, but rather to decide which sampling algorithm is being used. Superficially, this is very different from how we defined problems, but it is essentially the same thing.

6.2 STATISTICAL DISTANCE

We shall sometimes need to consider sampling algorithms that are slightly different. The answer to the Decision Diffie-Hellman problem is not unique, and sometimes it would be convenient if it was.

E *Example* 6.7. The $\text{DDH}'_{G,g}$ is identical to the $\text{DDH}_{G,g}$ problem, except that the sampling algorithm samples $z_1 \xleftarrow{r} G \setminus \{z_0\}$, not $z_1 \xleftarrow{r} G$.

We need tools to reason about such small differences.

D **Definition 6.1.** Let $\mathcal{X}_0, \mathcal{X}_1$ be probability spaces on a finite set S. The *statistical distance* between $\mathcal{X}_0$ and $\mathcal{X}_1$ is

$$\Delta(\mathcal{X}_0, \mathcal{X}_1) = \frac{1}{2} \sum_{s \in S} |\Pr[\mathcal{X}_0 = s] - \Pr[\mathcal{X}_1 = s]|.$$

Two probability spaces are ϵ-*close* if the statistical distance is at most ϵ.

E *Exercise* 6.8. Compute the statistical distance between the instance sampling algorithm for $\text{DDH}_{G,g}$ and $\text{DDH}'_{G,g}$.

E *Exercise* 6.9. Fix a finite set S. Show that the statistical distance Δ is a metric on the set of probability spaces on S.

If we consider a fixed algorithm and give it input sampled from two different probability spaces, the statistical distance between the outputs will be bounded by the statistical distance of the inputs.

E *Exercise* 6.10. Let $\mathcal{A}$ be an algorithm that takes as input elements of a finite set S and outputs elements of a finite set T, and let $\mathcal{X}_0, \mathcal{X}_1$ be probability spaces on S. Define $\mathcal{Y}_i$ to be the probability space on T induced by sampling

$s \xleftarrow{r} \mathcal{X}_i$ and computing $u \leftarrow \mathcal{A}(s)$. Show that

$$\Delta(\mathcal{Y}_0, \mathcal{Y}_1) \leq \Delta(\mathcal{X}_0, \mathcal{X}_1).$$

Two probability spaces are *statistically indistinguishable* if their statistical distance is small. The origin of the phrase lies in the following result.

Proposition 6.1. *Let* $\mathcal{X}_0, \mathcal{X}_1$ *be probability spaces on a finite set* S, *and let* $\mathcal{A}$ *be a distinguisher for* $\mathcal{X}_0$ *and* $\mathcal{X}_1$. *Then*

$$\mathbf{Adv}^{\mathrm{dist}}_{\mathcal{X}_0, \mathcal{X}_1}(\mathcal{A}) \leq 2\Delta(\mathcal{X}_0, \mathcal{X}_1).$$

Exercise 6.11. Prove Proposition 6.1.

Example 6.8. Let α, β be integers such that $1 < \alpha < \beta$. Let $S = \{0, 1, \ldots, \alpha - 1\}$, and let $\mathcal{X}_0$ be the uniform probability space on S. Let $\mathcal{X}_1$ be the probability space on S induced by sampling $r \xleftarrow{r} \{0, 1, \ldots, \beta - 1\}$ and taking $r \bmod \alpha$ as the result. This distribution will be uniform if α divides β. Otherwise, write $\beta = k\alpha + \nu$ with $0 \leq \nu < \alpha$, and we get that

$$\Pr[\mathcal{X}_1 = s] = \begin{cases} (k+1)/\beta & s < \nu, \\ k/\beta & \text{otherwise.} \end{cases}$$

A useful general bound on the statistical distance is α/β.

Exercise 6.12. Let S be a finite set, and let $T \subseteq S$. Let $\mathcal{X}_0$ be the uniform distribution on S, and let $\mathcal{X}_1$ be the uniform distribution on T. Compute the statistical distance $\Delta(\mathcal{X}_0, \mathcal{X}_1)$ as a function of the number of elements in S and T. Show that the statistical difference is $(|S| - |T|)/|S|$.

Exercise 6.13. With α, β, S, k and ν as in the previous example, let $T = \{0, 1, \ldots, k\} \times \{0, 1, \ldots, \alpha - 1\}$. Let $\mathcal{X}_0$ be the uniform probability space on T. Let $\mathcal{X}_1$ be the probability space induced by the following sampling algorithm: Sample $r \xleftarrow{r} \{0, 1, \ldots, \beta - 1\}$ and output $(\lfloor r/\alpha \rfloor, r \bmod \alpha)$. Show that

$$\Delta(\mathcal{X}_0, \mathcal{X}_1) \leq \frac{\alpha}{\beta}.$$

Example 6.9. The $\mathsf{DLog}'_{G,g,\beta}$ is identical to $\mathsf{DLog}_{G,g}$, except that the sampling algorithm samples $r \xleftarrow{r} \{0, 1, \ldots, \beta - 1\}$ and computes the instance $x \leftarrow g^r$.

Exercise 6.14. Compute the statistical distances between the instance sampling algorithms for $\mathsf{DLog}_{G,g}$ and $\mathsf{DLog}'_{G,g,\beta}$. Are the problems equivalent when $\beta \ggg n$? What if $\beta \lll n$? What if $\beta \approx n$?

Remark. Even though we cannot say that the problems are equivalent when $\beta \ll n$, that does not imply that the modified discrete logarithm problem is easy. Unless β is sufficiently small, of course.

6.3 DIFFIE-HELLMAN

We have used the Diffie-Hellman problem and its variants as examples. One question about reductions remain, however. Does there exist a reduction from the discrete logarithm problem to the Diffie-Hellman problem? Suprisingly, the answer is yes in some cases.

Let G be a cyclic group of prime order p with generator g. Define $\boxplus$: $G \times G \to G$ by $x \boxplus y = xy$, and $\boxdot : G \times G \to G$ by $x \boxdot y = x^{\log_g y}$. Then $\mathbb{K} = (G, \boxplus, \boxdot)$ is a finite field with p elements, and hence isomorphic to $\mathbb{F}_p$.

E *Exercise 6.15.* Show that the isomorphism $\mathbb{K} \to \mathbb{F}_p$ is given by $\log_g$, while its inverse is the map $a + \langle p \rangle \mapsto g^a$.

When we think of an element $x \in G$ as an element in $\mathbb{K}$, we shall denote it by $[x]$. Suppose we have an integer k such that $[x] = [y]^k$ in $\mathbb{K}$. Then mapping the relation to $\mathbb{F}_p$, we get that

$$\log_g x \equiv (\log_g y)^k \pmod{p}.$$

If we happen to know $\log_g y$, we can compute $\log_g x$.

So far, this is not actually interesting. However, suppose we have access to an oracle that on input of (x, y) outputs $x^{\log_g y}$. This means that $\mathbb{K}$ is not just an imagined mathematical structure, but a mathematical structure we can actually compute in. We compute the addition as $[x] \boxplus [y] = [xy]$ and use the oracle to compute multiplications $[x] \boxdot [y]$.

So how would we find the relation $[x] = [y]^k$? Recall that $\mathbb{K}^*$ is a cyclic group, so in particular any of the generic discrete logarithm algorithms we studied in Section 2.2 could work. In particular, if $p - 1$ has no large prime divisors, Pohlig-Hellman in combination with Shanks' BSGS or Pollard's rho method would work well. To find the relation, all we need to do is find a generator $[y]$ for $\mathbb{K}^*$ and compute $\log_{[y]}[x]$.

E *Exercise 6.16.* Let G be a group of prime order p with generator g, where $p \approx 2^{256}$. Suppose also that $p - 1$ is square-free and has no prime divisor larger than 2^8.

Estimate how many multiplications in $\mathbb{K}^*$ are required to compute discrete logarithms in $\mathbb{K}^*$.

Suppose we have a 2^{60}-solver for $\mathsf{CDH}_{G,g}$ that always outputs the correct answer. Estimate the time cost of computing a discrete logarithm using the reduction from DLog to CDH.

E *Exercise* 6.17. Solvers may be unreliable, in that their advantage may be smaller than one. Based on the numbers from Exercise 6.16, estimate how reliable a CDH solver must be for the discrete logarithm computation to be correct reasonably often.

Assume you have a CDH solver with advantage $1/2^{16}$. Use the techniques from Exercise 6.7 to estimate how many times you must run the CDH solver to make it sufficiently reliable. Redo the time estimate for computing a single discrete logarithm using the reduction from DLog to CDH.

Generally, we do not want our primes to have too much special structure. Therefore, we cannot expect $p - 1$ to be smooth. However, just as Pollard's $p - 1$ method can be turned into a general factoring method by (essentially) replacing $\mathbb{F}_p$ with an elliptic curve E over $\mathbb{F}_p$, we can replace the computation in $\mathbb{K}^*$ by a computation in $E(\mathbb{K}^*)$ for some suitable elliptic curve E defined over $\mathbb{F}_p$. We leave the development of these ideas to the interested reader.

The conclusion is that there is quite often a somewhat costly reduction from the discrete logarithm problem to the Diffie-Hellman problem. This is not very useful, but it suggests that the Diffie-Hellman problem is not too easy to solve unless the discrete logarithm problem is easy to solve.

6.4   RSA

The factoring problem is a classical mathematical problem. It nicely illustrates that the factoring instances we see in ordinary mathematics are different from the ones we use in cryptography, where we choose only the most difficult instances.

E *Example* 6.10. The *factoring* problem has instance set $\mathbb{Z}$ and answer set $\mathbb{Z} \times \mathbb{Z}$. The relation is $n \; R \; (p, q)$ if and only if $n = pq$ and $1 < p, q < n$. We shall only be interested in the distribution on $\mathbb{Z}$ induced by the RSA cryptosystem key generation algorithm from Section 3.3.

The factoring problem can be used effectively in cryptography, but we shall use two related problems, namely the RSA problem and a variant called the *strong* RSA problem.

E *Example* 6.11. The *RSA problem* has instance set $\mathbb{Z} \times \mathbb{Z} \times \mathbb{Z}$ and answer set $\mathbb{Z}$. The relation is $(n, e, x) \; R \; y$ if and only if $y^e \equiv x \pmod{n}$.

We shall only be interested in the probability space on $\mathbb{Z}^3$ induced by the RSA cryptosystem key generation algorithm and the uniform distribution on the set $\{0, 1, \ldots, n - 1\}$.

E *Exercise* 6.18. Show that the RSA problem reduces to the factoring problem.

Example 6.12. The *strong RSA problem* has instance set $\mathbb{Z} \times \mathbb{Z}$ and answer set $\mathbb{Z}$. The relation is $(n, x)\ R\ (e, y)$ if and only if $e > 1$ and $y^e \equiv x \pmod{n}$.

We shall only be interested in the probability space on $\mathbb{Z}^3$ induced by the RSA cryptosystem key generation algorithm generating n and the uniform distribution on the set $\{0, 1, \ldots, n - 1\}$.

We may create variants of the strong RSA problem by adding restrictions to what exponents e are allowed as a solution.

The strong RSA problem morally reduces to the RSA problem. The issue is that we need to be able to choose an exponent that the RSA solver can deal with. For most RSA modulus generation strategies, this is not difficult, but in principle the exponent e could give the solver crucial information that is hard to guess. Restrictions on the exponent for the strong RSA problem may also make the reduction non-trivial.

When convenient, we may assume that x is sampled from the uniform distribution on $\{i \mid 0 < i < n \wedge \gcd(i, n) = 1\}$ for both RSA problems. If $x = 0$ the solution 0 works for any exponent. Or if $\gcd(x, n) \neq 1, n$ we can factor n and trivially find some eth root.

There is evidence that factoring does not reduce to the RSA problem. However, the best known method to solve the RSA problem is to factor the RSA modulus.

6.5 LATTICE PROBLEMS

We defined a number of problems related to lattices in Sections 3.5 and 3.6. While these are fundamentally important, we shall only consider the problem we will actually use in the design and analysis of cryptosystems.

Learning With Errors   We first introduced the learning with errors problem in Section 3.6.2, related to q-ary lattices. There is very strong evidence that this problem is hard, even against adversaries who can do large quantum computations. (Essentially, it is possible to turn a solver for learning with errors into a solver for fundamental problems related to lattices. We shall not explore these results.)

Example 6.13. Let q be a prime, χ_s a probability space on $\mathbb{F}_q^l$ and χ a probability space on $\mathbb{F}_q^n$. The *(search) learning with errors* (LWE) problem has instance set $\mathbb{F}_q^{n \times l} \times \mathbb{F}_q^n$ and answer set $\mathbb{F}_q^l$. The sampling algorithm samples $\mathbf{G} \xleftarrow{r} \mathbb{F}_q^{n \times l}$, $\mathbf{b} \xleftarrow{r} \chi_s$ and $\mathbf{f} \xleftarrow{r} \chi$, computes $\mathbf{y} \leftarrow \mathbf{Gb} + \mathbf{f}$ and outputs $(\mathbf{G}, \mathbf{y}, \mathbf{b})$.

There is a large number of parameters in the LWE problem, including the prime q, the two dimensions n and l, and the exact shape of the two probability distributions. Their relationship with the difficulty of solving LWE is quite complicated. Also, there are sometimes functional requirements on the parameters as well.

As usual, there is also a decision variant of the learning problem. We shall phrase this in terms of a sampling oracle problem.

Example 6.14. Let q be a prime, χ_s a probability space on $\mathbb{F}_q^l$ and χ a probability space on $\mathbb{F}_q$. The *decision learning with errors* problem $\mathsf{LWE}_{q,l,\chi_s,\chi}$ provides the solver with the following oracle: Initially, it samples $\beta \xleftarrow{r} \{0,1\}$ and $\mathbf{b} \xleftarrow{r} \chi_s$. When it is asked for a sample, the oracle samples $\mathbf{g} \xleftarrow{r} \mathbb{F}_q^l$, $f \xleftarrow{r} \chi$ and $y_1 \xleftarrow{r} \mathbb{F}_q$, computes $y_0 \leftarrow \mathbf{g} \cdot \mathbf{b} + f$, and returns $(\mathbf{g}, y_b)$. The answer is β.

Observe that this sampling formulation reveals the rows of $\mathbf{G}$ and the coordinates of $\mathbf{y}$ one by one.

Under some circumstances, the decision LWE problem reduces to the search LWE problem. In particular, this means that variants of LWE that make decision oracles available are often easy.

One way to change the LWE problem is to add structure to the matrix $\mathbf{G}$. This has the advantage that the matrix description requires less space and some computational operations are faster. In principle, the added structure may make the problems easier, but this seems not to be the case when the problems are properly tuned.

We add structure by working over a ring $\mathbb{F}_q[X]/\langle f(X)\rangle$ for some polynomial of degree n. This polynomial need not be irreducible. The effect is that the lattice is not just a q-ary lattice but also an ideal lattice. We only define the decision variant of the LWE problem.

Example 6.15. Let R be a finite ring, and let $\mathcal{X}_s$ and $\mathcal{X}_n$ be probability spaces on R. The *decision ring learning with errors* (Ring-LWE) problem $\mathsf{RLWE}_{R,\mathcal{X}_s,\mathcal{X}_n}$ provides the solver with the following oracle: Initially, it samples $b \xleftarrow{r} \{0,1\}$ and $a \xleftarrow{r} \mathcal{X}_s$. When it is asked for a sample, the oracle samples $g \xleftarrow{r} R$, $e \xleftarrow{r} \mathcal{X}_n$ and $z_1 \xleftarrow{r} R$, computes $z_0 \leftarrow ga + e$, and returns (g, z_b). The answer is b.

Short Integer Solution   The learning with errors problem is in some sense related to q-ary lattices and finding the closest point to a vector. We can use these lattices in a different way, by asking for short vectors in q-ary lattices.

Example 6.16. Let q be a prime and S_0 a subset of $\mathbb{F}_q^l$. The *short integer solution* (SIS) problem SIS_{q,n,l,S_0} has instance set $\mathbb{F}_q^{n \times l}$ and answer set $\mathbb{F}_q^l$, with the relation that $\mathbf{G}\ R\ \mathbf{b}$ if and only if $\mathbf{Gb} = 0$, $\mathbf{b} \neq 0$ and $\mathbf{b} \in S_0$. The instance distribution is the uniform distribution on the instance set.

Again, there are many parameters in the SIS problem, and their relationship with the difficulty of solving SIS is complicated. As we shall see, there are also a number of functional requirements.

We shall not be interested in decision problems related to SIS, nor shall we care about sampling formulations.

We want to add structure to the SIS problem, and as for Ring-LWE we shall look at ideal lattices.

Example 6.17. Let R_p be a finite ring and S_0 a subset of R_p^l. The *ring short integer solution* (Ring-SIS) problem $\mathsf{RSIS}_{R_p, l, S_0}$ has instance set R_p^l and answer set R_p^l, with the relation that $\mathbf{g} \, R \, \mathbf{a}$ if and only if $\mathbf{g} \cdot \mathbf{a} = 0$, $\mathbf{a} \neq 0$ and $\mathbf{a} \in S_0$. The instance distribution is the uniform distribution on the instance set.

CHAPTER 7

Symmetric Cryptography

This chapter will deal with the problem of how to define security for symmetric cryptography and the corresponding primitives such as block ciphers and key stream generators, and how to reason about the security of the various constructions we saw in Chapter 1.

Our general approach to analysis of cryptographic constructions is to build a *reduction* that uses an adversary against one construction to perform some other task. Then we analyse the reduction to determine its success rate at its task, relative to the success rate of the underlying adversary.

This seems like we just replace one unknown quantity with a different unknown quantity. But this is superficial. The main problem is while cryptographic constructions can be quite complicated, the context in which we have to analyse them will be even more complicated. Our goal is to reduce the analysis of complex cryptographic constructions in complex contexts to the analysis of simpler cryptographic constructions or even natural mathematical problems in simpler contexts. Eventually, we can use the techniques sketched in Chapters 1–5 to estimate how hard it is to attack the primitives, and through our reductions establish an estimate for how hard it is to attack the cryptographic construction.

Our primary task in this chapter is to analyse a wide variety of security goals, a topic that is suprisingly complex. Then we discuss constructions and underlying primitives. One important general theorem allows us to discuss confidentiality and integrity separately.

Our secondary task is to give an introductory discussion of channels. This is in some sense the original cryptographic topic: how can two parties protect a conversation? We shall return to this topic in depth later.

Hash functions were first introduced in Section 4.2. Strictly speaking, hash functions are not related to encryption and decryption, but we discuss the concept in this chapter because the design of common hash functions is in some sense related to the design of symmetric primitives. We also cover ideal models for hash functions and other primitives.

On language   The modern style in cryptography is to define exactly what an adversary against a scheme is and how to measure how good that adversary is. In a sense, therefore, we only define precisely a notion of *insecurity level*, and we prove relations among insecurity levels.

We could define security levels as the negation of the insecurity level and express theorems in terms of security levels. However, since the actual proof relates insecurity levels, this quickly becomes complicated and awkward.

Instead, we adopt the convention that a statement that something is secure is informal, which is somehow appropriate since almost all positive statements about security levels are conjectural. The relations among insecurity relations then turn into opposite informal statements about informal security claims. The reader should compare the informal statements in the early chapters with the corresponding theorems in the later chapters.

7.1 DEFINING SECURITY

Before we begin defining security, we shall first extend the functionality of symmetric cryptosystems. Ciphertexts are quite often not processed in splendid isolation, but appear in some context. It is often inconvenient to include this context in the message encrypted, but it would often be convenient to somehow involve the context in the encryption and decryption of ciphertext, in order to tie the message to its context. In fact, not connecting the decryption of ciphertexts to their proper context is a consistent source of failures in secure systems design. We encode the context as *associated data*.

Definition 7.1. A *symmetric cryptosystem with associated data* consists of a set $\mathfrak{K}$ of *keys*, a set $\mathfrak{P}$ of *plaintexts*, a set $\mathfrak{F}$ of *associated data*, a set $\mathfrak{C}$ of *ciphertexts*, and two algorithms:

- an *encryption algorithm* $\mathcal{E}$ that on input of a key, associated data and a plaintext outputs a ciphertext; and

- a *decryption algorithm* $\mathcal{D}$ that on input of a key, associated data and a ciphertext outputs either a plaintext or the special symbol $\perp$ (indicating an invalid ciphertext).

For any key k, associated data ad and any plaintext m, we have that

$$\mathcal{D}(k, ad, \mathcal{E}(k, ad, m)) = m.$$

Associated data is only relevant for integrity, not confidentiality against eavesdroppers. We shall design and analyse a number of cryptosystems that do not support associated data. Technically, we can treat $\mathfrak{F}$ as a singleton set, but to simplify the presentation we ignore associated data in these cases.

7.1.1 Confidentiality

The meaning of *confidentiality* should be that the adversary cannot learn anything about the contents of a ciphertext. To do this, we must define precisely what it means to learn something. We must also somehow define the context in which this happens. In the following, we shall define a number of variants of confidentiality: one of them will define what we intuitively mean by confidentiality, one will be practical to work with and one will be convenient when using symmetric cryptography in other constructions.

Before we begin, we list two principles for security definitions.

- We want to design cryptosystems that are usable by everyone. Therefore, we do not know exactly what security the cryptosystem's users will need. Hence, we need to design for as widely useful security as possible.

- We want as strong security as possible with a reasonable effort. Security is defined in terms of what an adversary should not be able to do, so we get stronger security by making the adversary's job easier. We must be careful not to make the adversary's job trivial.

A Rather Long Discussion   What follows is a rather long discussion explaining one line of reasoning that eventually leads to a useful definition of confidentiality for symmetric cryptosystems. We include this to give an idea of the thought processes behind modern security definitions.

A Game   One useful way to define security is in terms of a *game* between an *adversary* and an entity that in some sense plays the role of the honest users of the cryptosystem. The adversary sends *queries* to the entity and the entity responds. In some sense, we are conducting an experiment with the adversary as subject, so we call the entity playing the role of the honest users an *experiment*. This experiment is used to define security.

For a given definition of security, the adversary is variable, while the experiment is fixed. The word *experiment* may include the adversary, while *game* always refers to the conversation between the experiment and the adversary.

We want a game where the experiment creates a ciphertext and gives it to the adversary, who then learns something about the decryption of the ciphertext and tells the experiment what it has learned.

Adversary Goal   What should the adversary learn? One possibility is that the adversary should learn the entire decryption of the ciphertext. However, this is clearly too strong, since there are many real-world examples where partial plaintext recovery has been useful for an adversary. There are even situations where not even partial plaintext has been needed, but rather some other useful information about the plaintext has leaked.

Whatever information the adversary wants to learn can be encoded as an integer, so for any adversary there is a function $f : \mathfrak{P} \to \mathbb{Z}$ that defines what the adversary wants to learn.

Figuring out which of many possible values is correct is harder than figuring out which of a few possible values is correct. This suggests that the adversary would be better off if the function has a smaller range, such as $\{0, 1\}$.

We want the experiment to create a ciphertext and then the adversary must correctly answer a single yes/no question about the decryption.

Message Length   We should now discuss a fundamental obstacle: the ciphertext cannot be shorter than the message in the average. If we want to hide the message length, the ciphertext will have to be at least as long as the longest plaintext, which may be hugely expensive. In other words, it is expensive to hide the length of a message.

We can partially hide the length by adding random-length padding, or padding the message so that the length is divisible by some fixed number, but since this tends to increase message length, it will increase the cost of the cryptosystem. We do not know the needs of the cryptosystem's users, which suggests that it is not the job of the cryptosystem designer to obscure the message length. Instead, the cryptosystem user will most likely have more information about the cost-benefit tradeoff and can do more intelligent padding.

Therefore, we shall not try to hide the length of the message. That means that we shall only be interested in functions f where the answer cannot be inferred from the length of the message. This requirement is hard to formalise. However, this will turn out to be easy to solve, so we ignore it for now.

Message Choice   We want to define security in terms of a game between an experiment that creates a ciphertext and gives it to an adversary.

The experiment must encrypt a message, but how should it choose the message? One possibility is that it should simply sample a message from some fixed probability distribution. There are a number of issues with this.

First of all, this is probably not how the cryptosystem will be used. Messages sent by actual people are rarely random. While any interesting message will be somewhat unpredictable, an adversary will often have partial knowledge of the message, even before the ciphertext is created.

Also, the way a cryptosystem is used may affect its security, and historically this has often been the case. As discussed above, the designers of a cryptosystem cannot easily predict how the cryptosystem will be used. In fact, we would like to capture the possibility that a cryptosystem can be used in a way that actually helps the adversary.

Ideally we would actually like the adversary to choose the message, but this obviously does not work. Instead, we are going to let the adversary choose *how to choose* the message. Technically, the adversary gives the experiment an algorithm $\mathcal{X}$ for sampling from the set of plaintexts.

The adversary's sampling algorithm must be somewhat well-behaved. First, every plaintext sampled should have the same length, which neatly solves the problem of most cryptosystems leaking the length of the plaintext. The second requirement relates the function f to the sampling algorithm $\mathcal{X}$.

Adversary Goal, Again We would like our cryptosystem to be secure regardless of which question the adversary tries to answer. However, it is very hard to both quantify over all possible questions and prove anything. A better choice is simply to let the adversary decide the question.

In other words, the adversary must decide both how the message is chosen (by providing the sampling algorithm $\mathcal{X}$) and which question (the function f) to answer about the message. We should think about the sampling algorithm $\mathcal{X}$ as the adversary's prior knowledge about the message, and f as what the adversary wants to learn by observing the ciphertext.

Recall that we want the adversary's job to be as easy as possible, without making it trivial. In order for the adversary's job to be non-trivial, we must require that the sampling algorithm $\mathcal{X}$ samples a message m such that $f(m)$ can be both 0 and 1. At this point, we shall be strict and require that when m has been sampled by $\mathcal{X}$, the probability that $f(m) = 0$ should be 1/2.

Associated Data We have defined symmetric cryptosystems to included associated data. The idea was that the associated data encodes the context the ciphertext appears in. We shall see that this is useful for system designers.

Usually, the context is public, which means that there is no need to protect the confidentiality of the associated data. However, sometimes it would be useful to ensure the confidentiality of the associated data, since that would give system designers even more useful functionality. Treating associated data as secret cause a number of technical problems in the presentation for little value, so we let associated data be public.

Summary So Far We define security as a game between an experiment and an adversary. The adversary chooses associated data, a message distribution and a question to answer, the latter two given by a sampling algorithm $\mathcal{X}$ and a function $f : \mathfrak{P} \to \{0,1\}$. The experiment chooses a key k, samples a message $m \xleftarrow{r} \mathcal{X}$, encrypts the message $c \leftarrow \mathcal{E}(k, ad, m)$, and sends the *challenge ciphertext* c to the adversary. The adversary responds with a *guess* $b' \in \{0,1\}$. We say that the adversary *wins* the game if $b' = f(m)$.

One observation is that any adversary can win the game with probability 1/2, simply by guessing 0. This means that we are only interested in adversaries that win significantly more often than half the time. Adversaries that win significantly *less* often than half the time are also interesting, just as someone who reliably predicts the wrong outcome of a coin toss is interesting.

We shall now discuss an alternative way to define this. Sometimes computing the function f is hard, so did the adversary win? Such functions can be used, provided $\mathcal{X}$ samples both m and $f(m)$ at the same time. Then the experiment would sample $(m, b) \xleftarrow{r} \mathcal{X}$, and we can compare b' against b.

In this definition, the function f disappears. Instead, the experiment essentially samples the message from one of two distinct sample distributions, and the adversary's job is to guess which distribution the message was sampled from. The chosen distribution is given by the sampled bit b, which we call the *challenge bit*. In some cases, it is actually convenient to have the adversary present two distinct sampling algorithms.

This also allows us to relax the conditions on the game. The answer to the yes/no question need no longer be determined by the message itself. Obviously, if a single message could be sampled from both distributions, this makes the adversary's job harder, since it is now impossible for the adversary to be correct with probability 1. But relaxing our requirements in this way makes other tasks easier, which we shall see later.

More Challenge Ciphertexts In practice, an adversary may not be interested in a single message, but in multiple messages. In this case, we would want the message sampler to sample a sequence of messages, and the messages need not be independent. The number of messages sampled and their length should be independent of the challenge bit eventually output.

However, messages may not only depend on each other but also other factors, such as ciphertexts. At this point, modelling becomes more complicated. Instead of presenting a single multi-message sampling algorithm, the adversary should present a sequence of single-message sampling algorithms. Each algorithm should output a message and a *state*. The experiment runs the sampling algorithm, encrypts the sampled message and gives the ciphertext to the adversary. The state output by the sampling algorithm is hidden from the adversary. Instead, the next sampling algorithm gets the state as input before sampling. The final sampling algorithm should also output the challenge bit.

The above idea is fairly simple, but ensuring that the adversary cannot trivially win is somewhat more tricky. Obviously, the adversary must promise that the number of messages sampled and their length will be independent of the challenge bit. We would also want the challenge bit to be independent of the ciphertexts created by the experiment. Since the ciphertexts may be correlated to the messages and the messages are correlated to the ciphertexts, this is too strong. Instead, we must ensure that the challenge bit is sampled in a way that does not depend on the observed ciphertexts. The only reasonable option is for the challenge bit to be sampled before any ciphertext is created.

Remark. This does rule out some adversarial strategies (such as having the challenge bit be the parity of all the messages sampled), but it seems hard to distinguish these adversarial strategies from strategies that would allow an adversary to win trivially, for example by having the challenge bit depend on a ciphertext, possibly in a very complicated fashion. Also, such adversarial strategies seem not to be too useful.

The game then proceeds as follows. The experiment chooses a key k. The adversary sends l_c sampling algorithms $\mathcal{X}_1, \mathcal{X}_2, \ldots, \mathcal{X}_{l_c}$ to the experiment, one at a time. For sampling algorithm i, the experiment samples $(ad_i, m_i, \sigma_i, b_i) \xleftarrow{r} \mathcal{X}_i(\sigma_{i-1})$, with $\sigma_0 = \bot$, encrypts $c_i \leftarrow \mathcal{E}(k, ad_i, m_i)$ and sends c_i to the adversary. Eventually, the adversary outputs a guess b'. We say that the adversary wins if $b' = b_1$.

Chosen Plaintext Attack Once we allow the adversary access to multiple challenge ciphertexts, we can simplify the discussion a bit by allowing the adversary to get encryptions of messages chosen by the adversary, in addition to the challenge ciphertexts. The adversary could obviously submit a sampling algorithm that would always output a single message, but being able to encrypt chosen plaintexts simplifies some arguments. We stress that this is not an increase in adversarial capability. This is why the attack we have so far defined is called a *chosen plaintext attack (CPA)*.

It is sometimes interesting to distinguish between allowing the adversary to see encryptions of chosen plaintexts only before getting the first challenge ciphertext, or only after getting the final challenge ciphertext. We shall not discuss such variants of the chosen plaintext attack further.

We must sometimes consider *deterministic* encryption schemes (where the ciphertext is fully determined by the key and the message). In this case, we would need to prevent the adversary from getting the same message encrypted twice, since that would usually make the adversary's job too easy.

Chosen Ciphertext Attack Sometimes an adversary is able to convince one of the honest users to accept an adversarially created ciphertext as honest, and therefore try to decrypt it. If the decryption is successful, the adversary may learn something about the decryption.

If the adversary is allowed to get a challenge ciphertext decrypted and learn something about the decryption, the adversary may learn the challenge bit. This would make the adversary's job too easy. We solve this in the simplest possible manner, by refusing to decrypt any of the challenge ciphertexts.

The augmented game works as follows: The experiment keeps track of the challenge ciphertexts. The adversary may send ciphertexts to the experiment. If the ciphertext is not a challenge ciphertext, the experiment runs the decryption algorithm on the ciphertext with the experiment's key and sends the result to the adversary. This is called a *chosen ciphertext attack (CCA)*.

We must be careful with associated data in this game, but the idea is that getting a ciphertext successfully decrypted in the wrong context is bad and therefore non-trivial. Consequently, we reject challenge ciphertexts only when paired with the associated data they were encrypted with.

Just as for the chosen plaintext attack, there are variants of the chosen ciphertext attack where the adversary is not allowed to ask for decryptions before getting the first challenge ciphertext, or after getting the final challenge ciphertext. One may also consider non-adaptive variants of the attack, where the adversary has to submit all the ciphertexts together. We shall not discuss these and other variants of the chosen plaintext attack further.

The experiment for semantic security proceeds as follows:

1. Let $\sigma = b = \perp$, and let $C = \emptyset$.

2. Choose a key $k \xleftarrow{r} \mathfrak{K}$.

3. When the adversary sends a query $(ad, \mathcal{X})$ (*challenge*), do:
   (a) Sample $(m, \sigma', b'') \xleftarrow{r} \mathcal{X}(\sigma)$. $\sigma \leftarrow \sigma'$. If $b = \perp$, then $b \leftarrow b''$.
   (b) $c \leftarrow \mathcal{E}(k, ad, m)$. $C \leftarrow C \cup \{(ad, c)\}$.
   (c) Send c to the adversary.

4. When the adversary sends a query (ad, m) (*chosen plaintext*), do:
   (a) $c \leftarrow \mathcal{E}(k, ad, m)$. Send c to the adversary.

5. When the adversary sends a query (ad, c) (*chosen ciphertext*), do:
   (a) If $(ad, c) \in C$, send $\perp$ to the adversary.
   (b) Otherwise, $m \leftarrow \mathcal{D}(k, ad, c)$ and send m to the adversary.

Eventually, the adversary outputs $b' \in \{0, 1\}$. If $b = \perp$ at that time, the experiment samples $b \xleftarrow{r} \{0, 1\}$.

FIGURE 7.1 The experiment $\mathbf{Exp}_{\Sigma}^{\mathsf{sem}}(\mathcal{A})$ for the semantic security game for a symmetric cryptosystem $\Sigma = (\mathfrak{K}, \mathfrak{P}, \mathfrak{F}, \mathfrak{C}, \mathcal{E}, \mathcal{D})$ with adversary $\mathcal{A}$.

Semantic Security We can now define confidentiality for a symmetric cryptosystem is defined using a game between an experiment and an adversary.

Definition 7.2. An (τ, l_c, l_e, l_d)-*adversary against semantic security* for a symmetric cryptosystem Σ is an interactive algorithm $\mathcal{A}$ that interacts with the experiment in Figure 7.1 making at most l_c challenge queries (where the length of messages sampled from the adversary-specified message sampler algorithms is independent of the experiment's state), l_e chosen plaintext queries and l_d chosen ciphertext queries, and where the runtime of the adversary and the experiment is at most τ.

The *advantage* of this adversary is defined to be

$$\mathbf{Adv}_{\Sigma}^{\mathsf{sem}}(\mathcal{A}) = 2|\Pr[E] - 1/2|,$$

where E is the event that b' output by $\mathcal{A}$ equals the experiment's b.

The adversary is a *chosen plaintext adversary* if it makes no chosen cipher-text queries. Otherwise it is a *chosen ciphertext adversary*. We may denote the advantage in these cases by $\mathbf{Adv}_{\Sigma}^{\text{sem-cpa}}(\mathcal{A})$ and $\mathbf{Adv}_{\Sigma}^{\text{sem-cca}}(\mathcal{A})$.

Remark. We use the word *advantage* to indicate that our adversary has an advantage over an adversary that just guesses. Some authors prefer to distinguish advantage from the success probability, using the latter when we do not measure our adversary relative to something else. We shall not do so.

Remark. Real-world adversaries are not interested in attacking cryptography. They want to attack a system that uses cryptography. One way to attack such a complicated system is to attack the cryptography. How do we distinguish attacks on the system through the cryptography from attacks that are incidental to the cryptography?

The idea is that we model the system and the adversary. When modelling, we identify the use of the cryptosystem, which we model as interactions with the experiment. The remainder of the system model is then merged with the adversary to create a single adversary against the cryptosystem.

If this adversary against the cryptosystem has a significant advantage, this is a cryptographic attack. Otherwise, the adversary against the system did not attack the cryptosystem. We shall see examples of larger systems later.

Remark. Another way to model the game is to give the adversary access to certain *oracles*, instead of letting it talk to an experiment. In our case, the adversary would get access to an encryption oracle, a decryption oracle and a challenge oracle. The experiment setup would then be in charge of setting up keys and common values. The different oracles will have to communicate amongst themselves. The adversary would send messages to the oracles, and the oracles would respond according to their programming.

Defining games in this way is essentially equivalent to what we have done, and there is no real difference in expressive power. Which style to choose is largely a matter of taste and temperament.

It is usually convenient to define a bit more machinery before proving schemes secure, which we will do in Section 7.2.1 and 7.2.2. However, there is one simple scheme that we can prove secure without introducing more machinery.

E │ *Exercise 7.1.* Show that for any $(\tau, 1, 0, 0)$-adversary $\mathcal{A}$ against the one-time pad from Section 1.2.8, $\mathbf{Adv}_{\text{OTP}}^{\text{sem-cpa}}(\mathcal{A}) = 0$, regardless of τ.

E │ *Exercise 7.2.* In this exercise, we shall make the adversary's job harder. It is sometimes interesting to consider the security of encrypting random messages. Partial leakage of the message is often unproblematic in these applications,

The left-or-right experiment is identical to the experiment from Figure 7.1, except that two steps are changed as follows:

1. $b \xleftarrow{r} \{0,1\}$. $C \leftarrow \emptyset$.
3. When the adversary sends a query (ad, m_0, m_1) (*challenge*), do:
   (a) $c \leftarrow \mathcal{E}(k, ad, m_b)$.
   (b) $C \leftarrow C \cup \{(ad, c)\}$.
   (c) Send c to the adversary.

FIGURE 7.2 The experiment $\mathbf{Exp}_{\Sigma}^{\text{ind}}(\mathcal{A})$ for the left-or-right security game for a symmetric cryptosystem $\Sigma = (\mathfrak{K}, \mathfrak{P}, \mathfrak{F}, \mathfrak{C}, \mathcal{E}, \mathcal{D})$ with an adversary $\mathcal{A}$, based on the experiment from Figure 7.1.

so the adversary's goal is to recover the entire message. Define a notion of *one-way* security for some subset of plaintexts that captures this idea.

[E] *Exercise 7.3.* This exercise continues the discussion on deterministic encryption. Let Σ be a deterministic symmetric cryptosystem. Give a $(\tau, 1, 1, 0)$-adversary against semantic security with advantage 1 and trivial bound τ.

Indistinguishability, or Left-or-Right Security The definition of semantic security is somewhat hard to work with. We shall now define a notion that is easier to work with, but which does not immediately appear to be the "correct" security notion for a symmetric cryptosystem.

The idea in this game is that the adversary's challenge queries will be pairs of equal-length messages. The experiment will either always encrypt the left message, or always encrypt the right message. Otherwise, the experiment is identical to the semantic security game experiment.

[D] **Definition 7.3.** An (τ, l_c, l_e, l_d)-*adversary against left-or-right security* (or *indistinguishability*) for a symmetric cryptosystem Σ is an interactive algorithm $\mathcal{A}$ that interacts with the experiment in Figure 7.2 making at most l_c challenge queries (pairs of equal-length messages), l_e chosen plaintext queries and l_d chosen ciphertext queries, and where the runtime of the adversary and the experiment is at most τ.

The *advantage* of this adversary is defined to be

$$\mathbf{Adv}_{\Sigma}^{\text{ind}}(\mathcal{A}) = 2|\Pr[E] - 1/2|,$$

where E is the event that b' output by $\mathcal{A}$ equals the experiment's b.

The adversary is a *chosen plaintext adversary* if it makes no chosen ciphertext queries. Otherwise it is a *chosen ciphertext adversary*. We may denote the advantage in these cases by $\mathbf{Adv}_{\Sigma}^{\text{ind-cpa}}(\mathcal{A})$ and $\mathbf{Adv}_{\Sigma}^{\text{ind-cca}}(\mathcal{A})$.

Remark. Some authors use the word *indistinguishability* only when the number of challenge queries is 1, and otherwise use the name *left-or-right security*. We shall consider indistinguishability and left-or-right as synonyms.

We begin by repeating Exercise 7.1. Compare the difficulty of the two proofs.

E | *Exercise 7.4.* Show that for any $(\tau, 1, 0, 0)$-adversary $\mathcal{A}$ against the one-time pad from Section 1.2.8, $\mathbf{Adv}_{\text{OTP}}^{\text{ind-cpa}}(\mathcal{A}) = 1/2$, regardless of τ.

Now we shall continue with the relationship between semantic security and indistinguishability, proving that any indistinguishability adversary can be turned into an equally good adversary against semantic security. Informally, semantic security implies indistinguishability. Note that we want something that is easier to work with than semantic security, but still provides the same security. So this is not the implication we want! What we want will come later.

The proof idea is that a left-or-right challenge query consisting of a pair of equal-length messages can be turned into sampling algorithm that either samples the left or the right message. With a bit of care, we can make sure that the sequence of sampling algorithms we create always encrypt left or always encrypt right. With this addition, the semantic security experiment will create a ciphertext with the exact same distribution as the indistinguishability experiment would have created in response to the original query.

The technical implementation of this idea takes the form of a *reduction*, in the sense that we shall reduce the problem of attacking semantic security to the problem of attacking indistinguishability. Between the adversary against indistinguishability and the semantic security experiment, we insert an intermediate whose job is to reinterpret $\mathcal{A}$'s queries and the experiment's responses. The intermediate makes the semantic security experiment $\mathbf{Exp}_{\Sigma}^{\text{sem}}$ look like the indistinguishability experiment $\mathbf{Exp}_{\Sigma}^{\text{ind}}$ to $\mathcal{A}$, while it makes $\mathcal{A}$ look like a semantic security adversary to the experiment $\mathbf{Exp}_{\Sigma}^{\text{sem}}$.

This intermediate is in some sense a reduction. However, this is a very simple form of reduction, and since the intermediate's job is to simulate other types of players, we shall call it a *simulator*.

T | **Proposition 7.1.** *Let $\mathcal{A}$ be a (τ, l_c, l_e, l_d)-adversary against indistinguishability for the symmetric cryptosystem Σ. Then $\mathcal{B}$ given in Figures 7.3 and 7.4 is a (τ', l_c, l_e, l_d)-adversary against semantic security for Σ, where τ' is essentially equal to τ, and*
$$\mathbf{Adv}_{\Sigma}^{\text{sem}}(\mathcal{B}) = \mathbf{Adv}_{\Sigma}^{\text{ind}}(\mathcal{A}).$$

Proof. We first prove that the combination of the semantic security experiment $\mathbf{Exp}_{\Sigma}^{\text{sem}}$ and the simulator $\mathbf{Sim}$ from Figure 7.4 behaves exactly the same

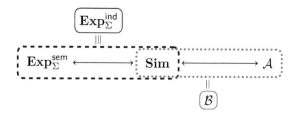

FIGURE 7.3 The idea used in the proof of Proposition 7.1. The part inside the dashed line behaves like the indistinguishability experiment, which means that the adversary should behave as expected. The part inside the dotted line becomes an adversary against semantic security.

as the indistinguishability experiment. Then we prove that the correct guess is the same in both cases. This will prove that the advantages are identical.

Since **Sim** just forwards chosen plaintext and ciphertext queries, these will be processed in the same way by $\mathbf{Exp}_\Sigma^{\mathsf{sem}}$ and **Sim** as by $\mathbf{Exp}_\Sigma^{\mathsf{ind}}$. By inspection, we see that **Sim** and $\mathbf{Exp}_\Sigma^{\mathsf{sem}}$ will always encrypt m_b in a challenge query, just as the indistinguishability experiment will in the challenge query. Therefore, challenge queries will be processed the same, given b.

The indistinguishability experiment samples the challenge bit at the start. The semantic security experiment samples the challenge bit when processing the first challenge query or at the end of the game. But until the first challenge query, nothing depends on the challenge bit, and in both cases the challenge bit is sampled from the uniform distribution on $\{0, 1\}$.

The adversary $\mathcal{B}$ runs a copy of $\mathcal{A}$ and the following simulator **Sim**:

- When $\mathcal{A}$ sends a challenge query (ad, m_0, m_1), **Sim** creates $\mathcal{X}$ which:
  - on input of $\bot$ samples $b \xleftarrow{r} \{0, 1\}$ and outputs (m_b, b, b); and
  - on input of $b \in \{0, 1\}$ outputs (m_b, b, b).

  The simulator sends $(ad, \mathcal{X})$ to the semantic security experiment.

  When the experiment responds with c, **Sim** sends c to $\mathcal{A}$.

- **Sim** forwards other queries to the experiment and the responses to $\mathcal{A}$.

When $\mathcal{A}$ outputs b', $\mathcal{B}$ outputs b'.

FIGURE 7.4 The simulator and adversary used in the proof of Proposition 7.1.

By inspection, we see that $\mathcal{A}$'s guess is correct if and only if $\mathcal{B}$'s guess is correct. Therefore, the advantage of $\mathcal{B}$ interacting with $\mathbf{Exp}_\Sigma^{\text{sem}}$ equals the advantage of $\mathcal{A}$ interacting with $\mathbf{Exp}_\Sigma^{\text{ind}}$.

By construction, the two adversaries make the same number of challenge, chosen plaintext and chosen ciphertext queries. With regard to runtime, there is a small amount of overhead for each query, since something has to be forwarded by **Sim**, but for any non-trivial adversary, this overhead can be safely ignored. It follows that τ' is essentially equal to τ. □

The statement on runtime in the above proposition, *essentially equal*, is remarkably imprecise. It can be made precise if we choose an underlying computational model. With some optimisations, the overhead will probably be negligible. However, the detail required will be considerable.

Alternatively, we could choose an approximate solution, saying that the cost is linear in the number of queries, typically denoted by $O(l_c + l_e + l_d)$. But for any reasonable underlying computational model, this cost will not just be linear in the number of queries but also in the length of the individual messages, so we would have to include the total message length in the above approximation. So the cost is roughly linear in the number of queries plus the total volume of data encrypted or decrypted.

In general, we shall have to assume that the adversary will be able to devote far more resources to attacking a system than the users of the system will want to devote to running the system. It follows that the runtime bound will be significantly larger than the total volume of data encrypted and decrypted. It then follows that difference between the two runtimes will be relatively small.

There will be obscure cases where this difference cannot be ignored, for instance if we for some reason only consider adversaries with brief runtimes.

The conclusion is that as a compromise between precision and economy, we write *essentially equal* to indicate that these two terms are not equal, but that the difference is relatively small and can usually be ignored. We do have to be careful in each proof that we understand what the difference is, that it really is relatively small and that it can usually be ignored.

Remark. Another important thing about the runtime mentioned in the above statement is that it is the *global* runtime, that is the runtime of both the adversary and the experiment, not just the runtime of the adversary. As discussed in the previous remark, the experiment usually models the honest users whose total runtime will be much smaller than the adversary's runtime, and as such this could be ignored. But we will later run into proof-technical issues where a *local* runtime bound would cause significant accounting issues. Using a global runtime is easier.

E *Exercise* 7.5. This exercise follows up on the theory of deterministic encryption and the result from Exercise 7.3. Define a variant of indistinguishability that could hold for deterministic encryption.

The real-or-random experiment is identical to the experiment from Figure 7.1, except that two steps are changed as follows:

1. $b \xleftarrow{r} \{0,1\}$. $C \leftarrow \emptyset$.

3. When the adversary sends a query (ad, m_0) (*challenge*), do:
   (a) Sample $m_1 \xleftarrow{r} \{m \in \mathfrak{P} \mid m$ has the same length as $m_0\}$.
   (b) $c \leftarrow \mathcal{E}(k, ad, m_b)$.
   (c) $C \leftarrow C \cup \{(ad, c)\}$.
   (d) Send c to the adversary.

FIGURE 7.5 The experiment for the real-or-random security game for a symmetric cryptosystem $\Sigma = (\mathfrak{K}, \mathfrak{P}, \mathfrak{F}, \mathfrak{C}, \mathcal{E}, \mathcal{D})$, based on the experiment from Figure 7.1.

Real-or-Random Security   We shall now define a notion that perhaps seems even farther from the "correct" security notion for a symmetric cryptosystem, but it will be much easier to work with later on.

The idea in this game is that the adversary's challenge queries will be single messages. The experiment will either always encrypt the adversary's chosen message, or always encrypt a randomly chosen message of the same length. Other than the changes to the accounting and the challenge query processing, the experiment is identical to the semantic security experiment and the indistinguishability experiment.

Contradicting our initial remark, this is some sense reasonable notion of security on its own. If the adversary cannot decide if a ciphertext contains a particular message or random nonsense, it can hardly expect to glean any useful information from a ciphertext. This approach will be very important.

Definition 7.4. An (τ, l_c, l_e, l_d)-*adversary against real-or-random security* for a symmetric cryptosystem Σ is an interactive algorithm $\mathcal{A}$ that interacts with the experiment in Figure 7.5 making at most l_c challenge queries, l_e chosen plaintext queries and l_d chosen ciphertext queries, and where the runtime of the adversary and the experiment is at most τ.

The *advantage* of this adversary is defined to be

$$\mathbf{Adv}_\Sigma^{\text{ror}}(\mathcal{A}) = 2|\Pr[E] - 1/2|,$$

where E is the event that b' output by $\mathcal{A}$ equals the experiment's b.

The adversary is a *chosen plaintext adversary* if it makes no chosen ciphertext queries. Otherwise it is a *chosen ciphertext adversary*. We may denote the advantage in these cases by $\mathbf{Adv}_\Sigma^{\text{ror-cpa}}(\mathcal{A})$ and $\mathbf{Adv}_\Sigma^{\text{ror-cca}}(\mathcal{A})$.

Remark. Note that an adversary cannot get advantage 1 against real-ord-random, since it could happen that the random message is the same as the adversarially chosen message. The probability is usually not large, but it does explain why some bounds we get later are strictly smaller than 1.

We first prove a result on the relationship between indistinguishability and real-or-random security, which is that any adversary against real-or-random security can be turned into an equally good adversary against indistinguishability. This means that if we have indistinguishability, we must also have real-or-random security.

The proof idea is much the same as the proof of Proposition 7.1, that a real-or-random challenge query consisting of a single message can be turned into a left-or-right challenge query simply by sampling a random message of the correct length. With this addition to the query, the left-or-right experiment will create a ciphertext with the exact same distribution as the real-or-random experiment would have created in response to the original query.

T **Proposition 7.2.** *Let $\mathcal{A}$ be a (τ, l_c, l_e, l_d)-adversary against real-or-random security for the symmetric cryptosystem Σ. Then there exists a (τ', l_c, l_e, l_d)-adversary $\mathcal{B}$ against indistinguishability for Σ, where τ' is essentially equal to τ, and*

$$\mathbf{Adv}_{\Sigma}^{\mathrm{ind}}(\mathcal{B}) = \mathbf{Adv}_{\Sigma}^{\mathrm{ror}}(\mathcal{A}).$$

E *Exercise 7.6.* Prove Proposition 7.2.

The next result proves that semantic security, indistinguishability and real-or-random security are essentially equivalent security notions. This means that we can choose the security notion that is most convenient when working with symmetric cryptosystems, and this is good.

This result nicely illustrates why real-or-random security is useful when working with symmetric cryptosystems. Replacing encryptions of meaningful messages with encryptions of random messages is a very powerful technique.

It is convenient first to prove a small lemma related to advantages. The power of this lemma is that it relates the advantage of an adversary with the difference in the adversary's behaviour in two different settings (as measured by the probability of the adversary's output being 1). This means that we can place an adversary in two carefully related settings and immediately say something about the possible difference in behaviour.

T **Lemma 7.3.** *Let $\mathcal{A}$ and $\mathbf{Exp}$ be interactive algorithms. Suppose $\mathbf{Exp}$ samples a bit $b \xleftarrow{r} \{0,1\}$, $\mathbf{Exp}$ and $\mathcal{A}$ interact, and then $\mathcal{A}$ outputs a bit $b' \in \{0,1\}$. Denote by $\mathbf{Exp}^i$ a variant of $\mathbf{Exp}$ that always chooses $b = i$. Let E be the event that $b = b'$ after the interaction, and let F_i be the event that $b' = 1$ after an interaction between $\mathbf{Exp}^i$ and $\mathcal{A}$. Then*

$$2|\Pr[E] - 1/2| = |\Pr[F_0] - \Pr[F_1]|.$$

The adversary $\mathcal{B}$ runs a copy of $\mathcal{A}$ and the following simulator **Sim**:

- At the start, **Sim** lets $\sigma = \bar{b} = \bot$.

- When $\mathcal{A}$ sends the challenge query $(ad, \mathcal{X})$, **Sim** samples $(m_0, \sigma', b'') \xleftarrow{r} \mathcal{X}(\sigma)$. If $\bar{b} = \bot$, $\mathcal{B}$ sets $\bar{b} = b''$. Then **Sim** sends the challenge query (ad, m_0) to the real-or-random experiment.

  When the experiment responds with c, **Sim** sends c to $\mathcal{A}$.

- **Sim** forwards other queries to the experiment and the responses to $\mathcal{A}$.

If $\mathcal{A}$ outputs b', $\mathcal{B}$ outputs $b' = 1$ if $\bar{b}' = \bar{b}$, and otherwise outputs $b' = 0$. If $\mathcal{A}$ exceeds its bounds, then $\mathcal{B}$ samples $b' \xleftarrow{r} \{0, 1\}$ and outputs b'.

FIGURE 7.6 The simulator and adversary used in the proof of Proposition 7.4.

Proof. Let E_i be the event that $b = b'$ after $\mathbf{Exp}^i$ and $\mathcal{A}$ interacts. We compute

$$\Pr[E] = \Pr[E_0]\Pr[b = 0] + \Pr[E_1]\Pr[b = 1] = \frac{1}{2}\left((1 - \Pr[F_0]) + \Pr[F_1]\right)$$

$$= \frac{1}{2} - \frac{1}{2}(\Pr[F_1] - \Pr[F_0]),$$

from which the claim immediately follows. □

Proposition 7.4. *Let $\mathcal{A}$ be a (τ, l_c, l_e, l_d)-adversary against semantic security for the symmetric cryptosystem Σ. Then $\mathcal{B}$ given in Figure 7.6 is a (τ', l_c, l_e, l_d)-adversary against real-or-random security for Σ, where τ' is essentially equal to τ, and*

$$\mathbf{Adv}_{\Sigma}^{\mathrm{ror}}(\mathcal{B}) = \frac{1}{2}\mathbf{Adv}_{\Sigma}^{\mathrm{sem}}(\mathcal{A}).$$

Proof. Since $\mathcal{A}$ can run for time at most τ, the maximal runtime of $\mathcal{B}$ and the experiment is essentially the same as τ for any non-trivial adversary.

We need to compute the probability that $\Pr[b = b']$ when $\mathcal{B}$ interacts with the real-or-random experiment, and Lemma 7.3 gives us

$$2|\Pr[b = b'] - 1/2| = |\Pr[b' = 1 \mid b = 0] - \Pr[b' = 1 \mid b = 1]|.$$

If $b = 0$ (which means that the real-or-random experiment will always encrypt the given challenge message), then since **Sim** chooses the challenge message exactly the same way that the semantic security experiment chooses the message to encrypt, $\mathcal{A}$ does not exceed its bounds, and we have that

$$2|\Pr[\bar{b}' = \bar{b} \mid b = 0] - 1/2| = \mathbf{Adv}_{\Sigma}^{\mathrm{sem}}(\mathcal{A}).$$

Also, $\Pr[\bar{b}' = \bar{b} \mid b = 0] = \Pr[b' = 1 \mid b = 0]$.

On the other hand, if $b = 1$ (which means that the real-or-random experiment will always encrypt messages that are independent of the challenge messages), then since $\bar{b}$ is only involved in the choice of challenge messages, $\bar{b}$ is independent of anything observed by $\mathcal{A}$. It follows that

$$\Pr[\bar{b}' = \bar{b} \mid b = 1] = 1/2.$$

In other words, since $\mathcal{A}$ has no information about the bit $\bar{b}$ chosen by $\mathcal{B}$ when $b = 1$, we get that $\Pr[\bar{b}' = \bar{b} \mid b = 1] = \Pr[b' = 1 \mid b = 1] = 1/2$. (This also holds if $\mathcal{A}$ exceeds its bounds.)

Combining the above, we get that

$$\mathbf{Adv}_{\Sigma}^{\text{ror}}(\mathcal{B}) = |\Pr[b' = 1 \mid b = 0] - \Pr[b' = 1 \mid b = 1]|$$

$$= \frac{1}{2} \cdot 2|\Pr[\bar{b}' = \bar{b} \mid b = 0] - 1/2| = \frac{1}{2}\mathbf{Adv}_{\Sigma}^{\text{sem}}(\mathcal{A}),$$

which proves the claim $\qquad\qquad\qquad\qquad\qquad\qquad\qquad\qquad\qquad\square$

Remark. Note again that the $1/2$ is somehow essential, since in the real-or-random experiment, the random case might encrypt the same message that the real case would encrypt. It follows that the real-or-random adversary cannot achieve advantage equal to 1.

Remark. This proof shows that we need to be careful about runtime when constructing adversaries. The reason is that the algorithms we are dealing with here are only guaranteed to terminate within the given time limit if they receive the expected input. If we give an algorithm input that it does not expect (such as encryptions of uniformly random messages when encryptions of messages sampled from specific distribution were expected), we cannot expect the algorithm to keep its runtime promise, or stay within other bounds. If we had not added the bounds check, the adversary could have failed to terminate when $b = 1$, in which case we would not have had an adversary at all.

Propositions 7.1, 7.2 and 7.4 mean that the three security notions defined are essentially equivalent. The factor $1/2$ is mostly unimportant.

7.1.2 A Single Challenge Suffices – Maybe

In the long discussion in Section 7.1.1, we began by defining security in terms of a single challenge query, before allowing multiple challenge queries. We shall now show that in a certain sense, it is sufficient to prove security for a single challenge query. But we shall immediately state a warning, namely that proving security for a single challenge query gives a weaker overall result. All other things being equal, a proof for multiple challenge queries is better than a proof for a single challenge query.

The adversary $\mathcal{B}$ runs a copy of $\mathcal{A}$ and the following simulator **Sim**:

- At the start, **Sim** samples $j \xleftarrow{r} \{1, 2, \ldots, l_c\}$.
- When $\mathcal{A}$ makes its ith challenge query (ad, m_0, m_1), then:
  - if $i < j$, **Sim** sends (ad, m_1) as a chosen plaintext query;
  - if $i > j$, **Sim** sends (ad, m_0) as a chosen plaintext query; and
  - if $i = j$, **Sim** sends (ad, m_0, m_1) as a challenge query.
  - When the experiment responds with a ciphertext c, **Sim** records (ad, c) and forwards the ciphertext to $\mathcal{A}$.
- If $\mathcal{A}$ makes a chosen ciphertext query (ad, c) that has been recorded, **Sim** immediately replies with $\bot$. Otherwise, **Sim** forwards the query to its experiment and forwards the response to $\mathcal{A}$.
- **Sim** forwards other queries to the experiment and the responses to $\mathcal{A}$.

If $\mathcal{A}$ outputs b', $\mathcal{B}$ outputs the same value b'. If $\mathcal{A}$ exceeds its bounds, then $\mathcal{B}$ samples $b' \xleftarrow{r} \{0, 1\}$ and outputs b'.

FIGURE 7.7 A simulator and a $(\tau', 1, l_e + l_c - 1, l_d)$-adversary $\mathcal{B}$ against indistinguishability based on a (τ, l_c, l_e, l_d)-adversary $\mathcal{B}$ against indistinguishability.

This theorem also nicely illustrates an important proof technique, namely the *hybrid argument*. The idea is that we construct a sequence of very similar games, and then connect the difference between consecutive games to one thing and difference between the end-points of the sequence to another thing.

Proposition 7.5. *Let $\mathcal{A}$ be a (τ, l_c, l_e, l_d)-adversary against indistinguishability for Σ. Then $\mathcal{B}$ given in Figure 7.7 is a $(\tau', 1, l_e + l_c - 1, l_d)$-adversary against indistinguishability for Σ, where τ' is essentially τ, and*

$$\mathbf{Adv}_\Sigma^{\mathrm{ind}}(\mathcal{A}) \leq l_c \mathbf{Adv}_\Sigma^{\mathrm{ind}}(\mathcal{B}).$$

Proof. Since $\mathcal{A}$ can run for time at most τ, the maximal runtime of $\mathcal{B}$ and the game is essentially the same as τ for any non-trivial adversary.

Let F_{j_0} be the event that **Sim** samples the value j_0 for j. Let E_{j_0, b_0} be the event that the indistinguishability experiment samples b_0 for b, **Sim** samples the value j_0 for j and then $\mathcal{B}$ outputs $b' = 1$.

First, observe that if $b = 0$ and **Sim** samples the value 1 for j, then $\mathcal{A}$ will get an encryption of the left-hand message for all its challenge queries. Likewise, observe that if $b = 1$ and **Sim** samples the value l_c for j, then $\mathcal{A}$ will get an encryption of the right-hand message for all its challenge queries. In these cases, we know that the runtime of the indistinguishability experiment and the adversary is at most τ and that $\mathcal{A}$ will never exceed its bounds. This means that $\mathcal{B}$ never stops with a random output. It follows that

$$\mathbf{Adv}_\Sigma^{\mathsf{ind}}(\mathcal{A}) = |\Pr[E_{1,0}] - \Pr[E_{l_c,1}]|. \tag{7.1}$$

If $b = 1$ and $\mathbf{Sim}$ samples the value j_0 for j, then $\mathcal{A}$ will get encryptions of the right-hand message for the first j_0 challenge queries, and encryptions of the left-hand message for the remaining challenge queries. However, if $b = 0$ and $\mathbf{Sim}$ samples the value $j_0 + 1$ for j, then again $\mathcal{A}$ will get encryptions of the right-hand message for the first j_0 challenge queries, and encryptions of the left-hand message for the remaining challenge queries. Since there are no other possible differences between these two cases, we have that

$$\Pr[E_{j_0,1}] = \Pr[E_{j_0+1,0}].$$

Finally, observe that

$$\mathbf{Adv}_\Sigma^{\mathsf{ind}}(\mathcal{B}) = \frac{1}{l_c} \sum_{j=1}^{l_c} |\Pr[E_{j,0}] - \Pr[E_{j,1}]|.$$

Now we use (7.1) and a telescoping sum to compute

$$\mathbf{Adv}_\Sigma^{\mathsf{ind}}(\mathcal{A}) = |\Pr[E_{1,0}] - \Pr[E_{1,1}] + \Pr[E_{2,0}] - \cdots - \Pr[E_{l_c,1}]|$$

$$\leq \sum_{j=1}^{l_c} |\Pr[E_{j,0}] - \Pr[E_{j,1}]| = l_c \mathbf{Adv}_\Sigma^{\mathsf{ind}}(\mathcal{B}). \qquad \square$$

Remark. The advantage loss factor l_c is probably essential for a general theorem. That has no implication for what can be proven with multiple challenge queries, and we should strive to prove statements without this loss.

The hybrid proof uses a single adversary. We could have used l_c single-challenge adversaries, parameterised by j. We would then prove that their average advantage must be at least $1/l_c$ times the advantage of the multi-challenge adversary. It would follow that at least one of the adversaries must have advantage greater than or equal to the average. The choice of proof structure is to some extent a question of style and preference, but sometimes one approach is easier to deal with than the other.

The proof for real-or-random security is essentially the same as for indistinguishability. But doing the proof for semantic security directly would be somewhat more complicated, because we cannot easily control the challenge bit. Instead, to get the corresponding result for semantic security, given a multi-challenge adversary against semantic security, we use Propositions 7.4 and 7.2 to get an adversary against indistinguishability with half the advantage. Then we use Proposition 7.5 to get a single-challenge adversary against indistinguishability with a smaller advantage. Finally, we use Proposition 7.1 to get a single-challenge adversary against semantic security. This illustrates the power of theorems proving reductions between security notions.

The experiment for the R-rnd game with an adversary $\mathcal{A}$ is identical to the experiment from Figure 7.1, except that two steps are changed as follows:

1. $b \xleftarrow{r} \{0, 1\}$. $C \leftarrow \emptyset$.

3. When the adversary sends a query (ad, m) (*challenge*), do:
   (a) If $b = 0$, encrypt $c \leftarrow \mathcal{E}(k, ad, m)$.
   (b) If $b = 1$, sample $c \leftarrow R_l$, where l is the length of m.
   (c) $C \leftarrow C \cup \{(ad, c)\}$.
   (d) Send c to the adversary.

FIGURE 7.8 The experiment for the R-random security game for a symmetric cryptosystem $\Sigma = (\mathfrak{K}, \mathfrak{P}, \mathfrak{F}, \mathfrak{C}, \mathcal{E}, \mathcal{D})$ and noise family R on $\mathfrak{C}$, based on the experiment from Figure 7.1.

7.1.3 Random-Looking Ciphertexts

Some cryptosystems actually provide a stronger notion of real-or-random than the one defined above. The idea is that encryptions of chosen messages are not only hard to distinguish from encryptions of random messages but also hard to distinguish from randomness that is completely independent not only of the chosen message and the associated data, but of the secret key itself.

This property is quite convenient when symmetric cryptography is used as part of a larger system, but it does also have direct applications. One application is to hide a ciphertext in random noise. Another is to show that ciphertexts are pseudo-random, which means that results requiring randomness (such as the left-over hash lemma) can be used.

Definition 7.5. Let $\Sigma = (\mathfrak{K}, \mathfrak{P}, \mathfrak{F}, \mathfrak{C}, \mathcal{E}, \mathcal{D})$ be a symmetric cryptosystem. A *noise family* R is a family of sampling algorithms indexed by the non-negative integers. An (τ, l_c, l_e, l_d)-*adversary against R-random* for a symmetric cryptosystem Σ is an interactive algorithm $\mathcal{A}$ that interacts with the experiment in Figure 7.8 making at most l_c challenge queries, l_e chosen plaintext queries and l_d chosen ciphertext queries, and where the runtime of the adversary and the experiment is at most τ.

The *advantage* of this adversary is defined to be

$$\mathbf{Adv}_{\Sigma}^{\text{R-rnd}}(\mathcal{A}) = 2|\Pr[E] - 1/2|,$$

where E is the event that b' output by $\mathcal{A}$ equals the experiment's b.

Exercise 7.7. Let Σ be a cryptosystem with ciphertext set $\mathfrak{C}$ and let R be a noise family on $\mathfrak{C}$. Prove that if $\mathcal{A}$ is any (τ, l_c, l_e, l_d)-adversary against real-or-random security, then there exists a (τ', l_c, l_e, l_d)-adversary against R-random security for the cryptosystem where τ' is essentially the same as τ and their advantage are roughly the same (up to a small multiple).

The most interesting case is where ciphertexts are bit strings and the noise family R_l is just the uniform distribution on the set of bit strings of some length related to l. If we have R-random security, then ciphertexts look like random bit strings, which can be quite useful in security proofs.

The following exercise disproves the converse claim of Exercise 7.7. If ciphertexts are random-looking, then we have secure encryption. But secure encryption does not imply random-looking ciphertexts.

E *Exercise* 7.8. Let Σ be any cryptosystem where ciphertexts are bit strings. Construct a new cryptosystem Σ' where ciphertexts are also bit strings, satisfying the two claims:

- ciphertexts are trivially distinguishable from random bit strings; and

- for any adversary $\mathcal{A}$ against real-or-random security for Σ', there exists an adversary $\mathcal{B}$ against real-or-random security for Σ that has the same advantage and essentially the same runtime.

Remark. Random-looking ciphertexts is, and to a lesser extent real-or-random, is related to a general approach to security often called *simulatability*. The idea is that it should be possible to simulate the honest parties' actions without any actual knowledge of what they are doing, except for some designed-in *leakage*. We will need this approach in Chapters 11 and 12. As a general approach this allows for some very powerful theorems on *composition* that could be useful, but we shall not develop this approach in full.

7.1.4 Integrity

For many applications, integrity is more important than confidentiality. Informally, we have integrity if the adversary is unable to create valid ciphertexts that decrypt to new messages. We shall discuss variants of these notions.

We shall define two integrity notions. The first, plaintext integrity, says that the adversary cannot come up with a ciphertext that decrypts to a new message, that is, one not previously submitted as a chosen plaintext query. This intuitively seems to match the sort of integrity we want in applications.

The second integrity notion, ciphertext integrity, says that the adversary cannot come up with a new valid ciphertext, that is, one not previously returned by a chosen plaintext query. This intuitively seems too strong for applications, but this notion is easier to work with and is what we will use for proofs. It also turns out that for many applications the stronger security notion is safer and easier to work with.

D **Definition 7.6.** An (τ, l_e, l_d)-*adversary against integrity* for a symmetric cryptosystem Σ is an interactive algorithm $\mathcal{A}$ that interacts with the experiment in Figure 7.9 making at most l_e chosen plaintext queries and l_d test

The integrity experiment $\mathbf{Exp}_\Sigma^{\mathrm{int}}$ proceeds as follows:

1. Choose a key $k \xleftarrow{r} \mathfrak{K}$.
2. Let $M = C = \emptyset$.
3. When the adversary sends a query (ad, m) (*chosen plaintext*), do:
   (a) $c \leftarrow \mathcal{E}(k, ad, m)$.
   (b) $M \leftarrow M \cup \{(ad, m)\}$, $C \leftarrow C \cup \{(ad, c)\}$.
   (c) Send c to the adversary.
4. When the adversary sends a query (ad, c) (*test*), do:
   (a) Compute $m \leftarrow \mathcal{D}(k, ad, c)$. Send m to the adversary.

FIGURE 7.9 Experiment for the integrity games for a symmetric cryptosystem $\Sigma = (\mathfrak{K}, \mathfrak{P}, \mathfrak{F}, \mathfrak{C}, \mathcal{E}, \mathcal{D})$.

queries, and where the runtime of the adversary and the experiment is at most τ.

The *plaintext and ciphertext (integrity) advantages* for this adversary are

$$\mathbf{Adv}_\Sigma^{\mathrm{int\text{-}ptxt}}(\mathcal{A}) = \Pr[E] \qquad \text{and} \qquad \mathbf{Adv}_\Sigma^{\mathrm{int\text{-}ctxt}}(\mathcal{A}) = \Pr[F],$$

where E is the event that for some test query (ad, c), the decryption $m \neq \bot$ and $(ad, m) \notin M$, and F is the event that for some test query $(ad, c) \notin C$, the decryption is not $\bot$. The ciphertexts in events E and F are called *forgeries*.

Informally, we say that a scheme has *plaintext integrity* if we have some reasonable argument for why any feasible integrity adversary has no significant plaintext integrity advantage. *Ciphertext integrity* carries the corresponding informal meaning. If we do know about feasible adversaries with significant advantage, we say that the scheme has *no plaintext/ciphertext integrity*.

Consider the events E and F in the above definition. Since the event E cannot happen unless F happens, it is clear that for any adversary against integrity, its plaintext advantage is not smaller than its ciphertext advantage. We shall now prove that the converse is not true, which shows that unlike our confidentiality notions, these two integrity notions are not equivalent.

Proposition 7.6. Let Σ_0 be any symmetric cryptosystem, and let Σ be the cryptosystem based on Σ_0 given in Figure 7.10. There exists a $(\tau, 1, 1)$-adversary $\mathcal{A}$ against integrity for Σ that has ciphertext advantage 1 where τ is trivial. Also, for any (τ', l_e, l_c)-adversary $\mathcal{B}$ against integrity for Σ, there exists a (τ'', l_e, l_c)-adversary $\mathcal{B}'$ against integrity for Σ_0 with τ'' essentially

The symmetric cryptosystem $\Sigma = (\mathfrak{K}, \mathfrak{P}, \mathfrak{F}, \mathfrak{C}, \mathcal{E}, \mathcal{D})$ has the same key set and plaintext set as Σ_0. The ciphertext set is $\mathfrak{C} = \mathfrak{C}_0 \times \{0, 1\}$. The encryption and decryption algorithms work as follows:

- On input of k, ad and m, $\mathcal{E}$ computes $c \leftarrow \mathcal{E}_0(k, ad, m)$, samples $i \xleftarrow{r} \{0, 1\}$ and outputs (c, i).

- On input of k, ad and (c, i), $\mathcal{D}$ computes and outputs $m \leftarrow \mathcal{D}_0(k, ad, c)$.

FIGURE 7.10 Symmetric cryptosystem $\Sigma = (\mathfrak{K}, \mathfrak{P}, \mathfrak{F}, \mathfrak{C}, \mathcal{E}, \mathcal{D})$ with no ciphertext integrity, based on a symmetric cryptosystem $\Sigma_0 = (\mathfrak{K}, \mathfrak{P}, \mathfrak{F}, \mathfrak{C}_0, \mathcal{E}_0, \mathcal{D}_0)$.

equal to τ and such that

$$\mathbf{Adv}_{\Sigma_0}^{\text{int-ptxt}}(\mathcal{B}') = \mathbf{Adv}_{\Sigma}^{\text{int-ptxt}}(\mathcal{B}).$$

Proof. The adversary $\mathcal{A}$ against ciphertext integrity for Σ is easy. It first makes a chosen plaintext query for an arbitrary message, getting (c, i) as a response. It then sends $(c, 1 - i)$ as a test query. It is clear that this has ciphertext integrity advantage 1, that its runtime is completely insignificant and that it needs a single chosen plaintext query and a single test query.

Now let $\mathcal{B}$ be an integrity adversary against Σ. Our adversary $\mathcal{B}'$ runs a copy of $\mathcal{B}$ and a simulator **Sim** that works as follows:

- When $\mathcal{B}$ makes a chosen plaintext query (ad, m), **Sim** forwards (ad, m) to the integrity experiment and receives c as the response. It then samples $i \xleftarrow{r} \{0, 1\}$ and sends (c, i) to $\mathcal{B}$.

- When $\mathcal{B}$ makes a test query (ad, c, i), **Sim** forwards (ad, c) to the integrity experiment and receives m as response. It then sends m to $\mathcal{B}$.

There is a small amount of overhead for each query, since something has to be forwarded by **Sim**, but for any non-trivial adversary, this overhead can be safely ignored. It follows that τ'' is essentially equal to τ'.

By inspection, we see that the combined processing of chosen plaintext and test queries by **Sim** and the integrity experiment for Σ_0 is identical to the processing of the queries done by the integrity experiment for Σ.

It follows that if $\mathcal{B}$ creates a ciphertext that decrypts to a new plaintext for Σ, we can derive a ciphertext that decrypts to a new plaintext for Σ_0. Therefore, the plaintext advantages of $\mathcal{B}'$ and $\mathcal{B}$ are equal. □

7.1.5 Chosen Plaintext Confidentiality and Ciphertext Integrity Suffice

The main theorem on ciphertext integrity is an important theorem that simplifies the analysis of schemes, since it allows us to deal with chosen ciphertext queries separately from challenge and chosen plaintext queries.

We shall first prove a small lemma on the probability of dependent events. This is a powerful lemma, since it allows us to bound the divergence of two games by isolating a single exceptional event that could cause them to diverge.

Lemma 7.7. *Let E_0, E_1, F_0 and F_1 be events such that $\Pr[F_0] = \Pr[F_1]$ and $\Pr[E_0 \mid \neg F_0] = \Pr[E_1 \mid \neg F_1]$. Then*

$$|\Pr[E_0] - \Pr[E_1]| \leq \Pr[F_0].$$

Proof. We compute

$$
\begin{aligned}
|\Pr[E_0] - \Pr[E_1]| &= |\Pr[E_0 \mid F_0]\Pr[F_0] + \Pr[E_0 \mid \neg F_0]\Pr[\neg F_0] - \\
&\quad \Pr[E_1 \mid F_1]\Pr[F_1] - \Pr[E_1 \mid \neg F_1]\Pr[\neg F_1]| \\
&= |\Pr[E_0 \mid F_0]\Pr[F_0] - \Pr[E_1 \mid F_1]\Pr[F_1]| \\
&= \Pr[F_0] \cdot |\Pr[E_0 \mid F_0] - \Pr[E_1 \mid F_1]| \leq \Pr[F_0]. \qquad \square
\end{aligned}
$$

Theorem 7.8. *Let Σ be a symmetric cryptosystem. Let $\mathcal{A}$ be a (τ, l_c, l_e, l_d)-adversary against indistinguishability. Then there exists a $(\tau_1', l_c, l_e, 0)$-adversary $\mathcal{B}_1$ against indistinguishability and a $(\tau_2', l_c + l_e, l_d)$-adversary $\mathcal{B}_2$ against integrity, with τ_1' and τ_2' essentially equal to τ, such that*

$$\mathbf{Adv}_\Sigma^{\mathsf{ind\text{-}cca}}(\mathcal{A}) \leq \mathbf{Adv}_\Sigma^{\mathsf{ind\text{-}cpa}}(\mathcal{B}_1) + 2\mathbf{Adv}_\Sigma^{\mathsf{int\text{-}ctxt}}(\mathcal{B}_2).$$

Proof. We first describe the adversaries. The indistinguishability adversary $\mathcal{B}_1$ runs a copy of $\mathcal{A}$ and the following simulator $\mathbf{Sim}_1$.

- When $\mathcal{A}$ makes a challenge or a chosen plaintext query, $\mathbf{Sim}_1$ forwards the query to the indistinguishability experiment. It keeps a record (ad, m, c) of chosen plaintext queries with responses.

- When $\mathcal{A}$ makes a chosen ciphertext query (ad, c), $\mathbf{Sim}_1$ checks if it has a record (ad, m, c) for some m. If it does, $\mathbf{Sim}_1$ sends m to $\mathcal{A}$. Otherwise it sends $\perp$ to $\mathcal{A}$.

If $\mathcal{A}$ outputs b', $\mathcal{B}_1$ outputs b'. If $\mathcal{A}$ exceeds its bounds, $\mathcal{B}_1$ samples $b' \xleftarrow{r} \{0,1\}$ and outputs b'.

The integrity adversary $\mathcal{B}_2$ runs a copy of $\mathcal{A}$ and the following simulator $\mathbf{Sim}_2$.

- $\mathbf{Sim}_2$ samples $b \xleftarrow{r} \{0,1\}$.

- When $\mathcal{A}$ sends a challenge query (ad, m_0, m_1), $\mathbf{Sim}_2$ sends a chosen plaintext query (ad, m_b). When it gets a response c, it records (ad, c). Then it forwards c to $\mathcal{A}$.

- When $\mathcal{A}$ sends a chosen ciphertext query (ad, c), then if (ad, c) has been recorded, $\mathbf{Sim}_2$ sends $\perp$ in response. Otherwise, $\mathbf{Sim}_2$ sends a test query (ad, c). When it gets a response m, it sends m to $\mathcal{A}$.

- Any other queries and their responses are forwarded unchanged.

The adversaries satisfy the claimed bounds. The first because we explicitly ensure that it does. The second because the simulator $\mathbf{Sim}_2$ and $\mathbf{Exp}_\Sigma^{\text{int}}$ perfectly simulate the indistinguishability experiment. There is some forwarding, but for any non-trivial adversary $\mathcal{A}$ this overhead is insignificant.

We begin by defining some events. Let E_0 be the event that $b = b'$ when $\mathcal{B}_2$ interacts with the integrity experiment, while E_1 is the event that $b = b'$ when $\mathcal{B}_1$ interacts with the indistinguishability experiment. Note that

$$\mathbf{Adv}_\Sigma^{\text{ind-cpa}}(\mathcal{B}_1) = 2|\Pr[E_1] - 1/2|.$$

In the same way, let F_0 be the event that a ciphertext submitted as a test query both decrypts to a message (not $\perp$) and is not in the experiment's ciphertext set C. Let F_1 be the event that $\mathcal{B}_2$ responds with $\perp$ to a chosen ciphertext c, but $\mathcal{D}(k, c) \neq \perp$. Note that

$$\mathbf{Adv}_\Sigma^{\text{int-ctxt}}(\mathcal{B}_2) = \Pr[F_0].$$

Finally, let E be the event that $b = b'$ when $\mathcal{A}$ interacts with the indistinguishability experiment. Then

$$\mathbf{Adv}_\Sigma^{\text{ind-cca}}(\mathcal{A}) = 2|\Pr[E] - 1/2|.$$

Now, we shall analyse these events. First, we see that F_0 and F_1 are corresponding events, so $\Pr[F_0] = \Pr[F_1]$. Further, if we look at the total interactions of $\mathcal{B}_1$ and $\mathcal{B}_2$ with their respective experiments, the only difference is that $\mathcal{B}_2$ may reject some extra ciphertexts, which is exactly F_1. Provided F_1 never happens, the two games behave the same, which means that

$$\Pr[E_0 \mid \neg F_0] = \Pr[E_1 \mid \neg F_1].$$

By inspection, we see that

$$\Pr[E] = \Pr[E_0].$$

Which means that

$$\begin{aligned}
\mathbf{Adv}_\Sigma^{\text{ind-cca}}(\mathcal{A}) &= 2|\Pr[E_0] - 1/2| \\
&= 2|\Pr[E_0] - \Pr[E_1] + \Pr[E_1] - 1/2| \\
&\leq 2|\Pr[E_0] - \Pr[E_1]| + 2|\Pr[E_1] - 1/2|
\end{aligned}$$

By the earlier arguments, we can apply Lemma 7.7 to get that

$$|\Pr[E_0] - \Pr[E_1]| \leq \Pr[F_0],$$

and the claim follows. $\qquad\qquad\square$

This theorem neatly illustrates why we need ciphertext integrity for proofs. The chosen plaintext adversary simulates the answers to chosen ciphertext queries by rejecting challenge ciphertexts and simply repeating the question when the ciphertext was obtained via a chosen plaintext query. This simulation works with ciphertext integrity. It would not work with plaintext integrity.

Benign Malleability   The symmetric cryptosystem from Figure 7.10 trivially has no ciphertext integrity, even if the underlying cryptosystem has ciphertext integrity. However, it is easy to recognise that a ciphertext (c, i) is a trivial forgery, since $(c, 1 - i)$ must then have been returned in response to a chosen plaintext query. This is often called *benign malleability*. While it is undesirable in some situations, the proof-technical issues can often be worked around, as the following exercise shows.

E | *Exercise* 7.9. Consider symmetric cryptosystems with ciphertext set $\mathfrak{C}$, and let R be any relation on $\mathfrak{C}$ that is computationally easy to decide.

   (a) Define *R-ciphertext integrity* as a variant of ciphertext integrity where a test query ciphertext is accepted as a forgery only if it is not R-related to any ciphertext returned in response to a chosen plaintext query.

   (b) Define *R-indistinguishability* as a variant of indistinguishability where the experiment will also reject a chosen ciphertext query if the ciphertext is related to any challenge ciphertext.

   (c) State and prove a version of Theorem 7.8 for R-ciphertext integrity and R-indistinguishability.

Non-malleability and Ciphertext Expansion   One measure that historically has been considered important, and which still is important in some contexts is *ciphertext expansion*, which is defined to be the difference in length (typically in bits) of the message and the encryption.

For long messages, ciphertext expansion is typically relatively small and matters little. However, for systems that exchange many small messages, ciphertext expansion can be relatively large (or even huge!). There are also systems where ciphertext expansion would be impossible, typically when retrofitting encryption onto a system not designed with encryption in mind.

One simple result is that under our confidentiality notions, deterministic encryption is insecure. Non-deterministic encryption requires ciphertext expansion. Also, integrity requires additional ciphertext expansion. This means that if we want to avoid ciphertext expansion, we have to have deterministic encryption, and we cannot have integrity.

We can have indistinguishability as long as the adversary queries a message at most once. Furthermore, we can have a property where any change to any ciphertext causes an unpredictable change to the decryption. This is called *non-malleability*. In this case, the adversary's chosen ciphertext queries should not tell the adversary anything not already known.

For fixed-length messages, a deterministic encryption scheme with no ciphertext expansion is essentially a bijection from the message set to the ciphertext set. If the message set and the ciphertext set are identical, the encryption and decryption algorithms must simply compute inverse permutations.

In other words, the cryptosystem defines a family of permutations indexed by message length. A cryptosystem with no ciphertext expansion is then a keyed family of indexed families of permutations. This is an expansion of the concept of block ciphers. We shall not study this concept further.

7.1.6 Authenticated Encryption with Associated Data

We have seen that randomness is vital for secure encryption. Faulty randomness generation is a major threat in cryptography. Bad (pseudo-)random generation have caused a number of disasters, and there also seem to have been deliberate attempts to sabotage pseudo-random number generators.

Therefore, it is useful to make cryptosystems that are more misuse resistant (or user-friendly, where the user is a designer of systems using cryptography), so that if randomness generation should somehow fail the cryptography will not fail completely. To deal with this we shall need another cryptographic object. We stress that this is something that can be used to build a symmetric cryptosystem, but it is not by itself a symmetric cryptosystem.

Definition 7.7. A *authenticated encryption scheme with associated data (AEAD)* consists of a set $\mathfrak{K}$ of *keys*, a set $\mathfrak{P}$ of *plaintexts*, a set $\mathfrak{F}$ of *associated data*, a set $\mathfrak{N}$ of *nonces*, a set $\mathfrak{C}$ of *ciphertexts*,

- a deterministic *encryption* algorithm $\mathcal{E}$ that on input of a key, a nonce, associated data and a plaintext outputs a ciphertext, and

- a deterministic *decryption* algorithm $\mathcal{D}$ that on input of a key, a nonce, associated data and a ciphertext outputs a plaintext or $\perp$.

For any key k, nonce no, associated data ad and plaintext m, we have that

$$\mathcal{D}(k, no, ad, \mathcal{E}(k, no, ad, m)) = m.$$

It is tempting to design a real-or-random-like security game for AEAD, but this does not work. When the adversary may specify the nonce, the encryption function becomes deterministic. The adversary could then ask for encryptions of several messages of very short length for a fixed nonce and associated data, and expect a collision when random messages are encrypted. An indistinguishability security game could be made to work, but the preferred security notion for AEAD is similar to random-looking ciphertexts from Section 7.1.3.

The R-rnd-aead experiment $\mathbf{Exp}_\Sigma^{R\text{-rnd-aead}}$ proceeds as follows:

1. Choose a key $k \xleftarrow{r} \mathfrak{K}$.
2. Sample $b \xleftarrow{r} \{0,1\}$, and let $C = \emptyset$.
3. When the adversary sends a query (no, ad, m) (*challenge*), do:
   (a) If $(no, ad, m, c) \in C$ for some c, send c to $\mathcal{A}$ and stop processing.
   (b) If $b = 0$, encrypt $c \leftarrow \mathcal{E}(k, no, ad, m)$.
   (c) If $b = 1$, sample $c \xleftarrow{r} R(l)$, where l is the length of m.
   (d) $C \leftarrow C \cup \{(no, ad, m, c)\}$.
   (e) Send c to the adversary.
4. When the adversary sends a query (no, ad, c) (*chosen ciphertext*), do:
   (a) If $(no, ad, m, c) \in C$ for some m, send m to the adversary.
   (b) Otherwise, if $b = 0$, send $\mathcal{D}(k, no, ad, c)$ to the adversary.
   (c) Otherwise, if $b = 1$, send $\perp$ to the adversary.

Eventually, the adversary outputs $b' \in \{0,1\}$.

FIGURE 7.11 The experiment for the R-random game for an AEAD cryptosystem $\Sigma = (\mathfrak{K}, \mathfrak{P}, \mathfrak{C}, \mathfrak{F}, \mathfrak{N}, \mathcal{E}, \mathcal{D})$, where R is a noise family for $\mathfrak{C}$.

D **Definition 7.8.** Let R be a noise family for $\mathfrak{C}$. A (τ, l_e, l_d)-*adversary against* R-*random* for an AEAD cryptosystem Σ with ciphertext set $\mathfrak{C}$ is an interactive algorithm $\mathcal{A}$ that interacts with the experiment in Figure 7.11 making at most l_e chosen plaintext queries and l_d chosen ciphertext queries, and where the runtime of the adversary and the experiment is at most τ.

The *advantage* of this adversary is defined to be

$$\mathbf{Adv}_\Sigma^{R\text{-rnd-aead}}(\mathcal{A}) = 2|\Pr[E] - 1/2|,$$

where E is the event that b' output by $\mathcal{A}$ equals the experiment's b.

It is easy to turn an AEAD into a cryptosystem. Simply sample a random nonce and encrypt the message using that nonce. Note that the ciphertext output by the AEAD encryption algorithm must be accompanied by the nonce, since without it the intended recipient cannot decrypt.

There can be other options for choosing the nonce or varying the associated data to achieve the required effect. In a conversation, the associated data might simply be a message counter, which would obviate the need for a nonce.

E *Exercise* 7.10. Describe the symmetric cryptosystem with associated data that we get from an AEAD system by choosing nonces at random. Prove that for any adversary against confidentiality or integrity, there exists an adversary against the AEAD scheme with roughly the same advantage, up to a small constant factor and for confidentiality a term related to the collision probability for randomly chosen nonces.

E *Exercise* 7.11. Plot the probability of nonce collisions for values of $|\mathfrak{N}|$ equal to 2^{32}, 2^{48}, 2^{64}, 2^{96} and 2^{128}, and l_e equal to $2, 2^2, 2^3, \ldots, 2^{|\mathfrak{N}|/2}$. What is the maximal value of l_e if the adversary's advantage should be at most 2^{-20}?

7.1.7 Multiple Keys

In practice a system that uses symmetric cryptosystems is unlikely to confine itself to a single key. Usually, there is a huge number of keys, even when there is not a huge number of users. Studying systems with more than one key is therefore important.

It is possible to design variants of the security games where the experiment has multiple independent keys, and the adversary may choose which key the experiment should use when answering a query.

As usual, these multi-key notions contain the single-key notions as special cases. Conversely, we can prove that any adversary against the multi-key notions can be turned into an adversary against a single-key notion, and the advantage of the multi-key adversary is at most that of the single-key adversary times the number of keys.

E *Exercise* 7.12. Define a multi-key variant of ror-cca, state a precise variant of the above informal claim and use a hybrid argument to prove the statement.

Another multi-key variant is to allow *key reveal*, where the adversary may learn a subset of the keys, chosen adaptively. The immediate problem is that the adversary cannot first ask for any challenge ciphertexts under some key, and then later ask for the key, since this will immediately reveal the challenge bit. The underlying problem is that revealing ciphertexts commits the experiment to a certain key, which is difficult to reveal. Most of the natural generalisations of the theorems we have proven for the single-key case are hard to prove for the multi-key case with key compromise. Stateful encryption is one approach to achieve multi-key security with key compromise which we shall investigate later.

7.2 CONFIDENTIALITY AND UNDERLYING PRIMITIVES

The previous section concerned itself only with the meaning of security for symmetric cryptosystems (and extended our notion of symmetric cryptosystems). This section shall discuss how to construct symmetric cryptosystems that achieve confidentiality, and what kind of primitives we need to construct them. Because of Theorem 7.8, we only have to care about chosen plaintext attacks (and consequently we can also ignore associated data for the moment). We shall achieve integrity in the next section.

The keystream generator experiment proceeds as follows:

1. Sample $b \xleftarrow{r} \{0,1\}$ and $k \xleftarrow{r} \mathfrak{K}$.

2. When the adversary sends a query $l < N$, the experiment samples $iv \xleftarrow{r} \mathfrak{V}$, computes $f(k, iv, l) = (z_{01}, z_{02}, \ldots, z_{0,l})$ and samples $(z_{11}, z_{12}, \ldots, z_{1,l})$ from the uniform distribution on G^l.

3. Then the experiment sends $(iv, z_{b,1}, z_{b,2}, \ldots, z_{b,l})$ to $\mathcal{A}$.

Eventually, the adversary outputs $b' \in \{0,1\}$.

FIGURE 7.12 The experiment for the security game for a key stream generator $f : \mathfrak{K} \times \mathfrak{V} \to G^N$.

7.2.1 Stream Ciphers

We shall only consider what is often called synchronious or additive stream ciphers, where a key stream generator expands a key and an initialisation vector into a string of symbols which is then added to the message (which is interpreted as a string of symbols). Traditionally, stream ciphers were bit oriented, but we can take the alphabet to be any group.

Definition 7.9. Let $f : \mathfrak{K} \times \mathfrak{V} \to G^N$ be a key stream generator. A (τ, l_c)-*adversary against* f is an interactive algorithm $\mathcal{A}$ that interacts with the experiment in Figure 7.12 making at most l_c queries to the experiment, and where the runtime of the adversary and the experiment is at most τ.

The *advantage* of this adversary is defined to be

$$\mathbf{Adv}_f^{\mathsf{ksg}}(\mathcal{A}) = 2|\Pr[E] - 1/2|,$$

where E is the event that b' output by $\mathcal{A}$ equals the experiment's b.

We will only compute as many key stream elements as is needed. The key stream must be computed by some algorithm whose cost is essentially linear in the number of key stream elements computed.

Remark. Sometimes we want a *pseudo-random generator* $f : \mathfrak{K} \to G^N$. Since there is no initialisation vector, each key expands into a single key stream. The security game is the single-query variant of the key stream security game.

Remark. There is a stronger notion of security for key stream generators, where the adversary is allowed to specify the initialisation vector to be used (a *pseudo-random function*). This is usually too strong a requirement, since it is not needed and may make key stream generator design harder. An intermediate variant is to specify some fixed sequence of initialisation vectors, which is often easy to design for and has advantages in many applications.

E *Example* 7.1. Let $f : \mathfrak{K} \times \mathfrak{V} \to G^N$ be a key stream generator. The *additive stream cipher* based on f is $(\mathfrak{K}, \cup_{l \leq N} G^l, \mathfrak{V} \times \cup_{l \leq N} G^l, \mathcal{E}, \mathcal{D})$, where:

- The encryption algorithm $\mathcal{E}$ takes as input a key $k \in \mathfrak{K}$ and message $m = m_1 m_2 \ldots m_l$ for some $l \leq N$. It samples $iv \xleftarrow{r} \mathfrak{V}$, computes $(z_1, z_2, \ldots, z_l) \leftarrow f(k, iv, l)$ and $w_i = m_i + z_i$ for $i = 1, 2, \ldots, l$, and outputs $c = (iv, w_1 w_2 \ldots w_l)$.

- The decryption algorithm $\mathcal{D}$ takes as input a key $k \in \mathfrak{K}$ and a ciphertext $c = (iv, w_1 w_2 \ldots w_l)$. It computes $(z_1, z_2, \ldots, z_l) \leftarrow f(k, iv, l)$ and $m_i = w_i - z_i$ for $i = 1, 2, \ldots, l$. It outputs $m = m_1 m_2 \ldots m_l$.

E *Exercise* 7.13. Let Σ be the additive stream cipher based on the key stream generator $f : \mathfrak{K} \times \mathfrak{V} \to G^N$. Let $\mathcal{A}$ be a $(\tau, l_c, l_e, 0)$-adversary against indistinguishability for Σ. Show that there exists a $(\tau', l_c + l_e)$-adversary $\mathcal{B}$ against the key stream generator f, where τ' is essentially equal to τ and

$$\mathbf{Adv}_{\Sigma}^{\text{ind-cpa}}(\mathcal{A}) \leq \mathbf{Adv}_f^{\text{ksg}}(\mathcal{B}) + \frac{2(l_c + l_e)^2}{|\mathfrak{V}|}.$$

Remark. The extra term in the bound on the adversary's advantage comes from possible collisions in the initialisation vectors. If we use stronger notions of security for the key stream generator, such as in the remark before the example, we can create stream ciphers with tighter security by using a fixed sequence of initialisation vectors, but this means that we have a *stateful* cipher. Sometimes this is quite practical, but sometimes it is not.

E *Exercise* 7.14. Repeat the previous exercise, but for R-random security as defined Section 7.1.3.

7.2.2 Pseudo-random Permutations and Functions

Block ciphers, also called pseudo-random permutations, are extremely useful primitives in symmetric cryptography. A block cipher is a family of permutations on a set.

A related concept is the pseudo-random function, which we have already seen in weak form, namely the key stream generator. A pseudo-random function (family) is simply a family of functions indexed by a set of keys.

D **Definition 7.10.** A *pseudo-random function (PRF) family* is a function $f : \mathfrak{K} \times S \to T$.

The idea in these security games is to have the experiment give the adversary access to either a function/permutation from the pseudo-random family,

The PRF experiment proceeds as follows:

1. Sample $b \xleftarrow{r} \{0,1\}$ and $k \xleftarrow{r} \mathfrak{K}$. Let $C_0 = C_1 = \emptyset$.
2. When the adversary sends a query $s \in S$ (*evaluate*), then:
   (a) If $s \notin C_1$, sample $u \xleftarrow{r} T$, and add s to C_1 and (s,u) to C_0.
   (b) Find $(s,u_1) \in C_0$. Compute $u_0 \leftarrow f(k,s)$. Send u_b to $\mathcal{A}$.

Eventually, the adversary outputs $b' \in \{0,1\}$.

FIGURE 7.13 The experiment for the PRF security game for a pseudo-random function $f : \mathfrak{K} \times S \to T$.

The experiment for the block cipher security game with an adversary $\mathcal{A}$ proceeds as follows:

1. Sample $b \xleftarrow{r} \{0,1\}$ and $k \xleftarrow{r} \mathfrak{K}$. Let $C_0 = C_1 = C_2 = \emptyset$.
2. When the adversary sends a query $s \in S$ (*evaluate*), then:
   (a) If $s \notin C_1$, sample $u \xleftarrow{r} S \setminus C_2$ and add (s,u) to C_0, s to C_1 and u to C_2. If $s \notin C_2$, sample $v \xleftarrow{r} S \setminus C_1$ and add (v,s) to C_0, s to C_2 and v to C_1.
   (b) Find $(s,u_1) \in C$ and $(v_1,s) \in C$. Compute $u_0 \leftarrow \pi^{-1}(k,s)$ and $v_0 \leftarrow \pi(k,s)$. Send (u_b, v_b) to $\mathcal{A}$.

Eventually, the adversary outputs $b' \in \{0,1\}$.

FIGURE 7.14 The experiment for the block cipher security game for a block cipher $\pi, \pi^{-1} : \mathfrak{K} \times S \to S$.

or a random function/permutation. For simplicity, the permutation experiment computes both the permutation and its inverse for each query.

Definition 7.11. Let $f : \mathfrak{K} \times S \to T$ be a PRF. A (τ, l_c)-*adversary against f* is an interactive algorithm $\mathcal{A}$ that interacts with the experiment in Figure 7.13 making at most l_c queries to the experiment, and where the runtime of the adversary and the experiment is at most τ.

The *advantage* of this adversary is defined to be

$$\mathbf{Adv}_f^{\mathrm{prf}}(\mathcal{A}) = 2|\Pr[E] - 1/2|,$$

where E is the event that b' output by $\mathcal{A}$ equals the experiment's b.

D **Definition 7.12.** Let $\pi, \pi^{-1} : \mathfrak{K} \times S \to S$ be a block cipher. A (τ, l_c)-*adversary against* (π, π^{-1}) is an interactive algorithm $\mathcal{A}$ that interacts with the experiment in Figure 7.14 making at most l_c queries to the experiment, and where the runtime of the adversary and the experiment is at most τ.

The *advantage* of this adversary is defined to be

$$\mathbf{Adv}^{\text{prp}}_{(\pi, \pi^{-1})}(\mathcal{A}) = 2|\Pr[E] - 1/2|,$$

where E is the event that b' output by $\mathcal{A}$ equals the experiment's b.

Remark. We have a weaker notion of pseudo-random function (sometimes called *weak pseudorandom function*) where the adversary does not choose where the function is evaluated. This is essentially the same as a key stream generator, except that a key stream generator has a small domain and a very large range, while it is often the opposite for pseudo-random functions. The key stream generator's range also has a very special form, while a pseudo-random function could have a more complex set as its range.

Remark. The design of pseudo-random functions and permutations is beyond the scope of this book. However, pseudo-random permutations or block ciphers have a long history. We believe we know good block ciphers.

Pseudo-random functions have received less direct attention, though recent hash standards do provide some very useful work. But the compression functions from many common hash functions are suitable as pseudo-random functions for some purposes. The HMAC construction is also useful if a function with a larger domain is needed (see Section 7.3.3). Another common strategy follows forthwidth.

Using block ciphers as pseudo-random functions   A pseudo-random permutation looks like a random permutation. A pseudo-random function looks like a random function. But as long as we do not inspect too many values, a random permutation looks like a random function. The upshot is that if we have a pseudo-random permutation, then we can treat it as a pseudo-random function as long as we do not evaluate it at too many points.

Sampling a random function and then evaluating it at distinct points is equivalent to independently sampling elements from the uniform distribution. Likewise, sampling a random permutation and then evaluating it at distinct points is equivalent to independently sampling elements from the uniform distribution, subject to the condition that the sampled elements are all distinct.

T **Lemma 7.9.** *Let $\mathcal{A}$ be an interactive algorithm that interacts with the experiment from Figure 7.15 and makes at most l_c queries. Then*

$$2|\Pr[E] - 1/2| \leq \frac{l_c^2}{2|S|},$$

The experiment for distinguishing random functions from random permutations, with an adversary $\mathcal{A}$ proceeds as follows:

1. Sample $b \leftarrow \{0,1\}$. Let $C_0 = C_1 = C_2 = \emptyset$.
2. When the adversary sends a query $s \in S$ (*evaluate*), then:
   (a) If $s \notin C_1$, sample $u_0 \xleftarrow{r} S$ and $u_1 \xleftarrow{r} S \setminus C_2$. Add (s, u_b) to C_0, s to C_1 and u_b to C_2.
   (b) Find $(s, u) \in C_0$. Send u to $\mathcal{A}$.

Eventually, the adversary outputs $b' \in \{0,1\}$.

FIGURE 7.15 Experiment for Lemma 7.9.

where E is the event that $b' = b$ when the adversary ends.

Proof. The experiment essentially samples two sequences of elements from S, one with replacement and one without replacement. Let F be the event that the sequence sampled with replacement results in distinct elements.

Proposition 2.11 (the birthday paradox) says that the sequence of elements sampled with replacement will consist of distinct elements except with a probability bounded by $l_c^2/(2|S|)$ so $\Pr[\neg F] \leq l_c^2/(2|S|)$.

If the elements sampled with replacement are all distinct, they are distributed exactly as the elements sampled without replacement, so $\Pr[E \mid F] = 1/2$. We then compute

$$2|\Pr[E] - 1/2| = 2|\Pr[E \mid F](1 - \Pr[\neg F]) + \Pr[E \mid \neg F]\Pr[\neg F] - 1/2|$$
$$= 2|(\Pr[E \mid \neg F] - 1/2)\Pr[\neg F]| \leq \Pr[\neg F]. \qquad \square$$

The above lemma says that unless you evaluate at too many points, a randomly chosen function is indistinguishable from a randomly chosen permutation. Since a block cipher should be hard to distinguish from a random permutation (with an inverse), and a PRF should be hard to distinguish from a random function, it follows by transitivity that a block cipher (without the inverse function) should be hard to distinguish from a random function.

Theorem 7.10 (PRP/PRF Switching Lemma). *Let (π, π^{-1}) be a block cipher on a set S. Let $\mathcal{A}$ be a (τ, l_c)-adversary against π as a PRF. Then there exists a (τ', l_c)-adversary $\mathcal{B}$ against (π, π^{-1}) as a block cipher, where τ' is essentially equal to τ, and*

$$\mathbf{Adv}_\pi^{\mathrm{prf}}(\mathcal{A}) \leq \mathbf{Adv}_{(\pi,\pi^{-1})}^{\mathrm{prp}}(\mathcal{B}) + \frac{l_c^2}{2|S|}.$$

Proof. First, we define a simulator **Sim** as follows: When the simulator receives a query from $\mathcal{A}$, it forwards the query to the block cipher experiment. When it receives the response (u', u''), it forwards u' to $\mathcal{A}$. Our adversary $\mathcal{B}$

runs a copy of **Sim** and $\mathcal{A}$, and when $\mathcal{A}$ eventually outputs a bit b', $\mathcal{B}$ also outputs b'. If $\mathcal{A}$ exceeds its bounds, $\mathcal{B}$ samples $b' \xleftarrow{r} \{0, 1\}$ and outputs b'.

Next, we shall define some events. Let E_β be the event that $\mathcal{A}$ outputs 1 when interacting with the PRF experiment with $b = \beta$. Let F_β be the event that $\mathcal{B}$ outputs 1 when interacting with the block cipher experiment with $b = \beta$. By Lemma 7.3, we have that

$$\mathbf{Adv}_\pi^{\mathsf{prf}}(\mathcal{A}) = |\Pr[E_0] - \Pr[E_1]| \qquad \mathbf{Adv}_{(\pi, \pi^{-1})}^{\mathsf{prp}}(\mathcal{B}) = |\Pr[F_0] - \Pr[F_1]|.$$

By inspection, when $b = 0$, $\mathcal{A}$ interacting with the PRF experiment behaves identically to the $\mathcal{B}$ interacting with the block cipher experiment. Therefore,

$$\Pr[E_0] = \Pr[F_0].$$

Next, by inspection we see that $\mathcal{A}$ interacting with the PRF experiment with $b = 1$ behaves identically to $\mathcal{A}$ interacting with the experiment from Figure 7.15 with $b = 0$. Furthermore, by inspection we see that $\mathcal{A}$ embedded in $\mathcal{B}$ interacting with the block cipher experiment with $b = 1$ behaves identically to $\mathcal{A}$ interacting with the experiment from Figure 7.15 with $b = 1$. Lemmas 7.3 and 7.9 then say that

$$|\Pr[E_1] - \Pr[F_1]| \le \frac{l_c^2}{2|S|}.$$

Finally, we compute

$$\begin{aligned} \mathbf{Adv}_\pi^{\mathsf{prf}}(\mathcal{A}) &= |\Pr[E_0] - \Pr[E_1]| \\ &= |\Pr[E_0] - \Pr[F_0] + \Pr[F_0] - \Pr[F_1] + \Pr[F_1] - \Pr[E_1]| \\ &\le |\Pr[F_0] - \Pr[F_1]| + |\Pr[F_1] - \Pr[E_1]|. \qquad \square \end{aligned}$$

7.2.3 Two Constructions

We saw counter mode defined earlier for block ciphers, but the construction works equally well for PRFs.

Example 7.2. Let f be a function from $\mathfrak{K} \times \mathfrak{V} \times \{1, 2, \ldots, N\}$ to some finite group G. For any $iv \in \mathfrak{V}$ and $k \in \mathfrak{K}$, the *counter (CTR) mode* is the function $\mathrm{CTR}_\pi : \mathfrak{K} \times \mathfrak{V} \to G^N$ given by $\mathrm{CTR}_\pi(k, iv) = z_1 z_2 \ldots z_N$, with

$$z_i = \pi(k, (iv, i)), \text{ where } 1 \le i \le N.$$

Exercise 7.15. Prove that any adversary against the counter mode key stream generator can be turned into a PRF adversary against f with the same advantage and essentially the same runtime.

Remark. The above construction essentially shows how to construct a PRF with a larger co-domain, at the cost of a somewhat smaller domain. We shall later study how to make the domain of a PRF larger.

The block cipher mode *cipher-block chaining mode* from Example 1.9 is a convenient way to turn a block cipher into a symmetric cryptosystem.

E *Exercise 7.16.* Let (π, π^{-1}) be a block cipher on a group G. Let R be the family of noise distributions where R_l is the uniform distribution on G^{l+1}. Prove that any R-random chosen plaintext adversary against CBC mode can be turned into an adversary against the block cipher with essentially equal runtime. If l_c is the number of messages encrypted and l is their total length, then the difference in advantage is bounded by $(l_c + l)^2/|G|$.

7.3 MESSAGE AUTHENTICATION CODES

A message authentication code is a tool we can use to achieve integrity. The idea is to compute a *tag* that we can attach to message. This tag allows someone with the correct key to verify that the message is unchanged, and appears in the right context. We shall then consider a few ways to construct message authentication codes, before we show how to combine them with symmetric cryptosystems in order to achieve integrity.

D **Definition 7.13.** A *message authentication code* MAC consists of a set $\mathfrak{K}$ of *keys*, a set $\mathfrak{P}$ of *plaintexts*, a set $\mathfrak{T}$ of *tags*, and

- a *tag* algorithm $\mathcal{S}$ that on input of a key, associated data and a plaintext outputs a tag; and

- a *verify* algorithm $\mathcal{V}$ that on input of a key, associated data, a plaintext and a tag outputs either 0 or 1.

We require that for any key k and any plaintext m

$$\mathcal{V}(k, m, \mathcal{S}(k, m)) = 1.$$

Remark. We get a MAC from a simple MAC μ (see Definition 1.5) by letting the tag algorithm evaluate the μ function on its input. The verify algorithm recomputes $\mu(k, m)$ and compares this tag value with the input. We usually skip this trivial wrapper in discussions.

As for integrity, we get two natural security notions (which coincide for a simple MAC). The first is that it should be hard to find a tag on a message for which the adversary has not already seen a tag. The adversary should be able to see a number of MAC tags. The only reasonable option is to allow the adversary to choose the messages to get tags for. This notion captures our intuition about what a MAC should be and what we would want in applications.

The MAC security experiment proceeds as follows:

1. Sample $k \leftarrow \mathfrak{K}$. Set $C_0 = C_1 = \emptyset$.

2. When the adversary sends a query $m \in \mathfrak{P}$ (*chosen message*), compute $t \leftarrow \mathcal{S}(k, m)$, add m to C_0 and (m, t) to C_1 and send t to $\mathcal{A}$.

3. When the adversary sends a query $(m, t) \in \mathfrak{P} \times \mathfrak{T}$ (*test*), compute $i \leftarrow \mathcal{V}(k, m, t)$ and send i to $\mathcal{A}$.

FIGURE 7.16 The experiment for the MAC security game for a message authentication code $\text{MAC} = (\mathfrak{K}, \mathfrak{P}, \mathfrak{T}, \mathcal{S}, \mathcal{V})$.

A stronger notion says that the adversary should not have seen the particular message-tag pair before. This notion is the more practical for applications, similar to the situation with plaintext and ciphertext integrity.

Definition 7.14. A (τ, l_c, l_e)-*adversary* against a message authentication code MAC is an interactive algorithm $\mathcal{A}$ that interacts with the experiment in Figure 7.16 making at most l_c test queries and l_e chosen message queries, and where the runtime of the adversary and the experiment is at most τ.

The *existential unforgeability advantage* and the *strong unforgeability advantage* of this adversary are defined to be

$$\mathbf{Adv}_{\text{MAC}}^{\text{euf-cma}}(\mathcal{A}) = \Pr[E] \quad \text{and} \quad \mathbf{Adv}_{\text{MAC}}^{\text{suf-cma}}(\mathcal{A}) = \Pr[F],$$

where E is the event that the experiment responds with 1 for some test query (m, t) with $m \notin C_0$, while the event F is that $(m, t) \notin C_1$ for this test query. The pair (m, t) is often called a *forgery* and $\mathcal{A}$ is often called a *forger*.

Remark. These existential and strong notions coincide for simple message authentication codes, since any message has exactly one tag that will verify. They do not need to coincide if the tag algorithm is deterministic.

We begin with an immediate result, namely that pseudo-random functions are good message authentication codes.

Proposition 7.11. Let $f : \mathfrak{K} \times \mathfrak{P} \to \mathfrak{T}$ be a function and let $\mathcal{A}$ be a (τ, l_c, l_e)-adversary against f as a MAC. Then there exists a $(\tau', l_c + l_e)$-adversary $\mathcal{B}$ against f as a PRF, where τ' is essentially equal to τ, and

$$\mathbf{Adv}_f^{\text{suf-cma}}(\mathcal{A}) \leq \mathbf{Adv}_f^{\text{prf}}(\mathcal{B}) + \frac{l_c}{|\mathfrak{T}|}.$$

Proof. We define a simulator **Sim**. When $\mathcal{A}$ makes a chosen message query m, the simulator forwards the message to the PRF experiment and returns the response to $\mathcal{A}$. When $\mathcal{A}$ makes a test query (m, t), the simulator sends m to the PRF experiment and receives a response t'. If $t = t'$, it sends 1 to the adversary, otherwise it sends 0. The PRF adversary $\mathcal{B}$ runs a copy of the simulator **Sim** and $\mathcal{A}$. If **Sim** ever responds with 1 to a test query (m, t) without a corresponding chosen message query m, $\mathcal{B}$ stops with output 1. When $\mathcal{A}$ stops, or exceeds its bounds, $\mathcal{B}$ stops with output 0.

If the PRF experiment's challenge bit $b = 0$, the interaction of $\mathcal{A}$ and **Sim** with the PRF experiment proceeds exactly as the interaction of $\mathcal{A}$ with the MAC simulator, unless the adversary submits a forgery. It follows that if $b = 0$, then the runtime of $\mathcal{B}$ interacting with the PRF simulator is at most τ plus some overhead because $\mathcal{B}$ is forwarding messages.

Let E be the event that the MAC experiment's response to a test query is 1, and let F_0 be the event that $\mathcal{B}$ outputs 1 when interacting with the PRF experiment. We then have that

$$\Pr[F_0] = \Pr[E] = \mathbf{Adv}_f^{\text{suf-cma}}(\mathcal{A}).$$

If the experiment's challenge bit $b = 1$, then since $\mathcal{B}$ stops after runtime τ, we know that the total runtime τ' is τ plus the additional cost of forwarding messages and the overhead of the PRF experiment, which for any non-trivial adversary is insignificant.

Let F_1 be the event that $\mathcal{B}$ outputs 1 when interacting with the PRF experiment with $b = 1$. Note that if $\mathcal{B}$ outputs 1, then $\mathcal{A}$ correctly predicted the value t before the PRF experiment sampled it from the uniform distribution. For l_c test queries, this happens with probability at most $l_c/|\mathfrak{T}|$.

We compute

$$\Pr[E] = |\Pr[F_0] - \Pr[F_1] + \Pr[F_1]| \leq \mathbf{Adv}_f^{\text{prf}}(\mathcal{B}) + \frac{l_c}{|\mathfrak{T}|}. \qquad \square$$

E *Exercise 7.17.* Prove that a good MAC need not be a PRF. (That is, given a MAC, construct a new MAC that is easy to distinguish from a random function, and show that any adversary against the new MAC can be turned into an adversary against the original MAC that is just as good.)

7.3.1 Polynomial Evaluation MACs

We begin by designing a MAC that does not have associated data. Technically, $\mathfrak{F}$ should then be a singleton set, but we simply ignore it. It is actually easy to construct a good *one-time* MAC, as we saw in Example 1.16 in Section 1.3.1.

E *Exercise 7.18.* Prove that for any value of τ, any $(\tau, l_c, 1)$-adversary $\mathcal{A}$ against OTPE from Example 1.16 has advantage

$$\mathbf{Adv}_{\text{OTPE}}^{\text{suf-cma}}(\mathcal{A}) \leq \frac{l_c(l + 1)}{p},$$

where l is the length of the single chosen message.

Hint: You may use Theorem 1.2.

This can be turned into a many-time MAC by combining OTPE with a PRF. The cost is that we turn this into a randomised MAC.

Example 7.3. Let $f : \mathfrak{K} \times \mathfrak{V} \to \mathbb{F} \times \mathbb{F}$ be a PRF. Define the function PE : $\mathfrak{K} \times \mathfrak{V} \times \mathfrak{P} \to \mathbb{F}$ by

$$\mathrm{PE}(k, iv, m) = \mathrm{OTPE}(k_1, k_2, m) \qquad \text{where } (k_1, k_2) = f(k, iv).$$

The tag algorithm samples $iv \xleftarrow{r} \mathfrak{V}$ and the MAC tag is $t = (iv, \mathrm{PE}(k, iv, m))$.

The analysis is fairly straight-forward. We first assume that there are no repetitions among the sampled iv values. Then we replace the PRF by a random function. In this situation, every iv value sampled in response to a chosen message query gives a new, independent instance of the one-time MAC OTPE. Exercise 7.18 and Theorem 1.2 gives us a bound on the adversary winning against each of the sampled one-time keys. Next, a test query with a new iv value will only win if the adversary correctly guesses a random value that the experiment has revealed no information about.

Exercise 7.19. Write out a precise theorem statement about the security of the MAC in Example 7.3 and complete the above proof sketch.

Exercise 7.20. Consider the MAC from Exercise 1.40. State and prove a theorem about the security of this MAC, relative to the block cipher used.

7.3.2 PRF-based MACs

We can build MACs directly from pseudo-random functions with a large domain, but we can also use pseudo-random functions with smaller domains, in a construction very similar to CBC-MAC from Section 1.3.2.

Example 7.4. Let G be any group, let $\mathfrak{P}$ be the set of non-empty strings of group elements, and let $f : \mathfrak{K} \times G \to G$ be a function. We define the MAC CBC-MAC$'$: $\mathfrak{K} \times \mathfrak{P} \to G$ as follows: Let $k \in \mathfrak{K}$, let $t_0 = 0 \in G$ be the additive identity and let $h = f(k, 0)$. Given a message $m = m_1 m_2 \ldots m_l$, let

$$t_i = f(k, t_{i-1} + m_i), \text{ where } 1 \leq i \leq l - 1.$$

Then CBC-MAC$'(k, m) = f(k, h + t_{l-1} + m_l)$.

Adding h is a useful technique. The effect is essentially to separate the final queries from the earlier queries. We could have achieved the same effect by using a different key, which in some situations is inconvenient.

Adding h is essential. Otherwise an adversary could use the chosen message query to evaluate the function $f(k, \cdot)$ at chosen points, enabling forgeries.

E | *Exercise* 7.21. Write out a precise theorem statement about the security of the MAC in Example 7.4 and prove the theorem.

7.3.3 Hash-based MACs

HMAC is a clever construction using certain hash functions to create a message authentication code. Let $\mathfrak{P}$ be the set of strings of elements from G. We hash $m_1 m_2 \ldots m_l \in \mathfrak{P}$ using a compression function $f : G \times G \to G$ by computing

$$s_i = f(s_{i-1}, m_i) \text{ for } i = 1, 2, \ldots, l,$$

where s_0 is some fixed value and the hash value is s_l. Recall that hashing the string $k\, m_1 \ldots m_l$ is not a PRF because of Exercise 4.5.

We first use the compression function to construct a PRF NMAC : $G \times G \times \mathfrak{P} \to G$ based on a compression function as follows: We compute

$$s_1 = f(k_1, m_1), \qquad s_i = f(s_{i-1}, m_i) \text{ for } i = 2, 3, \ldots, l,$$

and finally $\mathrm{NMAC}(k_1, k_2, m_1 m_2 \ldots m_l) = f(k_2, s_l)$. Note how the final computation is different, just as for CBC-MAC′.

The idea is that s_1 looks like a random element. This means that s_2 will also be random-looking, and so it goes. Finally, as long as the final element s_l is unique, the MAC tag will also be random-looking. The final element will be unique as long as there are no collisions among the elements of G computed for all of the queried messages.

The simplest proof strategy is to first stop if a collision in f is ever encountered by the PRF experiment. Now consider two messages $m_{01} m_{02} \ldots m_{0l_0}$ and $m_{11} m_{12} \ldots m_{1l_1}$, where the experiment computes

$$s_{j1} = f(k_1, m_{j1}) \qquad s_{ji} = f(s_{j,i-1}, m_{ji}) \text{ for } j = 0, 1 \text{ and } i = 2, \ldots, l_j.$$

If no collisions are encountered, we know that $s_{0l_0} \neq s_{1l_1}$. Which in turn means that the PRF is evaluated at two distinct points, which means that the function values are independent.

Since we have not yet fully discussed collision-resistance, we do not formalise this proof. However, we note one drawback, which is that *we never used the first key k_1*. In fact, our proof applies equally well to the MAC $(k, m) \mapsto f(k, h(m))$. But there is a significant difference between this MAC and NMAC, namely that for the simpler MAC an adversary can find two messages m, m' such that $h(m) = h(m')$. Now a valid tag on m will also be a valid tag for m'. This attack is not possible for NMAC. Instead, the adversary must submit queries to find a collision, and the input to the compression function is only partially under the adversary's control. The upshot is that a more careful argument can bound the adversary's collision finding advantage in terms of the number of queries made, not in terms of the computational effort expended by the adversary.

We leave the full proof to the interested reader.

HMAC is now derived from NMAC by choosing two constants i, j and defining HMAC$(k, m) =$ NMAC$(h(k + i), h(k + j), m)$. Recall that since $k + i$ and $k + j$ are single blocks, $h(k + i) = f(s_0, k + i)$ and $h(k + j) = f(s_0, k + j)$. And under reasonable assumptions on common compression functions, evaluating the compression function at two distinct randomly chosen points result in independent-looking function values, even when the two points are related. Therefore, it is reasonable to claim that HMAC is secure if NMAC is secure.

Remark. The HMAC construction reduces the size of the key, but that is not the clever part. The clever part is that, referring to the iterative construction of the hash function h, we can evaluate HMAC as

$$\text{HMAC}(k, m) = h((k + j)\, h((k + i)\, m)).$$

At the time when HMAC was designed, many software libraries and hardware implementations did not give direct access to the compression function. Therefore, even if there was an implementation of the hash function on the system, NMAC was non-trivial to implement, while HMAC was trivial to implement.

7.3.4 Encrypt-then-MAC

We shall now show how to achieve integrity for a symmetric cryptosystem. We encrypt the message using the cryptosystem and authenticate the ciphertext using the MAC. The crucial point is that we use independent keys for the encryption and the MAC. This means that the MAC cannot ruin the chosen plaintext security of the cryptosystem. Which means that a chosen plaintext adversary against the new system is essentially also a chosen plaintext adversary against the original system. Likewise, appending a MAC to an encryption of the plaintext trivially gives us integrity. If we have a suitable MAC, we even get ciphertext integrity, and we can now apply Theorem 7.8.

Historically, this construction was defined without associated data. We define it with associated data by using a MAC on the set $\mathfrak{F} \times \mathfrak{C}$. Such a MAC can be derived from a MAC on some larger set $\mathfrak{P}'$ that we can inject $\mathfrak{F} \times \mathfrak{C}$ into. It is extremely important that this function is injective. If associated data and ciphertext can be confused, this sometimes allows very effective attacks.

Example 7.5. Let $\Sigma_0 = (\mathfrak{K}_e, \mathfrak{P}, \mathfrak{C}_0, \mathcal{E}_0, \mathcal{D}_0)$ be a symmetric cryptosystem and let MAC $= (\mathfrak{K}_m, \mathfrak{F} \times \mathfrak{C}_0, \mathfrak{T}, \mathcal{S}, \mathcal{V})$ be a message authentication code. We construct a new symmetric cryptosystem with associated data $\Sigma = (\mathfrak{K}_e \times \mathfrak{K}_m, \mathfrak{P}, \mathfrak{F}, \mathfrak{C}_0 \times \mathfrak{T}, \mathcal{E}, \mathcal{D})$ as follows:

- The *encryption* algorithm $\mathcal{E}$ takes as input a key $(k_1, k_2) \in \mathfrak{K}_e \times \mathfrak{K}_m$, a message $m \in \mathfrak{P}$ and associated data $ad \in \mathfrak{F}$. It computes $w \leftarrow \mathcal{E}_0(k_1, m)$ and $t \leftarrow \mathcal{S}(k_2, (ad, w))$. The ciphertext is (w, t).

- The *decryption* algorithm $\mathcal{D}$ takes as input a key $(k_1, k_2) \in \mathfrak{K}_e \times \mathfrak{K}_m$, associated data ad and a ciphertext (w, t). If $\mathcal{V}(k_2, (ad, w)) = 0$, the algorithm outputs $\bot$. Otherwise, the algorithm outputs $\mathcal{D}_0(k_1, w)$.

The next two exercises combined with Theorem 7.8 gives the desired result.

E *Exercise* 7.22. Show that for any ciphertext integrity adversary against Σ, there exists a strong unforgeability adversary against MAC with the same advantage and essentially the same runtime.

E *Exercise* 7.23. Show that for any chosen plaintext indistinguishability adversary against the Σ, there exists a chosen plaintext indistinguishability adversary against Σ_0 with the same advantage and essentially the same runtime.

Remark. As discussed in Section 1.4 there are three natural combinations of encryption and message authentication: encrypt-then-MAC (as described above), MAC-then-encrypt (where we tag the message and encrypt the message-tag pair) and encrypt-and-MAC (where we tag the message, but do not encrypt the tag). Encrypt-then-mac is always secure. Encrypt-and-mac is in general insecure (since the tag typically reveals information about the message). Mac-then-encrypt will give us plaintext integrity.

As a general rule, encrypt-then-mac should be preferred. In practice other solutions are often used, sometimes for sensible reasons.

Remark. This is our first *compositional* construction. The previous remark illustrates that even though cryptographic constructions compose in many seemingly natural ways to build new cryptographic constructions, compositional construction is a subtle field, and we cannot expect security properties to compose nicely.

There are a number of ways to achieve cryptosystems with integrity other than the encrypt-then-mac paradigm. Some of them go via AEAD, which the above construction does not.

E *Example* 7.6. We consider an AEAD scheme that is similar to the SIV block cipher mode (which is a more careful construction relying only on a block cipher). Let $\mathfrak{N}$ be a set of nonces, G be a finite group, $\mathfrak{P}$ be the set of strings of group elements of length at most N, $\mathfrak{F}$ be a set of associated data, $f_1 : \mathfrak{K}_1 \times \mathfrak{V} \to G^N$ be a key stream generator, and let $f_2 : \mathfrak{K}_2 \times \mathfrak{N} \times \mathfrak{F} \times \mathfrak{P} \to \mathfrak{V}$ be a pseudo-random function family. We construct an AEAD scheme $(\mathfrak{K}_1 \times \mathfrak{K}_2, \mathfrak{P}, \mathfrak{F}, \mathfrak{N}, \mathfrak{P}, \mathcal{E}, \mathcal{D})$ as follows:

- The encryption algorithm $\mathcal{E}$ takes as input $(k_1, k_2) \in \mathfrak{K}_1 \times \mathfrak{K}_2$, $no \in \mathfrak{N}$, $ad \in \mathfrak{F}$ and $m \in G^l$. It computes $iv \leftarrow f_2(k_2, no, ad, m)$, $z \in G^l$ as a prefix of $f_1(k_1, iv)$ and $w \leftarrow z + m$. It outputs the ciphertext $c = (iv, w)$.

- The decryption algorithm $\mathcal{D}$ takes as input $(k_1, k_2) \in \mathfrak{K}_1 \times \mathfrak{K}_2$, $no \in \mathfrak{N}$, $ad \in \mathfrak{F}$ and $c = (iv, w) \in \mathfrak{N} \times G^l$. It computes $z \in G^l$ as a prefix of $f_1(k_1, iv)$, $m \leftarrow w - z$ and $iv' \leftarrow f_2(k-2, no, ad, m)$. If $iv \neq iv'$, output $\perp$, otherwise output m.

E | *Exercise* 7.24. Let R be the noise distribution family where R_l is the uniform distribution on $\mathfrak{N} \times G^l$. Show that the above scheme is R-random secure.

7.4 CHANNELS

There are many applications for symmetric cryptosystems, but the most important is the secure communications channel. This is the classic cryptographic problem, indeed, the only problem for much of cryptographic history.

We begin by considering Alice and Bob. They have a shared key k and they have a symmetric cryptosystem with associated data. Alice wants to send messages to Bob, and Bob wants to send messages to Alice. While the messages they send certainly may depend on the messages they receive, they may also depend on outside factors. In particular, there is no reason to expect Alice and Bob to alternate sending messages. Alice may send several messages without receiving a response from Bob.

Informally. A *channel* is a protocol Π with two players. Each party may send and receive ciphertexts via an underlying network. The two players receive a shared key as input. During the protocol execution, each party may receive messages as input, and each party may output messages. Either party may also receive $\top$ as input, after which the party will not accept any further messages as input. Either party may output $-$, $\top$ or $\perp$, after which no further messages will be accepted as input, and no further messages will be output.

Remark. There is a notion of *stateful encryption*, where the key material is allowed to change after each encryption. Essentially, the encryption and decryption algorithms keep a *state*. In some sense stateful encryption is synonymous to our notion of channel. We develop this theory more fully in Chapter 13.

Usually, Alice and Bob also want to send their messages with as little overhead as possible, both in terms of message expansion, but in particular in terms of message round-trips. Typically, when Alice wants to send a message to Bob, she wants to send a single ciphertext over the network and then no more. This also implies that we want asynchronous communications.

We usually assume that the adversary controls the underlying communications channel, and therefore the underlying network is not so important for our cryptographic analysis. But the underlying network is still a network in the real world, and it has certain properties during *normal operation* that influences how we want our cryptography to work, and what kind of functionality and security we should aim for. There is a huge variety of networks in the real world, from radio channels with low bandwidth and little error correction (which means that data received may be different from data sent) to modern high-speed, low-latency reliable wired networks. The network may not even be a classical network, but rather a third-party server or a composition of several logical or physical networks.

We shall only consider two types of networks. The first is a reliable network where messages sent are delivered unchanged, and they are delivered in the order they were sent. The second is a network where, *under normal operation*, there is no expectation that messages will be delivered, nor that they will be delivered in the order they were sent, but that if messages are delivered, they are delivered unchanged. This is under normal operations. When the adversary controls the network, there are no guarantees for anything for either network.

Remark. We did not formally define any requirements for correctness for a channel, but we expect the channel to be as reliable as the underlying network.

We want a channel that provides confidentiality. As usual, this will be up to message length and is similar to our earlier work. In addition, we want to define a notion of channel integrity, which is more complicated. The challenge is that we may also care about knowing who sent the message, that messages arrive in the correct order, and that lost messages are noticed. Defining these notions is somewhat technical, but essentially rely on identifying some correspondence between the received messages and the sent messages.

The basic requirement is that there should be some correspondence, essentially saying that the received messages were all sent by the players. The usual requirement that the adversary cannot replay old messages then amounts to saying that the correspondence should be injective. Saying that every message has arrived amounts to saying that the correspondence is surjective.

We begin by defining the *message transcript* of a protocol execution. Player i has a sequence of message-bit pairs $(m_{i,1}, r_{i,1}), (m_{i,2}, r_{i,2}), \ldots, (m_{i,l_i}, r_{i,l_i})$. We interpret $r_{i,j}$ to be 1 if Player i sent the message, and 0 otherwise.

Define $I_r = \{(i,j) \mid r_{i,j} = r\}$, $r = 0, 1$. Note that I_1 identifies the messages that were sent by either player, while I_0 identifies the messages that were received. We now want to consider the set S of maps from I_0 to I_1 with the property that if $(i,j) \in I_0$ maps to $(i', j') \in I_1$ then $m_{i,j} = m_{i',j'}$. For any given transcript, such maps need not exist, or there may be many of them.

We now define a hierarchy of increasingly strict requirements on S, listed in Table 7.1. Note that given a transcript of the players' input and output, it is computationally easy to verify if each of these conditions hold. Our next task is to sketch solutions for each requirement.

sc-0: Basic security. It is trivial to get the most basic security level by using a symmetric cryptosystem with ciphertext integrity. To send a message, simply encrypt it and send the ciphertext to the other party.

sc-1: Correct origin. One way to ensure correct origin is to use two keys, one for each direction: Alice encrypts and Bob decrypts with k_0, while Bob

Name	Definition	Explanation
sc-0	There exists a map $f \in S$.	Any message received was sent by either Alice or Bob.
sc-1	For any $(i,j) \in I_0$, we have that $f(i,j) = (1-i, j')$ for some j'.	Any message received by Alice was sent by Bob, and vice versa.
sc-2	The map f is injective.	If a message was received several times by one party, it was sent at least that many times by the other party.
sc-3	For any $(i, j_0), (i, j_1) \in I_0$ with $f(i, j_0) = (1 - i, j_0')$ and $f(i, j_1) = (1 - i, j_1')$, then $j_0 < j_1$ implies $j_0' < j_1'$.	The messages were received in the same order as they were sent.
sc-4	If (i, j_1') is in the image of f, then (i, j_0') is in the image of f for any $j_0' < j_1'$.	If some message sent by Alice was received, any previously sent message by Alice was also received, and vice versa.
sc-5	If any player outputs $\top$, then f is surjective.	If Alice or Bob considers the channel properly closed, then any message sent was also received.

TABLE 7.1 Requirements on correspondence between sent and received message sequences in a secure channel. The requirements are cumulative, so the third requirement also includes all the previous requirements.

encrypts and Alice decrypts with k_1. If we must, we can use a PRF to derive the two keys from a single key.

There are sometimes reasons for using the same key for encryption and decryption. In this case, we can use a symmetric cryptosystem with associated data. If Alice associates 0 with any message she sends, while Bob associates 1 with his messages, Alice and Bob can both verify the origin.

sc-2: Replay protection. The natural choice is to use a message counter. When Alice sends her jth message, she uses $(0, j)$ as associated data. When Bob sends his jth message, he uses $(1, j)$ as the associated data.

For a reliable channel, Bob will use $(0, j)$ as associated data when he decrypts the jth message from Alice, and likewise for Alice. If any decryption fails, it may be reasonable to close the channel.

For an unreliable channel, messages may be lost or arrive out of order and this scheme does not work. We shall simply attach the counter to the ciphertext, so that Bob knows which message has arrived. However, we must prevent Bob from reusing old counters, which would allow replay of old ciphertexts. We could store every used counter, but this quickly becomes impractical. We could store the biggest received counter and the counters for messages that have not yet been received. This is more practical, but for long-lived channels, the number of lost messages may become large.

A cheaper solution is to only store counters that are close to the biggest received counter. This means that messages that arrive too far out of order will be discarded, even though they are genuine. This means that the channel will be somewhat more unreliable than the underlying channel. This may be a problem, but it may in some situations represent a sensible compromise.

Remark. It is sometimes undesirable to send the counter in the clear. With the above approach, one could simply not send the counter. The recipient will instead try to decrypt with the next few counters, or if that fails, with the unused counters. This will increase the cost incurred by the underlying channel unreliability. Other strategies are possible, and depending on the channel properties and the application, these strategies could be much more efficient.

Remark. For an unreliable channel, this seems to be the best we can do without essentially creating a reliable channel from the unreliable channel. While this can certainly be done, it often does not make sense to integrate network functionality into the cryptographic functionality in this way.

From now on, we shall assume a reliable underlying network.

sc-3: Correct order. We return to the previous solution: Bob uses $(0, j)$ as associated data when he decrypts the jth message from Alice, and vice versa. If decryption fails, the channel is closed. This solution will trivially ensure that messages are received in the correct order, if they are received.

sc-4: Every previous message delivered. The previous solution still works.

sc-5: Every message delivered before orderly close. The idea is to ensure that the channel is closed in an orderly fashion, and ensure that both parties agree on the number of messages sent and received, in both directions.

The general idea is to send a close message, and typically the close message is sent both in a ciphertext and in the clear. The closing partner sends the close message, which will contain the number of messages sent ν_0 and received ν_1 by the closing partner, and have associated data $(2, i)$, where 2 indicates a close message and i identifies the closing partner. The responding partner will reply with a close response message that contains the number of messages sent ν_2 and received ν_3 by the responding partner, and has associated data $(3, 1 - i)$. The responding partner will then:

- If $\nu_0 = \nu_3$ and $\nu_1 = \nu_2$, the responding partner outputs $\top$ and stops.
- If $\nu_0 > \nu_3$, the responding partner outputs $\bot$ and stops.
- Otherwise, it waits for a close finish message.

When the close response message is received by the closing partner:

- If $\nu_2 = \nu_1$, output $\top$ and stop.
- If the number of received messages now equals ν_2, send a close finish message that contains the number of messages sent ν_0 and received ν_2, and has associated data $(4, i)$. Then output $\top$ and stop.
- Output $\bot$ and stop.

When the responding partner receives the close finish message, it outputs $\top$.

The three messages sent at the end of the protocol ensure that the two parties agree on the number of messages sent and received. The associated data ensures that these special messages cannot be confused with ordinary messages. Also, after sending these special messages, the number of sent messages will never change. It follows that the protocol will either ensure that every message is delievered, or the protocol will signal failure.

Because the channel is asynchronous, both parties can initiate the close at the same time. In this case, both parties send close response and close finish messages. This can be further optimised, but the benefit may be small.

Remark. The requirement sc-4 implies agreement on arrival and order for each sender's messages separately, but does not imply agreement on the order on all sent messages. That is, Alice knows the order in which Bob sent messages, but she may not know if Bob sent a particular message before or after receiving a given message from Alice. (Obviously, the messages themselves may contain information allowing Alice to decide this, but they need not.)

This is a consequence of our decision to have asynchronous communications. In fact, our channel model does not include any functionality for Alice and Bob to share this information.

We now have two choices. The first is to say that this should be part of the functionality of a channel. The other is to say that this is better handled at the application layer. Both choices make sense.

However, it is easy to augment a channel to provide this information. When Alice sends a message, she includes the number of received messages. Likewise for Bob. It is now clear that if sc-4 holds, Alice and Bob have enough information to recreate the other's exact message order. We will therefore not change our channel definition.

Remark. Another feature that could be included in a channel is acknowledgement of receipt. It is possible to build this into the cryptographic protocol, but it will often have a significant cost, both in round-trip time and in complexity. This can often be built into the application layer at acceptable cost.

The above schemes are based straight-forward on generic constructions, and illustrate the obvious power and utility in our definitions of symmetric cryptosystem with associated data.

Multiple Channels   Usually for applications that require multiple channels, there will also be multiple keys. In this case, everything is simple, and the above theory will work quite well. However, sometimes it is convenient to have multiple channels with a single shared key.

One solution is for the initiating player to send a random value to the responding player, who responds with a random value. The two random values form a *session identifier*, which is included in the associated data. The two random values can be sent in the clear.

We could also derive a key from a master key and the random values.

7.5   HASH FUNCTIONS

A *hash function* is a function $h : S \to T$. This definition is not very specific, and in particular there is no requirement that the cardinality of T is smaller than the cardinality of S. But in some settings it makes sense to be more restrictive. When S is a set of strings, we require that the computational cost of computing the hash of a string is linear in the length of the string. But a hash function remains a very general concept.

Remark. Hash functions are *unkeyed* constructions, unlike all the other constructions discussed so far in this chapter. So what is this discussion on hash functions doing in this chapter? The simple answer is that the most commonly used hash functions are designed and analysed using very similar techniques to the other symmetric primitives we have discussed. Even though we do not study these underlying techniques, it makes sense to classify hash functions as symmetric cryptography.

One of the properties we often want from a hash function is that it should *behave* as if it is an injective function. In other words, it should be infeasible to find collisions as discussed in Section 4.2. This is most interesting when the cardinality of S is greater than that of T, in which case h cannot be injective. But there are also interesting cases where T is the set with greater cardinality, but where we cannot use a function that is provably injective.

A hash function will in general not be invertible, but finding preimages is possible. Again, we often want this to be infeasible. There are two flavours of this variant: For the first, the adversary must produce a preimage of a given element of T, while for the second, the adversary must produce a collision for a given element of S, a second preimage of its image.

Definition 7.15. Let $h : S \to T$ be a hash function. Table 7.2 defines τ-*preimage finder*, τ-*second preimage finder* and τ-*collision finder*. The first two also involve sampling their input from some probability space $\mathcal{X}$, and the time bound τ includes the time required to sample the input. The advantage

Notion	Input	Output	Success event E_{notion}
preimage	$u \in T$	$s \in S$	$h(s) = u$
sec. pre.	$s \in S$	$s' \in S$	$s \neq s', \ h(s) = h(s')$
collision	none	$(s, s') \in S \times S$	$s \neq s', \ h(s) = h(s')$

TABLE 7.2 A τ-notion *finder* for a hash function h is an algorithm that given input as specified in the table produces output as specified in the table. The final column in the table lists the finder's success criterion.

of a finder $\mathcal{A}$ is

$$\mathbf{Adv}_{h,\mathcal{X}}^{\text{preimage}}(\mathcal{A}) = \Pr[E_{\text{preimage}}], \qquad \mathbf{Adv}_{h}^{\text{collision}}(\mathcal{A}) = \Pr[E_{\text{collision}}] \text{ and}$$
$$\mathbf{Adv}_{h,\mathcal{X}}^{\text{sec. pre.}}(\mathcal{A}) = \Pr[E_{\text{sec. pre.}}].$$

Remark. We omit $\mathcal{X}$ if it is the uniform probability space on T. For the probability space induced by the uniform distribution on S and h, we write S.

There is one immediate problem. Applying our usual theorem statement style to hash functions and collision finders as defined, we would state a theorem saying that if a certain adversary against a cryptosystem exists, then there exists a collision finder for the hash function. However, this does not actually say anything. If the hash function is not injective, there exists a collision $(s_0, s_1) \in S \times S$. Trivially then, there also exists a collision finder, namely the algorithm that outputs the collision (s_0, s_1). Therefore, the statement that a collision finder exists is trivially true, and our theorem statement is empty.

One option is to consider families of functions, and require that the collision finder must be able to work with a function chosen at random from this family. This gives us meaningful theorem statements, but it does not actually model reality, since the function is often not chosen at random. In fact, it is often a design goal that all the functions are equally good, and then certain non-technical reasons make it desirable to choose a natural function from the family. But this natural choice is by its very nature not random.

However, existence is not knowledge. We know that trivial collision finders exist, but we do not know how to explicitly describe any such collision finder. We should require that a theorem not only shows that a collision finder exists but also gives an explicit description of such a collision finder. Immediately, we note that the description may rely on other hypothesised algorithms, but the hash function should be easy to construct from these algorithms.

We restate the Merkle-Damgård construction from Section 4.2.2.

Example 7.7. Let $\mathfrak{A}, S', T$ be a finite sets, let $\mathfrak{A}^*$ denote the set of all finite strings of symbols from $\mathfrak{A}$, and let $f : \{0, 1\} \times \mathfrak{A} \times T \to T$ be a compression function. For any $u_0 \in T$, we define a function $h_{u_0} : \mathfrak{A}^* \to T$ as follows: Given a string $s_1 s_2 \ldots s_l \in \mathfrak{A}^*$, define $h_{u_0}(s_1 s_2 \ldots s_l) = u_l$ where $u_1, \ldots, u_l$ are

$$u_1 = f(1, s_1, u_0) \qquad\qquad u_i = f(0, s_i, u_{i-1}), \qquad 2 \leq i \leq l$$

Observe that computing h_{u_0} on a string of length l costs l computations of f, so the computational cost is linear in the length of the input.

T **Proposition 7.12.** *Let $\mathcal{A}$ be any τ-collision finder for h_{s_0} from Example 7.7. Then there exists a τ'-collision finder $\mathcal{B}$ for f, with τ' essentially equal to τ, such that*

$$\mathbf{Adv}_h^{\mathrm{collision}}(\mathcal{A}) = \mathbf{Adv}_f^{\mathrm{collision}}(\mathcal{B}).$$

The algorithm $\mathcal{B}$ is easy to derive from $\mathcal{A}$.

The proposition follows directly from Theorem 4.1 and its proof.

For (second) preimage finders, we run into a similar problem. In the extreme case, when $\Pr[s = s_0 \mid s \xleftarrow{r} \mathcal{X}] = 1$ for some $s_0 \in T$, we may again have trivial (second) preimage finders. We could try to avoid this, say by subtracting some probability derived from $\mathcal{X}$ from the advantage, but this does not work very well. Essentially, we shall have to use our judgement and remember that having a large advantage against a hash function relative to some very special probability space $\mathcal{X}$ may not be very useful.

As discussed in Section 4.2 there are some seemingly very natural relations between these three notions, but not all of them hold. First, it is clear that in general a collision finder does not imply a second preimage finder, because there is no reason to assume that a given collision relates in any way to the preimage we get. The other direction, however, holds.

Again, a second preimage finder does not imply a preimage finder, since the second preimage finder has more information to work with. A preimage finder need not imply a second preimage finder, as shown by Exercise 4.1. However, if we require that the preimage finder works for a part of the hash function that is sufficiently non-injective, this obstacle goes away and any such preimage finder does result in a second preimage finder, and then also a collision finder.

E *Exercise 7.25.* Let $h : S \to T$ be a hash function. Let $\mathcal{A}$ be a τ-second preimage finder for h with respect to a probability space $\mathcal{X}$ on S. Show that there exists a τ'-collision finder $\mathcal{B}$ for h, with τ' essentially equal to τ, such that

$$\mathbf{Adv}_h^{\mathrm{collision}}(\mathcal{A}) = \mathbf{Adv}_{h,\mathcal{X}}^{\mathrm{sec.\ pre.}}(\mathcal{A}),$$

and that $\mathcal{B}$ is easy to derive from $\mathcal{A}$.

E *Exercise 7.26.* Let $h : S \to T$ be a hash function. Let $h' : S \times \{0,1\} \to T$ be

$$h'(s, b) = h(s).$$

Suppose $\mathcal{A}$ is a τ-preimage finder for h' with respect to a probability space $\mathcal{X}$ on T. Show that there exists a τ'-preimage finder $\mathcal{B}$ for h with respect to $\mathcal{X}$, with τ' essentially equal to τ, and

$$\mathbf{Adv}_{h,\mathcal{X}}^{\text{preimage}}(\mathcal{B}) = \mathbf{Adv}_{h',\mathcal{X}}^{\text{preimage}}(\mathcal{A}).$$

Also show that there exists a trivial second preimage finder for h' with respect to any probability space $\mathcal{X}$ with advantage 1.

E *Exercise 7.27.* Suppose $h : S \to T$ is a hash function, with $T \subseteq S$. Show that there exists a hash function $h' : S \to \{0,1\} \times T$, such that for any τ-second preimage finder $\mathcal{A}$ for h' with respect to some probability space $\mathcal{X}$, there exists a τ'-second preimage finder $\mathcal{B}_1$ for h with respect to $\mathcal{X}$, such that τ' is essentially equal to τ, and

$$\mathbf{Adv}_{h',\mathcal{X}}^{\text{sec. pre.}}(\mathcal{B}_1) = \mathbf{Adv}_{h,\mathcal{X}}^{\text{sec. pre.}}(\mathcal{A}).$$

Also show that there is a trivial preimage finder $\mathcal{B}_2$ for h' with respect to the uniform probability space $\mathcal{X}'$ on the range of h', such that

$$\mathbf{Adv}_{h',\mathcal{X}'}^{\text{preimage}}(\mathcal{B}_2) = 1/2.$$

Hint: Use the construction from Exercise 4.1.

E *Exercise 7.28.* Let $h : S \to T$ be a hash function, and let $V \subseteq S$ be a set such that $|V| \ggg |T|$. Let $\mathcal{X}_0$ be the uniform probability space on V, and let $\mathcal{X}$ be the probability space on T induced by sampling s from $\mathcal{X}_0$ and computing $u \leftarrow h(s)$. Show that for any τ-preimage finder $\mathcal{A}$ for h with respect to $\mathcal{X}$, there exists a τ'-second preimage finder $\mathcal{B}$ for h with respect to $\mathcal{X}_0$, with τ' essentially equal to τ, such that the advantage of $\mathcal{B}$ is roughly the same as the advantage of $\mathcal{A}$, and that the algorithm $\mathcal{B}$ is easy to derive from $\mathcal{A}$.

7.6 IDEAL MODELS

There are many cryptographic functions that are random-looking, in the sense that except for being computable functions, they do not have any properties that you would not expect an average function to have. While we have a very clear grasp of this notion intuitively, it turns out to be very hard to capture this notion mathematically. Unlike when we discussed pseudo-random functions and permutations, there is no key that we can leverage.

E *Example 7.8.* For many reasonable hash functions h, the HMAC construction with some fixed key is considered to be random-looking.

Instead of modelling this intuitive notion, which we cannot, we introduce an idealisation. We stop working in the real world, called the *standard model*, and instead work in an ideal world.

To analyse a construction using a random-looking function, we replace the random-looking function by a random function. We then analyse the modified, *ideal world* construction. Our hope is that the random-looking function has no special properties that interact badly with our cryptographic construction.

As the previous paragraph makes clear, design and analysis in the random oracle model is a *heuristic* approach. In fact, we know that there are constructions that are insecure regardless of which random-looking function we replace the random oracle with. However, even given these separations, the random oracle model has shown itself to be a powerful heuristic and is today more or less essential in many applications in order to find efficient constructions.

Arbitrary functions on sets of non-trivial size are of course too large objects to work with in practice. But we want to work with them, and we want to isolate this impracticality somehow. We use a concept called an *oracle*, which is simply an interactive machine that responds to queries in a specified way and is allowed to do computationally impossible things in order to respond.

Remark. The word *oracle* as used in the literature carries many different meanings. The above is one definition. Earlier we discussed an alternative way to define security experiments, using oracles. In the current meaning, the word oracle implies the ability to do impractical computations, while in the security experiment it merely alludes to the ability to answer certain queries.

Before our oracle does anything, it samples a function from the uniform distribution on the set of functions with the required domain and range. Any other interactive machine may then query the oracle with elements from the domain, to which the oracle responds with the function values. This is the *random oracle model (ROM)*. There may be more than one random oracle.

The random oracle can model many functions that we have worked with previously. Some of our problems become easy in the random oracle model.

Example 7.9. Consider a function $f : \mathfrak{K} \times S \to T$. If we model this as a random oracle, then f is a very good pseudo-random function family. In this case, any one of the functions $f(k, \cdot)$ is a random function, so the only way to distinguish $f(k, \cdot)$ from a random function is to evaluate f at the experiment's key.

The probability of finding the correct key after testing $l < |\mathfrak{K}|$ keys is $l/|\mathfrak{K}|$. Assume that if the adversary does test the correct key, it will output 0, and that if it does not find the correct key, it outputs 0 with probability p. We can then compute its advantage as

$$\mathbf{Adv}_f^{\mathrm{prf}}(\mathcal{A}) \leq \left| \frac{l}{|\mathfrak{K}|} + p\frac{|\mathfrak{K}| - l}{|\mathfrak{K}|} - p \right| = (1 - p)\frac{l}{|\mathfrak{K}|}.$$

It follows that the optimal adversary outputs 0 only if it finds the correct key, and otherwise outputs 1. This adversary has advantage $l/|\mathfrak{K}|$.

The random oracle does as follows:

- Let $C_0 = C_1 = \emptyset$.
- When it receives a query s (*hash*), do:
    - If $s \notin C_1$, sample $u_0 \xleftarrow{r} T$, and add s to C_1 and (s, u_0) to C_0.
    - Find $(s, u) \in C_0$ and respond with u.

FIGURE 7.17 Random oracle with domain S and range T.

Example 7.10. If we model a hash function $h : S \to T$ as a random oracle, the probability of collisions is bounded by Proposition 2.11. Also, it will be hard to find preimages and second preimages for any reasonable probability space on S. In other words, a random oracle would be a very good hash function.

We should not use the random oracle model like this. We replace careful analysis with a strong assumption that is harder to analyse. We should only use random oracles when we otherwise cannot prove anything useful.

Instead of sampling a function, we shall work with the equivalent formulation from Figure 7.17. This alternative maintains a description of a partial function and samples *function values* as needed. This formulation is easier to manipulate, which is something that we usually need to do in security proofs.

We will often turn algorithms in the random oracle model into general algorithms for solving some problems. Forget for a moment that an actual practical adversary against our cryptosystem would not work in the random oracle model, but assume that it could be expressed as such. Then in principle, we should be able to run these algorithms in order to solve these problems. However, these algorithms will typically work with an implementation of the random oracle similar to that in Figure 7.17. If there are many queries to the random oracle, which there may very well be, this actually requires significant amounts of storage and the total time required for sampling and lookups may become significant. Sometimes there are tricks that can be played with pseudo-random functions in order to get a faster implementation.

But the main argument for why this is not an issue is that we never expect to run such an adversary. First of all, we have to assume that the underlying problems are hard, in which case the issue will never arise. Remember that if the cryptography that defends our society collapses, it will probably be small consolation that we can now do some computations more efficiently. Practical cryptography is not a win-win scenario, in this sense.

However, assumptions may very well be wrong. In the event that a practical adversary is found against some cryptosystem, it is unlikely that the cryptosystem attacked contains some key insight into solving the underlying problem. It follows that it should be possible to extract a specialised algorithm for solving the underlying problem from the adversary, stripped of the cryptographic interface. Generating many random values may be part of the attack, but if not, most of the random oracle could probably be removed.

Remark. A word of warning. We should be very skeptical about rationalising like in the above two paragraphs. We are doing provable security in order to ensure that our cryptographic designs have a sound theoretical foundation. Discarding this foundation when we encounter inconvenient results would be unwise and dangerous. But we should keep the big picture in mind.

A pseudo-random function family can be thought of as a function from a set of keys to a set of functions, each key mapping to a function. A block cipher (pseudo-random permutation family) can likewise be thought of as a function from a set of keys to a set of permutations. Just as for random-looking functions, some block ciphers are properly random-looking, and it therefore seems plausible to model them using a random map from the set of keys to the set of permutations. This is the *ideal cipher model*.

The ideal cipher model is not as widely used as the random oracle model, but we could use it to understand the tightness loss in multi-key settings.

Remark. The Feistel cipher approach to block cipher construction can be adapted to use random-looking functions as round functions. In the random oracle model, this Luby-Rackoff construction can be used to construct an ideal object that is indistinguishable from an ideal cipher. Likewise, there are constructions that can be used in the ideal cipher model to create a function that looks like a random function. These constructions are interesting, but beyond the scope of this book.

A special case of the ideal cipher model is the *random permutation model*, where we have a single permutation and its inverse. It is a special case, since fixing the key used in the ideal cipher model produces one random permuation.

CHAPTER 8

Public Key Encryption

Definitions of security for public key encryption is in some sense similar to definitions of security for symmetric encryption, which is both good and not unexpected. However, since symmetric encryption has different functional properties than public key encryption, the security notions are still somewhat different. For instance, the notion of integrity does not make much sense for public key encryption. Instead, we shall expand on the notion of non-malleability discussed briefly for symmetric encryption. The reader is therefore encouraged to notice similarities in definitions and proof techniques, but we caution against reading too much into these similarities, or making light of the differences.

The main design paradigm for public key encryption is hybrid encryption, which uses symmetric cryptosystems to build practical public key encryption.

The class of so-called homomorphic public key cryptosystems is of particular interest for many applications. To a certain extent, these cryptosystems allow a limited form of computation on encrypted data, which is both useful in general and is also a key property for many interesting constructions.

Commitment schemes are not public key encryption schemes, but the design and analysis is very similar to design and analysis of public key encryption schemes. Discussing commitment schemes therefore makes sense in this chapter, although we will only use them in later chapters.

We shall also make a first attempt at designing and analysing voting schemes, since voting schemes are strongly related to public key encryption.

8.1   DEFINING SECURITY

The many issues involved in defining security notions for encryption have already been extensively discussed in Section 7.1.

The major difference between symmetric cryptography and public key cryptography is that there is no need for *chosen plaintext* queries in the security games, since public key encryption implies that the adversary must have the public encryption key, and can therefore encrypt ciphertexts without any

DOI: 10.1201/9781003149422-8

secret information. Obviously, the adversary must get the encryption key, so we simply start off the games by giving the encryption key to the adversary.

Before we discuss security, we shall define public key cryptosystems including associated data. Just like for symmetric cryptosystems, including associated data extends the functionality of public key cryptosystems and makes it easier to design larger systems.

Definition 8.1. A *public key encryption* scheme PKE consists of three algorithms $(\mathcal{K}, \mathcal{E}, \mathcal{D})$:

- The *key generation* algorithm $\mathcal{K}$ takes no input and outputs an *encryption key* ek and a *decryption key* dk. To each encryption key ek there is an associated message set $\mathfrak{M}_{ek}$ and set of associated data $\mathfrak{F}_{ek}$.

- The *encryption* algorithm $\mathcal{E}$ takes as input an encryption key, associated data and a message. It outputs a ciphertext.

- The *decryption* algorithm $\mathcal{D}$ takes as input a decryption key, associated data and a ciphertext and outputs either a message or the special symbol $\perp$ indicating decryption failure.

We require that for any key pair (ek, dk) output by $\mathcal{K}$, any associated data $ad \in \mathfrak{F}_{ek}$ and any message $m \in \mathfrak{M}_{ek}$

$$\mathcal{D}(dk, ad, \mathcal{E}(ek, ad, m)) = m.$$

While the concept does not matter much, it is convenient for bookkeeping reasons to define a value for a public key cryptosystem, namely the probability of getting a collision among a set of encryption keys, and the probability of getting a collision among a set of ciphertexts. This value must be small if our cryptosystem is to be secure. In most cases, it will be very small and easy to determine, so we shall not bother computing it for most cryptosystems.

Definition 8.2. Let PKE $= (\mathcal{K}, \mathcal{E}, \mathcal{D})$ be a public key encryption scheme. Let $f_1(l)$ denote the probability that there is a collision of encryption keys after generating l key pairs. For any ek that can be output by $\mathcal{K}$ and any $m \in \mathfrak{M}_{ek}$, let $f_2(l, ek, m)$ denote the probability that there is a collision of ciphertexts after creating l encryptions of m under ek. Similarily, for any set S_0 of l messages and any set S_1 of l ciphertexts, let $f_3(l, ek, S_0, S_1)$ denote the probability that any of l encryptions of messages from S_0 collide with any ciphertext in S_1. The *collision probability* for PKE is

$$\mathbf{Col}_{\text{PKE}}(l) = \max\{f_1(l), \max_{ek, m} f_2(l, ek, m), \max_{ek, S_0, S_1} f_3(l, ek, S_0, S_1)\}.$$

Exercise 8.1. Let PKE $= (\mathcal{K}, \mathcal{E}, \mathcal{D})$ be a public key encryption scheme where $\mathcal{E}$ is a deterministic algorithm. Show that $\mathbf{Col}_{\text{PKE}}(2) = 1$. Compute the collision probability for l for the ElGamal cryptosystem from Section 3.2.1.

8.1.1 Confidentiality

The public key encryption confidentiality experiments proceed as follows:

1. Let $C = \emptyset$ and
   - (sem) let $\sigma = b = \perp$.
   - (ind)/(ror) sample $b \xleftarrow{r} \{0, 1\}$.

2. Compute $(ek, dk) \leftarrow \mathcal{K}$. Send ek to the adversary.

3. When the adversary sends a query $(ad, \mathcal{X})$ / (ad, m_0, m_1) / (ad, m_0) (*challenge* for semantic security, indistinguishability and real-or-random, respectively), do:

   (a) (sem) Sample $(m, \sigma', b'') \xleftarrow{r} \mathcal{X}(\sigma)$, compute $c \leftarrow \mathcal{E}(ek, ad, m)$, and set $\sigma \leftarrow \sigma'$. If $b = \perp$, then $b \leftarrow b''$.

   (b) (ind) Compute $c \leftarrow \mathcal{E}(ek, ad, m_b)$.

   (c) (ror) Sample $m_1 \xleftarrow{r} \{m \in \mathfrak{M}_{ek} \mid m \text{ has the same length as } m_0\}$ and compute $c \leftarrow \mathcal{E}(ek, ad, m_b)$.

   Add (ad, c) to C and send c to the adversary.

4. When the adversary sends a query (ad, c) (*chosen ciphertext*), do:

   (a) If $(ad, c) \in C$, send $\perp$ to the adversary.

   (b) Otherwise, $m \leftarrow \mathcal{D}(dk, ad, c)$ and send m to the adversary.

Eventually, the adversary outputs $b' \in \{0, 1\}$. If $b = \perp$ at that time, the experiment samples $b \xleftarrow{r} \{0, 1\}$.

FIGURE 8.1 The security experiments $\mathbf{Exp}_{\mathrm{PKE}}^{\mathrm{xxx}}(\mathcal{A})$, $\mathrm{xxx} \in \{\mathsf{sem}, \mathsf{ind}, \mathsf{ror}\}$ for a public key cryptosystem $\mathrm{PKE} = (\mathcal{K}, \mathcal{E}, \mathcal{D})$ with adversary $\mathcal{A}$.

Just as for symmetric cryptosystems, we have three equivalent notions of confidentiality.

Definition 8.3. A (τ, l_c, l_d)-*adversary against semantic security (respectively indistinguishability, real-or-random security) for a public key cryptosystem is an interactive algorithm $\mathcal{A}$ that interacts with the experiment in Figure 8.1 making at most l_c challenge queries and l_d chosen ciphertext queries, and where the runtime of the adversary and the experiment is at most τ.

For $\mathrm{xxx} \in \{\mathsf{sem}, \mathsf{ind}, \mathsf{ror}\}$, the *advantage* of this adversary is defined to be

$$\mathbf{Adv}_{\mathrm{PKE}}^{\mathrm{xxx}}(\mathcal{A}) = 2|\Pr[E] - 1/2|,$$

where E is the event that b' output by $\mathcal{A}$ equals the experiment's b.

If the number of chosen ciphertext queries $l_d = 0$, we say that $\mathcal{A}$ is a *chosen plaintext adversary* and use the notation $\mathsf{sem\text{-}cpa}$ (respectively $\mathsf{ind\text{-}cpa}$, ror-cpa). Otherwise we may say that $\mathcal{A}$ is a *chosen ciphertext adversary* and use the notation sem-cca (respectively ind-cca, ror-cca).

Remark. Security for public key cryptosystems is often defined with a single challenge query. We shall study this further in Section 8.1.2.

As expected, these notions are equivalent. The first two results are straightforward adaptions of one form of challenge query to another form. The third result is different, showing that in the real case we can simulate the semantic security game and deduce the adversary's success probability, while in the random case the adversary cannot guess the challenge bit. Compare the results to the corresponding results in Section 7.1.1.

Ⓣ **Proposition 8.1.** *Let $\mathcal{A}$ be a (τ, l_c, l_d)-adversary against indistinguishability for the public key cryptosystem* PKE. *The there exists a (τ', l_c, l_d)-adversary against semantic security for* PKE, *where τ' is essentially equal to τ, and*

$$\mathbf{Adv}_{\mathrm{PKE}}^{\mathsf{sem}}(\mathcal{B}) = \mathbf{Adv}_{\mathrm{PKE}}^{\mathsf{ind}}(\mathcal{A}).$$

Ⓣ **Proposition 8.2.** *Let $\mathcal{A}$ be a (τ, l_c, l_d)-adversary against real-or-random for the public key cryptosystem* PKE. *The there exists a (τ', l_c, l_d)-adversary against indistinguishability for* PKE, *where τ' is essentially equal to τ, and*

$$\mathbf{Adv}_{\mathrm{PKE}}^{\mathsf{ind}}(\mathcal{B}) = \mathbf{Adv}_{\mathrm{PKE}}^{\mathsf{ror}}(\mathcal{A}).$$

Ⓣ **Proposition 8.3.** *Let $\mathcal{A}$ be a (τ, l_c, l_d)-adversary against semantic security for the public key cryptosystem* PKE. *The there exists a (τ', l_c, l_d)-adversary against real-or-random for* PKE, *where τ' is essentially equal to τ, and*

$$\mathbf{Adv}_{\mathrm{PKE}}^{\mathsf{ror}}(\mathcal{A}) = \frac{1}{2}\mathbf{Adv}_{\mathrm{PKE}}^{\mathsf{sem}}(\mathcal{B}).$$

Ⓔ *Exercise* 8.2. Prove Propositions 8.1, 8.2 and 8.3.

Remark. Historically, when considering single-challenge variants of the security notions, some emphasis was placed on a variant of a chosen ciphertext attack where the adversary only made chosen ciphertext queries before the first challenge query, and never after. This is a *non-adaptive chosen ciphertext attack*, in the sense that the adversary's chosen ciphertext queries cannot be adapted to the challenge ciphertexts. (Note that it is only the challenge ciphertexts that the adversary may not adapt to. The adversary makes queries one at a time, and may adapt queries to previous responses. In other situations, non-adaptive may have a different and perhaps more restrictive meaning.) This does not usually make sense for applications, but there are certain cases where it makes sense, and there is a certain theoretical interest in studying also this attack. Also, there are classes of public key cryptosystems that can only achieve security against non-adaptive chosen ciphertext attacks.

This also explains why the plain chosen ciphertext attack is often called an *adaptive chosen ciphertext attack* in the literature.

[E] *Exercise* 8.3. Show that if there is a public key cryptosystem that is indistinguishable under chosen plaintext attack, there is a cryptosystem that is indistinguishable under chosen plaintext attack, but not under non-adaptive chosen ciphertext attack.

Hint: Cheat. Let the decryption algorithm output the decryption key for certain invalid ciphertexts. The non-adaptive chosen ciphertext attack is now easy. Then prove that it remains a public key cryptosystem and that it remains indistinguishable under chosen plaintext attacks.

Example: Chosen Plaintext Security and ElGamal   As we have seen in Section 3.2 we can turn the Diffie-Hellman key agreement protocol into a public key cryptosystem. This cryptosystem is trivially not chosen ciphertext secure, but if we take care about which group we use, it can be chosen plaintext secure. As usual, associated data is not interesting when we do not consider chosen ciphertext security.

[E] *Example* 8.1. Consider ElGamal encryption from Section 3.2.1 based on a group G of prime order p with generator g. The ElGamal public key is $y = g^b$, and an ElGamal ciphertext is (x, w) where $x = g^r$ and $w = y^r m$.

It seems hard to prove anything useful about ElGamal relying only on the discrete logarithm problem. Instead, we will need a stronger property, related to breaking the Diffie-Hellman protocol: Essentially, an eavesdropper cannot distinguish the shared secret from random group elements.

We begin with the simplest possible result for ElGamal. We shall improve on this result later.

[T] **Proposition 8.4.** *Let $\mathcal{A}$ be a $(\tau, 1, 0)$-adversary against real-or-random for ElGamal. Then there exists a τ'-distinguisher $\mathcal{B}$ for Decision Diffie-Hellman, with τ' essentially equal to τ, such that*

$$\mathbf{Adv}^{\text{ror-cpa}}_{\text{ELGAMAL}}(\mathcal{A}) = \mathbf{Adv}^{\text{DDH}}_G(\mathcal{B}).$$

Proof. We create an adversary $\mathcal{B}$ against DDH. It gets (x, y, z) as input and runs a copy of $\mathcal{A}$. It gives $\mathcal{A}$ the encryption key y. When $\mathcal{A}$ chooses m, $\mathcal{B}$ gives $\mathcal{A}$ the ciphertext (x, zm). When $\mathcal{A}$ outputs b', $\mathcal{B}$ outputs b'.

In both cases, the encryption key y has been sampled from the uniform distribution on G, just like the ElGamal key generation algorithm would do.

If (x, y, z) is a DDH tuple, then x has been sampled from the uniform distribution on G, just like the ElGamal encryption algorithm would sample it. Also, there exists r, t such that $x = g^r$, $y = g^t$ and $z = g^{rt}$. Therefore,

$(x, zm) = (g^r, y^r m)$ is an encryption of m, distributed exactly like encryptions of m output by the ElGamal encryption algorithm.

If (x, y, z) is a random tuple, then zm is a random value multiplied by something chosen by the adversary, which is distributed uniformly. It follows that (x, zm) is uniformly distributed in $G \times G$, just like an ElGamal encryption of a random message.

We compute the advantage of $\mathcal{B}$ as

$$\mathbf{Adv}_G^{\mathsf{DDH}}(\mathcal{B}) = |\Pr[b' = 1 \mid \text{DDH tuple}] - \Pr[b' = 1 \mid \text{random tuple}]|$$
$$= |\Pr[b' = 1 \mid b = 0] - \Pr[b' = 1 \mid b = 1]| = \mathbf{Adv}_{ElGamal}^{\mathsf{ror\text{-}cpa}}(\mathcal{A}),$$

where the probability in the second line refers to the game where $\mathcal{A}$ interacts with the real-or-random experiment.

The runtime of $\mathcal{B}$ is essentially the same as the runtime of $\mathcal{A}$ and the real-or-random experiment, so we have a τ'-adversary $\mathcal{B}$ against DDH with the same advantage as $\mathcal{A}$, where τ' is essentially equal to τ. $\qquad\square$

Towards Chosen Ciphertext Security Understanding how to construct public key cryptosystems with confidentiality under chosen ciphertext attack took a long time. A number of proposals were analysed with respect to various chosen ciphertext notions, but at first constructions were expensive and difficult. We shall look at some ideas for constructing secure systems.

One approach was in some sense inspired by ciphertext integrity for symmetric cryptosystems, the idea that it is hard to create a valid ciphertext without knowing the key. Obviously, anyone can create a ciphertext, so this does not work directly, but instead the idea becomes that it should be hard to create a valid ciphertext without in some sense knowing the decryption of the ciphertext. If that is true, chosen ciphertext queries are pointless, and we only need to analyse a chosen plaintext attack.

For our first example, we shall restrict ourselves to non-adaptive chosen ciphertext attacks. In these attacks, the point of the chosen ciphertext queries is to gain some information that will uncover some information about the decryption key, or help in choosing clever challenge messages.

The inspiration comes from looking at Diffie-Hellman tuples, the subset $\{(g^r, g^t, g^{rt})\} \subseteq G^3$. A group element y defines the subgroup $H_y = \{(g^r, y^r)\} \subseteq G^2$. Clearly, (x, z) is in this subgroup, but knowing only y, can we find other elements in the group? Choosing r and computing (g^r, y^r) is one method that obviously works. However, this also seems to be the only way to find elements of the subgroup. The hypothesis that this is the only efficient method is known as the *knowledge of exponent assumption*.

This motivates one variant of ElGamal (often known as *Damgård's ElGamal*) where the encryption key is (y_1, y_2). An encryption of $m \in G$ is $(g^r, y_1^r, y_2^r m)$. When decrypting (x_1, x_2, w), the decryption algorithm verifies that (x_1, x_2) lies in the appropriate subgroup, and then decrypts as $w x_1^{-b_2}$.

The difference from ElGamal is the second coordinate in the ciphertext, and verifying that this value is correct during decryption.

Note that the first two coordinates (x_1, x_2) of a valid ciphertext will lie in the subgroup H_{y_1}. By the knowledge of exponent assumption, the only way to come up with a valid ciphertext (x_1, x_2, w) is to choose r and compute $x_1 = g^r$ and $x_2 = y_1^r$. But in this case, the decryption of the ciphertext is wy_2^{-r}.

If we accept the hypothesis, chosen ciphertext queries before the challenge ciphertexts are revealed do not help the adversary. And against a chosen plaintext adversary, the cryptosystem reduces to plain ElGamal.

However, the cryptosystem falls trivially against an adaptive chosen ciphertext attack. Several approaches were attempted in order to defend against adaptive approaches. The main idea is to attach a tag to the ciphertext, and it should be hard to modify the ciphertext and the tag in such a way that the pair is still valid. The obvious approach is to use a sensible MAC and encrypt some randomness together with the message. The randomness will serve as the key for the MAC. That is, compute $w \leftarrow \mathcal{E}(ek, (r, m))$ and $t \leftarrow \mu(r, m)$, and we get the ciphertext $c = (w, t)$.

This approach seems to work well in practice. However, proving that it works is harder, and it cannot be done purely based on the usual MAC requirements. The reason is that tampering with the ciphertext may change the key in a predictable fashion, which means that if the MAC is vulnerable to a *related key* attack, the entire construction will be insecure.

8.1.2 A Single Challenge Suffices – Maybe

We said that sometimes security is defined for a single challenge query. We shall now prove that in some sense, it is sufficient to prove security for a single challenge query. However, this generic theorem is not *tight* in the sense that the advantage bound contains a factor l_c. Proving security for multiple challenges directly, without this non-tightness, would be better better. We begin with the generic result and illustrate later with two examples.

Proposition 8.5. *Let $\mathcal{A}$ be a (τ, l_c, l_d)-adversary against indistinguishability for PKE. Then there exists a $(\tau', 1, l_d)$-adversary $\mathcal{B}$ against indistinguishability for PKE, where τ' is essentially τ, such that*

$$\mathbf{Adv}_{\mathrm{PKE}}^{\mathrm{ind}}(\mathcal{A}) \leq l_c \mathbf{Adv}_{\mathrm{PKE}}^{\mathrm{ind}}(\mathcal{B}).$$

Exercise 8.4. Prove Proposition 8.5. Hint: Look at Proposition 7.5.

Example 8.2. Propositions 8.4 and 8.5 say that any $(\tau, l_c, 0)$-adversary $\mathcal{A}$ against real-or-random security for ElGamal can be turned into a τ'-adversary $\mathcal{B}$ against DDH, where τ' is essentially equal to τ, and

$$\mathbf{Adv}_{\mathrm{ELGAMAL}}^{\mathrm{ror\text{-}cpa}}(\mathcal{A}) \leq l_c \mathbf{Adv}_{G}^{\mathrm{DDH}}(\mathcal{B}).$$

Example 8.3. Consider ElGamal encryption as in Example 8.1. Observe that if (x_1, w_1) and (x_2, w_2) decrypt to m_1 and m_2, respectively, then $(x_1 x_2, w_1 w_2)$ decrypts to $m_1 m_2$, and (x_2^r, w_2^r) decrypts to m_1^r.

Next, consider a tuple $(x, y, z) \in G^3$. If this is a DDH tuple, then with y as the ElGamal encryption key, both (g, y) and (x, z) are encryptions of 1. Then for r, t sampled from the uniform distribution on $\{0, 1, \ldots, p-1\}$ we have that

$$(g^r x^t, y^r z^t)$$

is an encryption of 1, distributed identically to the output of $\mathcal{E}$.

If the tuple is not a DDH tuple, then (x, z) is an encryption of a generator for the group (because its order is prime). Then for r, t sampled from the uniform distribution on $\{0, 1, \ldots, p-1\}$ we have that

$$(g^r x^t, y^r z^t)$$

is an encryption of a message sampled from the uniform distribution on G and the ciphertext is distributed identically to the output of $\mathcal{E}$.

Consider a $(\tau, l_c, 0)$-adversary $\mathcal{A}$ against real-or-random security for ElGamal. We create an adversary $\mathcal{B}$ against DDH. It gets (x, y, z) as input and runs a copy of $\mathcal{A}$. It gives $\mathcal{A}$ the encryption key y.

When the adversary makes a challenge query for m, $\mathcal{B}$ samples r, t from the uniform distribution on $\{0, 1, \ldots, p-1\}$ and computes the encryption as

$$(g^r x^t, y^r z^t m).$$

When the adversary $\mathcal{A}$ eventually outputs b', $\mathcal{B}$ outputs b'.

As discussed above, if (x, y, z) is a DDH tuple, $\mathcal{B}$ will create encryptions of the messages that $\mathcal{A}$ queried. If it is not a DDH tuple, $\mathcal{B}$ will create encryptions of random messages unrelated to the messages that $\mathcal{A}$ queried.

It follows that if the input to $\mathcal{B}$ is a DDH tuple, then $\mathcal{B}$ perfectly simulates a the real-or-random simulator with $b = 0$. If the input is not a DDH tuple, $\mathcal{B}$ perfectly simulates the real-or-random simulator with $b = 1$.

Let E_β be the event that $\mathcal{A}$ outputs 1 when interacting with the real-or-random simulator with $b = \beta$. Let F_D be the event that $\mathcal{B}$ outputs 1 when given a DDH tuple, and let $F_{\neg D}$ be the event that $\mathcal{B}$ outputs 1 when given a non-DDH tuple. We get that

$$\Pr[E_0] = \Pr[F_D] \qquad \text{and} \qquad \Pr[E_1] = \Pr[F_{\neg D}].$$

The remaining obstacle is that the DDH adversary faces two overlapping distributions, because a tuple sampled from the uniform distribution on G^3 will be a DDH tuple with probability $1/p$. Let F_R be the event that $\mathcal{B}$ outputs 1 when given a tuple sampled from the uniform distribution on G^3. Then

$$\Pr[F_R] = \Pr[F_D] \frac{1}{p} + \Pr[F_{\neg D}] \frac{p-1}{p}.$$

We compute

$$\mathbf{Adv}^{\text{ind-cpa}}_{\text{ELGAMAL}}(\mathcal{A}) = |\Pr[F_D] - \Pr[F_{\neg D}]|$$
$$= |\Pr[F_D] - \Pr[F_R] + \frac{1}{p}(\Pr[F_D] - \Pr[F_{\neg D}])| \le \mathbf{Adv}^{\text{DDH}}_G(\mathcal{B}) + \frac{1}{p}.$$

This is a significantly better result than in the previous example.

Remark. We rely on a property of the Decision Diffie-Hellman problem called *random self-reducibility*: we can turn all but a small fraction of DDH problem instances into random instances of the DDH problem with the same answer.

Example: Chosen Ciphertext Security and Cramer-Shoup   The Cramer-Shoup public key encryption scheme is in some sense is similar to ElGamal, augmented with ideas superficially similar to those in the previous section.

We begin by studying two maps $f_1 : V \times V \to \mathbb{F}$ and $f_2 : V \times V \times \mathbb{F} \times V \to \mathbb{F}$, where V is a vector space over a field $\mathbb{F}$ with a dot product, given by

$$f_1(\mathbf{b}, \mathbf{r}) = \mathbf{b} \cdot \mathbf{r} \quad \text{and} \quad f_2(\mathbf{b}', \mathbf{b}'', \alpha, \mathbf{r}) = f_1(\mathbf{b}', \mathbf{r}) + \alpha f_1(\mathbf{b}'', \mathbf{r}).$$

For any $\mathbf{b}, \mathbf{g} \in V$ and $r \in \mathbb{F}$, we have that

$$f_1(\mathbf{b}, r\mathbf{g}) = r f_1(\mathbf{b}, \mathbf{g}) \quad \text{and} \quad f_2(\mathbf{b}', \mathbf{b}'', \alpha, r\mathbf{g}) = r(f_1(\mathbf{b}', \mathbf{g}) + \alpha f_1(\mathbf{b}'', \mathbf{g})).$$

If we let $\beta_1 = f_1(\mathbf{b}, \mathbf{g})$, $\beta_2 = f_1(\mathbf{b}', \mathbf{g})$ and $\beta_3 = f_1(\mathbf{b}'', \mathbf{g})$, we get

$$f_1(\mathbf{b}, r\mathbf{g}) = r\beta_1 \quad \text{and} \quad f_2(\mathbf{b}', \mathbf{b}'', \alpha, r\mathbf{g}) = r(\beta_2 + \alpha\beta_3).$$

This means that β_1 determines the function f_1 on the subspace spanned by $\mathbf{g}$. But outside that subspace, β_1 does not determine the function.

E *Exercise 8.5.* Assume that $\dim V = l > 1$ and that $\mathbb{F}$ has q elements. Fix $\beta, \beta' \in \mathbb{F}$ and two linearly independent vectors $\mathbf{g}$ and $\mathbf{t}$. Show that there are exactly q^{l-2} vectors $\mathbf{b}$ in V such that $f_1(\mathbf{b}, \mathbf{g}) = \beta$ and $f_1(\mathbf{b}, \mathbf{t}) = \beta'$.

E *Exercise 8.6.* Assume that $\dim V = l > 1$ and that $\mathbb{F}$ has q elements. Fix $\beta_2, \beta_3, \beta', \beta'' \in \mathbb{F}$, two distinct elements α, α', two linearly independent vectors $\mathbf{g}$ and $\mathbf{t}$, and a vector $\mathbf{t}'$ not in the span of $\mathbf{g}$ and not equal to $\mathbf{t}$. Show that there are exactly q^{l-4} pairs of vectors $(\mathbf{b}, \mathbf{b}')$ in $V \times V$ such that

$$f_1(\mathbf{b}, \mathbf{g}) = \beta_2 \quad f_1(\mathbf{b}', \mathbf{g}) = \beta_3 \quad f_2(\mathbf{b}, \mathbf{b}', \alpha, \mathbf{t}) = \beta' \quad f_2(\mathbf{b}, \mathbf{b}', \alpha', \mathbf{t}') = \beta''.$$

E *Example 8.4.* Let G be a group of prime order p with generator g. Let $\mathfrak{F}$ be a set of associated data. Let $h : G^3 \times \mathfrak{F} \to \{0, 1, \ldots, p-1\}$ be a hash function. The *Cramer-Shoup public key encryption scheme* CS-PKE works as follows:

- The *key generation* algorithm samples $g_2 \xleftarrow{r} G\backslash\{1,g\}, b_1, b_2, b_3, b_4, b_5, b_6 \xleftarrow{r} \{0, 1, \ldots, p-1\}$. It computes $y_1 \leftarrow g^{b_1}g_2^{b_2}$, $y_2 \leftarrow g^{b_3}g_2^{b_4}$ and $y_3 \leftarrow g^{b_5}g_2^{b_6}$. It outputs $ek = (y_1, y_2, y_3)$ and $dk = (b_1, b_2, b_3, b_4, b_5, b_6)$.

- The *encryption* algorithm takes ek, $ad \in \mathfrak{F}$ and $m \in G$ as input. It samples $r \xleftarrow{r} \{0, 1, \ldots, p-1\}$, computes $x_1 \leftarrow g^r$, $x_2 \leftarrow g_2^r$, $w \leftarrow y_1^r m$ and

$$t \leftarrow (y_2 y_3^{h(x_1, x_2, w, ad)})^r,$$

  and outputs $c = (x_1, x_2, w, t)$.

- The *decryption* algorithm takes dk, ad and $c = (x_1, x_2, w, t)$ as input. If

$$t \neq x_1^{b_3} x_2^{b_4} (x_1^{b_5} x_2^{b_6})^{h(x_1, x_2, w, ad)}.$$

  it outputs $\perp$, otherwise it outputs $w x_1^{-b_1} x_2^{-b_2}$.

We can reason about the cryptosystem based on the arguments for the functions f_1 and f_2, because with $g_2 = g^\omega$ we get

$$y_1 = g^{f_1((b_1, b_2), (1, \omega))} \qquad y_2 = g^{f_1((b_3, b_4), (1, \omega))} \qquad y_3 = g^{f_1((b_5, b_6), (1, \omega))}$$

and $t = g^{r f_2((b_3, b_4), (b_5, b_6), h(x_1, x_2, w), (1, \omega))}$.

E | *Exercise* 8.7. Show that CS-PKE is a public key encryption scheme.

T | **Proposition 8.6.** *Let $\mathcal{A}$ be a $(\tau, 1, l_d)$-adversary against indistinguishability for the above public key encryption scheme CS-PKE. Then there exists a τ_1'-collision finder $\mathcal{B}_1$ for h and a τ_3'-adversary $\mathcal{B}_3$ against Decision Diffie-Hellman, where τ_1' and τ_3' are essentially equal to τ, and*

$$\mathbf{Adv}_{\text{CS-PKE}}^{\text{ind-cca}}(\mathcal{A}) \leq 2\mathbf{Adv}_h^{\text{collision}}(\mathcal{B}_1) + 2\mathbf{Adv}_G^{\text{DDH}}(\mathcal{B}_3) + \frac{2l_d + 6}{p}.$$

It is easy to construct $\mathcal{B}_1$ from $\mathcal{A}$.

This is the most complicated security proof yet, and in order to make the proof readable, we need to introduce some structure to the argument. The proof will proceed via a sequence of games, between the adversary $\mathcal{A}$ and an experiment. For each new game, we make some small modifications to the experiment, and then analyse the effect of these changes on the game.

The first change is to forbid chosen ciphertext queries that contain a hash function collision with the challenge ciphertext. This will later allow us to use Exercise 8.6, since α and α' will be hash values. Then we stop using the random value r for certain computations, instead doing the computations with the secret key. This will later allow us to use values for which we do not know r. Next, the experiment generates x_1, x_2 for the challenge ciphertext in a different way, which will hide the encrypted message by Exercise 8.5. Finally, we stop accepting chosen ciphertexts that could not have been created by the encryption algorithm. This is allowed because of Exercise 8.6, and it allows us to use Exercise 8.5 to show that the message is hidden.

Remark. We cannot use the random self-reducible trick from Example 8.3 because of the way we use Decision Diffie-Hellman in the proof. Thus we must rely on Proposition 8.5 to get a more useful result.

Proof of Proposition 8.6 In the following, let E_i be the event that the adversary's guess b' equals the challenge bit b of the experiment in Game i.

Ⓖ **Game 0** The initial game is $\mathcal{A}$ interacting with the indistinguishability experiment. Then

$$\mathbf{Adv}_{\text{CS-PKE}}^{\text{ind-cca}}(\mathcal{A}) = 2|\Pr[E_0] - 1/2|. \tag{8.1}$$

Let $\tau_0 = \tau$.

Ⓖ **Game 1** In this game, our experiment shall reject any chosen ciphertext query where the hash value is the same as the hash value from the single challenge ciphertext. If $\mathcal{A}$ exceeds its bounds, we stop.

The runtime cost of this change insignificant, so the runtime bound τ_1 on this game is essentially equal to τ_0.

Ⓣ **Lemma 8.7.** *It is easy to construct a τ_1'-collision finder $\mathcal{B}_1$ for h from $\mathcal{A}$, with τ_1' essentially equal to τ_1, and*

$$|\Pr[E_1] - \Pr[E_0]| \leq \mathbf{Adv}_h^{\text{collision}}(\mathcal{B}_1). \tag{8.2}$$

Proof. Let F_i be the event in Game i that a chosen ciphertext query with the same hash value as the challenge query happens. A rejection happens in this game that would not happen in the previous game only if F_1 happens. Until F_i happens, the two games proceed identically, which means that $\Pr[F_0] = \Pr[F_1]$. By Lemma 7.7, we have that

$$|\Pr[E_1 - \Pr[E_0]| \leq \Pr[F_1].$$

If F_1 happens, a collision for h has been found.

We construct a τ_1'-collision finder for h as follows. It runs this game, with the following modification. If the experiment in the game finds two matching hashes that would cause it to reject a ciphertext, the collision finder outputs the collision and stops. We have that τ_1' is essentially equal to τ_1, and

$$\mathbf{Adv}_h^{\text{collision}}(\mathcal{B}_1) = \Pr[F_1]. \qquad \square$$

Ⓖ **Game 2** In this game, we change how the experiment computes its single challenge ciphertext. It will now use the decryption key to compute the encryption and the tag, computing

$$w \leftarrow x_1^{y_1} x_2^{y_2} \qquad\qquad t \leftarrow x_1^{b_3} x_2^{b_4} (x_1^{b_5} x_2^{b_6})^{h(x_1,x_2,w,ad)}.$$

This changes nothing, so

$$\Pr[E_2] = \Pr[E_1]. \tag{8.3}$$

The extra runtime cost of this is insignificant, so the runtime bound τ_2 on this game is essentially equal to τ_1.

The value r sampled when creating x_1 and x_2 is not used afterwards.

G **Game 3** Again, we change how the experiment computes its single challenge ciphertext. It will now sample $r' \xleftarrow{r} \{0, 1, \ldots, p-1\} \setminus \{r\}$ and compute

$$x_2 \leftarrow g_2^{r'}.$$

If $\mathcal{A}$ exceeds its bounds, the experiment samples $b' \xleftarrow{r} \{0,1\}$ and stops.

Because we enforce the bounds and the extra work is insignificant, the runtime bound τ_3 on this game is essentially equal to τ_2.

Remark. We exclude the value r when sampling r'. This guarantees that (g_2, x_1, x_2) is not a DDH tuple, which we will need. But it does not match the DDH game, where we either sample a DDH tuple or a random tuple. This difference causes the $1/p$ terms in the next result.

T **Lemma 8.8.** *There exists a τ_3'-adversary $\mathcal{B}_3$ against Decision Diffie-Hellman such that*

$$|\Pr[E_3] - \Pr[E_2]| \leq \mathbf{Adv}_G^{\mathsf{DDH}}(\mathcal{B}_3) + \frac{3}{p}. \tag{8.4}$$

Proof. The adversary $\mathcal{B}_3$ gets a tuple (g_2, x_1, x_2) as input. If $g_2 \in \{1, g\}$, $\mathcal{B}_3$ outputs a random bit and stops. Otherwise, it runs a copy of $\mathcal{A}$ and a slightly modified experiment that uses g_2 from its input when creating the encryption key, and uses x_1 and x_2 from its input when creating the challenge. When $\mathcal{A}$ outputs b', $\mathcal{B}$ outputs 1 if $b' = b$ and 0 otherwise.

The runtime bound on $\mathcal{B}_3$ essentially equal to τ_3.

Let F_D be the event that $\mathcal{B}_3$ outputs 1 when its input is sampled from the uniform distribution on DDH tuples, let $F_{\neg D}$ be the event that $\mathcal{B}_3$ outputs 1 when its input is sampled from the uniform distribution on non-DDH tuples, and let F_R be the event that $\mathcal{B}_3$ outputs 1 when its input is sampled from the uniform distribution on G^3.

If the input is sampled from the uniform distribution on DDH tuples, then $\mathcal{B}_3$ simulates Game 2 perfectly, except with probability $2/p$. It follows that $\Pr[F_D] = \frac{p-2}{p}\Pr[E_1] + 1/p$.

If the input is sampled from the uniform distribution on non-DDH tuples, then $\mathcal{B}_3$ simulates Game 3 perfectly, except with probability $2/p$. It follows that $\Pr[F_{\neg D}] = \frac{p-2}{p}\Pr[E_2] + 1/p$.

The probability that a tuple sampled from the uniform distribution on G^3 is a DDH tuple is $1/p$, which means that

$$\Pr[F_R] = \frac{1}{p}\Pr[F_D] + \frac{p-1}{p}\Pr[F_{\neg D}].$$

We compute

$$\mathbf{Adv}_G^{\mathrm{DDH}}(\mathcal{B}_3) = |\Pr[F_D] - \Pr[F_R]|$$

$$= |\frac{p-1}{p}\Pr[F_D] + \frac{1}{p}\Pr[F_D] - \frac{1}{p}\Pr[F_D] - \frac{p-1}{p}\Pr[F_{\neg D}]|$$

$$\geq |\Pr[F_D] - \Pr[F_{\neg D}]| - \frac{1}{p} = |\Pr[E_2] - \Pr[E_3]| - \frac{3}{p}. \qquad \square$$

Remark. The technique used here is a powerful technique. We can change how we generate certain values, as long as the resulting values are indistinguishable, and as long as what we do with the value does not depend on how the values were generated. This explains why we changed the way the challenge ciphertext was created, because if we did not make this change, we could not change how we generate (x_1, x_2).

Game 4 We now change how the experiment generates g_2 and how it responds to chosen ciphertext queries. The experiment will sample $w \xleftarrow{r} \{2, 3, \ldots, p-1\}$ and compute $g_2 \leftarrow g^w$. When responding to a chosen ciphertext query with ciphertext (x_1, x_2, w, t), if $x_2 \neq x_1^w$ the experiment returns $\perp$ to $\mathcal{A}$. Otherwise, it proceeds as before.

The runtime cost of this is insignificant, so the runtime bound τ_4 on this game is essentially equal to τ_3.

This changes nothing unless the adversary $\mathcal{A}$ submits a valid ciphertext with $x_2 \neq x_1^w$. Let F_i' denote the event that this happens in Game i. Then $\Pr[F_3'] = \Pr[F_4']$ since the two games proceed identically unless F_i' happens.

Recall that no chosen ciphertext query with a hash equal to the hash from the challenge ciphertext is allowed, because of the change in Game 1.

Letting the decryption keys (b_3, b_4) and (b_5, b_6) play the role of the **b** and **b**′ vectors, the value $(1, \omega)$ play the role of the **g** vector, and the hash values play the role of α and α', then with reference to Exercise 8.6 we see that the encryption key y_2 and y_3 gives values for β_2 and β_3, while the tag from the challenge ciphertext defines β' and the query ciphertext defines β''.

Exercise 8.6 now says that for each chosen ciphertext query, there is a unique value of the secret key that satisfies the given values. There are a total of p secret keys that are consistent with the encryption key and the challenge ciphertext tag. For l_d chosen ciphertext queries, there are at most l_d distinct values of the secret key that satisfies at least one of the queries. In other words,

$$\Pr[F_4'] \leq \frac{l_d}{p}.$$

By Lemma 7.7, we have that

$$|\Pr[E_4] - \Pr[E_3]| \leq \Pr[F_4'] \leq \frac{l_d}{p}. \tag{8.5}$$

Finally, when we have disallowed queries with $x_2 \neq x_1^\omega$, the adversary $\mathcal{A}$ can get no information about (b_1, b_2) except that revealed by y_1. Then by Exercise 8.5 we know that from $\mathcal{A}$'s point of view, when creating the challenge ciphertext any value of $x_1^{b_1} x_2^{b_2}$ is equally likely. This means that the adversary has no information about which message was encrypted, and that

$$\Pr[E_4] = 1/2. \tag{8.6}$$

The two adversaries come from the two lemmas, and the bound on the adversary's advantage follows from equations (8.1)–(8.6). This concludes the proof of Proposition 8.6.

8.1.3 Non-Malleability

Since anyone can encrypt anything they want with a public key cryptosystem, there is no notion of security for public key cryptosystems corresponding to integrity. However, there are some underlying ideas that do translate.

One thing that integrity is designed to protect against is tampering with a ciphertext to create a new ciphertext whose decryption is somehow related to the decryption of the original ciphertext. This property is called *malleability*, which we usually do not want (except when we do, see Section 8.3).

The adversary's goal is now, based on a target ciphertext with an unknown decryption, to create a new ciphertext that decrypts to something related. The decryption of the target ciphertext must be unknown, because if it is known it is very easy to get a ciphertext with a related decryption.

It is convenient to define some language. Given any relation on the set of plaintexts, we define a relation on the set of ciphertexts by saying that the ciphertexts are related if and only if the decryptions are not $\perp$ and related.

The adversary's goal is then to present a relation on the set of plaintexts and a valid ciphertext that is related to the target ciphertext. When considering applications, there seems to be no reason why the adversary will only want to create a single related ciphertext. There could be applications where the goal is to create many ciphertexts that are related to the target ciphertext. Individually, each of the ciphertexts may not be meaningfully related, but as a collection, they could have a meaningful relationship. We therefore allow the target ciphertext to be related to a collection of ciphertexts.

To summarise, the adversary must first get an encryption of a target message that was at least partially chosen by the experiment. The adversary must then present a relation and a collection of ciphertexts. These ciphertexts must then be related to the target ciphertext.

Just as confidentiality, where anyone can answer correctly with probability $1/2$, for malleability anyone can come up with a relation and some ciphertexts such that any target ciphertext is related to the proffered ciphertext collection. It is too easy for the adversary unless we correct.

We correct by measuring how easy it was for the adversary to find the related ciphertexts in the first place. We do this by having the experiment choose a second message in the same way that the target message was chosen, but keep this second message secret. If this second message is often related to the adversary's ciphertext collection, then it is easy to create a related message. The adversary has no information about the second message, so it cannot choose the relation to take it into account.

If most messages are related, but the target message is not, that is also a sort of achievement. So we define the advantage to be the distance between the probability that the target message is related to the adversary's collection of ciphertexts, and the probability that the second message is related.

As usual, there is no reason for the adversary to only get a single challenge ciphertext, which means that we should allow a sequence of stateful message sampler algorithms, like for semantic security. But should the adversary also be allowed to submit many relations and ciphertext collections? It seems reasonable that this should be so. However, this makes it more difficult to decide the adversary's efficiency. It seems that the only option is to count the number of times the target messages are related, and how many times the second set of messages is related.

As usual, we allow ciphertexts decrypted with modified associated data (that is, decrypted outside their proper context) to count as modified ciphertexts. Also, we assume that the sampler algorithms always output messages with fixed length.

Definition 8.4. A (τ, l_c, l_d, l_t)-*adversary* $\mathcal{A}$ *against non-malleability* for a public key cryptosystem PKE is an interactive algorithm that interacts with the experiment from Figure 8.2 making at most l_c challenge queries and l_d chosen ciphertext queries, and includes at most l_t ciphertexts in total in the test queries, and where the runtime of the adversary and the simulator is at most τ.

The *advantage* of this adversary is defined to be

$$\mathbf{Adv}^{\mathrm{nm}}_{\mathrm{PKE}}(\mathcal{A}) = |\Pr[E_0] - \Pr[E_1]|,$$

where E_i is the event that $l_i \geq l_{1-i}$ at the end of the game, $i = 0, 1$.

If the number of chosen ciphertext queries $l_d = 0$, we say that $\mathcal{A}$ is a *chosen plaintext adversary* and use the notation nm-cpa. Otherwise we say that $\mathcal{A}$ is a *chosen ciphertext adversary* and use the notation nm-cca.

Exercise 8.8. Show that ElGamal (see Section 3.2.1) and textbook RSA (see Section 3.3.2) are malleable. (That is, provide efficient adversaries against non-malleability.)

We shall now investigate the relationships of non-malleability with the confidentiality properties. It is important to note that non-malleability does not correspond to integrity, so there's no immediate reason to expect a theorem where chosen plaintext indistinguishability and chosen ciphertext non-malleability implies chosen ciphertext indistinguishability. Instead, it turns

The non-malleability experiment proceeds as follows:

1. Let $\sigma_0 = \sigma_1 = \perp$, $l_0 = l_1 = 0$ and set $C = M_0 = M_1$ to be empty lists.

2. Compute $(ek, dk) \leftarrow \mathcal{K}$. Send ek to the adversary.

3. When the adversary sends a query $(ad, \mathcal{X})$ (*challenge*), do:

   (a) Sample $(m_0, \sigma_0') \xleftarrow{r} \mathcal{X}(\sigma_0)$ and $(m_1, \sigma_1') \xleftarrow{r} \mathcal{X}(\sigma_1)$.

   (b) Add (ad, m_0) to M_0 and (ad, m_1) to M_1.

   (c) $\sigma_0 \leftarrow \sigma_0'$, $\sigma_1 \leftarrow \sigma_1'$.

   (d) Compute $c \leftarrow \mathcal{E}(ek, ad, m_0)$, and add (ad, c) to C.

   (e) Send c to the adversary.

4. When the adversary sends a query $(R, (ad_1, c_1), \ldots, (ad_l, c_l))$ (*test*), do:

   (a) If any $(ad_i, c_i) \in C$, send $\perp$ to the adversary.

   (b) Compute $m_i \leftarrow \mathcal{D}(dk, ad_i, c_i)$, $i = 1, 2, \ldots, l$. If any decryption fails, send $\perp$ to the adversary.

   (c) If M_0 R $((ad_1, m_1), (ad_2, m_2), \ldots, (ad_l, m_l))$, then set $l_0 \leftarrow l_0 + 1$.

   (d) If M_1 R $((ad_1, m_1), (ad_2, m_2), \ldots, (ad_l, m_l))$, then set $l_1 \leftarrow l_1 + 1$.

   (e) Send $\top$ to the adversary.

5. When the adversary sends a query (ad, c) (*chosen ciphertext*), do:

   (a) If $(ad, c) \in C$, send $\perp$ to the adversary.

   (b) Otherwise, $m \leftarrow \mathcal{D}(dk, ad, c)$ and send m to the adversary.

FIGURE 8.2 The experiment $\mathbf{Exp}_{\mathrm{PKE}}^{\mathrm{nm}}(\mathcal{A})$ for the non-malleability game for a public key cryptosystem $\mathrm{PKE} = (\mathcal{K}, \mathcal{E}, \mathcal{D})$ with adversary $\mathcal{A}$.

out that non-malleability implies indistinguishability, and furthermore, chosen ciphertext indistinguishability implies chosen ciphertext non-malleability.

The result that non-malleability implies indistinguishability is not very surprising. If you can tell what is inside a ciphertext, then it is trivial to create a related ciphertext. There are some technicalities, though.

The result that chosen ciphertext indistinguishability implies chosen ciphertext non-malleability is more suprising, and crucially depends on being able to get decryptions of chosen ciphertexts. The idea is that we tamper with the ciphertext and get the new ciphertext decrypted. We will also need to decrypt the adversary's ciphertext collection. And then we check if the relation holds, which should give us some information about the target message.

Before we state the first result, we need a small discussion on plaintext sets. We shall ned an easy-to-compute, length-preserving bijection π on the plaintext set $\mathfrak{P}$ with no fixed points. The existence of such a bijection is reasonably clear (in principle, there might be only a single message of a given length, but except for the empty message, that is unreasonable; and we shall ignore the empty message), but the tricky part is that it should be easy to compute. In principle, it might be possible to design a plaintext space where it is easy to verify membership, but hard to manipulate elements. However, no reasonable cryptosystem would ever use a such a plaintext set, so this is not a real restriction.

Proposition 8.9. *Suppose that for any encryption key* ek, *there exists a* π *on the plaintext set* $\mathfrak{M}_{ek}$ *that is length-preserving with no fixed points, such that the cost of* $2l_c$ *computations of* π *is small compared to* τ.

Let $\mathcal{A}$ *be a* (τ, l_c, l_d)-*adversary against indistinguishability for the public key cryptosystem* PKE. *Then there exists a* (τ', l_c, l_d, l_c)-*adversary* $\mathcal{B}$ *against non-malleability, where* τ' *is essentially equal to* τ, *and*

$$\mathbf{Adv}^{\text{nm}}_{\text{PKE}}(\mathcal{B}) = \frac{1}{2}\mathbf{Adv}^{\text{ind}}_{\text{PKE}}(\mathcal{A}).$$

Proof. The adversary $\mathcal{B}$ runs a copy of $\mathcal{A}$ and the following simulator **Sim**:

- When **Sim** gets ek, it forwards it to $\mathcal{A}$.
- **Sim** forwards any chosen ciphertext query and its response.
- When $\mathcal{A}$ makes the ith challenge query (ad, m_{i0}, m_{i1}), **Sim** creates the sample algorithm $\mathcal{X}$ that:
    - If the input is $\perp$, sample $b \xleftarrow{r} \{0,1\}$. Otherwise, the input is b.
    - Output $(m_{i,b}, b)$.

  It sends $(ad, \mathcal{X})$ to the experiment and forwards the response to $\mathcal{A}$.

Eventually $\mathcal{A}$ outputs b'. If $b' = 0$, $\mathcal{B}$ creates a relation R such that $m \, R \, m'$ if and only if $\pi(m) = m'$. If $b' = 1$, let R be the negation of this relation. We extend the relation to tuples in the obvious way. Then it creates encryptions of $\pi(m_{1,b'}), \pi(m_{2,b'}), \ldots$ and sends the test query with R and these ciphertexts to the simulator.

The ciphertext response to a challenge query is created in exactly the same way by the indistinguishability experiment and **Sim** interacting with the non-malleability experiment.

From this it also follows that chosen ciphertext queries are handled in the same way by the indistinguishability experiment and **Sim** interacting with the non-malleability experiment.

Let E be the event that $b' = b$, and for $i = 0, 1$ let E_i be the event that the i message sequence is related to the ciphertexts.

Let F be the event that the non-malleability experiment's two sample states are identical (which means that the two message sequences will be identical). This happens with probability $1/2$, so

$$\Pr[F] = 1/2.$$

If F happens, then $l_0 = l_1$, so this contributes nothing to our adversary's non-malleability advantage. We therefore only need to analyse the case when the two states are distinct. We also note that F is independent of E.

When F does not happen, exactly one of the two message sequences will be related to $\mathcal{B}$'s ciphertext collection, which means that either $l_0 \geq l_1$ or $l_0 \leq l_1$, and $\Pr[l_0 \geq l_1] = \Pr[E_0]$, while $\Pr[l_0 \leq l_1] = \Pr[E_1]$.

If $b' = b$, then the left message sequence will be related to $\mathcal{B}$'s ciphertext collection. It then follows that if $b' \neq b$, the right message sequence will be related to $\mathcal{B}$'s ciphertext collection. It follows that

$$\Pr[E_0 \mid E \wedge \neg F] = 1 \quad \text{and} \quad \Pr[E_1 \mid \neg E \wedge \neg F] = 1.$$

We compute

$$\mathbf{Adv}^{\mathrm{nm}}_{\mathrm{PKE}}(\mathcal{B}) = |\Pr[l_0 \geq l_1] - \Pr[l_1 \geq l_0]| = |\Pr[E_0] - \Pr[E_1]|$$

$$= \frac{1}{2}|\Pr[E_0 \mid \neg F] - \Pr[E_1 \mid \neg F]| = \frac{1}{2}|\Pr[E] - \Pr[\neg E]|$$

$$= \frac{1}{2}|2\Pr[E] - 1| = \frac{1}{2}\mathbf{Adv}^{\mathrm{ind}}_{\mathrm{PKE}}(\mathcal{A}). \qquad \square$$

Remark. The loss of advantage seems inherent. Since the non-malleability simulator samples the two message sequences from the same probability space, it is always possible that identical message sequences could be sampled. This gives an upper bound on any adversary's advantage against non-malleability.

Theorem 8.10. *Let $\mathcal{A}$ be a (τ, l_c, l_d, l_t)-adversary against non-malleability for the public key cryptosystem* PKE. *Then there exists a $(\tau', l_c, l_d + l_t)$-adversary $\mathcal{B}$ against indistinguishability, where τ' is essentially equal to τ, and*

$$\mathbf{Adv}^{\mathrm{ind\text{-}cca}}_{\mathrm{PKE}}(\mathcal{B}) = \frac{1}{2}\mathbf{Adv}^{\mathrm{nm\text{-}cca}}_{\mathrm{PKE}}(\mathcal{A}).$$

Proof. The adversary $\mathcal{B}$ runs a copy of $\mathcal{A}$ and the following simulator **Sim**:

- Let $\sigma_0 = \sigma_1 = \bot$, $l_0' = l_1' = 0$ and set C, M_0 and M_1 to be empty lists.
- When **Sim** gets ek, it samples $b' \xleftarrow{r} \{0,1\}$ and forwards ek to $\mathcal{A}$.
- **Sim** forwards any chosen ciphertext query and its response.
- When $\mathcal{A}$ makes a challenge query $(ad, \mathcal{X})$, **Sim** does:
  - Sample $(m_0, \sigma_0') \leftarrow \mathcal{X}(\sigma_0)$ and $(m_1, \sigma_1') \leftarrow \mathcal{X}(\sigma_1)$. Add (ad, m_0) to M_0 and (ad, m_1) to M_1.
  - $\sigma_0 \leftarrow \sigma_0'$, $\sigma_1 \leftarrow \sigma_1'$.
  - Send (ad, m_0, m_1) to the experiment. Forward the response to $\mathcal{A}$.

- When $\mathcal{A}$ sends a test query $(R, (ad_1, c_1), \ldots, (ad_l, c_l))$, **Sim** does:
  - Send the chosen ciphertext (ad_i, c_i) to the experiment and receive the response m_i, for $i = 1, 2, \ldots, l$. If the response to any query is $\bot$, send $\bot$ to the adversary.
  - If $M_0 \; R \; ((ad_1, m_1), (ad_2, m_2), \ldots, (ad_l, m_l))$, then set $l_0' \leftarrow l_0' + 1$.
  - If $M_1 \; R \; ((ad_1, m_1), (ad_2, m_2), \ldots, (ad_l, m_l))$, then set $l_1' \leftarrow l_1' + 1$.
  - Send $\top$ to the adversary.

When $\mathcal{A}$ stops, $\mathcal{B}$ outputs 0 if $l_0 \geq l_1$, otherwise it outputs b'.

Regardless of the challenge bit b, **Sim** and the indistinguishability experiment will perfectly simulate the non-malleability experiment, so the runtime bound τ' of the system is essentially equal to τ.

Let F_0 be the event that $l_0 \geq l_1$ when $\mathcal{A}$ terminates, and let F_1 be the event that $l_0 \leq l_1$ when $\mathcal{A}$ terminates. Let E_0 and E_1 be the corresponding events when $\mathcal{A}$ interacts with the non-malleability experiment. We compute

$$\mathbf{Adv}_{\mathrm{PKE}}^{\mathrm{ind\text{-}cca}}(\mathcal{B}) = |\Pr[F_0 \mid b = 0] + \frac{1}{2}\Pr[\neg F_0 \mid b = 0]$$

$$- \Pr[F_0 \mid b = 1] - \frac{1}{2}\Pr[\neg F_0 \mid b = 1]|$$

$$= |\frac{1}{2}\Pr[F_0 \mid b = 0] + \frac{1}{2} - \frac{1}{2}\Pr[F_0 \mid b = 1] - \frac{1}{2}|$$

$$= \frac{1}{2}|\Pr[F_0 \mid b = 0] - \Pr[F_0 \mid b = 1]|.$$

When $b = 0$, the indistinguishability experiment and **Sim** perfectly simulate the non-malleability experiment, so F_0 then coincides with E_0. When $b = 1$ the two message sequences essentially change role. This means that F_0 then coincides with E_1. The claim follows. □

We proved that non-malleability implies indistinguishability, and that the converse is true for chosen ciphertext adversaries. The converse is not true for chosen plaintext and non-adaptive chosen ciphertext attacks.

E *Exercise* 8.9. Let $\mathrm{PKE} = (\mathcal{K}, \mathcal{E}, \mathcal{D})$ be a public key encryption system. We construct a second cryptosystem $\mathrm{PKE}' = (\mathcal{K}', \mathcal{E}', \mathcal{D}')$ as follows: $\mathcal{K}' = \mathcal{K}$, $\mathcal{E}'(ek, m) = (\mathcal{E}(ek, m), 0)$ and $\mathcal{D}'(dk, (c, b)) = \mathcal{D}(dk, c)$.

(a) Show that for any non-adaptive chosen ciphertext (τ, l_c, l_d)-adversary $\mathcal{A}$ against indistinguishability for PKE', there exists a (τ', l_c, l_d)-adversary $\mathcal{A}$ against indistinguishability for PKE. That is, if PKE is indistinguishable, so is PKE'. (In other words, changing the cryptosystem does not ruin indistinguishability.)

(b) Give a $(\tau, 1, 0, 1)$-adversary against PKE' non-malleability with trivial τ.

E | *Exercise* 8.10. Use Proposition 8.9, Theorem 8.10 and the result from the previous exercise to show that if there is a public key cryptosystem that is indistinguishable under non-adaptive chosen ciphertext attack, there is a public key cryptosystem that is indistinguishable under non-adaptive chosen ciphertext attack, but not under adaptive chosen ciphertext attack.

Remark. A major achievement in the study of public key encryption was the complete characterisation of the relations between the different indistinguishability and non-malleability variants. Proposition 8.9, Theorem 8.10 and Exercises 8.3, 8.9 and 8.10 complete most of this picture, but there are some remaining issues. These are left to the interested reader.

Remark. It seems like we only care about non-malleability in a chosen ciphertext attack setting. This makes sense. An attack on non-malleability is in some sense inherently a chosen ciphertext attack, since the ciphertexts presented get decrypted. In other settings, only indistinguishability seems interesting. Because non-malleability is essentially equivalent to indistinguishability under a chosen ciphertext attack, the only two public key encryption security notions that we care about are indistinguishability under chosen plaintext attack or chosen ciphertext attack. However, there are somewhat obscure settings where indistinguishability under non-adaptive chosen ciphertext attack makes sense.

8.1.4 Multiple Key Pairs

In practice a system that uses public key encryption will not confine itself to a single key pair. There will be many key pairs. Studying systems with more than one key pair is therefore important. The second exercise shows that it is in some sense sufficient to study a single key pair. The third exercise shows that sometimes we can do better than that.

E | *Exercise* 8.11. Define multi-key variants of the security notions semantic security, indistinguishability, real-or-random and non-malleability.

E | *Exercise* 8.12. Use a hybrid argument to prove that for any adversary against multi-key indistinguishability, there is an adversary against indistinguishability with essentially the same runtime whose advantage is equal to the multi-key adversary's advantage divided by the number of key pairs.

E | *Exercise* 8.13. Consider the ElGamal cryptosystem as discussed in Example 8.3. Use the techniques from that example to prove that for any chosen plaintext multi-key adversary against ElGamal, there is a solver for Decision Diffie-Hellman with the same advantage and essentially the same runtime.

Remark. The main difficulty with multi-key adversaries against public key cryptosystems is that it is natural to allow the adversary to ask for decryption keys. However, once the adversary has gotten a challenge ciphertext for some key pair, asking for that decryption key reveals the challenge plaintext, allowing the adversary to win trivially. This means that we must forbid the adversary from asking for the decryption key of a key pair after getting challenge ciphertexts for that key pair. In practice, we often use a stronger limitation and say that the adversary must ask for all the decryption keys before asking for any challenge ciphertexts. (This is often called *non-adaptive corruption*, and should not be confused with non-adaptive chosen ciphertext attack.)

This sounds reasonable, but in certain applications of public key cryptography, we would like to guarantee that encryptions are secure *until* the adversary asks for the decryption key. The security level changes with time.

One approach that has been used is to have multiple challenge bits, either one for each key pair or even one for each challenge ciphertext. In some sense, this captures the appropriate security notion for public key encryption. But in some applications, encryptions under different keys may contain related information. This is then not a satisfactory solution. There are many other approaches that are unsatisfactory for applications.

The above problem is real and serious, but it should probably not be seen as a problem with our definitions, but rather as indicative of a fundamental limitation of public key encryption: If a decryption key leaks, the secrecy of everything encrypted under that key is lost, retroactively. Preventing this loss of secrecy is called *forward secrecy*. A significant part of modern cryptographic research deals with this issue, and we shall return to it in Chapters 10 and 13.

8.1.5   Multiple Recipients

It is a common use case that some user wants to send the same message to multiple recipients. While the user could encrypt the message to multiple recipients separately, this may be wasteful, either of effort, bandwidth or storage. We shall study a fairly efficient solution, a construction we shall return to when we discuss hybrid encryption. But first, we must define what a multi-recipient public key encryption scheme is.

Definition 8.5. A *multi-recipient public key encryption* scheme $(\mathfrak{P}, \mathfrak{F}, \mathfrak{C}, \mathcal{K}, \mathcal{E}, \mathcal{D})$ consists of a plaintext set $\mathfrak{P}$, a ciphertext set $\mathfrak{C}$ and three algorithms:

- The *key generation* algorithm $\mathcal{K}$ outputs an *encryption key* ek and a decryption key dk.

- The *encryption* algorithm $\mathcal{E}$ takes as input a set of encryption keys $\{ek_1, \ldots, ek_l\}$, associated data $ad \in \mathfrak{F}$ and a message $m \in \mathfrak{P}$ and outputs a ciphertext $c \in \mathfrak{C}$.

- The *decryption* algorithm $\mathcal{D}$ takes as input a decryption key dk, associate data $\mathfrak{F}$ and a ciphertext c and outputs a message $m \in \mathfrak{P}$ or the special symbol $\perp$ indicating decryption failure.

We require that for any set of key pairs $\{(ek_i, dk_i)\}$ output by $\mathcal{K}$, any associated data $ad \in \mathfrak{F}$ and any message $m \in \mathfrak{P}$ and any of the decryption keys dk_j, we have that
$$\mathcal{D}(dk_j, ad, \mathcal{E}(\{ek_i\}, ad, m)) = m.$$

E *Exercise* 8.14. Define semantic security, indistinguishability and real-or-random security for multi-recipient public key encryption schemes. Prove analogues of Propositions 8.1, 8.2 and 8.3.

Note that an adversary against a multi-recipient scheme needs to be characterised also by the number of key pairs needed.

E *Exercise* 8.15. Define non-malleability for multi-recipient public key encryption schemes. Prove that non-malleability implies indistinguishability and that indistinguishability under chosen ciphertext attack implies non-malleability (that is, prove analogues of Proposition 8.9 and Theorem 8.10).

Remark. There may also be a need for stronger security. For instance, if Alice receives a ciphertext and believes that it was also encrypted to Bob, she would like to be certain that Bob will be able to decrypt the ciphertext *and that it will decrypt to the same message*. While it is fairly straight-forward to capture the requirement, we will not explore the design of such schemes.

E *Example* 8.5. Let $\Sigma = (\mathfrak{K}_s, \mathfrak{P}, \mathfrak{F}_s, \mathfrak{C}, \mathcal{E}_s, \mathcal{D}_s)$ be a symmetric cryptosystem with associated data, and let $\mathrm{PKE}_0 = (\mathcal{K}_0, \mathcal{E}_0, \mathcal{D}_0)$ be a public key encryption scheme such for any key pair (ek, dk) output by $\mathcal{K}_0$ we have that $\mathfrak{K}_s \subseteq \mathfrak{M}_{ek}$. We shall also assume that $\mathfrak{F}_s = \mathfrak{F}_1 \times \mathfrak{F}_2$, and that there is an injection from the set of finite sets of pairs of encryption keys and ciphertexts to $\mathfrak{F}_2$.

We construct a multi-recipient scheme $\mathrm{PKE} = (\mathcal{K}, \mathcal{E}, \mathcal{D})$ as follows:

- The key generation algorithm $\mathcal{K}$ computes $(ek, dk) \leftarrow \mathcal{K}_0$ and outputs $(ek, (ek, dk))$.

- The encryption algorithm $\mathcal{E}$ takes as input a set of encryption keys $\{ek_i\}$, associated data $ad \in \mathfrak{F}_1$ and a message $m \in \mathfrak{P}$. It samples $k \xleftarrow{r} \mathfrak{K}_s$, computes $x_i \leftarrow \mathcal{E}_0(ek, k)$ for every encryption key ek_i and $w \leftarrow \mathcal{E}_s(k, (ad, \{(ek_i, x_i)\}), m)$. The ciphertext is $c = (\{(ek_i, x_i)\}, w)$.

- The decryption algorithm $\mathcal{D}$ takes as input a decryption key (ek, dk), associated data ad and a ciphertext $c = (\{(ek_i, x_i)\}, w)$. If $ek_i = ek$, it computes $k \leftarrow \mathcal{D}_0(dk, x_i)$ and $m \leftarrow \mathcal{D}_s(k, (ad, \{(ek_i, x_i)\}), w)$. If either decryption fails or $ek_i \neq ek$ for all i, then the algorithm outputs $\perp$, otherwise it outputs m.

E *Exercise* 8.16. Prove that the above scheme is a multi-recipient public key encryption scheme.

E *Exercise* 8.17. Show that for any adversary $\mathcal{A}$ against PKE, there exists a multi-key adversary $\mathcal{B}_1$ against Σ (multi-key in the sense of Section 7.1.7) and a multi-key adversary $\mathcal{B}_2$ against PKE_0 such that the sum of the advantages of $\mathcal{B}_1$ and $\mathcal{B}_2$ bound the advantage of $\mathcal{A}$, and the runtimes of $\mathcal{B}_1$ and $\mathcal{B}_2$ are essentially the same as the runtime of $\mathcal{A}$.

Hint: We are running somewhat ahead of ourselves here, but looking at the proof of Theorem 8.14 may make sense.

Remark. Some care must be taken to protect the recipient list and the list of encryptions of the symmetric key, because otherwise an adversary may be able to add or remove recipients from the list, or tamper with symmetric key encryptions that the current recipient cannot decrypt (and therefore cannot detect modifications), which trivially makes the entire ciphertext different from the challenge ciphertext, which in turn would allow an adversary to get it decrypted through a chosen ciphertext query.

8.2 KEY ENCAPSULATION MECHANISMS

The ElGamal and Cramer-Shoup cryptosystems discussed in Section 8.1.1 are not actually very practical, since the message space is very small. Most of the public key encryption schemes from Chapter 3 encrypt the actual message using a more or less standard symmetric cryptosystem. The public key part of the system is used to encrypt a key for the symmetric cryptosystem, not the message. The idea that what we actually just want to encrypt a random key for a symmetric cryptosystem is quite powerful, for two reasons: It gives us a design framework for practical public key cryptosystems, and more importantly, encrypting random messages gives us new design options and allows us to design simpler systems.

We shall first define security for key encapsulation mechanisms. Then we shall show how to use such a mechanism to create a general and practical public key encryption scheme.

8.2.1 Key Encapsulation Mechanisms

As usual, we begin by augmenting our definition of key encapsulation mechanism to include associated data. While we saw in Section 8.1.5 that we could include associated data in the symmetric encryption, it is still convenient to be able to include associated data with the encapsulated key.

D **Definition 8.6.** A *key encapsulation mechanism* (KEM) consists of three algorithms $(\mathcal{KK}, \mathcal{KE}, \mathcal{KD})$ and an associated symmetric key set $\mathfrak{K}_s$:

- The *key generation* algorithm $\mathcal{KK}$ takes no input and outputs an *encapsulation key* ek and a *decapsulation key* dk. To each encapsulation key ek there is an associated set of associated data $\mathfrak{F}_{ek}$.

- The *encapsulation* algorithm $\mathcal{KE}$ takes as input an encapsulation key ek and associated data $ad \in \mathfrak{F}_{ek}$, and outputs an encapsulation (ciphertext) c and a key $k \in \mathfrak{K}_s$.

- The *decapsulation* algorithm $\mathcal{KD}$ takes as input a decapsulation key dk, associated data ad and an encapsulation (ciphertext) c and outputs either a key k or the special symbol $\perp$ indicating decapsulation failure.

We require that for any key pair (ek, dk) output by $\mathcal{KK}$, any associated data $ad \in \mathfrak{F}_{ek}$ and any pair (c, k) output by $\mathcal{KE}(ek, ad)$, we get that $\mathcal{KD}(dk, ad, c) = k$.

The goal for a key encapsulation mechanism is just to encapsulate random messages. Semantic security or indistinguishability therefore does not make much sense for a key encapsulation mechanism. Instead, we shall use a variant of real-or-random security, asking if an adversary can decide if a given key is the real key or a random key.

The real-or-random experiment proceeds as follows:

1. Sample $b \xleftarrow{r} \{0, 1\}$ and let $C = \emptyset$.

2. Compute $(ek, dk) \leftarrow \mathcal{KK}$. Send ek to the adversary.

3. When the adversary sends a query ad (*challenge*), do:
   (a) $(c, k_0) \leftarrow \mathcal{KE}(ek, ad)$.
   (b) Sample $k_1 \xleftarrow{r} \mathfrak{K}_s$. Add (ad, c) to C. Send (c, k_b) to the adversary.

4. When the adversary sends a query (ad, c) (*chosen ciphertext*), do:
   (a) if $c \in C$, send $\perp$ to the adversary.
   (b) Otherwise, compute $k \leftarrow \mathcal{KD}(dk, ad, c)$ and send k to the adversary.

Eventually, the adversary outputs $b' \in \{0, 1\}$.

FIGURE 8.3 The experiment $\mathbf{Exp}^{\text{ror}}_{\text{KEM}}(\mathcal{A})$ for the real-or-random security game for a key encapsulation mechanism $\text{KEM} = (\mathfrak{K}_s, \mathcal{KK}, \mathcal{KE}, \mathcal{KD})$.

🄳 **Definition 8.7.** A (τ, l_c, l_d)-*adversary against real-or-random security* for a key encapsulation mechanism is an interactive algorithm $\mathcal{A}$ that interacts with the experiment in Figure 8.3 making at most l_c challenge queries and l_d chosen ciphertext queries, and the runtime of the adversary and the experiment is at most τ.

The *advantage* of this adversary is defined to be

$$\mathbf{Adv}^{\text{ror}}_{\text{PKE}}(\mathcal{A}) = 2|\Pr[E] - 1/2|,$$

where E is the event that b' output by $\mathcal{A}$ equals the experiment's b.

If the number of chosen ciphertext queries $l_d = 0$, we say that $\mathcal{A}$ is a *chosen plaintext adversary*, and use the notation ror-cpa. Otherwise we say that $\mathcal{A}$ is a *chosen ciphertext adversary* and use the notation ror-cca.

We begin with a simple result, namely that any public key cryptosystem with a reasonable message space can be used as a KEM.

E *Exercise 8.18.* Let PKE $= (\mathcal{K}, \mathcal{E}, \mathcal{D})$ be a public key encryption scheme, and suppose $\mathfrak{K}_s$ is a subset of the plaintext space for any encryption key output. Explain how to construct a KEM from PKE, and show that for any adversary against ror for the KEM, there is an adversary against ind with the same advantage and essentially equal runtime.

E *Example 8.6.* Let G be a finite cyclic group with generator g and order p. Let $\mathfrak{K}_s = G$ and consider the following ElGamal-based scheme. The key generation algorithm is the same as in ElGamal. The encapsulation algorithm samples $r \xleftarrow{r} \{0, 1, \ldots, p-1\}$, and computes $c \leftarrow g^r$ and $k \leftarrow y^r$. The decapsulation algorithm computes $k \leftarrow c^b$.

There is also a hashed variant, where $\mathfrak{K}_s$ is some sensible set and we compute the key as $k \leftarrow h(c, y^r)$ and $k \leftarrow h(c, c^b)$, for some hash function $h : G \times G \rightarrow \mathfrak{K}_s$.

E *Exercise 8.19.* Prove that the two ElGamal variants are chosen plaintext secure, that is, show that for any adversary against a chosen plaintext attack, there is a Decision Diffie-Hellman distinguisher with essentially the same advantage and runtime. Show that for the first variant, there is a chosen ciphertext adversary with advantage close to 1.

8.2.1.1 Example: RSA-KEM

We define the key encapsulation mechanism RSA-KEM first introduced in Section 3.3.4. We shall prove it secure in the random oracle model introduced in Section 7.6.

We shall implicitly consider only the RSA problem relative to whichever RSA key generation algorithm we use.

E *Example 8.7.* For any RSA modulus n, let $h : \mathbb{Z}_n \times \mathbb{Z}_n \times \mathfrak{F} \rightarrow \mathfrak{K}_s$ be a hash function. The RSA-KEM scheme works as follows:

- The *key generation* algorithm is the standard RSA key generation algorithm from Section 3.3.

- The *encapsulation* algorithm on input of (n, e) and $ad \in \mathfrak{F}$ samples $r \xleftarrow{r} \{1, 2, \ldots, n - 1\}$, computes $c \leftarrow r^e \bmod n$ and $k \leftarrow h(c, r, ad)$.
- The *decapsulation* algorithm on input of (n, d), ad and $c \in \{1, 2, \ldots, n - 1\}$ computes $k \leftarrow h(c, c^d \bmod n, ad)$.

E *Exercise* 8.20. Show that RSA-KEM is a key encapsulation mechanism.

T **Proposition 8.11.** *Let A be a (τ, l_c, l_d)-adversary against real-or-random for the above key encapsulation mechanism RSA-KEM in the random oracle model, making at most l_h hash queries. Then there exists a τ'-solver B for the RSA problem, where τ' is essentially equal to a small multiple of τ, and*

$$\mathbf{Adv}_{\text{RSA-KEM}}^{\text{ror-cca}}(A) \leq \mathbf{Adv}^{\text{RSA}}(B).$$

The main idea is that since the random oracle values are independent, the adversary cannot learn anything about the decryption of challenge ciphertexts unless it queries the random oracle with the eth root of a challenge ciphertext. The main difficulty is to answer decryption queries consistently.

Remark. In principle, the adversary may spend all its time computing hash values. The number of hash queries l_h can therefore only be bounded by the number of hash computations that can be done in time τ. This is not a small bound, so we need to be careful about how much work we do for each query.

Proof of Proposition 8.11 The proof proceeds as a sequence of games, between the adversary A and an experiment. Let E_i be the event that the adversary's guess b' equals the experiment's challenge bit b in Game i.

G **Game 0** The initial game is A interacting with the real-or-random experiment and the random oracle. Then

$$\mathbf{Adv}_{\text{RSA-KEM}}^{\text{ror-cca}}(A) = 2|\Pr[E_0] - 1/2|. \tag{8.7}$$

Let $\tau_0 = \tau$.

G **Game 1** We modify the random oracle so that it keeps a more careful list of queries and the properties of the queries and does as follows:

- For a query (c, r, ad) where (ad, c, r, k, i) is stored, it responds with k.
- When it receives a query (c, r, ad) where (ad, c, r, k, i) is not stored for any k, i, sample $k \xleftarrow{r} \mathfrak{K}_s$, store (ad, c, r, k, i), where $i = 1$ if $r^e \equiv c \pmod{n}$ and 0 otherwise, and output k.

This does not change the observable behaviour of the random oracle, so

$$\Pr[E_1] = \Pr[E_0]. \tag{8.8}$$

For every unique hash query, the random oracle in this game does a computation and stores a bit more data, compared to the oracle in the previous game. The cost of this extra work is linear in the number of hash queries, and the constant is small. It follows that τ_1 is at most a small multiple of τ_0.

Ⓖ **Game 2** We again modify the random oracle as follows:

- For a query $(c, \perp, ad)$ *from the experiment* where $(ad, c, r, k, 1)$ is stored for any r, it responds with k.
- For a query $(c, \perp, ad)$ *from the experiment* where no tuple $(ad, c, \cdot, \cdot, \cdot)$ is stored, sample $k \xleftarrow{r} \Re_s$, store $(ad, c, \perp, k, 1)$ and respond with k.
- When it receives a query (c, r, ad) *from $\mathcal{A}$* where $r^e \equiv c \pmod{n}$, and where $(ad, c, \perp, k, 1)$ is stored, replace the stored tuple by $(ad, c, r, k, 1)$.

We also modify the experiment so that it makes queries $(c, \perp, ad)$ to the random oracle instead of (c, r, ad).

Since raising to the eth power is a bijection on $\mathbb{Z}_n$, the ciphertext c uniquely determines a hash query with the correct eth root, even if the root is not explicitly recorded. This change to the random oracle and the experiment is therefore not observable, and we get

$$\Pr[E_2] = \Pr[E_1]. \tag{8.9}$$

The extra work done in this game is linear in the number of random oracle queries, so τ_2 is essentially equal to at most a small multiple of τ_1.

Ⓖ **Game 3** In this game, we modify two steps of the experiment. The first change chooses ciphertexts as random integers, which ensures that the experiment does not know the decryptions. The second change stops decrypting ciphertexts. In particular, the experiment now no longer uses the decryption key. The changes are as follows:

3(a) The experiment samples $c \xleftarrow{r} \{1, 2, \ldots, n-1\}$ and queries $(c, \perp, ad)$ to the random oracle to get k_0.

4(b) The experiment queries $(c, \perp, ad)$ to the random oracle to get k and sends k to the adversary.

Again, because raising to the eth power is a bijection on $\mathbb{Z}_n$, it does not matter if the experiment first samples r and computes c, or if it just samples c directly. Because of the changes to the way the experiment queries the random oracle, the r value computed by the decryption algorithm will never be used. Not computing it therefore changes nothing.

It follows that the changes to the experiment are not observable so

$$\Pr[E_3] = \Pr[E_2]. \tag{8.10}$$

The changes in this game involve doing less work, but the amount of work is linear in the number of queries made to the experiment, which for any non-trivial adversary will be insignificant compared to τ_2. We therefore get that τ_3 is essentially equal to τ_2.

Let F be the event that the adversary queries the random oracle with (c, r, ad) for a challenge ciphertext c with $r^e \equiv c \pmod{n}$.

If F never happens, the adversary has no information about the decryption of any challenge ciphertext. It follows that

$$\Pr[E_3 \mid \neg F] = 1/2.$$

From this it follows that

$$
\begin{aligned}
2|\Pr[E_3] - 1/2| &= 2|\Pr[E_3 \mid F]\Pr[F] + \Pr[E_3 \mid \neg F](1 - \Pr[F]) - 1/2| \\
&= 2|\Pr[E_3 \mid F] - 1/2|\Pr[F] \leq \Pr[F].
\end{aligned}
\tag{8.11}
$$

The following lemma completes the argument, by combining equations (8.7)–(8.12).

Lemma 8.12. *There exists a τ'-solver $\mathcal{B}$ for the RSA problem, with τ' essentially equal to a small multiple of τ_3, and*

$$\Pr[F] \leq \mathbf{Adv}^{\mathsf{RSA}}(\mathcal{B}).\tag{8.12}$$

Proof. The solver $\mathcal{B}$ takes (n, e, c_0) as input, runs a copy of the adversary $\mathcal{A}$ together with the modified random oracle and experiment, further modified as follows:

- If c_0 is not relatively prime to n, use c_0 to factor n and compute the eth root of c_0 in the usual way.

- For the ith challenge query, the experiment samples $r_i \xleftarrow{r} \{1, 2, \ldots, n-1\}$ and computes $c_i \leftarrow c_0 r_i^e \bmod n$.

- If the adversary queries the random oracle with (c_i, r, ad) with c_i being the ciphertext from the ith challenge query, and $r^e \equiv c_i \pmod{n}$, then $\mathcal{B}$ outputs $rr_i^{-1} \bmod n$. If r_i is not invertible, use r_i to factor n and compute the eth root of c_0 in the usual way.

If $\mathcal{B}$ does not stop before $\mathcal{A}$ stops, $\mathcal{B}$ outputs 1.

If c_0 is relatively prime to n, then multiplication by c_0 modulo n is a bijection, so $\mathcal{B}$ samples ciphertexts from the correct distribution. This means that $\mathcal{B}$ perfectly simulates the experiment and random oracle that $\mathcal{A}$ expects.

Therefore, if c_0 is relatively prime to n, the probability that $\mathcal{B}$ outputs the correct eth root is equal to $\Pr[F]$. If c_0 is not relatively prime to n, then $\mathcal{B}$ outputs the correct eth root with probability 1. We get that

$$\Pr[F] \leq \mathbf{Adv}^{\mathsf{RSA}}(\mathcal{B}).$$

The adversary $\mathcal{B}$ creates challenge ciphertexts in a slightly more expensive way. For every random oracle query, $\mathcal{B}$ must compare the ciphertext against the table of challenge ciphertexts, an operation comparable to the lookup anyway done for every hash query. It follows that the runtime of $\mathcal{B}$ is essentially equal to at most a small multiple of τ_3. □

Remark. This proof relies on the random self-reducibility of the RSA problem.

8.2.1.2 Example: DH-KEM

We shall now look again at the second ElGamal variant from Example 8.6.

Example 8.8. Let G be a group of order p with generator g, and let $h : \mathfrak{F} \times G \times G \to \mathfrak{K}_s$ be a hash function. The *Diffie-Hellman KEM* DH-KEM scheme works as follows:

- The *key generation* algorithm is the ElGamal key generation algorithm.
- The *encapsulation* algorithm on input of y and $ad \in \mathfrak{F}$ samples $r \xleftarrow{r} \{0, 1, \dots, p-1\}$ and computes $c \leftarrow g^r$ and $k \leftarrow h(ad, c, y^r)$.
- The *decapsulation* algorithm on input of b, ad and $c \in G$ computes $k \leftarrow h(ad, c, c^b)$.

Exercise 8.21. Show that DH-KEM is a key encapsulation mechanism.

The DH-KEM scheme is very similar to the RSA-KEM scheme, and it is tempting to simple copy the proof of that scheme. However, a problem appears when the modified random oracle must recognise queries for (ad, c, c^b) after the modified experiment has stopped submitting c^b to the oracle. In fact, recognising this particular tuple is exactly the Decision Diffie-Hellman problem, which means that we cannot easily recognise the correct query.

It gets worse. The adversary's chosen ciphertext query essentially gives the adversary the capability to recognise pairs (c, z) such that $z = c^b$: the adversary submits c as a chosen ciphertext query and hashes (c, z); if the keys are equal, $z = c^b$ except with very small probability.

The answer lies in introducing a new variant of the Diffie-Hellman problem where the adversary has this capability, in the sense that the adversary may query an experiment with pairs of group elements (c, z) and learn if $c^a = z$. We then prove that this capability is sufficient to simulate the chosen ciphertext query response, which means that in the random oracle model, the chosen ciphertext query gives the adversary exactly this capability, and nothing else. (Immediately, we should emphasise that a single decryption query for c gives the adversary the ability to test with many different z, which means that the capabilities are essentially the same, but not numerically identical.)

[T] **Proposition 8.13.** *Let $\mathcal{A}$ be a (τ, l_c, l_d)-adversary against real-or-random for the above key encapsulation mechanism DH-KEM in the random oracle model, making at most l_h hash queries. Then there exists a (τ', l_h)-solver $\mathcal{B}$ for the Strong Diffie-Hellman problem, where τ' is essentially equal to a small multiple of τ, and*

$$\mathbf{Adv}^{\mathrm{ror}}_{\mathrm{DH\text{-}KEM}}(\mathcal{A}) \leq \mathbf{Adv}^{\mathrm{SDH}}_{G}(\mathcal{B}).$$

The proof closely follows that of Proposition 8.11 and is based on the same ideas, but where the adversary in the RSA case can easily check if an adversarial hash query corresponds to a experiment query, our adversary will have to query its oracle.

[E] *Exercise* 8.22. Prove Proposition 8.13.

[E] *Exercise* 8.23. Define a multi-key variant of real-or-random for key encapsulation mechanisms. Use the techniques from Example 8.3 to show that for any multi-key adversary against DH-KEM, there is an solver for Strong Diffie-Hellman with the same advantage and essentially the same runtime.

Remark. Another way to turn Diffie-Hellman into a key encapsulation method is to start with ElGamal and then apply Proposition 8.15 below. This results in a scheme with larger ciphertexts and a more expensive decryption algorithm, which sounds bad. However, it is secure if the Decision Diffie-Hellman problem is hard, while the above construction is secure if the Strong Diffie-Hellman problem is hard. Since these are distinct problems, one scheme may turn out to be secure while the other may turn out to be insecure.

8.2.2   Hybrid Encryption

Let $\mathrm{KEM} = (\mathcal{KK}, \mathcal{KE}, \mathcal{KD})$ be a KEM for $\mathfrak{K}_s$, and let $\Sigma = (\mathfrak{K}_s, \mathfrak{P}, \mathfrak{F}, \mathfrak{C}, \mathcal{E}_s, \mathcal{D}_s)$ be a symmetric cryptosystem.

The $\mathrm{KEM} + \Sigma$-*hybrid* public key encryption scheme works as follows:

- The *key generation algorithm* is $\mathcal{KK}$.

- The *encryption algorithm* takes as input an encapsulation key ek, associated data $ad \in \mathfrak{F}$ and a message $m \in \mathfrak{P}$. It computes $(x, k) \leftarrow \mathcal{KE}(ek)$ and $w \leftarrow \mathcal{E}_s(k, ad, m)$. The ciphertext is $c = (x, w)$.

- The *decryption algorithm* takes as input a decapsulation key dk, associated data ad and a ciphertext $c = (x, w)$. It first computes $k \leftarrow \mathcal{KD}(dk, x)$. If $k = \bot$, the decryption algorithm outputs $\bot$. Otherwise, it outputs $\mathcal{D}_s(k, ad, w)$.

[E] *Exercise* 8.24. Show that the above scheme is a public key encryption scheme.

Theorem 8.14. *Let $\mathcal{A}$ be a (τ, l_c, l_d)-adversary against the KEM $+\Sigma$-hybrid. Then there exists a (τ'_1, l_c, l_d)-adversary $\mathcal{B}_1$ against KEM and a $(\tau'_2, 1, 0, l_d)$-adversary $\mathcal{B}_2$ against Σ, where τ'_1 and τ'_2 are essentially equal to τ, and*

$$\mathbf{Adv}^{\mathrm{ind}}_{\mathrm{KEM}+\Sigma}(\mathcal{A}) \leq 2\mathbf{Adv}^{\mathrm{ror}}_{\mathrm{KEM}}(\mathcal{B}_1) + l_c\mathbf{Adv}^{\mathrm{ind}}_{\Sigma}(\mathcal{B}_2).$$

The strategy is to first use the real-or-random security game to replace the encapsulated keys by random keys. What remains is a number of encryptions under independent keys, and here we use a hybrid argument.

Proof of Theorem 8.14   The proof proceeds as a sequence of games, between the adversary $\mathcal{A}$ and an experiment. Let E_i be the event that the adversary's guess b' equals the challenge bit b of the experiment in Game i.

Game 0   The initial game is $\mathcal{A}$ interacting with the indistinguishability experiment. Then

$$\mathbf{Adv}^{\mathrm{ind}}_{\mathrm{KEM}+\Sigma}(\mathcal{A}) = 2|\Pr[E_0] - 1/2|. \qquad (8.13)$$

Let $\tau_0 = \tau$.

Game 1   We add bookkeeping to avoid decrypting certain encapsulations.

- When the experiment creates a challenge ciphertext c with encapsulation x and encapsulated key k, the experiment records (x, k).
- If the adversary submits a chosen ciphertext query $c = (x, w)$, where (x, k) has been recorded by the experiment, the experiment does not decapsulate x to k, but instead decrypts w using k.

Since the KEM is correct, this change is not observable, so

$$\Pr[E_1] = \Pr[E_0].$$

The experiment needs to keep track of the KEM ciphertexts created in response to challenge queries, but this work will be insignificant compared to any non-trivial adversary. It follows that τ_1 is essentially equal to τ_0.

Game 2   In this game, instead of using the encapsulated keys for challenge ciphertexts, we use a completely random key.

- When the experiment creates a challenge ciphertext c for a message m with associated data ad, it computes $(x, k') \leftarrow \mathcal{KE}(ek)$, samples $k \xleftarrow{r} \mathfrak{K}_s$ and computes $w \leftarrow \mathcal{E}_s(k, ad, m)$. It records (x, k).

Exercise 8.25. Show that there exists a (τ'_1, l_c, l_d)-adversary $\mathcal{B}_1$ against KEM, where τ'_1 is essentially equal to τ_2, such that

$$|\Pr[E_2] - \Pr[E_1]| \leq \mathbf{Adv}^{\mathrm{ror}}_{\mathrm{KEM}}(\mathcal{B}_1). \qquad (8.14)$$

With this change, the encryption of the message is decoupled from the key encapsulation mechanism. This means that the only way the adversary can decide what has been encrypted is to break the symmetric key cryptosystem.

E *Exercise 8.26.* Show that there exists a $(\tau_2', 1, 0, l_d)$-adversary $\mathcal{B}_2$ against Σ, where τ_2' is essentially equal to τ_2, such that

$$2|\Pr[E_2] - 1/2| \le l_c \mathbf{Adv}_\Sigma^{\text{ind}}(\mathcal{B}_2). \tag{8.15}$$

With the two adversaries from the two exercises, the claim now follows from equations (8.13)–(8.15). This concludes the proof of Theorem 8.14.

Remark. If we had defined a notion of multiple-key security for a symmetric cryptosystem, we could have used that notion and removed the term l_c from the theorem. However, while there may be interesting constructions that achieve interesting multiple-key security results, for any standard construction the best argument is typically the same hybrid argument that is used above.

There are many alternative hybrid constructions, in particular with respect to associated data. We put the associated data in the symmetric ciphertext, but it could just as well be put in the key encapsulation.

E *Exercise 8.27.* Prove a variant of Theorem 8.14 for a hybrid construction that includes associated data in x, not in w.

A second alternative for integrating the associated data is to use a PRF to derive the symmetric key, from the encapsulated key and the associated data. We leave this construction for the interested reader.

8.2.3 Chosen Plaintext Security is Sufficient – Maybe

It is hard to design public key encryption schemes that are secure against chosen ciphertext. We have seen some concrete examples. Schemes secure against chosen plaintext attack are in some sense much more natural and also much more plentiful. It is therefore interesting to study how to use them to construct chosen ciphertext secure schemes from chosen plaintext secure schemes.

Remark. One of the insights required is that the encryption algorithm of a public key scheme can be considered as computing a function from the message space and a set of randomness to the ciphertext space. In theory this can actually be a bit tricky, but in practice it is straight-forward.

The Fujisaki-Okamoto construction that we shall now study is based on the idea that if the randomness used to encrypt a message is derived from the message itself, this allows the decryption algorithm to verify that the encryption was properly created. If used carefully, this actually allows us to achieve non-malleability in the random oracle model.

Directly constructing a public key encryption scheme with this technique would not work, since it would result in a deterministic encryption scheme, which would be insecure. But we can use this approach to construct a key encapsulation mechanism, which we have seen is sufficient.

Example 8.9. Let PKE$' = (\mathcal{K}, \mathcal{E}', \mathcal{D}')$ be a public key encryption scheme. Suppose that for each encryption key ek output by $\mathcal{K}$, there exists a set T_{ek} and a function from $\mathfrak{M}_{ek} \times T_{ek}$ to the ciphertext space, denoted $\mathcal{E}'_{ek}(\cdot, \cdot)$, such that for all messages $m \in \mathfrak{M}_{ek}$, the encryption algorithm on input of ek and m induces the same probability distribution on the ciphertext space as sampling $t \xleftarrow{r} T_{ek}$ and evaluating $\mathcal{E}'_{ek}(m, t)$. We shall also assume that it is easy to compute $\mathcal{E}'_{ek}(m, t)$ for all ek output by $\mathcal{K}$ and all $m \in \mathfrak{M}_{ek}$ and $t \in T_{ek}$.

Let $\mathfrak{F}$ and $\mathfrak{K}_s$ be finite sets and let $h_{ek} : \mathfrak{F} \times \mathfrak{M}_{ek} \to T_{ek} \times \mathfrak{K}_s$ be a family of hash functions.

The key encapsulation mechanism KEM $= (\mathcal{KK}, \mathcal{KE}, \mathcal{KD})$ works as follows:

- The *key generation* algorithm $\mathcal{KK}$ computes $(ek, dk) \leftarrow \mathcal{K}$ and outputs the encapsulation key ek and the decapsulation key (ek, dk).

- The *encapsulation* algorithm $\mathcal{KE}$ takes as input an encapsulation key ek and associated data $ad \in \mathfrak{F}$. It samples $r \xleftarrow{r} \mathfrak{M}_{ek}$, computes $(t, k) \leftarrow h_{ek}(ad, r)$ and $c \leftarrow \mathcal{E}'_{ek}(r, t)$, and outputs (c, k).

- The *decapsulation* algorithm $\mathcal{KD}$ takes as input a decapsulation key (ek, dk), associated data ad and an encapsulation c. It computes $r \leftarrow \mathcal{D}'(dk, c)$, $(t, k) \leftarrow h_{ek}(ad, r)$ and $c' \leftarrow \mathcal{E}'_{ek}(r, t)$. If decryption fails or $c \neq c'$, the algorithm outputs $\perp$. Otherwise it outputs k.

Remark. This construction has one potential problem, which both potentially serious and interesting. Unless great care is taken, the implementation of the decapsulation algorithm could leak information about the result of decrypting c. Recall that the underlying cryptosystem is not secure under a chosen ciphertext attack, and it may be that even knowledge about failed decryption attempts could be enough to break security.

One way in which information could leak could be through a *timing attack*, for instance if the implementation does not compute c' if the decryption fails.

Proposition 8.15. *Let $\mathcal{A}$ be a (τ, l_c, l_d)-adversary against* KEM *in the random oracle model, making at most l_h hash queries. Then there exists a $(\tau', l_c, 0)$-adversary $\mathcal{B}$ against* PKE$'$, *with τ' essentially equal to a small multiple of τ, such that*

$$\mathbf{Adv}^{\mathrm{ror}}_{\mathrm{KEM}}(\mathcal{A}) \leq 2\mathbf{Adv}^{\mathrm{ror}}_{\mathrm{PKE}'}(\mathcal{B}) + 2\mathbf{Col}_{\mathrm{PKE}'}(2l_d) + \frac{4l_c l_h}{l_m},$$

where l_m is a lower bound on $|\mathfrak{M}_{ek}|$ for any ek output by $\mathcal{K}$.

The proof idea is that anyone who has a list of queries to the random oracle can compute which ciphertexts could result from those queries. Any ciphertext that is not among those queries will almost certainly decrypt to $\perp$. This can be used to simulate responses to chosen ciphertext queries, which leaves us with a chosen plaintext adversary. Since no information about the seeds leaks out of the ciphertexts, it follows that we have a secure KEM.

Proof of Proposition 8.15 The proof proceeds as a sequence of games, between the adversary $\mathcal{A}$ and an experiment. Let E_i be the event that the adversary's guess b' equals the experiment's challenge bit b in Game i. As usual, the number of hash queries is bounded by a constant multiple of τ.

Ⓖ **Game 0** The initial game is $\mathcal{A}$ interacting with the real-or-random experiment and the random oracle. Then

$$\mathbf{Adv}^{\text{ror-cca}}_{\text{KEM}}(\mathcal{A}) = 2|\Pr[E_0] - 1/2|. \tag{8.16}$$

Let $\tau_0 = \tau$.

Ⓖ **Game 1** We modify the random oracle so that it keeps a more careful list of queries and the properties of queries:

- For a query (ad, r) where (ad, r, t, k, c, i) is stored, output (t, k).
- When it receives a query (ad, r) where no tuple $(ad, r, \cdot, \cdot, \cdot, \cdot)$ is stored, sample $t \xleftarrow{r} \mathfrak{M}_{ek}$ and $k \xleftarrow{r} \mathfrak{K}_s$, compute $c \leftarrow \mathcal{E}'_{ek}(r, t)$, store $(ad, r, t, k, c, 0)$ and output (t, k).

This does not change the observable behaviour of the random oracle, so

$$\Pr[E_1] = \Pr[E_0]. \tag{8.17}$$

For every unique hash query, the random oracle in this game does a computation and stores a bit more data, compared to the oracle in the previous game. The cost of this extra work is linear in the number of hash queries l_h, and the constant is small. Since l_h is at most a small multiple of τ_0, it follows that τ_1 is at most a small multiple of τ_0.

Ⓖ **Game 2** In this game we modify the experiment so that it rejects ciphertexts that have not been generated by hashing a random nonce. If the experiment receives a chosen ciphertext query (ad, c) for which no tuple $(ad, \cdot, \cdot, \cdot, c, \cdot)$ is stored by the random oracle, it responds with $\bot$.

This changes the observable behaviour of the experiment only if $\mathcal{D}'(dk, ad, c) = t$ and the random oracle returns a value t such that $\mathcal{E}_{ek}(r, t) = c$. This cannot happen if there is a tuple $(ad, r, \cdot, \cdot, c', \cdot)$ stored, with $c \neq c'$, so it follows that no tuple $(ad, r, \cdot, \cdot, \cdot, \cdot)$ is stored.

This means that when the experiment runs the decryption procedure, the random oracle will sample $t \xleftarrow{r} T_{ek}$ and then compute $c'' \leftarrow \mathcal{E}_{\mathcal{E}}(r, t)$. The resulting ciphertext distribution is exactly the same as for the encryption algorithm itself, which means that the probability that $c'' = c$ for at least one of the l_d chosen ciphertext queries can be bounded (somewhat loosely) by the collision probability for the cryptosystem, and we get that

$$|\Pr[E_2] - \Pr[E_1]| \leq \mathbf{Col}_{\text{PKE}'}(2l_d). \tag{8.18}$$

The extra cost of dealing with each chosen ciphertext query is comparable to the cost of a hash query, and since $l_d \leq l_h$ we get that τ_2 is at most a small multiple of τ_1. (In practice, l_d will be very small compared to l_h.)

Game 3 In this game we modify the experiment so that it uses the random oracle's records when responding to chosen ciphertext queries, instead of using the decryption key. The experiment responds to a chosen ciphertext query (ad, c) (Step 4(b) in Figure 8.3) as follows:

- If a tuple $(ad, \cdot, \cdot, k, c, \cdot)$ is stored by the random oracle, send k to the adversary. Otherwise, send $\bot$ to the adversary.

Since we disallowed the decapsulation of ciphertexts for which no such tuple exists, and the decapsulation of c with associated data ad can only be k, this change to the experiment is not observable. We get

$$\Pr[E_3] = \Pr[E_2]. \tag{8.19}$$

The cost of answering a chosen ciphertext query is slightly smaller in this game, and otherwise the games are identical, so τ_3 is not greater than τ_2.

Note that in this and any later game, the decryption key is never used.

Game 4 In this game we modify the experiment so that the it stops if a nonce for a challenge ciphertext has been queried previously. The random oracle is modified as follows:

- When the experiment responds to a challenge query and queries the random oracle with r, if a tuple $(\cdot, r, \cdot, \cdot, \cdot, \cdot)$ is already stored by the random oracle, the game samples $b' \xleftarrow{r} \{0, 1\}$ and stops. Otherwise, the random oracle proceeds as before, but stores the tuple $(ad, r, t, k, c, 1)$.

Since r is sampled from the uniform distribution on $\mathfrak{M}_{ek}$ and there has been at most l_h random oracle queries, the probability that the nonce sampled for any single challenge query has been queried before is at most $l_h/|\mathfrak{M}_{ek}|$. Since there are at most l_c challenge queries, all independent, we get that

$$|\Pr[E_4] - \Pr[E_3]| \leq \frac{l_c l_h}{l_m}, \tag{8.20}$$

where l_m is a lower bound on $|\mathfrak{M}_{ek}|$ for any ek output by $\mathcal{K}$.

Game 5 In this game we modify the game so that it disallows hashing the nonces used for challenge ciphertexts. We modify the random oracle as follows:

- When the adversary $\mathcal{A}$ queries the random oracle with (ad, r) and a tuple $(ad, r, t, \cdot, \cdot, 1)$ is stored by the random oracle, the game samples $b' \xleftarrow{r} \{0, 1\}$ and stops. Otherwise, the random oracle proceeds as before.

Note that in this game, the t value that is stored in response to challenge ciphertext queries is only ever used to create a challenge ciphertext, and no other information about the value leaks to the adversary.

Let F_i be the event that the adversary $\mathcal{A}$ queries the random oracle with r as used in the random oracle. If F_i never happens, the games proceed in an identical fashion, so $\Pr[F_5] = \Pr[F_4]$ and

$$|\Pr[E_5] - \Pr[E_4]| \le \Pr[F_5]. \tag{8.21}$$

Also, if F_5 never happens, the adversary has no information about the decapsulation of any challenge ciphertext, so $\Pr[E_5 \mid \neg F_5] = 1/2$. If F_5 happens, a random bit is chosen as the adversary's guess and $\Pr[E_5 \mid F_5] = 1/2$. Therefore,

$$\Pr[E_5] = 1/2. \tag{8.22}$$

We bound $\Pr[F_5]$ using an adversary against the underlying cryptosystem. The claim follows from equations (8.16)–(8.22) and the following lemma.

Lemma 8.16. *There exists an $(\tau', l_c, 0)$-adversary $\mathcal{B}$ against* PKE′, *with τ' essentially equal to τ_5, and*

$$\Pr[F_5] \le \mathbf{Adv}_{\text{PKE}'}^{\text{ror}}(\mathcal{B}) + \frac{l_c l_h}{l_m},$$

where l_m is a lower bound on $|\mathfrak{M}_{ek}|$ for any ek output by $\mathcal{K}$.

Proof. The adversary $\mathcal{B}$ runs a copy of $\mathcal{A}$ and a simulator **Sim** that runs the experiment and random oracle from Game 5, with the following modifications:

- When $\mathcal{B}$ gets ek, **Sim** sends ek to $\mathcal{A}$ instead of running $\mathcal{K}$.
- When responding to a challenge ciphertext query from $\mathcal{A}$, then **Sim** samples r as in Game 5, but **Sim** does not compute c. Instead, **Sim** makes a challenge query with r and then uses the ciphertext response as c. The simulator **Sim** then proceeds as in Game 5.

If a random oracle query by the adversary $\mathcal{A}$ causes the simulated Game 5 to stop, $\mathcal{B}$ outputs 1. If $\mathcal{A}$ outputs a bit, or $\mathcal{A}$ and **Sim** runs for time greater than τ_5 or makes more than l_c challenge queries, then $\mathcal{B}$ outputs 0.

First of all, since $\mathcal{A}$ and the simulator **Sim** can run for time at most τ_5, the runtime of $\mathcal{B}$ is at most τ_5 and it makes at most l_c challenge queries. The additional runtime required by message passing to the ror-cpa experiment is very small, and for any interesting adversary l_c is very small compared to τ, it follows that τ' is essentially equal to τ_5.

If the challenge ciphertexts **Sim** gets are all encryptions of the messages chosen by **Sim**, then **Sim** and its experiment perfectly simulate Game 5 for $\mathcal{A}$. If G_0 is the event that $\mathcal{B}$ outputs 1 in this case, we get that

$$\Pr[G_0] = \Pr[F_5].$$

If the challenge ciphertexts returned to **Sim** are encryptions of random messages, then the adversary $\mathcal{A}$ has no information about the messages that **Sim** uses to decide whether to stop the game. If G_1 is the event that $\mathcal{B}$ outputs 1 in this case, we get that every random oracle query stops the game with probability at most $l_c/|\mathfrak{M}_{ek}|$, and

$$\Pr[G_1] \leq \frac{l_c l_h}{|\mathfrak{M}_{ek}|}.$$

The claim follows from

$$\Pr[F_5] \leq \Pr[G_0] - \Pr[G_1] + \frac{l_c l_h}{|\mathfrak{M}_{ek}|} \leq |\Pr[G_0] - \Pr[G_1]| + \frac{l_c l_h}{|\mathfrak{M}_{ek}|}. \qquad \square$$

This construction can be applied to ElGamal, but as we shall see there are alternative constructions with smaller ciphertexts and less computation. This construction cannot be applied to textbook RSA, since it is not chosen plaintext secure. However, there are techniques one can use to make even weak schemes like textbook RSA chosen plaintext secure.

E *Exercise 8.28.* This exercise adapts Exercise 7.2 to public key encryption. Define a notion of *one-way* security for public key encryption. Prove that any one-way chosen plaintext adversary against textbook RSA can be turned into a solver for the RSA problem.

E *Exercise 8.29.* Let PKE$' = (\mathcal{K}, \mathcal{E}', \mathcal{D}')$ be any public key encryption scheme. Let $\mathfrak{N}$ be a large set, and suppose there is an easy-to-compute bijection between $\mathfrak{N} \times \mathfrak{N}$ and a subset of $\mathfrak{M}_{ek}$ for any ek output by $\mathcal{K}$. Suppose also that we have a family of easy-to-compute invertible permutations π_{ek} on $\mathfrak{M}_{ek}$. Show that if we model the permutations as random permutations, you can construct a public key encryption scheme PKE $= (\mathcal{K}, \mathcal{E}, \mathcal{D})$ with message space $\mathfrak{N}$, such that any ind-cpa adversary against PKE can be turned into an equally efficient one-way adversary against PKE$'$.

8.2.3.1 Example: RLWE-KEM

We have seen many KEM constructions that do not need the Fujisaki-Okamoto construction. However, it is a useful tool, and we shall study one application now. This is also the first lattice-based example.

We first construct a public key cryptosystem that looks like a "noisy" version of ElGamal. This scheme is not secure against chosen ciphertext attacks. Therefore, we apply the Fujisaki-Okamoto construction to the public key cryptosystem, to get a chosen ciphertext secure key encapsulation mechanism.

E *Example 8.10.* Let q be prime. For any integer i, denote by $i \bmod^{\pm} q$ the integer congruent to i modulo q with least absolute value. The field $\mathbb{F}_q$ is isomorphic to $\mathbb{Z}/\langle q \rangle$. For any coset $x = i + \langle q \rangle$, let $x \bmod^{\pm} q$ denote $i \bmod^{\pm} q$. We extend this map to $\mathbb{F}_q^n$ in the natural way.

Let R be a ring, which is isomorphic to $\mathbb{F}_q^n$ as a group. We extend the mod$^\pm$ operation to R. Let $R_{01} \subseteq R$ be the subset corresponding to $\{0,1\}^n \subseteq \mathbb{Z}^n$. We identify these sets.

Let $\mathcal{X}_n$ be a probability space on R, with the property that if we sample $e_1, e_2, e_3 \xleftarrow{r} \mathcal{X}_n$, then $\|(e_3 - e_1 e_2) \bmod^\pm q\|_\infty < q/4$.

The *RLWE* public key encryption scheme RLWE-PKE is:

- The *key generation* algorithm $\mathcal{K}_0$ samples $g \xleftarrow{r} R$, $a \xleftarrow{r} \mathcal{X}_n$ and $e_0 \xleftarrow{r} \mathcal{X}_n$, computes $y \leftarrow ga + e_0$, and outputs $ek = (g, y)$ and $dk = a$.

- The *encryption* algorithm $\mathcal{E}_0$ takes as input an encryption key $ek = (g, y)$ and a message $m \in R_{01}$. It samples $r \xleftarrow{r} \mathcal{X}_n$, $e_1, e_2 \xleftarrow{r} \mathcal{X}_n$, computes $x \leftarrow gr + e_1$ and $w \leftarrow yr + e_2 + \lfloor q/2 \rfloor m$, and outputs $c = (x, w)$.

- The *decryption* algorithm $\mathcal{D}_0$ takes as input a decryption key $dk = a$ and a ciphertext $c = (x, w)$. It outputs $\lfloor (w - xa \bmod^\pm q)/(q/2) \rceil$.

E| *Exercise 8.30.* Show that RLWE-PKE is a public key encryption scheme.

We now want to apply the construction from Example 8.9 to the above public key cryptosystem. In order for Proposition 8.15 to apply, we need to prove that RLWE-PKE is secure under a chosen plaintext attack.

T| **Proposition 8.17.** *Let $\mathcal{A}$ be a $(\tau, 1, 0)$-adversary against real-or-random for the public key cryptosystem* RLWE-PKE *from Example 8.10, making at most a single challenge query. Then there exists a $(\tau_1', 1)$-distinguisher $\mathcal{B}_1$ and a $(\tau_2', 2)$-distinguisher $\mathcal{B}_2$ for Ring-LWE, with τ_1' and τ_2' essentially equal to τ, such that*

$$\mathbf{Adv}^{\mathsf{ror}}_{\text{RLWE-PKE}}(\mathcal{A}) \leq 2\mathbf{Adv}^{\mathsf{RLWE}}_{R, \mathcal{X}_n}(\mathcal{B}_1) + 2\mathbf{Adv}^{\mathsf{RLWE}}_{R, \mathcal{X}_n}(\mathcal{B}_2).$$

The proof relies on the fact that the Ring-LWE samples look random, and that both x, y and w are computed as Ring-LWE samples. The first step is to replace x by a random ring element. The next step is to replace y and w (taking the message into account) by random ring elements. We cannot do all the replacements at the same time, because the randomness is different in the two cases. Also, we cannot do the second replacement unless we first do the first replacement, since x does not (yet) have a uniform distribution.

It is absolutely crucial that we have only a single challenge query in this proof, which means that we must use Proposition 8.5 to get a useful result.

Proof of Proposition 8.17   The proof proceeds as a sequence of games, between the adversary $\mathcal{A}$ and an experiment. Let E_i be the event that the adversary's guess b' equals the experiment's challenge bit b in Game i.

Ⓖ **Game 0** The initial game is $\mathcal{A}$ interacting with the real-or-random experiment and the random oracle. Then

$$\mathbf{Adv}_{\text{RLWE-PKE}}^{\text{ror}}(\mathcal{A}) = 2|\Pr[E_0] - 1/2|. \tag{8.23}$$

Let $\tau_0 = \tau$.

Ⓖ **Game 1** In this game, we modify the key generation algorithm so that it samples $y \xleftarrow{r} R$ instead of computing y. If $\mathcal{A}$ exceeds its bounds, the experiment samples $b' \xleftarrow{r} \{0, 1\}$ and stops.

Because we enforce the bounds, we get that τ_1 is essentially equal to τ_0.

Ⓔ *Exercise 8.31.* Show that there exists a $(\tau_1', 1)$-distinguisher $\mathcal{B}_1$ for RLWE, with τ_1' essentially equal to τ_1, such that

$$|\Pr[E_1] - \Pr[E_0]| \leq \mathbf{Adv}_{R,\mathcal{X}_n}^{\text{RLWE}}(\mathcal{B}_1). \tag{8.24}$$

Ⓖ **Game 2** We modify the computation of the single challenge ciphertext:

- Sample $x, z \xleftarrow{r} R$ and compute $w \leftarrow z + \lfloor q/2 \rfloor m_b$.

If $\mathcal{A}$ exceeds its bounds, the experiment samples $b' \xleftarrow{r} \{0, 1\}$ and stops.

Because we enforce the bounds, we get that τ_2 is essentially equal to τ_1.

Ⓔ *Exercise 8.32.* Show that there exists a $(\tau_2', 2)$-distinguisher $\mathcal{B}_2$ for RLWE, with τ_2' essentially equal to τ_2, such that

$$|\Pr[E_2] - \Pr[E_1]| \leq \mathbf{Adv}_{R,\mathcal{X}_n}^{\text{RLWE}}(\mathcal{B}_2). \tag{8.25}$$

The single challenge ciphertext in this game consists of two random ring elements, from which it follows that

$$\Pr[E_2] = 1/2. \tag{8.26}$$

The claim then follows from equations (8.23)–(8.26).

8.3 HOMOMORPHIC ENCRYPTION

We begin with an example that should give some intuition about what *homomorphic* means. Afterwards, we shall define the abstract concept of homomorphism in great detail, and then simplify to two forms of homomorphism: group-homomorphism and ring-homomorphism.

Ⓔ *Example 8.11.* Consider ElGamal encryption based on a group G of prime order p with generator g. The ElGamal encryption key is $y = g^b$, and an ElGamal ciphertext is $(x, w) \in G \times G$ where $x = g^r$ and $w = y^r m$. The decryption algorithm computes $\mathcal{D}(b, (x, w)) = wx^{-b}$.

Note that the ElGamal ciphertext space $G \times G$ is a group with identity $(1, 1)$ and the obvious group operation. The decryption algorithm computes a map from $G \times G$ to G. Since for any $(x_1, w_1), (x_2, w_2) \in G \times G$ we have that

$$\mathcal{D}(b, (x_1, w_1)(x_2, w_2)) = w_2 w_1 (x_1 x_2)^{-b} = \mathcal{D}(b, (x_1, w_1))\mathcal{D}(b, (x_2, w_2)).$$

The decryption algorithm computes a group homomorphism. In particular, if c_1 is an encryption of m_1 and c_2 is an encryption of m_2, we get

$$\mathcal{D}(b, c_1 c_2) = \mathcal{D}(b, c_1)\mathcal{D}(b, c_2) = m_1 m_2.$$

Since the decryption algorithm computes a group homomorphism, we say that ElGamal is a *group homomorphic* cryptosystem.

The intuitive idea about a cryptosystem being homomorphic is that there is some common algebraic structure on the ciphertext space and the message space and that the decryption algorithm is a homomorphism. However, while it is always possible to define some isomorphic structure on the ciphertext space, this is not always the best way to look at cryptosystems, so we shall have to relax our intuitive notion.

Let *ops* be a tuple of operations on some set S. An *ops-algebraic circuit* (or *algebraic circuit over ops*) C is a labeled directed graph with no loops and no cycles where every internal vertex (a vertex with both incoming and outgoing edges) is labeled with an operation from *ops* and an ordering on the incoming edges, the number of which is equal to the arity of the operation. The vertices without incoming edges are called *inputs*, and the vertices without outgoing edges are called *outputs*. An output is only allowed to have one incoming edge. A circuit with l_{in} inputs and l_{out} outputs is an (l_{in}, l_{out})-*circuit*.

An (l_{in}, l_{out})-circuit C with an ordering on its inputs and outputs defines a function $C : S^{l_{in}} \to S^{l_{out}}$ in the following fashion:

- Label each edge out of an input vertex with the input element.

- Choose an unlabeled internal node whose incoming edges are labeled. Compute the operation in the vertex label on the incoming values in specified order. Label every outgoing edge with the result.

- Repeat the previous step until every outgoing edge has been labeled.

- The output is the labels on the output vertices' incoming edges.

We need the order on incoming edges to handle non-commutative operations.

Remark. Any (l_{in}, l_{out})-circuit can be considered to be a collection of l_{out} $(l_{in}, 1)$-circuits. Evaluating the former will often be much faster than l_{out} separate evaluations, but this can be considered an optimisation. When it simplifies the presentation, we shall consider circuits with a single output vertex.

Example 8.12. Consider the set of algebraic circuits over an abelian group G. Any such circuit defines a function of the form $(x_1, \ldots, x_{l_{in}}) \mapsto x_1^{a_1} \ldots x_{l_{in}}^{a_{l_{in}}}$ for some integers $(a_1, \ldots, a_{l_{in}})$.

Example 8.13. Consider the set of algebraic circuits over a field $\mathbb{F}$. A $(l_{in}, 1)$-algebraic circuit over $(+, \cdot)$ defines a particular method for computing a multivariate polynomial from $\mathbb{F}[X_1, \ldots, X_{l_{in}}]$.

The *length* of a path in an algebraic circuit is the number of internal vertices on the path. The *op-length* of a path in an algebraic circuit is the number of internal vertices on the path labeled with the operator *op*. The *depth* of an algebraic circuit is the maximum length of any path in the circuit, and the *op-depth* is the maximum *op*-length of any path in the circuit.

Let *ops*, *ops'* be matching tuples of operations (corresponding operators have the same arity). For any *ops*-algebraic circuit C, we can easily replace the labels on the internal vertices with the corresponding operations from *ops'*. We denote the resulting *ops'*-algebraic circuit by $C^{ops/ops'}$.

We are now finally ready to define what it means for a cryptosystem to be homomorphic.

Definition 8.8. Let $\text{PKE} = (\mathcal{K}, \mathcal{E}, \mathcal{D})$ be a public key encryption scheme, and let $\mathfrak{P}_{ek}$ and $\mathfrak{C}_{ek}$ denote the plaintext and ciphertext spaces corresponding to an encryption key ek, and let $\mathfrak{C}_{ek,f}$ be the set of ciphertexts with a non-zero probability of being output by $\mathcal{E}(ek, \cdot)$, the *fresh ciphertexts*.

Suppose for every ek, there is a tuple of operations ops_{ek} on $\mathfrak{P}_{ek}$, and that there exists a matching tuple of operations ops'_{ek} on $\mathfrak{C}_{ek}$. Let $\mathcal{H}_{ek}$ be a set of ops_{ek}-algebraic circuits, $\mathcal{H}$ be the map taking ek to $\mathcal{H}_{ek}$, and let *ops* be the map taking ek to ops_{ek}. We say that PKE is $\mathcal{H}$-*homomorphic* if for any (ek, dk) output by $\mathcal{K}$, any $(l_{in}, 1)$-circuit $C \in \mathcal{H}_{ek}$ and fresh ciphertexts $c_1, c_2, \ldots, c_{l_{in}}$, we have that

$$\mathcal{D}(dk, C^{ops_{ek}/ops'_{ek}}(c_1, \ldots, c_{l_{in}})) = C(\mathcal{D}(dk, c_1), \ldots, \mathcal{D}(dk, c_{l_{in}})).$$

We say that PKE is *ops-homomorphic* if $\mathcal{H}_{ek}$ contains every ops_{ek}-algebraic circuit for any ek output by $\mathcal{K}$.

We say that PKE is *group-homomorphic* if it is *ops*-homomorphic and ops_{ek} defines a group structure on $\mathfrak{P}_{ek}$.

We say that PKE is *ring-homomorphic* if it is *ops*-homomorphic and ops_{ek} defines a ring structure on $\mathfrak{P}_{ek}$.

Remark. It is very easy to get non-interesting homomorphic cryptosystems. Consider the tuple $(C, c_1, \ldots, c_{l_{in}})$ to be a ciphertext and then modify the decryption algorithm to decrypt the ciphertexts and apply the circuit to the result.

Remark. Symmetric cryptosystems can also be homomorphic, and the above definition is easy to generalise. In fact, many other primitives can be homomorphic, such as message authentication codes and signatures.

Remark. For unstructured and limited circuit sets $\mathcal{H}_{ek}$, we sometimes say that a $\mathcal{H}$-homomorphic cryptosystem is a *somewhat homomorphic* cryptosystem.

We can let the key generation algorithm take as input a positive integer. Suppose such a cryptosystem PKE is a $\mathcal{H}$-homomorphic cryptosystem for some family of operations *ops*, and suppose that for any (ek, dk) output by $\mathcal{K}$ on input of L, the set $\mathcal{H}_{ek}$ contains every circuit of depth L. Then we say that PKE is *leveled homomorphic*.

The difference between a leveled and a somewhat homomorphic scheme is that for a leveled homomorphic scheme, we can choose how large circuits the scheme should be able to deal with. For a somewhat homomorphic scheme, the ciphertext set is fixed.

Remark. Sometimes some extra data is needed to compute the operations on $\mathfrak{C}_{ek}$. This is either included in the encryption key, but sometimes (for instance for symmetric cryptosystems) it is convenient to have an *evaluation key*. The key generation algorithm will create and output this key.

Remark. A somewhat more general formulation replaces the ciphertext operations *ops'* with an *evaluation* algorithm $\mathcal{EV}$, which takes as input an encryption key ek, a $(l_{in}, 1)$-circuit C over ops_{ek} and fresh ciphertexts $c_1, c_2, \ldots, c_{l_{in}}$, and outputs a ciphertext. The correctness requirement then becomes

$$\mathcal{D}(dk, \mathcal{EV}(ek, C, c_1, \ldots, c_{l_{in}})) = C(\mathcal{D}(dk, c_1), \ldots, \mathcal{D}(dk, c_{l_{in}})).$$

The evaluation algorithm will often use discrete operations like ops'_{ek} on the ciphertext space, but may do other operations as well to get better overall performance. The information that $\mathcal{EV}$ uses to organise these extra computations could be encoded in the ciphertexts, but this becomes increasingly contrived.

It follows that Definition 8.8 is an ideal definition that only works in certain cases, while having an evaluation algorithm is better in many cases.

8.3.1 Group Homomorphic Cryptosystems

There is a fairly large class of cryptosystems that have essentially the same abstract structure, even though the underlying group structures are very different. These schemes have important practical applications.

Example 8.14. ElGamal as defined in Example 8.11 is a group-homomorphic cryptosystem. The ciphertext space is $G \times G$, which has a natural group structure, and the decryption algorithm computes a group homomorphism.

Example 8.15. Textbook RSA as defined in Section 3.3.2 with invertible elements as the plaintext set is group homomorphic. The group is $\mathbb{Z}_n^*$ for an encryption key (n, e), and the decryption map is a group automorphism.

Example 8.16. Let n be a product of two primes p and q, such that $(p-1)/2$ and $(q-1)/2$ are both odd. Then -1 has Legendre symbol modulo both p and q, so -1 is a quadratic non-residue modulo n.

The Goldwasser-Micali cryptosystem works as follows:

- The *key generation* algorithm generates an RSA modulus $n = pq$ as above, and outputs $ek = n$ and $dk = (p, q)$.
- The *encryption* algorithm takes $ek = n$ and $m \in \{-1, 1\}$ as input, samples $r \xleftarrow{r} \mathbb{Z}_n^*$ and computes the ciphertext as $c \leftarrow r^2(-1)^m$.
- The *decryption* algorithm takes $dk = p$ and $c \in \mathbb{Z}_n^*$ as input. If $\left(\frac{c}{n}\right) = 1$, it outputs $\left(\frac{c}{p}\right)$, otherwise $\perp$.

Exercise 8.33. Show that the Goldwasser-Micali cryptosystem is group homomorphic, and identify the plaintext group.

Example 8.17. Let n be a product of two primes p and q, such that $\gcd(n, (p-1)(q-1)) = 1$. Then n is invertible modulo $(p-1)(q-1)$.

The Paillier cryptosystem works as follows:

- The *key generation* algorithm generates an RSA modulus $n = pq$ as above, computes $d \leftarrow (p-1)(q-1)$, and outputs $ek = n$ and $dk = (n, d)$.
- The *encryption* algorithm takes $ek = n$ and $m \in \mathbb{Z}_n$ as input, samples $r \xleftarrow{r} \mathbb{Z}_{n^2}^*$ and computes the ciphertext as $c \leftarrow r^n(1+n)^m$.
- The *decryption* algorithm takes $dk = (n, d)$ and $c \in \mathbb{Z}_{n^2}^*$ as input and computes $t \leftarrow c^d \bmod n^2$, $t' \leftarrow (t-1)/n$ (over the integers) and $m \leftarrow t'd^{-1} \bmod n$.

Exercise 8.34. Show that the Paillier cryptosystem is group homomorphic, and identify the plaintext group.

The above schemes are group homomorphic, and for all of them the ciphertext space and the plaintext space have group structures, and the decryption map computes a group homomorphism. These schemes all have very nice algebraic structures, and in fact generalise in a single abstract cryptosystem.

Example 8.18. Consider a family $\mathfrak{F} = \{(G, H, \mathfrak{P}, \phi_0, \phi_1)\}$, where G and $\mathfrak{P}$ are finite abelian groups, H is a subgroup of G, $\phi_0 : \mathfrak{P} \to G$ and $\phi_1 : G \to \mathfrak{P}$ are homomorphisms such that $\phi_1 \circ \phi_0$ is the identity on $\mathfrak{P}$, and $\ker \phi_1 = H$. We assume that the descriptions are sufficient to sample from the uniform probability spaces on G and H. We also assume that there is a sampling algorithm $\mathcal{K}_0$ that outputs an element of $\mathfrak{F}$.

The group-homomorphic public key cryptosystem $\text{PKE}_{\mathfrak{F}} = (\mathcal{K}, \mathcal{E}, \mathcal{D})$ is:

- The *key generation* algorithm $\mathcal{K}$ samples $(G, H, \mathfrak{P}, \phi_0, \phi_1) \xleftarrow{r} \mathcal{K}_0$ and outputs $ek = (G, H, \mathfrak{P}, \phi_0)$ and $dk = (G, \mathfrak{P}, \phi_1)$.

- The *encryption* algorithm takes as input $ek = (G, H, \mathfrak{P}, \phi_0)$ and $m \in \mathfrak{P}$. It samples $r \xleftarrow{r} H$ and computes $c = r\phi_0(m)$. The output is c.

- The *decryption* algorithm takes as input $dk = (G, \mathfrak{P}, \phi_1)$ and $c \in G$. It computes $m = \phi_1(c)$ and outputs m.

Since $r \in H = \ker \phi_1$ we have that

$$\phi_1(r\phi_0(m)) = (\phi_1 \circ \phi_0)(m) = m.$$

In other words, we have a public key cryptosystem, and since ϕ_0 and ϕ_1 are homomorphisms, it is also group-homomorphic.

We will need the family $\mathfrak{F}' = \{(G, H, \mathfrak{P}, \phi_0)\}$, which is the same as $\mathfrak{F}$, except that ϕ_1 has been forgotten. The corresponding sampling algorithm $\mathcal{K}_0'$ uses $\mathcal{K}_0$ to sample $(G, H, \mathfrak{P}, \phi_0, \phi_1)$, forgets ϕ_1 and outputs $(G, H, \mathfrak{P}, \phi_0)$.

Exercise 8.35. Show that Examples 8.11, 8.15, 8.16 and 8.17 are special cases of the abstract group homomorphic cryptosystem from Example 8.18. For each example, describe the family $\mathfrak{F}$ is, how to sample from the group and subgroup and how to evaluate the homomorphisms.

Textbook RSA fits technically, but not morally. In particular, H is trivial, so the following discussion will show that it is trivially insecure. But we already knew that from the discussion in Section 3.3.3.

The idea in this cryptosystem is that an element of $H = \ker \phi_1$ acts as noise that hides the message, and that this noise can be removed by the homomorphism ϕ_1. If we can distinguish elements of H from elements not in the subgroup, we can also distinguish encryptions of the identity element from encryptions of non-identity elements. The converse also holds.

Definition 8.9. Let $\mathfrak{F} = \{(S, T, aux)\}$ be a family of tuples where S is a set and T is a proper subset of S, and aux is some additional information. We assumpe that the descriptions are sufficient to sample from the uniform probability spaces on S and T. Let $\mathcal{K}_0$ be an algorithm that outputs an element of $\mathfrak{F}$. A τ-*distinguisher* $\mathcal{A}$ for the *subset membership problem* $(\mathfrak{F}, \mathcal{K}_0)$ is an algorithm that interacts with the experiment from Figure 8.4, and where the runtime of the distinguisher and the experiment is at most τ.

The *advantage* of the distinguisher is defined to be

$$\mathbf{Adv}_{\mathfrak{F}, \mathcal{K}_0}^{\text{smp}}(\mathcal{A}) = \Pr[E],$$

The distinguishing experiment for a subset membership problem with an adversary $\mathcal{A}$ proceeds as follows:

1. Compute $(S, T, aux) \leftarrow \mathcal{K}_0$ and sample $b \xleftarrow{r} \{0, 1\}$. If $b = 0$, sample $s \xleftarrow{r} S$. Otherwise, sample $s \xleftarrow{r} T$.

2. Send (S, T, aux, s) to $\mathcal{A}$.

Eventually the distinguisher outputs $b' \in \{0, 1\}$.

FIGURE 8.4 Experiment $\mathbf{Exp}_{\mathfrak{F}, \mathcal{K}_0}^{smp}(\mathcal{A})$ for the distinguishing game for a subset membership problem.

where E is the event that $\mathcal{A}$'s output b' equals the challenge bit b.

We say that $(\mathfrak{F}, \mathcal{K}_0)$ is a *subgroup membership problem* if for any (S, T, aux) sampled by $\mathcal{K}_0$, we have that S is a group and T is a subgroup of S.

Theorem 8.18. *Let $\mathfrak{F}'$, $\mathcal{K}_0'$ and* PKE *be as in Example 8.18. Let $\mathcal{A}$ be a $(\tau, 1, 0)$-real-or-random adversary against* PKE. *Then there exists a τ'-distinguisher $\mathcal{B}$ for the subgroup membership problem $(\mathfrak{F}', \mathcal{K}_0')$, where τ' is essentially equal to τ, and*

$$\mathbf{Adv}_{\mathfrak{F}, \mathcal{K}_0}^{smp}(\mathcal{B}) = \mathbf{Adv}_{PKE}^{ror}(\mathcal{A}).$$

Also, let $\mathcal{B}'$ be a τ''-distinguisher for $(\mathfrak{F}, \mathcal{K}_0)$. Then there exists a $(\tau''', 1, 0)$-real-or-random adversary $\mathcal{A}'$ against PKE, *with τ''' essentially equal to τ'', and*

$$\mathbf{Adv}_{PKE}^{ror}(\mathcal{A}') = \mathbf{Adv}_{\mathfrak{F}, \mathcal{K}_0}^{smp}(\mathcal{B}').$$

The following result is convenient to establish the structure of the ciphertext group G, which is the key to proving the theorem.

Lemma 8.19. *Let $\mathfrak{F}$ be as in Example 8.18 and let $(G, H, \mathfrak{P}, \phi_0, \phi_1) \in \mathfrak{F}$. Then for any $c \in G$, there exists a unique pair $(r, m) \in H \times \mathfrak{P}$ such that $c = r\phi_0(m)$. (In other words, G is isomorphic to $H \times \mathfrak{P}$, and the isomorphism is multiplication.)*

Proof. Let $c \in G$. Define $m = \phi_1(c)$ and $r = c/\phi_0(m)$. Then $r \in H$ since

$$\phi_1(r) = \phi_1(c/\phi_0(m)) = m/m.$$

Let (r', m') be any other such pair. Then $r = r'$ because

$$m = \phi_1(r\phi_0(m)) = \phi_1(c) = \phi_1(r'\phi_0(m')) = m'. \qquad \square$$

E | *Exercise* 8.36. Use the above lemma to prove Theorem 8.18.

Hint: Observe that an encryption of the identity is a random element of the noise subgroup H, while an encryption of a random message is a random element of the ciphertext group G. The converse also holds.

8.3.2 Somewhat Homomorphic Cryptosystems

We close this section with an interesting example of a somewhat homomorphic cryptosystem. It is very similar to RLWE-PKE from Example 8.10.

E | *Example* 8.19. The *BGV* (Brakerski-Gentry-Vaikuntanathan) public key encryption scheme RLWE-HOM based on a ring R as in Example 8.10 is the following scheme:

- The *key generation* algorithm $\mathcal{K}$ samples $g \xleftarrow{r} R$, $a \xleftarrow{r} \mathcal{X}_n$ and $e_0 \xleftarrow{r} \mathcal{X}_n$, computes $y \leftarrow ga + pe_0$, and outputs $ek = (g, y)$ and $dk = a$.

- The *encryption* algorithm $\mathcal{E}$ takes as input an encryption key $ek = (g, y)$ and a message $m \in R_{01}$. It samples $r \xleftarrow{r} \mathcal{X}_n$ and $e_1, e_2 \xleftarrow{r} \mathcal{X}_n$, computes $x \leftarrow gr + pe_1$ and $w \leftarrow yr + pe_2 + m$, and outputs $c = (x, w)$.

- The *decryption* algorithm $\mathcal{D}$ takes as input a decryption key $dk = a$ and a ciphertext $c = (x, w)$. It outputs $((w - xa) \bmod^{\pm} q) \bmod p$.

The parameters from this scheme need careful calibration, just as in Example 8.10, but with quite different goals in mind, as we shall now discuss.

This system is somewhat additively homomorphic, simply by adding ciphertexts coordinatewise. It is somewhat, because if we do too many additions, the sums of the noise terms become too big and decryption fails. But the interesting trick with this cryptosystem is that it also gives us a form of multiplication. Consider ciphertexts (x_1, w_1) and (x_2, w_2), and compute $x_3 = x_1 x_2$, $w_3 = w_1 w_2$ and $w_3' = x_1 w_2 + x_2 w_1$. Then

$$(w_1 - x_1 a)(w_2 - x_2 a) = w_1 w_2 - (x_1 w_2 + x_2 w_1)a + x_1 x_2 a^2 = w_3 - w_3' a + x_3 a^2.$$

If we are careful with the parameters, we can now lift the result to the integers $(\bmod^{\pm})$ and reduce modulo p, to get the product of the two decryptions on the left-hand side and a kind of decryption of (x_3, w_3', w_3) on the right-hand side. Which means that (x_3, w_3', w_3) in some sense is a product ciphertext.

Together with a number of other tricks, this gives us a useful somewhat homomorphic cryptosystem. We can also apply so-called *bootstrapping*, where we essentially use the somewhat homomorphic cryptosystem to evaluate its own decryption circuit, using an encryption of its own key. If correctly tuned, bootstrapping provides a way to reduce the ciphertext noise to a fixed level, which gives us *fully homomorphic encryption*. We shall not develop this theory.

8.4   COMMITMENT SCHEMES

Sometimes Alice will need to make a decision at some point in time. Alice will need to keep her decision secret, but at the same time Bob must be able to ensure that she does not change her mind.

We do this with *commitment schemes*, which will allow Alice to commit to some value by giving Bob a commitment, and then later convince Bob that she really committed to that value by opening the commitment. For some schemes, Alice and Bob need to get a commitment key from a common reference string.

Definition 8.10. A *commitment scheme* consists of three algorithms $(\mathcal{CK}, \mathcal{CC}, \mathcal{CO})$:

- The *key generation* algorithm $\mathcal{CK}$ takes no input and outputs a *commitment key* ck. A plaintext set $\mathfrak{P}$ is associated to a commitment key.

- The *commit* algorithm $\mathcal{CC}$ takes as input a commitment key ck and a message $m \in \mathfrak{P}$, and outputs a *commitment* u and an *opening* o.

- The *verification* algorithm $\mathcal{CO}$ takes as input a commitment key ck, a commitment u, a message $m \in \mathfrak{P}$ and an opening o and outputs 0 or 1.

For any commitment key ck output by $\mathcal{CK}$, any message $m \in \mathfrak{P}$ and any commitment u and opening o output by $\mathcal{CC}$ on input of ck and m we have

$$\mathcal{CO}(ck, u, m, o) = 1.$$

Remark. Some commitment schemes do not require a commitment key, in which case $\mathcal{CK}$ outputs an empty string. In this case, we will usually omit the commitment key and the key generation algorithm.

We first consider five commitment schemes, three based on a hash function, one based on a public key cryptosystem and one based on a cyclic group.

Example 8.20. Let $h : \mathfrak{P} \to S$ be a hash function. There is no commitment key. To commit to a message m, we compute the commitment $u \leftarrow h(m)$.
  To verify that u is a commitment to m, check that $u = h(m)$.

Example 8.21. Let $h : \mathfrak{P} \to \mathbb{Z}_n^+$ be a hash function. There is no commitment key. To commit to a message m, sample $o \leftarrow \mathbb{Z}_n^+$ and compute the commitment $u \leftarrow h(m) + o$. The opening is o.
  To verify that o is an opening of u to m, check that $u = h(m) + o$.

Example 8.22. Let $h : T \times \mathfrak{P} \to S$ be a hash function, where T is a finite set. There is no commitment key. To commit to a message m, sample $o \leftarrow T$ and compute the commitment $u \leftarrow h(o, m)$. The opening is o.
  To verify that o is an opening of u to m, check that $u = h(o, m)$.

E *Example 8.23.* Let $(\mathcal{K}, \mathcal{E}, \mathcal{D})$ be a public key cryptosystem. Recall that a probabilistic algorithm like $\mathcal{E}$ can be considered as a deterministic algorithm taking as its input the original input ek and m and also its randomness r. We denote this deterministic algorithm by $\mathcal{E}_{ek}(m, r)$.

To generate a commitment key, compute $(ek, dk) \leftarrow \mathcal{K}$ and let $ck = ek$.

To commit to a message m, sample randomness o and compute the commitment $u \leftarrow \mathcal{E}_{ck}(m, o)$. The opening is o.

To verify that o is an opening of u to m, check that $u = \mathcal{E}_{ck}(m, o)$.

E *Example 8.24.* This is the *Pedersen* commitment scheme. Let G be a cyclic group of prime order n. To generate a commitment key, sample two non-identity group elements g and h. The commitment key is $ck = (g, h)$.

To commit to a message $m \in \{0, 1, \ldots, n-1\}$, sample $o \leftarrow \{0, 1, \ldots, n-1\}$ and compute the commitment $u \leftarrow g^o h^m$. The opening is r.

To verify that o is an opening of u to m, check that $u = g^o h^m$.

E *Exercise 8.37.* Prove that the above five examples are all commitment schemes.

A commitment scheme has two-sided security requirements. The commitment must not reveal any information about the message to Bob. Also, it must not be possible for Alice to change her mind about what she committed to.

E *Example 8.25.* Consider the commitment scheme from Example 8.20. When Bob gets a commitment u, it is trivial for him to decide if Alice committed to a message m or not, simply by checking if u is equal to $h(m)$.

E *Example 8.26.* Consider the commitment scheme from Example 8.21. When Alice has made a commitment u to a message m with opening o, then for any message m' the value $o' = o - h(m) + h(m')$ will be an opening of u to m'.

D **Definition 8.11.** Consider a commitment scheme COM $= (\mathcal{CK}, \mathcal{CC}, \mathcal{CO})$. A (τ, l_c)-*adversary against hiding* is an interactive algorithm $\mathcal{A}$ that interacts with the experiment $\mathbf{Exp}^{\text{hide}}_{\text{COM}}(\mathcal{A})$ from Figure 8.5 making at most l_c challenge queries, and the runtime of the adversary and the experiment is at most τ. A τ-*adversary against binding* is an interactive algorithm $\mathcal{A}$ that interacts with the experiment $\mathbf{Exp}^{\text{bind}}_{\text{COM}}(\mathcal{A})$ from Figure 8.5, and the runtime of the adversary and the experiment is at most τ.

The *advantages* of adversaries against hiding and binding are

$$\mathbf{Adv}^{\text{hide}}_{\text{COM}}(\mathcal{A}) = 2|\Pr[E] - 1/2|, \qquad \mathbf{Adv}^{\text{bind}}_{\text{COM}}(\mathcal{A}) = \Pr[F],$$

where E is the event that $b = b'$ in $\mathbf{Exp}^{\text{hide}}_{\text{COM}}(\mathcal{A})$, and F is the event that $m_0 \neq m_1$ and $\mathcal{CO}(ck, u, m_0, o_0) = \mathcal{CO}(ck, u, m_1, o_1) = 1$ in $\mathbf{Exp}^{\text{bind}}_{\text{COM}}(\mathcal{A})$.

The experiment $\mathbf{Exp}_{\mathrm{COM}}^{\mathrm{hide}}(\mathcal{A})$ with an adversary $\mathcal{A}$ proceeds as follows:

1. Sample $b \xleftarrow{r} \{0, 1\}$.
2. Compute $ck \leftarrow \mathcal{CK}$ and send ck to $\mathcal{A}$.
3. When the adversary sends a query m_0, m_1 (*challenge*), do:

   (a) Compute $(u, o) \leftarrow \mathcal{CC}(ck, m_b)$ and send u to $\mathcal{A}$.

Eventually, the adversary outputs b'.

The experiment $\mathbf{Exp}_{\mathrm{COM}}^{\mathrm{bind}}(\mathcal{A})$ with an adversary $\mathcal{A}$ proceeds as follows:

1. Compute $ck \leftarrow \mathcal{CK}$ and send ck to $\mathcal{A}$.

Eventually, the adversary outputs u, m_0, m_1, o_0 and o_1.

FIGURE 8.5 Experiments for defining security notions for a commitment scheme $\mathrm{COM} = (\mathcal{CK}, \mathcal{CC}, \mathcal{CO})$ against an adversary $\mathcal{A}$.

We have defined *computational* hiding. If for any commitment key ck and any two distinct messages $m_0, m_1 \in \mathfrak{P}$ we have that the probability spaces induced by $\mathcal{CC}(ck, m_i)$ are ϵ-close for some usefully small ϵ, we have *statistical* hiding. If they are 0-close, we have *perfect* hiding.

We have also defined *computational* binding. If any adversary's advantage is 0 regardless of τ we have *unconditional* binding.

All other things being equal, unconditional is better than statistical is better than computational.

Exercise 8.38. Show that if a commitment scheme is perfectly hiding, any adversary has advantage 0. Also, if a commitment scheme is statistically hiding with bound ϵ, then any adversary has advantage 2ϵ.

Exercise 8.39. Show that a commitment scheme is unconditionally binding if and only for any commitment key there exists a function f_{ck} from the set of commitments to the set of messages such that $f_{ck}(\mathcal{CC}(ck, m)) = m$ for all messages m. This function does not have be easy to compute.

Example 8.27. Consider the commitment scheme from Example 8.22.

This scheme is hiding for reasonable hash functions, since the randomness will mask the message in the hash computation. For most hash functions we want to use, we cannot expect statistical hiding, much less prove it.

Any adversary against binding can trivially be turned into a collision finder for the hash function, which means that if we use any collision resistant hash function the scheme is binding. However, since most hash functions we would want to use are many-to-one, binding is not unconditional.

E *Exercise* 8.40. Consider the commitment scheme from Example 8.23. Explain why this scheme is unconditionally binding.

Also show that any adversary against hiding can be turned into an adversary against indistinguishability for the cryptosystem.

E *Exercise* 8.41. Show that the Pedersen scheme from Example 8.24 is perfectly hiding. Also show that any adversary against binding can be turned into a discrete logarithm solver for the cyclic group G.

It is not a coincidence that the perfectly hiding scheme is computationally binding, and the unconditionally binding scheme is computationally hiding.

E *Exercise* 8.42. Prove that no commitment scheme can be both perfectly hiding and unconditionally binding.

We shall now discuss some commitment schemes with special properties.

8.4.1 Equivocable and Extractable

This commitment scheme is based on Paillier encryption from Example 8.17, and the commitment algorithm is a variant of Paillier encryption. The commitment key contains a Paillier ciphertext, and by varying the message encrypted we can "trapdoor" commitment keys, so that either we can open a commitment to any value, or extract the committed message from a commitment.

These properties are interesting because of their application in security proofs. Great care must be taken using any such commitment scheme.

Let p, q be distinct large primes such that $(p-1)/2$ and $(q-1)/2$ are also primes distinct from p and q. Let $n = pq$. Let k be a positive integer and let $S = \{0, 1, \ldots, n^2 2^k\}$. Note that n is relatively prime to the order of $\mathbb{Z}_n^*$.

E *Exercise* 8.43. Let $-n^2 < \nu < n^2$. Compute the statistical distance between the uniform distribution on $\{0, 1, \ldots, n^2 2^k\}$ and the uniform distribution on $\{\nu, \nu+1, \ldots, n^2 2^k + \nu\}$. How does it behave as k grows?

For almost all $r \in \mathbb{Z}_{n^2}^*$, $-r^2$ has maximal order. To generate a commitment key, first generate an RSA modulus n satisfying the requirements. Sample $r \xleftarrow{r} \mathbb{Z}_{n^2}^*$ and compute $g \leftarrow -r^2$. The commitment key is (n, g).

To commit to a message $m \in \mathbb{Z}_n$, sample $r \xleftarrow{r} S$, and compute the commitment $u \leftarrow g^r (1+n)^m$. The opening is r.

To verify that o is an opening of u to m, check that $u = g^o (1+n)^m$.

Equivocable A commitment scheme is *equivocable* if we can generate the commitment key in a particular way, which will give us a "trapdoor" that allows us to open a commitment to any message, in such a way that the opening is indistinguishable from the "correct" opening.

Formally, we have two additional algorithms: an *equivocable commitment key generator* $\mathcal{CK}_{eq}$ that outputs an *equivocable* commitment key and a "trapdoor"; and an *equivocate* algorithm $\mathcal{CQ}$ that on input of a commitment key, a "trapdoor", a commitment, an opening to a message and a target message, outputs an opening of the commitment to the target message. An equivocable commitment key must provide statistical hiding.

Sample $r \xleftarrow{r} \mathbb{Z}_{n^2}^*$ and $b \xleftarrow{r} \{0, 1, \ldots, n-1\}$ such that $\gcd(b, n) = 1$, and let

$$g = -r^{2n}(1 + n)^b.$$

Let the "trapdoor" $a \in \{0, 1, \ldots, n-1\}$ be an inverse of b modulo n.

E | *Exercise 8.44.* Prove that the above method of generating g samples from a distribution that is identical to the output distribution of the commitment key generation algorithm $\mathcal{CK}$.

If o is an opening of a commitment u to m, such that $u = g^o(1+n)^m$, then

$$g^{o-(m-m')a}(1 + n)^{m'} = u.$$

Observe that for any two messages m, m', we have that $-n^2 < (m-m')a < n^2$. Therefore, the result of Exercise 8.43 applies. It follows first of all that the commitment scheme is statistically hiding, and second that we can open any commitment to any value, since if o is an opening of u to m, then $o' = o - (m - m')a$ is an opening of u to m', except with very small probability.

We have now proved first of all that an equivocable commitment key is indistinguishable from the ordinary commitment key, and that the opening output by the equivocate algorithm is indistinguishable from an ordinary opening.

Extractable A commitment scheme is *extractable* if we can generate the commitment key in a particular way, which will give us a "trapdoor" that allows us to extract the message committed to from the commitment.

Formally we have two additional algorithms: an *extractable commitment key generator* $\mathcal{CK}_{ex}$ that outputs an *extractable* commitment key and a "trapdoor"; and an *extraction* algorithm $\mathcal{CX}$ that on input of a commitment key, a "trapdoor" and a commitment, outputs a message. An extractable commitment key must almost always be unconditionally binding.

Sample $r \xleftarrow{r} \mathbb{Z}_{n^2}^*$ and let

$$g = -r^{2n}.$$

The "trapdoor" is $a = (p-1)(q-1)/2$, essentially the Paillier decryption key.

E $\quad$ *Exercise* 8.45. Let $u = g^r(1+n)^m$. Show that $u^a = (1+n)^{am}$, and if $u = g^{r'}(1+n)^{m'}$, then $m \equiv m' \pmod{n}$. Show also that

$$(1+n)^e \equiv 1 + en \pmod{n^2}.$$

The exercise shows that the commitment scheme with such a generator is unconditionally binding. If we have a commitment u, then we know that

$$u^a - 1 \equiv amn \pmod{n^2} \qquad \Rightarrow \qquad \frac{u^a - 1}{n} \equiv am \pmod{n},$$

which means that it is easy to recover m if we know the "trapdoor" value a.

The extractable commitment key is essentially a Paillier encryption of 0. It follows that any adversary against hiding for the extractable commitment key is also an adversary against Paillier encryption.

We have now proved first of all that an extractable commitment key is indistinguishable from the ordinary commitment key, and that the extraction algorithm will always succeed in extracting the correct message (in the sense that it is impossible to open the commitment to any other message).

8.4.2 Group Homomorphic

We shall now consider a class of commitment schemes where the set of commitments, the message set and the randomness set have group structures and the commit algorithm is essentially a group homomorphism. This gives us a group homomorphic commitment scheme, which is useful.

The two commitment schemes we shall consider are very similar, and both rely on a cyclic group of prime order p. The commitment key contains one or two group elements. A commitment to a message in $\mathbb{Z}_p$ uses a single random value from $\mathbb{Z}_p$, which is sampled from the uniform distribution. The opening of the commitment is the random value. We shall use the notation $\mathcal{CC}(m; o)$ to denote a commitment to m with opening o.

These schemes are group homomorphic in the sense that for any commitments u_1, u_2 with openings (m_1, o_1) and (m_2, o_2), then for any $\alpha_1, \alpha_2 \in \mathbb{Z}_p$ we have that $\mathcal{CO}(u_1^{\alpha_1} u_2^{\alpha_2}, \alpha_1 m_1 + \alpha_2 m_2, \alpha_1 o_1 + \alpha_2 o_2) = 1$. Also, for any commitment u to a message m, there is a unique opening o such that $\mathcal{CO}(ck, u, m, o) = 1$.

ElGamal-based construction $\quad$ The first commitment scheme is a small modification of the commitment scheme from Example 8.23 using the ElGamal encryption scheme as the underlying public key cryptosystem. The commitment key is a group element y. To commit to a value $m \in \mathbb{Z}_p$, the commitment algorithm chooses a random value $o \in \mathbb{Z}_p$ and computes $x \leftarrow g^o$ and $w \leftarrow y^o g^m$. The commitment is $u = (x, w)$.

The only difference is that we encrypt g^m, not m as in Exercise 8.40.

By Exercise 8.40 the commitment scheme is unconditionally binding. and an adversary against hiding for the commitment scheme is an adversary against ElGamal, which becomes a distinguisher for Decision Diffie-Hellman.

Pedersen commitments   The second commitment scheme is the Pedersen commitment scheme from Example 8.24. By Exercise 8.41 the commitment scheme is perfectly hiding, and an adversary against binding for the commitment scheme can be turned into discrete logarithm solver.

8.5   CRYPTOGRAPHIC VOTING

An *election* is a process in which a set of *voters* come together to agree upon a *result*. Usually, this process consists of voters *casting ballots*, which somebody then *counts* to produce the result. We call this process a *voting system*.

How the result is computed from the cast ballots is encoded in a *counting function*. We usually represent the cast ballots as a string of ballots, so the counting function is a function from a set of strings of ballots to the set of results (which we conveniently leave undefined for the moment).

We shall assume that there is a total order on the ballot set. This allows us to sort a string of ballots. Counting should be independent of the order in which ballots are counted. For any string of ballots $v_1 v_2 \ldots v_l$, let $v'_1 v'_2 \ldots v'_l$ be the corresponding sorted string of ballots. Then $f(v_1 \ldots v_l) = f(v'_1 \ldots v'_l)$.

It is convenient to consider not only strings of ballots but also strings of ballots and the special symbol $\perp$. We extend any counting function to such strings, by first removing the $\perp$ symbols from the string and then evaluating the counting function on the result. With respect to sorting, we shall arbitrarily declare that $\perp$ is sorted before any ballot.

We will sometimes need our counting function to have a somewhat technical property, which essentially says that if the counting function agrees for two equal-length ballot strings, then regardless of what ballots we add to the two initial ballot strings, the counting function will continue to agree. When combined with the fact that the counting function does not care about the order of ballots in the string, this becomes a strong property of counting functions.

Definition 8.12. A counting function f is *additive* if for any two ballot strings v, v' of equal length, if $f(v) = f(v')$, then $f(vv'') = f(v'v'')$ for any string v''.

Example 8.28. The simplest possible election is the *yes/no* election. Typically, we encode these values as 1 and 0, and the counting function can be

$$f_1(v_1 v_2 \ldots v_l) = \begin{cases} 0 & \sum_i v_i < l/2, \\ 1 & \sum_i v_i > l/2, \text{ and} \\ \perp & \text{otherwise.} \end{cases}$$

Often, a counting function that gives the number of yes votes, $f_2(v_1 v_2 \ldots v_l) = \sum_i v_i$, is used instead. This reveals more information about the ballots than the precise result, but this may be desired information.

Note that f_1 is not additive, but f_2 is.

Example 8.29. A more complicated election is the *first past the post* election, where there are multiple candidates and the one with the most votes wins.

There does not have to be a winner in such contests. In practice, we order the ν candidates and use the additive counting function $f(v_1 v_2 \ldots v_l) = (k_1, k_2, \ldots, k_\nu)$, where k_i is the number of votes for candidate i.

Example 8.30. An even more complicated election is proportional representation, where there are multiple seats and multiple parties and the parties are awarded seats approximately in proportion to their fraction of the ballots cast. Some rule for approximation must be used, such as d'Hondt or Sainte-Laguë.

A counting function that gives out the number of seats allocated to each party will not be additive.

Remark. There are even more complicated counting functions, such as voting systems using transferable votes, or with some combination of party and candidate voting. It is therefore tempting to say that a voting system should simply output the sorted list of ballots cast as its result, and leave the actual counting to a non-cryptographic process. This is not always safe.

For most elections it is crucial that voters are free to cast the ballot they want without undue influence. In order to achieve this we need *ballot secrecy* or *confidentiality*. Traditionally, we use paper ballots prepared privately by each voter, collected in ballot boxes and eventually counted. Most modern democracies have established reasonable procedures to ensure confidentiality for political elections, but for many other elections these procedures are too costly. Therefore, we want to study how to use cryptographic techniques to simplify election procedures, in particular for fully electronic elections.

We begin with a simple definition for a voting scheme. This essentially amounts to public key encryption, but with a few additions that will allow us achieve a simple form of security. We extend this definition in Chapter 14.

The scenario motivating this definition is that some trusted authority runs the election with the help of a *ballot box*. The trusted authority first prepares the election, making some information public. The voters use the public information to cast ballots, by creating *encrypted ballots* that they give to the ballot box. Eventually, the ballot box forwards the encrypted ballots to the authority, which decrypts them and computes the counting function.

For most elections it is important that any one voter casts at most one ballot. In our scenario the trusted third party ensures that it counts at most one ballot per voter by having each encrypted ballot tied to a voter identity. As a general rule, we do not want to deal with how to identify voters, so we enable this functionality using associated data attached to each cast ballot. The assumption is that the election authority somehow specifies how to encode voter identities and perhaps other data into the associated data.

Definition 8.13. A *simple cryptographic voting scheme* VOTE for a counting function f consists of a (totally ordered) set of ballots $\mathfrak{P}$, a set of associated data $\mathfrak{F}$ and three algorithms that work as follows:

- The *setup* algorithm $\mathcal{VS}$ runs with no input and outputs a *ballot casting key* and a *counting key*.

- The *casting* algorithm $\mathcal{CB}$ takes as input a bk, $ad \in \mathfrak{F}$ and a ballot $v \in \mathfrak{P}$, and outputs an encrypted ballot c.

- The *counting* algorithm $\mathcal{VC}$ takes as input a counting key ck and encrypted ballots $c_1, c_2, \ldots, c_l$, and outputs a result.

A voting scheme is l_0-*correct* if for any $l < l_0$, any (bk, ck) output by $\mathcal{VS}$, any list of distinct associated data $(ad_1, ad_2, \ldots, ad_l)$ and any list of ballots $(v_1, v_2, \ldots v_l)$, then with $c_i \leftarrow \mathcal{CB}(bk, ad_i, v_i)$, $i = 1, 2, \ldots, l$ and

$$res \leftarrow \mathcal{VC}(ck, c_1, c_2, \ldots, c_l),$$

then $res = f(v_1, \ldots, v_l)$. It is *correct* if it is l_0-correct for any l_0.

With respect to the motivating scenario, the setup algorithm models the preparation step, where the casting key is public information. The casting algorithm models creating the encrypted ballot. The counting algorithm models the trusted authority decrypting the ballots and counting them.

For the moment, we do not want to care about how to decide which ballots to count. Note that there is no algorithm corresponding to the ballot box, which is just a conduit for ballots. In our modelling the ballot box is the adversary, who controls which encrypted ballots get counted.

As usual, we need some challenge encrypted ballots for which the adversary does not immediately know the decryption. The simplest mechanism is to give the adversary a left-or-right challenge query. Essentially, the adversary adaptively provides a sequence of left hand ballots and right hand ballots, and one of these sequences are encrypted.

The adversary's control of the ballot box is modelled by a count query in the experiment, which allows the adversary to specify which ballots get counted. We also need to allow the adversary to insert arbitrary ballots into the ballot box, which is modelled by the chosen ciphertext query. We also include the restriction that the associated data attached to the encrypted ballots must be distinct. For typical elections, the voter identity will be encoded in the associated data and at most one ballot should be counted for each voter.

Note that the adversary specifies the encrypted ballots to be counted indirectly, first creating a list of encrypted ballots either via challenge or chosen ciphertext queries. The adversary then selects encrypted ballots from this list to be counted. We could have allowed the adversary to specify the encrypted ballots to be counted directly, but our indirect approach gives the adversary the same capabilities and simplifies the accounting.

There is one problem with the above description: If the left hand challenge ballots give a different result than the right hand challenge ballots, the adversary can trivially learn the challenge bit by looking at the result. We must therefore require that the two lists of ballots give identical results. The adversary's job is to decide which ballots were cast, not what the count is.

Technically, this requirement is not enforced. Instead, we introduce a new technique that will become important later, where the adversary's answer does not count towards the advantage unless the requirement is satisfied.

There are counting functions for which no cryptosystem will give security under this definition. For instance, consider a *veto counting function* where ballots are 0 and 1, and the counting function says 1 if and only if two or more ballots are 1. If the left hand ballots are 0 and 0, the right hand ballots are 0 and 1, and the adversary casts an encryption of 1, the result will be a veto if and only if the right hand ballots were cast.

Note that the experiment cannot take the adversary's ballot into account when evaluating the left-hand and the right-hand ballots. There is no general way to get those particular ballots decrypted, since a voting scheme does not contain a decryption algorithm. In principle, we could create fresh encryptions of both the left-hand and the right-hand ballots and count both, but we shall not go in this direction. It would work in the present simple case, but would become increasingly unworkable when we move to more realistic scenarios.

This security definition does work without further qualifications for additive counting functions or when there are no chosen ciphertext queries.

D **Definition 8.14.** A (τ, l_v, l_c, l_d)-adversary against *indistinguishability* for a voting scheme VOTE is an interactive algorithm $\mathcal{A}$ that interacts with the experiment in Figure 8.6 making count queries with at most l_v encrypted ballots, l_c challenge queries and l_d chosen ciphertext queries, and where the runtime of the adversary and the experiment is at most τ.

We say that an execution is *fresh* if the execution either never reaches Step 5(c), or if $f(L'_0) = f(L'_1)$ whenever this step is reached.

The *advantage* of this adversary is defined to be

$$\mathbf{Adv}^{\mathrm{ind}}_{\mathrm{VOTE}}(\mathcal{A}) = 2|\Pr[E] - 1/2|,$$

where E is the event that $b' = b$ given that the execution is fresh, or that b' equals a fresh coin toss if the execution is not fresh.

Remark. The computation of the partial results in Step 5(b) is not actually used in the experiment. But these values are used to determine if an execution is fresh. Computing the value here allows us to use freshness in later games, without incurring an increase in computational cost. This is mostly a technical

The experiment for the left-or-right security game with an adversary $\mathcal{A}$ proceeds as follows:

1. Sample $b \xleftarrow{r} \{0,1\}$, and let L_C, L_0, L_1 be empty lists.

2. Compute $(bk, ck) \leftarrow \mathcal{VS}$ and send bk to $\mathcal{A}$.

3. When the adversary sends a query (ad, v_0, v_1) (*challenge*), do:

    (a) Compute $c \leftarrow \mathcal{CB}(bk, ad, v_b)$.

    (b) Append (ad, c) to L_C, v_0 to L_0 and v_1 to L_1.

    (c) Send c to $\mathcal{A}$.

4. When the adversary sends a query (ad, c) (*chosen ciphertext*), do:

    (a) If the pair (ad, c) is in L_C, send $\perp$ to the adversary and stop.

    (b) Append (ad, c) to L_C and $\perp$ to L_0 and L_1.

    (c) Send $\top$ to $\mathcal{A}$.

5. When the adversary sends a query $(i_1, i_2, \ldots, i_{l'})$ (*count*), do:

    (a) Let $L_C = ((ad_1, c_1), (ad_2, c_2), \ldots, (ad_l, c_l))$. If $i_j = i_{j'}$ or $ad_{i_j} = ad_{i_{j'}}$ for some $j \neq j'$, or $i_j > l$ or $i_j < 1$ for any j, stop.

    (b) With $L_\nu = (v_{\nu,1}, v_{\nu,2}, \ldots, v_{\nu,l})$, define $L'_\nu = (v_{\nu,i_1}, v_{\nu,i_2}, \ldots, v_{l,i_{l'}})$ for $\nu = 0, 1$. Compute $f(L'_0)$ and $f(L'_1)$.

    (c) Compute $res \leftarrow \mathcal{VC}(ck, c_{i_1}, c_{i_2}, \ldots, c_{i_{l'}})$ and send res to $\mathcal{A}$.

Eventually, the adversary outputs $b' \in \{0, 1\}$.

FIGURE 8.6 The experiment $\mathbf{Exp}^{\mathsf{ind}}_{\mathrm{VOTE}}(\mathcal{A})$ for the left-or-right security game for a simple voting scheme VOTE with adversary $\mathcal{A}$.

trick which allows us to ignore obscure, theoretical counting functions with significant computational costs. It has no practical significance, since it is not computationally costly to compute counting functions used in practice, which were mostly invented when humans did any computations required.

Instead of introducing the notion of freshness and complicating the definition of the event E, we could instead have the experiment respond with $\perp$ to any non-fresh query. This would essentially match how we treat the decryption query in the public key encryption experiment. In our present simple case, this would in fact be easier to work with. However, when we later introduce more complicated types of voting schemes (and other cryptographic schemes), the amount of accounting required becomes significant. The analysis is simplified by separating this accounting from the experiment through freshness.

For any execution that is not fresh, the event E occurs with probability $1/2$. The adversary can easily determine if an execution is fresh or not. This captures our requirement that the adversary should not be able to learn the challenge bit from the result, either because the adversary does not learn the result or because the challenge bit is independent of the result.

With the simplest possible counting function, namely returning the string of ballots sorted, this security notion essentially reduces to hiding who cast which ballot. For counting functions that reveal less about the string of ballots counted, indistinguishability hides more information about exactly which ballots were cast. However, these counting functions are often not additive, in which case the security notion may not be applicable.

Since the adversary controls the notional ballot box, we do not attempt to hide the number of ballots cast, even if the result does hide that number. If we want to hide the number of ballots cast (which could be a legitimate goal), we would need to restrict the adversary significantly. We shall not attempt this, but we return to a similar problem in Chapter 14.

It is non-trivial to define analogues of semantic security and real-or-random security notions. The reason is that the adversary submits challenge queries separately. We could define an analogue of real-or-random by allowing the adversary to specify one list of ballots, and then sample a second list of ballots from the set of lists of ballots of equal length that give the same result as the first list. Likewise, we could do semantic security by defining sample spaces over lists of ballots. Obviously, this is only possible if we can sample from the resulting sets and spaces. The cost is that the adversary loses the ability to choose ballots adaptively, in particular the ability to select encrypted ballots after the fact. We shall not attempt such a definition.

One way to get a more workable semantic security game would be to relax the freshness condition, so that the challenge ballots that are decrypted need not have the same effect on the result. This allows the adversary to encode some information about the challenge bit in the result. However, we can measure that information by computing the statistical distance between the two probability spaces that we sample the ballots from, and subtract this distance from the adversary's advantage. Such definitions have been attempted. We shall not attempt such a definition.

One way to get a more workable real-or-random security game would be to change the counting procedure, so that the encrypted ballots from the challenge queries are replaced by fresh encryptions of the real ballots from those queries before running the counting algorithm. The replacement of the challenge query ciphertexts is often called a *recovery step*, and has been used in a number of security definitions. We shall not discuss this further.

If we wanted to relax the freshness condition as described above, such a real-or-random security game is a natural way to prove that the statistical distance is the only advantage the adversary has.

While this approach to real-or-random security will work, it will also introduce a number of technical difficulties. We shall not attempt such a definition.

Remark. Again, we are not usually trying to hide the ballots cast, but rather who cast which ballots. This is one reason why semantic security and real-or-random seem to be incorrect modelling choices for voting.

We shall consider one extremely simple voting scheme, based on public key encryption alone, and a more cryptographically sophisticated scheme based on homomorphic encryption, with somewhat restricted functionality. The two schemes will be secure only in a very limited sense. Instead of using more realistic security notions and schemes, we shall use these two very simple schemes to motivate a set of questions that we will answer in Chapter 14.

8.5.1 Simple Voting Scheme

The following is in some sense the cryptographic analogue of the traditional postal voting system, where the ballot is placed inside an envelope and identifying information is written on the outside of the envelope. The idea is that the encryption plays the role of the envelope.

We immediately note that unlike a ballot enclosed in an envelope, an encrypted ballot can be trivially and perfectly duplicated. This will cause confidentiality attacks, where an adversary attempts to use a chosen ciphertext query to duplicate a challenge ballot and thereby reveal the challenge bit. We use associated data to prevent this sort of attack.

Example 8.31. Let PKE $= (\mathcal{K}, \mathcal{E}, \mathcal{D})$ be a public key cryptosystem with message space $\mathfrak{P}$ and associated data $\mathfrak{F}$.

The simple cryptographic voting scheme VOTE $= (\mathfrak{P}, \mathfrak{F}, \mathcal{VS}, \mathcal{CB}, \mathcal{VC})$ is:

- The *setup* algorithm $\mathcal{VS}$ computes $(ek, dk) \leftarrow \mathcal{K}$ and outputs (ek, dk).

- The *casting* algorithm $\mathcal{CB}$ takes as input a casting key ek, associated data ad and a ballot v, computes $c \leftarrow \mathcal{E}(ek, ad, v)$, and outputs (ad, c).

- The *count* algorithm $\mathcal{VC}$ takes as input a counting key dk and encrypted ballots $(ad_1, c_1), (ad_2, c_2), \ldots, (ad_l, c_l)$. It computes $v_i \leftarrow \mathcal{D}(dk, ad_i, c_i)$ for $i = 1, 2, \ldots, l$ and outputs the sorting of the string $v_1 v_2 \ldots v_l$.

Exercise 8.46. Prove that the above scheme is correct with respect to the counting function that simply sorts its argument.

Exercise 8.47. Suppose the encrypted ballot does not include the associated data. Give an adversary with trivial runtime and advantage 1.

Proposition 8.20. *Suppose every ballot in $\mathfrak{P}$ has the same length. Let $\mathcal{A}$ be a (τ, l_v, l_c, l_d)-adversary adversary against indistinguishability for VOTE. Then there exists a (τ', l_c, l_d)-adversary $\mathcal{B}$ against indistinguishability for PKE, with*

τ' *essentially equal to* τ, *and*

$$\mathbf{Adv}^{\text{ind}}_{\text{VOTE}}(\mathcal{A}) = \mathbf{Adv}^{\text{ind-cca}}_{\text{PKE}}(\mathcal{B}).$$

The proof begins by observing that if the execution is determined to be non-fresh, the game can immediately stop and pretend that the adversary output a random bit. This simplifies the later analysis, since there will only be count query responses if the execution is fresh.

Now, in the count query, instead of actually decrypting the ciphertexts from the challenge queries, we instead use the corresponding left hand ballots. By correctness for the public key cryptosystem, we know that the decryption will either equal the left hand ballots or the right hand ballots. If the execution is fresh, the left hand and right hand ballots have the same effect on the count, so always using the left hand ballots does not change anything.

Then we create a simulator from the experiment which replaces key generation, challenge ballot encryptions and chosen ciphertext decryptions with queries to the indistinguishability experiment for the cryptosystem. This turns any voting scheme adversary into a cryptosystem adversary.

Proof of Proposition 8.20   The proof proceeds as a sequence of games, between the adversary $\mathcal{A}$ and an experiment. Let E_i be the event in Game i that the adversary's guess b' equals the challenge bit b of the experiment given that the execution is fresh.

Ⓖ **Game 0**   The initial game is $\mathcal{A}$ interacting with the indistinguishability experiment $\mathbf{Exp}^{\text{ind}}_{\text{VOTE}}(\mathcal{A})$. Then

$$\mathbf{Adv}^{\text{ind}}_{\text{VOTE}}(\mathcal{A}) = 2|\Pr[E_0] - 1/2|. \tag{8.27}$$

Let $\tau_0 = \tau$.

Ⓖ **Game 1**   We change Step 5(b) so that if the execution is not fresh, the game immediately stops and samples $b' \xleftarrow{r} \{0,1\}$.

This game is identical to the previous game until a count query is made that renders the execution non-fresh, so

$$\Pr[E_1] = \Pr[E_0]. \tag{8.28}$$

The runtime cost of this change is insignificant, so the runtime bound τ_1 on this game is essentially equal to τ_0.

Ⓖ **Game 2**   We change the counting step so that ciphertexts from challenge queries are not decrypted, but the corresponding ballots are used directly.

5(c). For $j = 1, 2, \ldots, l'$, compute

$$v_j''' \leftarrow \begin{cases} v_{0,i_j} & v_{0,i_j} \neq \perp, \\ \mathcal{D}(dk, ad_{i_j}, c_{i_j}) & \text{otherwise.} \end{cases}$$

and $res \leftarrow f(v_1''', v_2''', \ldots, v_{l'}''')$. Send res to $\mathcal{A}$.

Consider ballots defined by

$$v'_j = \mathcal{D}(dk, ad_{i_j}, c_{i_j}) \qquad v''_j = \begin{cases} v_{b,i_j} & v_{0,i_j} \neq \perp, \\ \mathcal{D}(dk, ad_{i_j}, c_{i_j}) & \text{otherwise..} \end{cases}$$

for $j = 1, 2, \ldots, l'$. Observe that v'_j are the ballots that the counting algorithm $\mathcal{VC}$ would have computed in the previous game. By correctness of the public key cryptosystem, we see that $v''_j = v'_j$ for $j = 1, 2, \ldots, l'$. It follows that

$$f(v'_1 v'_2 \ldots v'_{l'}) = f(v''_1 v''_2 \ldots v''_{l'}).$$

By design, $v_{0,i_j} = \perp$ if and only if $v_{b,i_j} = \perp$, indicating that the encrypted ballot is a chosen ciphertext. For these indexes we have that $v''_j = \mathcal{D}(dk, ad_{i_j}, c_{i_j}) = v'''_j$. For indexes with $v_{0,i_j} \neq \perp$, we have that $v''_j = v_{b,i_j}$ and $v'''_j = v_{0,i_j}$, which means that these ballots coincide with the ballots distinct from $\perp$ in L'_b and L'_0, respectively. If the execution does compute the result of the count, the execution will also be fresh in this game, which means that $f(L'_0) = f(L'_1)$. The counting function is additive, so we get that

$$f(v''_1 v''_2 \ldots v''_{l'}) = f(v'''_1 v'''_2 \ldots v'''_{l'}).$$

It follows that this modified counting query has the same behaviour as the counting query in the previous game, so

$$\Pr[E_2] = \Pr[E_1]. \tag{8.29}$$

For each ballot there is now an extra lookup. For some ballots, there is one decryption operation less. This means that τ_2 is essentially the same as τ_1.

Also observe that in this game, no pair (ad, c) that originated with a challenge query is later decrypted.

The proposition now follows from (8.27)–(8.29) and the following lemma.

Lemma 8.21. *There is a (τ', l_c, l_d)-adversary $\mathcal{B}$ against indistinguishability for* PKE, *with τ' essentially equal to τ_1, and*

$$\mathbf{Adv}_{\mathrm{PKE}}^{\mathrm{ind\text{-}cca}}(\mathcal{B}) = 2|\Pr[E_1] - 1/2|.$$

Proof. The adversary $\mathcal{B}$ runs a copy of $\mathcal{A}$ and the simulator **Sim**, which equals the experiment in Game 1 modified as follows:

2. Send ek to $\mathcal{A}$.

3(a). Make a challenge query (ad, v_0, v_1) and receive the response c.

5(c). For each $j = 1, 2, \ldots, l'$, if $v_{0,i_j} \neq \perp$, set $v'''_j \leftarrow v_{0,i_j}$, otherwise make a chosen ciphertext query (ad_{i_j}, c_{i_j}) and receive v'''_j as response. Compute $res \leftarrow f(v'''_1, v'''_2, \ldots, v'''_{l'})$. Send res to $\mathcal{A}$.

Eventually, $\mathcal{A}$ outputs b' (or b' is a fair coin toss if the game terminates because the execution became non-fresh). Then $\mathcal{B}$ outputs b'. Let τ' be an upper bound on the runtime of $\mathcal{B}$ interacting with $\mathbf{Exp}_{\mathrm{PKE}}^{\mathrm{ind\text{-}cca}}(\mathcal{B})$.

By inspection, we see that **Sim** interacting with $\mathbf{Exp}_{\mathrm{PKE}}^{\mathrm{ind\text{-}cca}}(\mathcal{B})$ behaves in the same way as the experiment in Game 1, with the challenge bit in $\mathbf{Exp}_{\mathrm{PKE}}^{\mathrm{ind\text{-}cca}}(\mathcal{B})$ playing the same role as the challenge bit b in Game 1. In particular, no chosen ciphertext query is forbidden, because we require distinct associated data. It follows that

$$\Pr[E] = \Pr[E_1],$$

where E is the event that b' equals the challenge bit in $\mathbf{Exp}_{\mathrm{PKE}}^{\mathrm{ind\text{-}cca}}(\mathcal{B})$.

Finally, we observe that every challenge and count query involves the same amount of work, but there is some additional forwarding of messages. It follows that τ' is essentially equal to τ_1. $\qquad\square$

Discussion   Confidentiality relies on the trusted authority doing the counting. We shall sketch one way to *distribute* counting, so that no single player will know who cast which ballot. Better methods will appear in Chapter 14.

This is motivated by the traditional postal voting process, where the ballot is inside an inner envelope, which is again inside an outer envelope. The outer envelope lists the voter's name, while the inner envelope is anonymous. It is now possible to organise counting such that no single counter can correlate ballots and voters. This can be approximated using *nested encryption*.

There is however a second problem, which is that of *election integrity*, that the result matches the cast ballots. Obviously, since the adversary controls the ballot box and may insert arbitrary encrypted ballots, we have no integrity anyway. But even if the adversary makes no chosen ciphertext queries and includes every cast ballot in the count, election integrity does not hold if either of the two counting parties is controlled by the adversary. The reason is that either party can lie about the decryptions of its ciphertexts. We need further tools in order to solve this problem.

8.5.2   Simple Homomorphic Counting

We shall now consider a different way to do counting, using a group-homomorphic cryptosystem. Consider first a yes/no election, where the result is the number of yes ballots cast. The idea is to encode yes and no as group elements. To count the encrypted ballots, we use the cryptosystem's operation to combine all the ciphertexts into a single ciphertext and then decrypt it. The decryption is then a product of powers of the two elements used to encode yes and no, with the powers being the number of yes and no ballots cast.

Example 8.32. Let l_0 be a positive integer, and let $\mathfrak{F}$ be as in Example 8.18, but with the additional requirement that for each tuple, there are distinct public elements $g_0, g_1 \in \mathfrak{P}$ such that one of them has order greater than l_0. Let $\text{PKE}_{\mathfrak{F}} = (\mathcal{K}, \mathcal{E}, \mathcal{D})$ be the corresponding group-homomorphic cryptosystem.

The voting scheme $\text{VOTE}_{\mathfrak{F}}$ with ballot set $\{0, 1\}$ and associated data set $\mathfrak{F}$, for the counting function $f(v_1 v_2 \ldots v_l) = \sum_i v_i$ that sums the ballots, is:

- The *setup* algorithm $\mathcal{VS}$ computes $(ek, dk) \leftarrow \mathcal{K}$ and outputs (ek, dk).

- The *casting* algorithm $\mathcal{CB}$ takes as input a casting key ek, associated data ad and a ballot v, computes $c \leftarrow \mathcal{E}(ek, g_v)$, and outputs c.

- The *counting* algorithm $\mathcal{VC}$ takes as input a counting key dk and a list of encrypted ballots $c_1, c_2, \ldots, c_l$. It computes $x \leftarrow \mathcal{D}(dk, \prod_i c_i)$, finds k such that $x = g_0^{l-k} g_1^k$ and outputs k.

This scheme is l_0-correct, since if (ek, dk) was output by $\mathcal{K}$ and $c_i \leftarrow \mathcal{E}(ek, g_{v_i})$ for $i = 1, 2, \ldots, l$, then

$$\mathcal{D}(dk, \prod_i c_i) = \prod_i \mathcal{D}(dk, c_i) = \prod_i g_{v_i} = g_0^{l - \sum_i v_i} g_1^{\sum_i v_i}.$$

One possible cryptosystem is ElGamal over some group G, as in Example 8.11. Typically, g_0 would be 1 and g_1 would be a generator for the group. Of course, in order to recover the election result, we need to compute $\log_{g_1} x$, which is easy since the discrete logarithm will be small.

It would be nice to generalise this scheme to a one-out-of-ν election, where you can vote for one out of ν candidates. The obvious idea is to have ν group elements $g_1, g_2, \ldots, g_\nu$ chosen so that if $0 \le k_i < l_0$ for $i = 1, 2, \ldots, \nu$, then $g_1^{k_1} g_2^{k_2} \ldots g_\nu^{k_\nu} = 1$ implies $k_1 = k_2 = \cdots = k_\nu = 0$. In general, when ν becomes large it becomes hard to recover $k_1, k_2, \ldots, k_k$ from the group element. However, for some specific group structures ν can be usefully large.

The absolutely simplest solution is to use multiple ciphertexts. If we have a group homomorphic cryptosystem with plaintext group $\mathfrak{P}$, we can trivially extend this to a cryptosystem with plaintext group $\mathfrak{P}^\nu$.

There are other, more efficient solutions. One possibility is multi-ElGamal, with the ciphertext space $G^{\nu+1}$ and message space G^ν. The encryption key is ν group elements $y_1, y_2, \ldots, y_\nu$, and an encryption of $(m_1, \ldots, m_\nu) \in G^\nu$ has the form $(g^r, y_1^r m_1, \ldots, y_\nu^r m_\nu)$. Our generators could take the form $(1, \ldots, 1, g, 1, \ldots, 1)$. Further trade-offs are possible.

A completely different group structure is used by the Paillier cryptosystem, where the plaintext group structure is $\mathbb{Z}_n^+$. In this case, we can use $g_i = l_0^{i-1}$, as long as $l_0^\nu < n$. This is often a practical solution, though arithmetic in the ciphertext group $\mathbb{Z}_{n^2}^*$ is quite expensive.

Of course, these schemes can also be used for elections where a voter may cast a ballot for more than one candidate.

Ⓣ **Proposition 8.22.** Let $\mathcal{A}$ be a $(\tau, l_0, l_c, 0)$-*adversary against indistinguishability for* $\text{VOTE}_{\mathfrak{F}}$. *Then there exists a* $(\tau', l_c, 0)$-*adversary* $\mathcal{B}$ *against indistin-*

guishability for $\text{PKE}_{\mathfrak{F}}$, *with* τ' *essentially equal to* τ, *and*

$$\mathbf{Adv}^{\text{ind}}_{\text{VOTE}_{\mathfrak{F}}}(\mathcal{A}) = \mathbf{Adv}^{\text{ind-cpa}}_{\text{PKE}_{\mathfrak{F}}}(\mathcal{B}).$$

As for Proposition 8.20, the proof begins by observing that if the execution is determined to be non-fresh, the game can immediately stop and pretend that the adversary output a random bit. This simplifies the later analysis, since there will only be a count query response if the execution is fresh.

Now, in the count query, instead of actually multiplying the ciphertexts from the challenge queries and decrypting the result, we can instead simply multiply the corresponding left hand ballots in a suitable fashion. By correctness for the group-homomorphic cryptosystem, we know that the decryption will either equal the product of the left hand ballots or the right hand ballots. If the execution is fresh, the left hand and right hand ballots have the same product, so always using the left hand ballots does not change anything.

It is now easy to create a simulator from the experiment, which replaces key generation and challenge ballot encryptions with queries to the indistinguishability experiment for the cryptosystem. This turns any adversary against the voting scheme into an adversary against the cryptosystem.

E | *Exercise* 8.48. Prove Proposition 8.22.

E | *Exercise* 8.49. Adapt the simple voting scheme from Example 8.32 to use the somewhat homomorphic cryptosystem from Example 8.19. State and prove a result similar to Proposition 8.22 for your construction.

Discussion   As for the scheme from Section 8.5.1, a problem with this scheme is the trusted authority, which can easily decrypt individual encrypted ballots, not just the ciphertext containing the election result. We shall study efficient tools that allow us to distribute the decryption, so this problem is solvable.

The second problem is election integrity, emphasised by the fact that the adversary is not allowed to make chosen ciphertext queries (which we can sort of justify by using digital signatures). In this case, too, the authority may choose to lie about the decryption. Again, we shall study efficient tools that allow us to verify that this computation was done correctly.

However, the real problem with this scheme is that a dishonest voter may submit an encryption of $g_0^{-1}g_1^2$ which increases the result by 2, while staying consistent even when $g_0 \neq 1$. Yet again, we shall study various tools that allow us to ensure that only valid ballots have been submitted.

CHAPTER 9

Digital Signatures

There are many ways to look at digital signatures, but for defining security the natural counterpart is the message authentication code. In fact, the experiment for message authentication codes could have worked for digital signatures as well. But since digital signatures are verifiable by anyone that has the verification key, we can dispense with the test query.

We could have studied constructions that essentially turn message authentication codes into signature schemes, to further this parallel. But this is not the best way to approach the design of signature schemes. Instead, we shall study two well-known design approaches: the hash-and-sign paradigm exemplified by RSA-FDH, and the Fiat-Shamir approach with the Schnorr signature scheme as its most famous example. The identification schemes used by this approach are related to interactive arguments, which we discuss in Chapter 11.

Unlike for symmetric cryptosystems and public key encryption, the study of signature schemes contains fewer generic theorems. However, we do spend some time investigating what exactly provable security means.

We extend the discussion on secure channels from Chapter 7 to a discussion about messaging, a discussion that will continue in Chapter 13.

9.1 DEFINING SECURITY

Before we can arrive at a security definition for digital signatures, we must first discuss the adversary's goal and the adversary's capabilities. Note the similarity between signatures and message authentication codes.

We begin with the adversary's capabilities. The adversary will be able to see valid signatures on some messages, but the question is how the messages are chosen. The adversary can probably influence the messages signed, so we allow the adversary to choose the messages, a *chosen message attack*.

The natural adversarial goal is to forge a valid signature on some message. Yet again, we have to decide how this message is chosen, and again it makes sense to allow the adversary to choose the message. Like for message authentication codes, we do have one choice left. Is it sufficient for the adversary to

The chosen message experiment proceeds as follows:

1. Set $C_0 = C_1 = \emptyset$. Compute $(vk, sk) \leftarrow \mathcal{K}$ and send vk to the adversary.

2. When the adversary sends a query m (*chosen message*), compute $\sigma \leftarrow \mathcal{S}(sk, m)$, add m to C_0 and (m, σ) to C_1, and send σ to the adversary.

Eventually, the adversary outputs (m_0, σ_0).

FIGURE 9.1 The experiment $\mathbf{Exp}^{\mathsf{cma}}_{\mathsf{SIG}}(\mathcal{A})$ for the chosen message security game for a digital signature scheme $\mathsf{SIG} = (\mathcal{K}, \mathcal{S}, \mathcal{V})$ with adversary $\mathcal{A}$.

come up with a new signature on a message it already has a signature for, or must it be able to forge a signature for a new message? Again, as for message authentication codes, the latter seems sufficient for most applications, while for security proofs we will often need the former.

Definition 9.1. A (τ, l_s)-*adversary* against a signature scheme SIG is an interactive algorithm $\mathcal{A}$ that interacts with the experiment in Figure 9.1 making at most l_s chosen message queries, and where the runtime of the adversary and the experiment is at most τ.

The *existential unforgeability advantage* and the *strong unforgeability advantage* of this adversary are defined to be

$$\mathbf{Adv}^{\mathsf{euf\text{-}cma}}_{\mathsf{SIG}}(\mathcal{A}) = \Pr[E] \quad \text{and} \quad \mathbf{Adv}^{\mathsf{suf\text{-}cma}}_{\mathsf{SIG}}(\mathcal{A}) = \Pr[F]$$

where E is the event that $m_0 \notin C_0$ and $\mathcal{V}(vk, m_0, \sigma_0) = 1$, and F is the event that $(m_0, \sigma_0) \notin C_1$ and $\mathcal{V}(vk, m_0, \sigma_0) = 1$.

We say that the pair (m_0, σ_0) is a *forgery*, and also that $\mathcal{A}$ is a *forger*.

Exercise 9.1. Let SIG be any signature scheme. Construct a new signature scheme SIG' such that:

- any existential unforgeability adversary against SIG' can be turned into an existential unforgeability adversary against SIG; and

- there exists a trivial strong unforgeability adversary against SIG'.

9.1.1 Example: One-time Signatures

We begin with Lamport's *one-time* signature scheme. The reason is that it is very simple and illustrates one design approach, based on one-way functions.

We define Lamport's signature scheme for single bit messages. It is one-time, so we can sign bit strings by signing each bit with a different key.

Example 9.1. Let $h : S \to T$ be a function. Lamport's one-time signature scheme LOTS $= (\mathcal{K}_1, \mathcal{S}_1, \mathcal{V}_1)$ works as follows:

- The *key generation* algorithm $\mathcal{K}_1$ samples $s_0, s_1 \in S$ and computes $u_0 \leftarrow h(s_0)$, $u_1 \leftarrow h(s_1)$. It outputs the verification key $vk = (u_0, u_1)$ and signing key $sk = (s_0, s_1)$.

- The *signing* algorithm $\mathcal{S}_1$ takes as input $sk = (s_0, s_1)$ and $m \in \{0, 1\}$, and outputs $\sigma = s_m$.

- The *verification* algorithm $\mathcal{V}_1$ takes as input $vk = (u_0, u_1)$, $m \in \{0, 1\}$ and σ. It outputs 1 if $h(\sigma) = u_m$, otherwise 0.

Since a signature in Lamport's scheme is a preimage, it is more or less obvious that any adversary against Lamport's scheme can be turned into a preimage finder, but even so we state the result and prove it.

Proposition 9.1. *Let $\mathcal{A}$ be a $(\tau, 1)$-adversary against LOTS. Then there exists a τ'-preimage finder $\mathcal{B}$ for h, with τ' essentially equal to τ, such that*

$$\mathbf{Adv}_{\text{LOTS}}^{\text{euf-cma}}(\mathcal{A}) \leq \frac{1}{2} \mathbf{Adv}_{h,S}^{\text{preimage}}(\mathcal{B}).$$

Exercise 9.2. Prove Proposition 9.1.

Exercise 9.3. As in Section 4.5.1, Lamport's scheme can be defined for longer messages. State and prove a similar result to Proposition 9.1 for this variant.

Exercise 9.4. Lamport's scheme can be extended from $\{0, 1\}$ to $\{0, 1, \ldots, k-1\}$ using two hash chains of length k as described in Section 4.5.1. State and prove a similar result to Proposition 9.1 for the resulting signature scheme.

Exercise 9.5. Merkle signatures use Merkle trees to extend one-time signatures to (stateful) many-time signatures without increasing the size of the public verification key, at the cost of increased signature size and key generation time. State and prove a similar result to Proposition 9.1 for Merkle signatures.

9.1.2 Example: Cramer-Shoup Signatures

There are many ways to turn a one-time (or a few-time) signature scheme into a normal signature scheme. One approach is to use trees as discussed in Section 4.5.2. This can be achieved using only hash functions, but one interesting approach is based on more standard asymmetric cryptography. This allows us to make trees that are so wide that they have very short height.

E *Example* 9.2. Let p and q be primes such that $(p-1)/2$ and $(q-1)/2$ are also prime. Let $n = pq$, and let $Q_n \subseteq \mathbb{Z}_n^*$ be the subgroup of *quadratic residues* (squares modulo n). Knowing p and q, it is easy to compute eth roots in Q_n for any small e.

Consider the following signature scheme:

- The key generation algorithm chooses $n = pq$ as a standard RSA modulus, subject to the above requirements, a small prime $e \in \{0, 1, \ldots, \min\{p, q\} - 1\}$, and samples $u, v \xleftarrow{r} Q_n$. The verification key is $vk = (n, u, v, e)$, while the signing key is $sk = (p, q, u, v, e)$.

- To sign a message $m \in \{0, 1, \ldots, e - 1\}$, let σ be an eth root of uv^m. The signature is σ.

- To verify a signature σ on a message m under verification key (n, u, v, e), verify that $\sigma^e v^{-m} = u$.

We claim that this is a one-time signature, and that a very specific adversary against the system can be turned into an adversary against the RSA problem. We shall sketch the proof by showing that any two signatures on distinct elements can be used to compute an eth root of some element in Q_n. And then we shall show that we can embed an RSA challenge value in this element, and at the same time create a verification key and a signature on a message of our choice. (Note that the message is chosen by us, not the adversary, which is what we meant by *very specific adversary* above.)

We shall need a lemma that will allow us to compute roots in $\mathbb{Z}_n^*$.

T **Lemma 9.2.** *Let G be a group and $s, u \in G$, and let k, k' be relatively prime integers such that $s^k = u^{k'}$. Then if α, β are integers such that $k\alpha + k'\beta = 1$, we get that*

$$(s^\beta u^\alpha)^k = u.$$

Proof. We compute $(s^\beta u^\alpha)^k = (s^k)^\beta u^{k\alpha} = (u^{k'})^\beta u^{k\alpha} = u.$ □

First, suppose we have two signatures σ, σ' on two distinct messages m, m', and assume without loss of generality that $m' > m$. Then we have that

$$\sigma^e v^{-m} = u = (\sigma')^e v^{-m'}, \quad \text{or} \quad (\sigma'\sigma^{-1})^e = v^{m-m'}.$$

Since $0 < m - m' < e$ and e is prime, $\gcd(m - m', e) = 1$. With valid signatures on two distinct messages, Lemma 9.2 recovers an eth root of v.

Next, suppose we have an RSA problem (n, e, x) where n and e are of the form from the example. We want to find y such that $y^e = x$ in $\mathbb{Z}_n$. We may assume that $x \in Q_n$, since this happens with probability about $1/4$ and we will know if our procedure fails (because the answer is incorrect or because the procedure requires too much time). We choose some $m \in \{0, 1, \ldots, e - 1\}$ and random $\sigma \in Q_n$. Then we let $v = x$ and $u = \sigma^e v^{-m}$. We have created a verification key (n, u, v, e) and a signature σ on a message m.

Since x has been sampled from the uniform distribution on Q_n, we get that v has the correct distribution. Likewise, σ has been sampled from the uniform distribution on Q_n and raising to the eth power is a bijection on Q_n, we have that u has been sampled from the uniform distribution on Q_n. Also, u and v are independent. Finally, signatures on messages are fully determined by the verification key and the message, so the signature σ on m that we generate is both valid and correctly distributed.

It follows that under the assumption that $x \in Q_n$, our very specific adversary against the signature scheme will get the input it expects, and therefore be able to create a forgery. As discussed above, this will give us an eth root of x, which is the y we were looking for.

Example 9.3. Let p, q and n be as in Example 9.2. Let k be an upper bound on messages.

Consider the following signature scheme:

- The key generation algorithm chooses $n = pq$ as a standard RSA modulus, subject to the above requirements, and samples $u, v \xleftarrow{r} Q_n$. The verification key is $vk = (n, u, v)$, while the signing key is $sk = (p, q, u, v)$.

- To sign a message $m \in \{0, 1, \ldots, k-1\}$, choose a small prime $e > k$ and let s be an eth root of uv^m. The signature is $\sigma = (e, s)$.

- To verify a signature $\sigma = (e, s)$ on a message m under verification key (n, u, v), verify that $s^e v^{-m} = u$.

This is a much better signature scheme, since we can use it multiple times, as long as we do not reuse primes. However, it is hard to prove that this system is secure when the adversary chooses the messages. If we choose the messages, we can prove it secure, much like the scheme above.

The idea is to use the above two signature schemes in a tree structure, where we create a very wide and shallow tree. We use the second scheme to sign verification keys for the first scheme, and then use the first scheme to sign actual messages. As long as the primes for the second scheme are chosen from a large enough set of primes, we do not have to worry about reuse of primes.

With some care, we can even reuse the RSA modulus and the element v between the two schemes, which makes it much more efficient. Also, when we want to sign a message with the second scheme, we actually know the message before we have to generate the key material, which means that instead of using the signing algorithm, we can generate the verification key and the signature at the same time, as in the discussion after the example.

The verification key of the first scheme is too large for the message space of the second scheme, so we use a hash function, as discussed in Section 9.1.3.

Example 9.4. Let $h : S \to \{0, 1, \ldots, k-1\}$ be a hash function. We shall suppose that for any RSA modulus n we care about, $\mathbb{Z}_n$ can be injected into S. Let $\mathcal{X}_p$ be some probability space on a set of primes contained in $\{k, k+1, \ldots, 2k\}$.

The *Cramer-Shoup* signature scheme CS-SIG works as follows:

- The *key generation* algorithm samples distinct random primes p and q, both larger than $2k$, such that $(p-1)/2$ and $(q-1)/2$ are both prime, computes $n \leftarrow pq$ and samples $u, v \xleftarrow{r} Q_n$ and $e \xleftarrow{r} \mathcal{X}_p$. The verification key $vk = (n, u, v, e)$, and the signing key is (p, q, u, v, e).

- The *signing* algorithm takes sk and $m \in \{0, 1, \ldots, k-1\}$ as input. It samples $\tilde{s} \xleftarrow{r} Q_n$ and $\tilde{e} \xleftarrow{r} \mathcal{X}_p$, computes $\tilde{u} = \tilde{s}^e v^{-m}$ and an $\tilde{e}$th root s of the value $uv^{h(\tilde{u})}$. The signature is $\sigma = (\tilde{e}, s, \tilde{s})$.

- The *verification* algorithm takes vk, $m \in \{0, 1, \ldots, k-1\}$ and σ as input. It verifies that $\tilde{e}$ is odd, that $\tilde{e} \neq e$, computes $\tilde{u}' = \tilde{s}^e v^{-m}$ and verifies that $u = s^{\tilde{e}} v^{-h(\tilde{u}')}$. If everything verifies, it outputs 1, otherwise 0.

The signing algorithm could have sampled $\tilde{u}$ and computed the eth root $\tilde{s}$, which would have been the signing algorithm from Example 9.2. But the computation in the signature algorithm produces an equivalent distribution for the generated signatures and is much faster. Since it does not actually use the signing key, it is also more convenient with respect to the security proof.

Compared to RSA-FDH from Section 4.3.2, the signing algorithm is not very expensive. It needs to find one fairly small prime, compute three exponentiations to fairly small exponents and compute one $\tilde{e}$th root in Q_n. There are fairly fast methods available to find small primes, which means that this scheme is just slightly slower than RSA-FDH. The verification algorithm is much slower than the RSA-FDH verification algorithm.

Note that we do not verify that the prime $\tilde{e}$ actually is prime in the verification algorithm. If we ensured that it was prime, this would simplify parts of the security proof, but verifying primality is somewhat expensive, so as long as we can make the proof work without verifying primality, we should do so.

The security of this signature scheme relies on the hardness of computing roots modulo an RSA modulus, but there is one issue, which is that an adversary is free to choose exactly which root should be used. This means that we will need a variant of the ordinary RSA problem.

Proposition 9.3. *Let $\mathcal{A}$ be a (τ, l_s)-adversary against strong unforgeability for the above signature scheme* CS-SIG. *Then there exists a τ_1'-collision finder $\mathcal{B}_1$ for h and τ''-adversaries $\mathcal{B}_{2,1}$, $\mathcal{B}_{2,2}$ and $\mathcal{B}_{2,2}$ against the Strong RSA problem, where τ_1' is essentially equal to τ, while τ'' is larger than τ by a factor of about l_s^2 signature computations, and*

$$\mathbf{Adv}_{\text{CS-SIG}}^{\text{suf-cma}}(\mathcal{A}) \leq \mathbf{Adv}_h^{\text{collision}}(\mathcal{B}_1) + \frac{3k+2}{n_0} + 2\epsilon(l_s)$$

$$+ \mathbf{Adv}^{\text{SRSA}}(\mathcal{B}_{2,1}) + l_s \mathbf{Adv}^{\text{SRSA}}(\mathcal{B}_{2,2}) + \frac{3}{2}\mathbf{Adv}^{\text{SRSA}}(\mathcal{B}_{2,3}),$$

where $\epsilon(l_s)$ is the probability that there is a collision among $l_s + 1$ primes sampled from $\mathcal{X}_p$, and n_0 is a lower bound on the RSA moduluses output by the key generation algorithm. The algorithm $\mathcal{B}_1$ is easy to derive from $\mathcal{A}$.

The first step of the security proof is to forbid collisions in the hash function h and the prime generation during signing. The next step is to introduce a new key generation and signing algorithm, which actually does not have to compute roots modulo n. We can do this by using the fact that if we know an $\alpha\beta$th root of something, we also know an αth root. Once this is done, we identify three possible types of forgery, and we bound each one in terms of various adversaries. Here we use similar tricks to the argument after Example 9.2.

Remark. We immediately note that this alternative stateful signing algorithm is not useful in practice, since it is not only stateful but also slower and in its more efficient variants also requires much more key material. But it is useful in the *security proof*, because the computational cost turns out to be manageable in this context, and it is easy to deal with the state.

Remark. With a bit more care, the first two Strong RSA solvers claimed in the theorem could be turned into solvers for the plain RSA problem, which may be a more difficult problem than the Strong RSA problem.

Proof of Proposition 9.3 In the following, let E_i be the event that the adversary outputs a valid forgery in Game i.

Ⓖ **Game 0** The initial game is $\mathcal{A}$ interacting with the chosen message experiment. Then
$$\mathbf{Adv}^{\mathsf{suf\text{-}cma}}_{\mathsf{CS\text{-}SIG}}(\mathcal{A}) = \Pr[E_0]. \tag{9.1}$$

Let $\tau_0 = \tau$.

Ⓖ **Game 1** In this game, we modify the chosen message experiment so that it immediately stops if during signature generation a prime $\tilde{e}$ or a quadratic residue $\tilde{u}$ is sampled that has been sampled before during signature generation or key generation. The game also stops if any hash values collide for any pair of generated signatures, or for a generated signature and the adversary's forgery.

Let $F_{c,i}$ be the event that a collision happens in Game i. Until such a collision happens, this game and the previous game proceed identically, so

$$\Pr[F_{c,1}] = \Pr[F_{c,0}] \quad \text{and} \quad \Pr[E_1 \mid \neg F_{c,1}] = \Pr[E_0 \mid \neg F_{c_0}].$$

By Lemma 7.7, it follows that

$$|\Pr[E_1] - \Pr[E_0]| \leq \Pr[F_{c,1}]. \tag{9.2}$$

Since this game stops when the first collision happens, the adversary's runtime remains bounded by τ_0. The extra computation in the experiment amounts to keeping a sorted list of primes, quadratic residues and hashes sampled. Compared to the runtime of interesting adversaries, this work will be insignificant. It follows that the runtime τ_1 is essentially equal to τ_0.

The probability of a collision in prime generation is $\epsilon(l_s)$. Since quadratic residues are sampled from the uniform distribution on a much larger set, $\epsilon(l_s)$ is also a useful upper bound on the collision probability for quadratic residues.

A collision finder $\mathcal{B}_1$ for h is immediate from the description of the game, and its runtime τ_1' will be essentially the same as this game.

Now it follows that

$$\Pr[F_{c,1}] \leq 2\epsilon(l_s) + \mathbf{Adv}_h^{\text{collision}}(\mathcal{B}_1). \tag{9.3}$$

Ⓖ **Game 2** In this game, we change how the experiment generates keys and signatures. Let the signature on the jth chosen message m_j generated in the game be $(\tilde{e}_j, s_j, \tilde{s}_j)$, with associated value $\tilde{u}_j$.

- Generating u and v during key generation is changed. The experiment samples exponents $\tilde{e}_1, \tilde{e}_2, \ldots, \tilde{e}_{l_s}$ and elements $\hat{u}, \hat{v} \xleftarrow{r} \mathbb{Z}_n^*$, and computes

$$u \leftarrow \hat{u}^2 \prod_\nu \tilde{e}_\nu, \qquad\qquad v \leftarrow \hat{v}^2 \prod_\nu \tilde{e}_\nu.$$

- For the jth chosen message, the $\tilde{e}_j$th root of $uv^{h(\tilde{u}_j)}$ is computed as

$$s_j \leftarrow (\hat{u}\hat{v}^{h(\tilde{u}_j)})^2 \prod_{\nu \neq j} \tilde{e}_\nu.$$

Squaring is a group homomorphism from $\mathbb{Z}_n^*$ onto Q_n, while exponentiation to any prime smaller than p and q is a permutation on Q_n. It follows that u and v are distributed as if they were sampled from the uniform distribution on Q_n. Also, since exponentiation to the $\tilde{e}$th power is a permutation, any $\tilde{e}$th root is unique. The verification key and the signatures computed in this game are therefore distributed exactly as in the previous game, and that

$$\Pr[E_2] = \Pr[E_1].$$

Since $\mathcal{A}$'s view is identical in this game and the previous game, its runtime is unchanged. The experiment does require greater computational resources. Each signature computation requires l_s exponentiations beyond what the ordinary signing algorithm requires, which is certainly larger than the computation required by l_s signature creations. This means that τ_2 is larger than τ_1 by a factor comparable to l_s^2 signature creations.

Remark. For almost any use case and any interesting adversary, τ_2 is essentially equal to τ_1, but it is conceivable that some system dedicates a huge amount of time for signature generation. For instance, if the honest users expect to generate about 2^{50} signatures, an adversary with runtime equal to 2^{80} signature creations would result in a game with runtime approximately equal to 2^{100} signature creations. A more careful analysis could improve upon these estimates, but this example does show that runtime accounting does need a lot of care.

The factors p and q are never used in Game 2. In Game 2 a forgery $(\tilde{e}_0, s_0, \tilde{s}_0)$ with associated value $\tilde{u}_0$ must fall into one of three disjoint classes:

1. There exists a j such that $\tilde{e}_0 = \tilde{e}_j$ and $\tilde{u}_0 = \tilde{u}_j$.
2. There exists a j such that $\tilde{e}_0 = \tilde{e}_j$, but $\tilde{u}_0 \neq \tilde{u}_j$.
3. For all j we have that $\tilde{e}_0 \neq \tilde{e}_j$.

These classes correspond to a forgery for the Example 9.2 scheme and two types of forgery for the Example 9.3 scheme. Let $E_{2,i}$ be the event that a forgery of class i is output. Since the forgery classes are disjoint, we get that

$$\Pr[E_2] = \Pr[E_{2,1}] + \Pr[E_{2,2}] + \Pr[E_{2,3}]. \tag{9.4}$$

We construct one adversary against the Strong RSA problem for each class and show that its advantage is equal to some multiple of the corresponding forgery probability. (In fact, two of these adversaries are strictly speaking adversaries against ordinary RSA, but the exponent is much larger than is usually the case with the RSA problem.) Proposition 9.3 now follows from equations (9.1)–(9.4) and the following three lemmas.

Lemma 9.4. *There exists a $\tau'_{2,1}$-adversary $\mathcal{B}_{2,1}$ against the Strong RSA problem, where $\tau'_{2,1}$ is essentially equal to τ_2, and*

$$\Pr[E_{2,1}] = \mathbf{Adv}^{\mathsf{SRSA}}(\mathcal{B}_{2,1}).$$

Proof. If $\tilde{e}_0 = \tilde{e}_j$ and $\tilde{u}_0 = \tilde{u}_j$, we have that

$$s_0^e v^{-m_0} = \tilde{u}_0 = s_j^e v^{-m_j}.$$

Since the signature-message pairs must differ and we have that $u_0 = u_j$, we know that $m_0 \neq m_j$ and $s_0 \neq s_j$. It follows that

$$(s_0/s_j)^e = v^{m_j - m_0} = \hat{v}^{(m-m')2} \prod_\nu \tilde{e}_\nu.$$

Since $|m - m'| < e$ and all the primes are distinct, we get that the exponents are relatively prime and Lemma 9.2 gives us an eth root of $\hat{v}$.

The adversary $\mathcal{B}_{2,1}$ gets (n, x) as input and runs a copy of Game 2 and $\mathcal{A}$, modified as follows:

- The key generation step does not generate an RSA modulus, but instead uses the input n. Also, this step does not sample $\hat{v}$, but instead sets $\hat{v} = x$.

When a forgery of class 1 is output, the above argument gives us an eth root of $\hat{v} = x$. The runtime of this Strong RSA solver is essentially equal to τ_2. $\square$

T **Lemma 9.5.** *There exists a $\tau'_{2,2}$-adversary $\mathcal{B}_{2,2}$ against the Strong RSA problem, where $\tau'_{2,2}$ is essentially equal to τ_2, and*

$$\Pr[E_{2,2}] \le l_s \mathbf{Adv}^{\mathsf{SRSA}}(\mathcal{B}_{2,2}).$$

Proof. If $\tilde{e}_0 = \tilde{e}_j$ and $\tilde{u}_0 \ne \tilde{u}_j$, we have that

$$s_0^{\tilde{e}_j} v^{-h(\tilde{u}_0)} = u = s_j^{\tilde{e}_j} v^{-h(\tilde{u}_j)}.$$

We have forbidden hash collisions, so $h(\tilde{u}_0) \ne h(\tilde{u}_j)$, and it follows that

$$(s_0/s_j)^{\tilde{e}_j} = v^{h(\tilde{u}_j)-h(\tilde{u}_0)}.$$

Again, since $|h(\tilde{u}_j) - h(\tilde{u}_0)| < \tilde{e}_j$, we will be able to use Lemma 9.2 to find an $\tilde{e}_j$th root of v. However, we need to modify the key generation so that we do not already know such a root, while also being able to sign the message.

The adversary $\mathcal{B}_{2,2}$ gets (n, x) as input and runs a copy of $\mathcal{A}$ and Game 2, modified as follows:

- At the start, the game samples $j_0 \xleftarrow{r} \{1, 2, \ldots, l_s\}$.
- The key generation step does not generate an RSA modulus, but instead uses the input n. Also, the computation of v and u is changed as follows: Instead of sampling, set $\hat{v} = x$ and

$$v \leftarrow \hat{v}^{2e} \prod_{\nu \ne j_0} \tilde{e}_\nu.$$

Sample $r_1, r_2 \xleftarrow{r} Q_n$ and compute

$$s_{j_0} \leftarrow r_1^{\prod_{\nu \ne j_0} \tilde{e}_\nu}, \qquad \tilde{u}_{j_0} \leftarrow r_2^e, \qquad u \leftarrow s_{j_0}^{\tilde{e}_{j_0}} v^{-h(\tilde{u}_{j_0})}.$$

- For the jth chosen message query, $j \ne j_0$, the $\tilde{e}_j$th root of the value $uv^{h(\tilde{u}_j)}$ is computed as

$$s_j \leftarrow r_1^{\prod_{\nu \ne j} \tilde{e}_\nu} \hat{v}^{-2h(\tilde{u}_{j_0})e} \prod_{\nu \ne j, j_0} \tilde{e}_\nu.$$

- For the j_0th chosen message query, s_{j_0} and $\tilde{u}_{j_0}$ have already been chosen, so the eth root $\tilde{s}_{j_0}$ of the value $\tilde{u}_{j_0} v^{m_j}$ is computed as

$$\tilde{s}_{j_0} \leftarrow r_2 \hat{v}^{2m_j} \prod_{\nu \ne j_0} \tilde{e}_\nu.$$

Observe that the verification key and all the signatures are generated with the correct distribution, since for most signatures all that changes is how the $\tilde{e}_j$th root is computed. For the j_0th signature, the elements are generated in a different order, but it does not matter if we sample the root and compute its eth power, or sample the power and compute its eth root. It follows that $\mathcal{A}$ behaves identically when running as part of $\mathcal{B}_{2,2}$ and as part of Game 2.

If a class 2 forgery is output and $\tilde{e}_0 = \tilde{e}_{j_0}$, the initial argument gives us an $\tilde{e}_{j_0}$th root of $\hat{v} = x$.

Finally, we can observe that the game simulated by $\mathcal{B}_{2,2}$ requires essentially the same amount of computation as Game 2. The lemma follows. □

Lemma 9.6. *There exists a $\tau'_{2,3}$-adversary $\mathcal{B}_{2,3}$ against the Strong RSA problem, where $\tau'_{2,3}$ is essentially equal to τ_2, and*

$$\Pr[E_{2,3}] \leq \frac{3}{2}\mathbf{Adv}^{\mathsf{SRSA}}(\mathcal{B}_{2,3}) + \frac{3k+2}{n_0}.$$

Before we begin with the proof, we first discuss the main obstruction in this case that is different from the previous cases. If $\tilde{e}_0 \neq \tilde{e}_j$ for all j, then

$$s_0^{\tilde{e}_0} = uv^{h(\bar{u}_0)}.$$

The natural approach is to make the right-hand side of the equation be some power of an RSA challenge and then apply Lemma 9.2. And the obvious way to achieve this is to let v be the RSA challenge and let u be some power of v. However, unless we are very careful, we cannot guarantee that the power of our RSA challenge will be divisible by $\tilde{e}_0$. The simple solution is to choose a very big exponent when creating u.

One more note before we begin: We may assume that any RSA challenge squared will be a generator for Q_n. If it is not, the challenge will either be ± 1, for which finding roots is trivial, or it will be congruent to ± 1 modulo either p or q, but not both, which allows us to factor n.

Proof. The adversary $\mathcal{B}_{2,3}$ gets (n, x) as input and runs a copy of Game 2 and $\mathcal{A}$, modified as follows.

- The key generation step does not generate an RSA modulus, but instead uses the input n. Also, this step does not sample $\hat{v}$, but instead sets $\hat{v} = x$. It then samples $r \xleftarrow{r} \{0, 1, \ldots, n^2 - 1\}$ and computes $\hat{u} = \hat{v}^r$.

When a forgery of class 3 is output, we get that

$$s_0^{\tilde{e}_0} = x^{2(r+h(\bar{u}_0))} \prod_\nu \tilde{e}_\nu.$$

The algorithm divides the exponents by $\gcd(\tilde{e}_0, r + h(\bar{u}_0))$ and then applies Lemma 9.2 to recover a root of x.

Let F be the event that a forgery of class 3 is output, and let F' be the event that $\mathcal{B}_{2,3}$ succeds in extracting a non-trivial root from a forgery. We have that $\mathbf{Adv}^{\mathsf{SRSA}}(\mathcal{B}_{2,3}) = \Pr[F']$.

First, as discussed above, we may assume that x generates Q_n. Since the probability space induced by taking r modulo $|Q_n| = (p-1)(q-1)/4$ is $1/n$-close to uniform, the distribution of $\hat{u}$ is $1/n$-close to uniform. It follows that the key generated is $1/n$-close to the correct distribution, which means that

$$\Pr[F] \geq \Pr[E_{2,3}] - \frac{2}{n}.$$

We need to estimate the probability that $\tilde{e}_0$ divides $2(r + h(\tilde{u}_0)) \prod_\nu \tilde{e}_\nu$, which we shall do through some small modifications to the game. We note that these modifications need not be efficient, since we are not going to use them to create an adversary, but just to estimate a probability. We still need to make sure that the adversary's view does not change too much, though. The argument is technical, but the idea is that the adversary cannot predict r completely, so cannot find $\tilde{e}_0$ and $\tilde{u}_0$ such that $\tilde{e}_0$ both fits the constraints and divides the required value. See also Examples 6.8 and 6.13.

Write $r = \alpha|Q_n| + \beta$. Let ℓ be a prime dividing $\tilde{e}_0$. The prime must be odd and smaller than k, so it cannot divide $2 \prod_\nu \tilde{e}_\nu$, which means that it must divide $\alpha|Q_n| + \beta + h(\tilde{u}_0)$. The prime ℓ does not divide $|Q_n|$, which means that multiplication by $|Q_n|$ is a permutation modulo ℓ. This means that ℓ divides $\alpha|Q_n| + \beta + h(\tilde{u}_0)$ if and only if α is congruent to one particular value modulo ℓ. We shall now estimate this probability.

If we sample α and β from an appropriate joint distribution and compute r, instead of sampling r and then determining α and β, that does not change the adversary's view. Then, if we instead sample α and β independently from appropriate ranges, the resulting distribution is $1/n$-close. Now the adversary has no information about α.

Since the adversary has no information about α, we can postpone sampling α until after the adversary has output the forgery. Then, instead of sampling α from the correct range, we can slightly increase the size of the range we sample from, so that the size is divisible by ℓ. The resulting distribution is at least ℓ/n-close to the previous distribution, but now we know that the distribution of α modulo ℓ is uniform.

When α is sampled such that α modulo ℓ is uniformly distributed, the probability that it is congruent to a particular value modulo ℓ is exactly $1/\ell$.

Therefore, the probability that we cannot extract a root from the signature is at most $1/\ell + 2/n + 2\ell/n$. We can bound this value by $1/3 + 2k/n$. Then

$$\Pr[F'] \geq (1 - 1/3 - 2k/n)\Pr[F] \geq \frac{2}{3}\Pr[F] - 2k/n.$$

The lemma follows. □

Remark. It may be possible to reduce the constant $3/2$ in the lemma almost to 1, by using a more careful argument at the end of the proof (taking into account that the entire $\tilde{e}_0$ must be a divisor). However, given the much larger factor l_s in the previous lemma, there seems to be no point to doing this.

9.1.3 Hash Functions Do No Harm – Maybe

We deliberately defined the Cramer-Shoup signature scheme in the previous section without hashing the message, which means that it can only be used to sign short messages. To give the scheme a more useful message space, we sign the hash of a message instead of the message itself.

We now prove the standard theorem about using hash functions to extend the message space. This is an essentially defensive use of hash functions, in that we prove that the hash function does not make the signature scheme worse, except if collisions can be found. We note that in many constructions, the hash function plays a greater offensive role, preventing attacks on the "underlying" signature scheme, and we return to this in Section 9.2.

Example 9.5. Let $\text{SIG}_0 = (\mathcal{K}, \mathcal{S}_0, \mathcal{V}_0)$ be a signature scheme and $\{h_{vk} : S \to T_{vk}\}$ a family of hash functions such that $T_{vk} \subseteq \mathfrak{M}_{vk}$ for any vk output by $\mathcal{K}$.

The composed signature scheme $\text{SIG} = (\mathcal{K}, \mathcal{S}, \mathcal{V})$ uses signature and verification algorithms modified as follows:

- The *signing* algorithm $\mathcal{S}$ takes as input a signature key sk and a message $m \in S$. It computes $\sigma \leftarrow \mathcal{S}_0(sk, h_{vk}(m))$ and outputs σ.

- The *verification* algorithm $\mathcal{V}$ takes as input a verification key vk, a message $m \in S$ and a signature σ. It outputs $\mathcal{V}_0(vk, h_{vk}(m), \sigma)$.

Proposition 9.7. *Let $h : S \to T$ be a hash function and let $\text{SIG}_0 = (\mathcal{K}, \mathcal{S}_0, \mathcal{V}_0)$ be a signature scheme such that $T \subseteq \mathfrak{M}_{vk}$ for any verification key vk output by $\mathcal{K}_0$. Let $\text{SIG} = (\mathcal{K}, \mathcal{S}, \mathcal{V})$ be the scheme from Example 9.5, and let $\mathcal{A}$ be a (τ, l_s)-adversary against SIG. Then there exists a τ_1-collision finder $\mathcal{B}_1$ for h and a (τ_2, l_s)-adversary $\mathcal{B}_2$ against SIG_0, where τ_1 and τ_2 are essentially equal to τ, such that*

$$\mathbf{Adv}^{\text{suf-cma}}_{\text{SIG}}(\mathcal{A}) \leq \mathbf{Adv}^{\text{collision}}_{h}(\mathcal{B}_1) + \mathbf{Adv}^{\text{suf-cma}}_{\text{SIG}_0}(\mathcal{B}_2), \text{ and}$$
$$\mathbf{Adv}^{\text{euf-cma}}_{\text{SIG}}(\mathcal{A}) \leq \mathbf{Adv}^{\text{collision}}_{h}(\mathcal{B}_1) + \mathbf{Adv}^{\text{euf-cma}}_{\text{SIG}_0}(\mathcal{B}_2).$$

The algorithm $\mathcal{B}_1$ is easy to derive from $\mathcal{A}$.

Remark. The proof applies both to existential and strong unforgeability.

Proof. The collision finder $\mathcal{B}_1$ runs a copy of the signature experiment and $\mathcal{A}$. When $\mathcal{A}$ outputs its forgery, it compares the hash of the message to the hashes of the messages in the chosen message queries. If it finds a collision, it outputs that collision. Otherwise it stops.

The collision finder runs a copy of the signature experiment interacting with the adversary, so the adversary runs as normal and requires runtime at most τ. Searching for a collision requires at most $l_s + 1$ hash computations, which means that the runtime τ_1 of $\mathcal{B}_1$ is essentially equal to τ.

The forger $\mathcal{B}_2$ runs a copy of the adversary $\mathcal{A}$ and the simulator **Sim**:

- When the simulator receives vk, it sends vk to the adversary $\mathcal{A}$.
- When the simulator receives a chosen message query m from the adversary, it computes $m' \leftarrow h(m)$ and sends a chosen message query m'. It forwards the response σ to the adversary $\mathcal{A}$.

When the adversary outputs (m_0, σ_0), $\mathcal{B}_2$ outputs $(h(m_0), \sigma_0)$.

The key generation algorithms are identical, and by inspection we see that the response to the chosen message query will be identical to the signature experiment's response. The adversary $\mathcal{A}$ therefore runs as normal and requires runtime at most τ. For each chosen message query the simulator must forward some messages, and also before it outputs the final forgery, but for any reasonable adversary this is negligible, so τ_2 is essentially equal to τ.

Let m_j be the jth chosen message. There are three disjoint forgery classes:

1. For some j, $m_0 = m_j$ (strong forgery).

2. For some j, $h(m_0) = h(m_j)$, but $m_0 \neq m_j$ (hash collision).

3. The hash $h(m_0)$ is not equal $h(m_j)$ for any j (existential forgery).

Let E_i be the event that the adversary's forgery is of class i, so that

$$\mathbf{Adv}_{\mathrm{SIG}}^{\mathsf{suf\text{-}cma}}(\mathcal{A}) = \Pr[E_1] + \Pr[E_2] + \Pr[E_3], \quad \mathbf{Adv}_{\mathrm{SIG}}^{\mathsf{euf\text{-}cma}}(\mathcal{A}) = \Pr[E_2] + \Pr[E_3],$$

We immediately see that

$$\mathbf{Adv}_h^{\mathsf{collision}}(\mathcal{B}_1) = \Pr[E_2].$$

If E_1 happens, then the signatures on the message will differ, so $\sigma_0 \neq \sigma_j$ and the forgery output by $\mathcal{B}_2$ will be a new signature on a message that was submitted as a chosen message query, which counts as a forgery for strong unforgeability for SIG_0. Finally, if E_3 happens, then the forgery output by $\mathcal{B}_2$ will be a signature on a message that was never submitted as a chosen message query, which counts as a forgery for SIG_0. In other words,

$$\mathbf{Adv}_{\mathrm{SIG}_0}^{\mathsf{suf\text{-}cma}}(\mathcal{B}_2) \geq \Pr[E_1] + \Pr[E_3], \qquad \mathbf{Adv}_{\mathrm{SIG}_0}^{\mathsf{euf\text{-}cma}}(\mathcal{B}_2) = \Pr[E_3].$$

The claim follows. □

9.1.4 Multiple Key Pairs

Signatures are rarely used in a context where there is only one signature key pair. There will be many users (key pairs), and it is important to consider the security of the scheme when used in such a context. Fortunately, this is easy, much easier than in the case of public key or symmetric encryption.

The multi-user chosen message experiment proceeds as follows:

1. Set $C_0 = C_1 = C_2 = \emptyset$.

2. When the adversary sends the ith query $\perp$ (*key generation*):

   (a) Compute $(vk_i, sk_i) \leftarrow \mathcal{K}$ and send (i, vk_i) to the adversary.

3. When the adversary sends a query i (*key reveal*):

   (a) Add i to C_2 and send (i, sk_i) to the adversary.

4. When the adversary sends a query (i, m) (*chosen message*):

   (a) Compute $\sigma \leftarrow \mathcal{S}(sk_i, m_i)$, add (i, m) to C_0 and (i, m, σ) to C_1, and send σ to the adversary.

Eventually, the adversary outputs (i, m_0, σ_0).

FIGURE 9.2   The multi-user chosen message experiment $\mathbf{Exp}_{\mathrm{SIG}}^{\mathsf{mu\text{-}cma}}(\mathcal{A})$ for a digital signature scheme $\mathrm{SIG} = (\mathcal{K}, \mathcal{S}, \mathcal{V})$ with adversary $\mathcal{A}$.

The main obstacle is revealing keys. This is a huge problem for encryption, where the adversary may do challenge queries for many different key pairs, but we cannot predict which key pairs. This typically makes reductions harder, since we must be able to reveal every decryption key.

For signatures, the adversary may create a forgery for any key pair, but the forgery must happen for a definite key pair, unlike breaking secrecy. The obvious approach is therefore to guess which key pair the forgery will happen for and simulate all the other keys.

Definition 9.2. A (τ, l_u, l_s)-*adversary* against a signature scheme SIG is an interactive algorithm $\mathcal{A}$ that interacts with the experiment in Figure 9.2 making at most l_u key generation queries and l_s chosen message queries, and where the runtime of the adversary and the experiment is at most τ.

The *existential unforgeability advantage* and *strong unforgeability advantage* of this adversary are defined to be

$$\mathbf{Adv}_{\mathrm{SIG}}^{\mathsf{euf\text{-}mu\text{-}cma}}(\mathcal{A}) = \Pr[E], \quad \text{and} \quad \mathbf{Adv}_{\mathrm{SIG}}^{\mathsf{suf\text{-}mu\text{-}cma}}(\mathcal{A}) = \Pr[F],$$

where E is the event that $i \notin C_2$, $(i, m_0) \notin C_0$ and $\mathcal{V}(vk_i, m_0, \sigma_0) = 1$, and F is the event that $i \notin C_2$, $(i, m_0, \sigma_0) \notin C_1$ and $\mathcal{V}(vk_i, m_0, \sigma_0) = 1$.

Exercise 9.6. Show that for any multi-user adversary against a signature scheme, there exists a single-user adversary against a signature scheme, and that the advantage of the multi-user adversary is at most the advantage of the single-user adversary times the number of key pairs. That is, prove that for any (τ, l_u, l_s)-adversary $\mathcal{A}$ against SIG, there exists a (τ', l_s)-adversary $\mathcal{B}$

The experiment for the random message attack with an adversary $\mathcal{A}$ proceeds as follows:

1. Set $C_0 = C_1 = C_2 = \emptyset$.

2. Compute $(vk, sk) \leftarrow \mathcal{K}$ and send vk to the adversary.

3. When the adversary sends a query $\perp$ (*target*):

   (a) Sample $m \xleftarrow{r} \mathfrak{M}_{vk}$, add m to C_2 and send m to the adversary.

4. When the adversary sends a query m (*chosen message*):

   (a) Stop if $m \notin C_2$.

   (b) Compute $\sigma \leftarrow \mathcal{S}(sk, m)$.

   (c) Add m to C_0 and (m, σ) to C_1, then send σ to the adversary.

5. When the adversary sends a query $\top$ (*random message*):

   (a) Sample $m \xleftarrow{r} \mathfrak{M}_{vk}$, compute $\sigma \leftarrow \mathcal{S}(sk, m)$, add m to C_0 and C_2, and (m, σ) to C_1. Then send (m, σ) to the adversary.

Eventually, the adversary outputs (m_0, σ_0).

FIGURE 9.3   The experiment $\mathbf{Exp}_{\mathrm{SIG}}^{\mathrm{rma}}(\mathcal{A})$ for the random message security game for a digital signature scheme $\mathrm{SIG} = (\mathcal{K}, \mathcal{S}, \mathcal{V})$ with adversary $\mathcal{A}$.

against SIG, with τ' essentially equal to τ, such that

$$\mathbf{Adv}_{\mathrm{SIG}}^{\mathrm{euf\text{-}mu\text{-}cma}}(\mathcal{A}) \leq l_u \mathbf{Adv}_{\mathrm{SIG}}^{\mathrm{euf\text{-}cma}}(\mathcal{A}), \quad \mathbf{Adv}_{\mathrm{SIG}}^{\mathrm{suf\text{-}mu\text{-}cma}}(\mathcal{A}) \leq l_u \mathbf{Adv}_{\mathrm{SIG}}^{\mathrm{suf\text{-}cma}}(\mathcal{A}).$$

9.2   HASH AND SIGN PARADIGM

We saw how to use hash functions to extend the message space of signature schemes in Section 9.1.3. We shall now consider how hash functions can be used more offensively, to make it harder to attack systems. Again, we shall use the construction from Example 9.5.

We first define a new notion of security, where the adversary receives a collection of *target messages* chosen at random from the message space, and must create a forgery for one of these messages. The adversary may ask for signatures, but the experiment will only sign target messages.

We emphasise that this notion is a purely synthetic notion and not intended as a model of real-world security requirements.

Definition 9.3. A (τ, l_h, l_s, l'_s)-*random message adversary* against a signature scheme SIG is an interactive algorithm $\mathcal{A}$ that interacts with the experiment in Figure 9.3 making at most l_h target queries, l_s chosen message queries and l'_s random message queries, and where the runtime of the adversary and the experiment is at most τ.

The *existential unforgeability advantage* and the *strong unforgeability advantage* of this adversary is defined to be

$$\mathbf{Adv}_{\mathrm{SIG}}^{\mathsf{euf\text{-}rma}}(\mathcal{A}) = \Pr[E], \qquad \mathbf{Adv}_{\mathrm{SIG}}^{\mathsf{suf\text{-}rma}}(\mathcal{A}) = \Pr[F],$$

where E is the event that $m_0 \in C_2$, $m_0 \notin C_0$ and $\mathcal{V}(vk, m_0, \sigma_0) = 1$, while F is the event that $m_0 \in C_2$, $(m_0, \sigma_0) \notin C_1$ and $\mathcal{V}(vk, m_0, \sigma_0) = 1$.

A random message query can be simulated by a target query and a chosen message query. Typically, an adversary will either make chosen message queries or random message queries, but not both, so we have essentially defined two distinct attacks. Informally, we shall call the former a *strong random message* adversary and the latter a *weak random message* adversary.

It is immediate that any random message adversary will also be a chosen message adversary. In fact, as the following exercise shows, this is a strictly weaker security notion. It also shows that the random message security notion is usually not very interesting on its own.

E *Exercise 9.7.* Let SIG be any signature scheme with a large message space. Design a scheme SIG′ such that:

- any random message adversary against SIG′ can be turned into a chosen message adversary against SIG with almost the same advantage; and
- there exists a trivial chosen message adversary against SIG′.

Now we show that if we model the hash function as a random function, the composition of any signature scheme with a hash function as given in Example 9.5 gives us proper security. In order to avoid some technical problems, we shall actually consider a family of hash functions $\{h_{vk} : S \to \mathfrak{M}_{vk}\}$.

T **Proposition 9.8.** *Let* SIG *be the composition of a signature scheme* $\mathrm{SIG}_0 = (\mathcal{K}, \mathcal{S}_0, \mathcal{V}_0)$ *and a family of hash functions* $\{h_{vk} : S \to \mathfrak{M}_{vk}\}$ *given in Example 9.5. Let* $\mathcal{A}$ *be a* (τ, l_s)-*adversary against* SIG *in the random oracle model, making at most* l_h *queries to the random oracle. Then there exists a* $(\tau', l_h + 1, l_s, 0)$-*random message adversary* $\mathcal{B}$ *against* SIG_0, *with* τ' *essentially equal to* τ, *such that*

$$\mathbf{Adv}_{\mathrm{SIG}}^{\mathsf{euf\text{-}cma}}(\mathcal{A}) \leq \mathbf{Adv}_{\mathrm{SIG}_0}^{\mathsf{euf\text{-}rma}}(\mathcal{B}) + \epsilon, \quad \mathbf{Adv}_{\mathrm{SIG}}^{\mathsf{suf\text{-}cma}}(\mathcal{A}) \leq \mathbf{Adv}_{\mathrm{SIG}_0}^{\mathsf{suf\text{-}rma}}(\mathcal{B}) + \epsilon,$$

where $\epsilon = \max_{vk}\{l_h^2/|\mathfrak{M}_{vk}|\}$ *and the maximum is taken over* vk *that could be output by* $\mathcal{K}$.

Proof. The adversary $\mathcal{B}$ runs a copy of $\mathcal{A}$ and the following simulator **Sim**:

- The simulator runs a copy of the random oracle from Figure 7.17, with the following modification: When the adversary queries the random oracle with a new message m, **Sim** does not sample a hash value, but sends a target query to its experiment and uses the response as the hash value.

- When the adversary makes a chosen message query m, **Sim** queries m to its random oracle and receives the response m'. It then sends m' as a chosen message query to its experiment, and forwards the response σ to the adversary $\mathcal{A}$.

When $\mathcal{A}$ outputs its forgery (m_0, σ_0), the adversary $\mathcal{B}$ queries m_0 to **Sim**'s random oracle and receives the response m'_0. It then outputs (m'_0, σ_0).

We compare $\mathbf{Exp}_{\mathrm{SIG}}^{\mathrm{cma}}$ in the random oracle model with the composition of $\mathbf{Exp}_{\mathrm{SIG}_0}^{\mathrm{rma}}$ and **Sim**. The key generation is unchanged. The simulation of the random oracle by **Sim**, since the target query samples a message uniformly at random. By inspection, we see that the simulated chosen message response is identical to the response that would have been computed by $\mathbf{Exp}_{\mathrm{SIG}}^{\mathrm{cma}}$ in the random oracle model. It follows that the view of $\mathcal{A}$ is identical in both cases, and therefore that the adversary achieves its goal with the requisite probability, and within the requisite time bound.

Let E and F be existential and strong unforgeability events, so that

$$\mathbf{Adv}_{\mathrm{SIG}}^{\mathrm{euf\text{-}cma}}(\mathcal{A}) = \Pr[E], \qquad \mathbf{Adv}_{\mathrm{SIG}}^{\mathrm{suf\text{-}cma}}(\mathcal{A}) = \Pr[F],$$

and let G be the event that two target queries result in the same response. Note that G corresponds to a collision in a random function, which we bound using Proposition 2.11. We get

$$\Pr[G] \leq \epsilon.$$

Suppose first that G does not happen. Then (m_0, σ_0) is a forgery for SIG if and only (m'_0, σ_0) is forgery for SIG$_0$. It follows that

$$\Pr[E \mid \neg G] \leq \mathbf{Adv}_{\mathrm{SIG}_0}^{\mathrm{euf\text{-}rma}}(\mathcal{B}_2), \qquad \Pr[F \mid \neg G] \leq \mathbf{Adv}_{\mathrm{SIG}_0}^{\mathrm{suf\text{-}rma}}(\mathcal{B}_2),$$

The claim follows by Lemma 7.7. □

In practice the message is hashed to a relatively small set using an ordinary hash function, and then a random-looking hash function maps the small set to $\mathfrak{M}_{vk}$. Technically, we should use Proposition 9.7 to model this case. If we model the composition as a single random-looking hash function, the birthday bound on the collision probability will be incorrect. But it is not important.

E | *Exercise* 9.8. It is often inconvenient to have hash functions tailored to every verification key. Suppose you have a single hash function $h : S \to T$ with the property that for any vk output by $\mathcal{K}$, $T \subseteq \mathfrak{M}_{vk}$ and $|\mathfrak{M}_{vk}|/|T| \leq 2$. State a variant of Proposition 9.8 that captures this situation and prove the claim.

Hint: The random message adversary will need to make more target queries than the number of hash queries from the chosen message adversary.

Remark. For this strategy, it makes sense to hash the verification key together with the message. This cryptographic separation may frustrate optimised attacks when the adversary attacks many verification keys, not just one.

There is a different strategy that creates a weaker adversary. The idea is to add some randomness before hashing the message. This means that the hash will be randomised, which means that the adversary will no longer be able to do chosen message attacks, only random message attacks. The main downside is that the randomness must be part of the signature, so signatures will be longer. The signature is no longer deterministic, which may be problematic.

Example 9.6. Let $\text{SIG} = (\mathcal{K}, \mathcal{S}_0, \mathcal{V}_0)$ be a signature scheme and $\{h_{vk} : V \times S \to T_{vk}\}$ a family of hash functions such that $T_{vk} \subseteq \mathfrak{M}_{vk}$ for any vk output by $\mathcal{K}$.

The composed signature scheme $\text{SIG} = (\mathcal{K}, \mathcal{S}, \mathcal{V})$ uses signature and verification algorithms modified as follows:

- The *signing* algorithm $\mathcal{S}$ takes as input a signature key sk and a message $m \in S$, samples $r \xleftarrow{r} V$ and outputs $(r, \mathcal{S}_0(sk, h_{vk}(r, m)))$.

- The *verification* algorithm $\mathcal{V}$ takes as input a verification key vk, a message $m \in S$ and a signature (r, σ). It outputs $\mathcal{V}_0(vk, h_{vk}(r, m), \sigma)$.

Proposition 9.9. *Let* SIG *be the composition of a signature scheme* SIG_0 *and a family of hash functions* $\{h_{vk}\}$ *given in Example 9.6. Let* $\mathcal{A}$ *be a* (τ, l_s)-*adversary against* SIG *in the random oracle model, making at most* l_h *queries to the random oracle. Then there exists a* $(\tau', l_h + 1, 0, l_s)$-*random message adversary* $\mathcal{B}$ *against* SIG_0, *with* τ' *essentially equal to* τ, *such that*

$$\mathbf{Adv}_{\text{SIG}}^{\text{euf-cma}}(\mathcal{A}) \leq \mathbf{Adv}_{\text{SIG}_0}^{\text{euf-cma}}(\mathcal{B}) + \epsilon + \frac{l_h l_s}{|V|},$$

$$\mathbf{Adv}_{\text{SIG}}^{\text{suf-cma}}(\mathcal{A}) \leq \mathbf{Adv}_{\text{SIG}_0}^{\text{suf-cma}}(\mathcal{B}) + \epsilon + \frac{l_h l_s}{|V|},$$

where $\epsilon = \max_{vk}\{l_h^2/|\mathfrak{M}_{vk}|\}$ *and the maximum is taken over* vk *that could be output by* $\mathcal{K}$.

Exercise 9.9. Prove Proposition 9.9.

With this result, we can now trivially construct a variant of RSA-FDH from Section 4.3.2 with randomised hashes, by proving that a weak random message adversary against textbook RSA can be turned into an RSA solver. The proof uses random self-reducibility, as used to prove Proposition 8.11.

Proposition 9.10. *Let* $\mathcal{A}$ *be a* $(\tau, l_h + 1, 0, l_s)$-*random message adversary against textbook RSA. Then there exists a* τ'-*solver* $\mathcal{B}$ *for the RSA problem,* *with τ' being essentially equal to a small multiple of τ, and*

$$\mathbf{Adv}^{\mathsf{RSA}}(\mathcal{B}) = \mathbf{Adv}^{\mathsf{suf\text{-}rma}}_{\mathsf{TB\text{-}RSA}}(\mathcal{A}).$$

Proof. The adversary $\mathcal{B}$ runs a copy of $\mathcal{A}$ and the following simulator **Sim**:

- The simulator gets (n, e, x) as input and sends $vk = (n, e)$ to $\mathcal{A}$.
- Given a target query, the simulator samples $r \xleftarrow{r} \mathbb{Z}_n^*$ and computes $m \leftarrow r^e x$. It records (r, m) and responds with m.
- Given a random message query, the simulator samples $\sigma \xleftarrow{r} \mathbb{Z}_n^*$ and computes $m \leftarrow \sigma^e$. It responds with (m, σ).

When $\mathcal{A}$ outputs its forgery (m_0, σ_0), $\mathcal{B}$ has (r, m_0) and outputs σ/r.

By inspection, an execution of $\mathcal{B}$ proceeds exactly as the game between the random message experiment and the adversary $\mathcal{A}$ as far as $\mathcal{A}$ is concerned.

The simulator records the randomness together with the message for randomness. The extra resources required does not make τ' significantly larger than τ. It follows that τ' is essentially equal to τ.

Finally, when the adversary $\mathcal{A}$ outputs a forgery (m_0, σ_0) and the simulator **Sim** has recorded (r, m_0), then $m_0 = r^e x$, so $\mathcal{B}$ outputs the correct answer to the challenge. The claim follows. $\qquad\square$

9.2.1 Example: RSA-FDH

The RSA-FDH scheme from Section 4.3.2 is the primary example of the sort of construction discussed in Proposition 9.8. The proposition says that any adversary against RSA-FDH in the random oracle model becomes a strong random message adversary against textbook RSA as in Section 4.3. We therefore need to study the security of this system.

Remark. To simplify the analysis, we shall work with a variant of textbook RSA that will never accept messages that are not invertible modulo the RSA modulus. Since any such message is either zero or immediately reveals the factors of the RSA modulus, disallowing them is not a practical problem.

Unlike the weak random message attack studied in Proposition 9.10, the security of textbook RSA under a strong random message attack is not exactly the same as solving an RSA problem. It is not a problem that the adversary can get many challenges (target messages) and choose which one to attack, but the adversary may also get roots for some of these challenges (which he would not in a weak random message attack), and that is difficult to model.

Immediately, there is one very easy result that we can make. At the start of the experiment, we sample $j_0 \xleftarrow{r} \{1, 2, \ldots, l_h\}$. For the j_0th target query, we return the RSA challenge. Since it is distributed exactly as the other targets, the adversary has no information about j_0. It follows that the adversary's forgery is for the j_0th target message with probability $1/l_h$.

Unfortunately, such a result would be very weak. The reason is that l_h is essentially the number of hash queries, which can be very large (it may make good sense for the adversary to look at many different hash values). The odds of our RSA challenge being the one that the adversary actually solves is then very small. Which means that we could in principle have a very good adversary against RSA-FDH turn into a very bad adversary against RSA.

Fortunately, a better result is possible by choosing a slightly more optimistic strategy. We use random self-reducibility as in the proof of Proposition 8.11 to embed the RSA challenge not just in the $j_0$the query, but in many target queries. Now we have to contend with two issues: The adversary does not forge a signature for any of the messages we have embedded the challenge in, or the adversary asks for a signature for a message we have embedded the challenge in. The second case is problematic, since we cannot respond correctly to the signature query without first solving the RSA challenge ourselves. And if we cannot respond correctly, there is no reason to expect the adversary to be able to do its computation. In other words, we want to embed the challenge in as many target messages as possible in order to increase the likelihood that the adversary's forgery will be useful, but we want to embed the challenge in as few target messages as possible in order to increase the likelihood that the adversary will produce a forgery. The answer turns out to be a compromise.

We first state a standard result from probability theory, and then apply it to prove a better result for RSA-FDH.

[E] *Exercise 9.10.* Consider an urn with l_h balls, of which a fraction δ are blue, and the remaining are red. Suppose we draw $l_s + 1$ balls from the urn with replacement. Show that the probability that we first draw l_s blue balls and then a red ball is $\delta^{l_s}(1 - \delta)$.

[E] *Exercise 9.11.* Show that the function $f : (0, 1) \to \mathbb{R}$ given by $f(\delta) = \delta^{l_s}(1-\delta)$ has its maximal value

$$\frac{1}{l_s}\left(\frac{l_s}{l_s + 1}\right)^{l_s+1} = \frac{1}{l_s}\left(1 - \frac{1}{l_s + 1}\right)^{l_s+1}$$

at $\delta = l_s/(l_s + 1)$. Also, $\exp(1)/l_s$ is an upper bound on this maximum, and $\exp(1)/l_s$ is a good approximation for the maximum when l_s is large.

[T] **Proposition 9.11.** *Let $\mathcal{A}$ be a (τ, l_h, l_s)-random message adversary against textbook RSA. Then there exists a τ'-solver $\mathcal{B}$ for the RSA problem, with τ' being essentially equal to a small multiple of τ, and*

$$\mathbf{Adv}_{\text{TB-RSA}}^{\text{suf-rma}}(\mathcal{A}) \leq l_s\left(1 - \frac{1}{l_s + 1}\right)^{-(l_s+1)}\mathbf{Adv}^{\text{RSA}}(\mathcal{B}).$$

The idea is to embed the RSA problem into many target query responses. Which is chosen at random such that we have a fair chance that the adversary never wants a signature on a message with the challenge embedded.

Proof of Proposition 9.11   The proof proceeds as a sequence of games, between the adversary $\mathcal{A}$ and an experiment. Let E_i be the event that the adversary outputs a forgery in Game i. Let $\delta = l_s/(l_s + 1)$.

Ⓖ **Game 0**   The initial game is $\mathcal{A}$ interacting the random message attack experiment. Then

$$\mathbf{Adv}^{\mathsf{suf\text{-}rma}}_{\mathsf{TB\text{-}RSA}}(\mathcal{A}) = \Pr[E_0]. \tag{9.5}$$

Let $\tau_0 = \tau$.

Ⓖ **Game 1**   In this game, we modify the experiment so that it samples target messages and keeps records in a slightly different way, as follows:

- Step 3 is processed as follows:
    (a) Sample $r \xleftarrow{r} \mathbb{Z}_n^*$ and sample $b \xleftarrow{r} \mathrm{Bernoulli}(1 - \delta)$.
    (b) Compute $m \leftarrow r^e$.
    (c) Add (r, m, b) to C_2.

- Step 4(a) is done as:
    4(a) Stop if there is no tuple $(\cdot, m, \cdot) \in C_2$.

This changes nothing, so

$$\Pr[E_1] = \Pr[E_0]. \tag{9.6}$$

The target query requires roughly an extra exponentiation, so τ_1 is larger than τ_0 by l_h exponentiations, which is at most a small multiple of τ_0.

Ⓖ **Game 2**   In this game, we arbitrarily stop if the bit b recorded with a message in a signature query is not 0. We thus modify Step 4(a) as follows:

4(a) Stop the game if there is no tuple $(\cdot, m, 0) \in C_2$.

We now define the event F_i to be the event that in Game i, $i \geq 2$, the adversary outputs a forgery (σ_0, m_0) and there is no tuple $(\sigma_0, m_0, 1) \in C_2$.

The adversary has no information about the bit b, and these bits are sampled independently, so we get that

$$\Pr[F_2] \geq \delta^{l_s}(1 - \delta)\Pr[E_1]. \tag{9.7}$$

(It will be greater if there are fewer than l_s chosen message queries.)

The experiment in this game must also look at a bit, but the cost is trivial and τ_2 is essentially equal to τ_1.

Ⓖ **Game 3**   In this game, we sample an element and embed this element in the target messages that the experiment will later refuse to sign. We modify the game as follows:

- At the start of the game, the experiment samples $x \xleftarrow{r} \mathbb{Z}_n^*$.

- In Step 3(b) we embed the element x in the target message as follows:

  3(b) Compute $m \leftarrow \sigma^e x^b$.

- In Step 4 we use the known eth root to compute the signature, instead of the signing exponent d, as follows:

  4(a) Stop if there is no tuple $(r, m, 0) \in C_2$.

  4(b) Compute $\sigma \leftarrow r$.

Since x is invertible, multiplication by x is a bijection, so this does not change the distribution of target messages. Also, since the experiment will not sign messages with $b = 1$, signatures are correct. It follows that

$$\Pr[F_3] = \Pr[F_2]. \tag{9.8}$$

The cost of a single multiplication is much smaller than the per-query cost we have already incurred, so τ_3 is essentially equal to a small multiple of τ_2.

Conclusion Note that in this game, the signing exponent d is never used. We can now construct an RSA solver from the final game. The proposition then follows from equations (9.5)–(9.8) and the following lemma.

Lemma 9.12. *There exists an τ'-solver $\mathcal{B}$ for the RSA problem, with τ' essentially equal to τ_3, such that*

$$\Pr[F_3] \leq \mathbf{Adv}^{\mathsf{RSA}}(\mathcal{B}).$$

Proof. The adversary $\mathcal{B}$ runs a copy of $\mathcal{A}$ and the following simulator **Sim**:

- The simulator gets (n, e, x) from $\mathcal{B}$, sets $vk = (n, e)$ and sends vk to $\mathcal{A}$.

- Given a target query, the simulator samples $r \xleftarrow{r} \mathbb{Z}_n^*$ and b from the Bernoulli distribution with parameter $1 - \delta$ and computes $m \leftarrow r^e x^b$. It records (r, m, b) and responds with m.

- Given a chosen message query m, where (r, m, b) has been recorded in response to the target query, the simulator stops the simulation if $b = 0$, and otherwise responds with r.

When $\mathcal{A}$ outputs its forgery (m_0, σ_0), $\mathcal{B}$ recovers its record (r, m_0, b) and outputs σ / r.

By inspection, we see that an execution of $\mathcal{B}$ proceeds exactly as Game 3 as far as the adversary $\mathcal{A}$ is concerned, and everything done by **Sim** except key generation is the same as is done by the experiment in Game 3. It follows that τ' is essentially equal to τ_3.

All that remains is to observe that when the adversary $\mathcal{A}$ outputs a forgery (m_0, σ_0), and the simulator **Sim** has recorded $(r, m_0, 1)$, then $m_0 = r^e x$, so $\mathcal{B}$ outputs the correct answer to the challenge. The claim follows. □

We can now get a good security result for RSA-FDH by combining Proposition 9.8 and Proposition 9.11. If we compare to the Cramer-Shoup signature scheme, this result is in some sense better since it relies on the plain RSA problem instead of the Strong RSA problem. However, the Cramer-Shoup scheme avoids idealisations like the random oracle model, which is in some other sense better. This means that the schemes are not quite comparable. From a practical point of view, RSA-FDH is somewhat faster, in particular since it can be used with a small exponent.

9.2.2 Example: Improved RSA-FDH – Maybe

We shall now consider a variant of textbook RSA. It is laughably insecure against a chosen message attack but is secure against a random message attack. The main proof obstacle for RSA-FDH was that when we choose target messages, we had to decide if we should be able to sign a particular message, or if we should embed a challenge in a hash value. In the variant scheme, we say that the message is a pair of ring elements, and a signature is an eth root of one of the ring elements. We choose which element to sign using a pseudorandom function, so that we can embed a challenge in one element and know an eth root of the other. Since the adversary does not know where the challenge is embedded, a forgery solves the RSA challenge with probability $1/2$. Proposition 9.8 then gives us a better security claim.

E *Example* 9.7. Let $f : \mathfrak{K} \times S \to \{0,1\}$ be a function such that $\mathbb{Z}_n \times \mathbb{Z}_n$ can be injected into S for any n output by the TB-RSA key generation algorithm.
The modified textbook RSA scheme (TB-RSA') works as follows:

- The *key generation* algorithm runs the textbook RSA key generation algorithm to get (n, e) and (n, d), and then samples $k \xleftarrow{r} \mathfrak{K}$. The verification key is $vk = (n, e)$, while the signing key is $sk = (n, d, k)$.

- The *signing algorithm* takes $sk = (n, d, k)$ and $(m_0, m_1) \in \mathbb{Z}_n^* \times \mathbb{Z}_n^*$ as input. It computes $b \leftarrow f(k, (m_0, m_1))$ and $\sigma \leftarrow m_b^d$. The signature is σ.

- The *verification algorithm* takes $vk = (n, e)$, $(m_0, m_1) \in \mathbb{Z}_n^* \times \mathbb{Z}_n^*$ and σ as input. It outputs 1 if $\sigma^e = m_0$ or $\sigma^e = m_1$, and otherwise outputs 0.

T **Proposition 9.13.** *Let $\mathcal{A}$ be a (τ, l_h, l_s)-random message adversary against* TB-RSA'*. Then there exists a (τ', l_h)-PRF distinguisher $\mathcal{B}_1$ for f and a τ''-solver $\mathcal{B}_2$ for the RSA problem, with τ' and τ'' being essentially equal to small multiples of τ, and*

$$\mathbf{Adv}_{\text{TB-RSA'}}^{\text{suf-rma}}(\mathcal{A}) \leq \mathbf{Adv}_f^{\text{prf}}(\mathcal{B}_1) + 2\mathbf{Adv}^{\text{RSA}}(\mathcal{B}_2) + \frac{3l_h^2}{n_0}.$$

Proof of Proposition 9.13   The proof proceeds as a sequence of games, between the adversary $\mathcal{A}$ and an experiment. Let E_i be the event that the adversary outputs a forgery in Game i.

G **Game 0**   The initial game is $\mathcal{A}$ interacting with the random message attack experiment. Then

$$\mathbf{Adv}^{\mathsf{suf\text{-}rma}}_{\mathrm{TB\text{-}RSA'}}(\mathcal{A}) = \Pr[E_0]. \tag{9.9}$$

Let $\tau_0 = \tau$.

G **Game 1**   In this game, we shall prevent trivial attacks on the scheme caused by collisions among the ring elements sampled for target messages. If the adversary outputs a forgery for the message (m_0, m_1), and there is a second pair (m'_0, m'_1) in C_2 such that $m_i = m'_j$ for $i, j \in \{0, 1\}$, we discard the forgery.

We can give a bound on the probability of this happening by bounding the probability that such a collision exists. The experiment essentially samples $2l_h$ random elements, and the birthday bound (Proposition 2.11) gives us a suitable upper bound on the probability of a collision. By Lemma 7.7, we get

$$|\Pr[E_1] - \Pr[E_0]| \leq \frac{2l_h^2}{n}. \tag{9.10}$$

Detecting a collision requires time at most linear in the number l_h of target queries, which means that τ_1 is essentially equal to a small multiple of τ_0.

G **Game 2**   Now we change how we sample elements. We also record some additional bookkeeping data and use this when responding to chosen message queries. In response to a target query, the experiment samples $r_0, r_1 \xleftarrow{r} \mathbb{Z}_n^*$, computes $m_0 \leftarrow r_0^e$, $m_1 \leftarrow r_1^e$ and $b \leftarrow f(k, (m_0, m_1))$, and records (r_0, r_1, m_0, m_1, b). When responding to a chosen message query, the recorded bit is used instead of recomputing b and the signature is computed as $\sigma \leftarrow r_b$.

As usual, these changes are not observable by $\mathcal{A}$, so

$$\Pr[E_2] = \Pr[E_1]. \tag{9.11}$$

For each target query, we need to record more data, we need to compute two exponentiations and we need to compute f. The total cost is at most linear in l_h, which means that τ_2 is essentially equal to a small multiple of τ_1.

G **Game 3**   In this game, we replace the pseudo-random function f by sampling random bits instead. This causes two potential problems. First, f may be distinguishable from a random function. Second, if we ever sample the same pair twice, we may now sample distinct bits, which could cause two distinct chosen message queries to return different signatures.

E *Exercise* 9.12. Show that a τ'-PRF distinguisher $\mathcal{B}_1$ for f exists, with τ' essentially equal to τ_3, such that

$$|\Pr[E_3] - \Pr[E_2]| \leq \mathbf{Adv}_f^{\mathsf{prf}}(\mathcal{B}_1) + \frac{l_h^2}{2n^2} \leq \mathbf{Adv}_f^{\mathsf{prf}}(\mathcal{B}_1) + \frac{l_h^2}{n}. \tag{9.12}$$

G **Game 4** We change the target query to sample b before sampling r_0 and r_1. This is a trivial bit of bookkeeping to prepare for more significant changes.

This changes nothing, so

$$\Pr[E_4] = \Pr[E_3]. \tag{9.13}$$

G **Game 5** We now sample a challenge and embed it into the ring elements that we will not sign. At the start of the game, the experiment samples $x \xleftarrow{r} \mathbb{Z}_n^*$. In response to a target query, it computes $m_b \leftarrow r_b^e$ and $m_{1-b} \leftarrow r_{1-b}^e x$.

As usual, multiplication by x is a bijection, and r_{1-b} will never be used in the experiment, so this is not observable by the adversary and

$$\Pr[E_5] = \Pr[E_4]. \tag{9.14}$$

For each target query, we have to do one more multiplication. We also have to sample one more element. The cost of this is linear in l_h, which means that τ_5 is at most essentially equal to a small multiple of τ_4.

The claim now follows from equations (9.9)–(9.15).

E *Exercise* 9.13. Show that a τ''-adversary $\mathcal{B}_2$ against the RSA problem exists, with τ'' essentially equal to τ_5, such that

$$\Pr[E_5] = 2\mathbf{Adv}^{\mathsf{RSA}}(\mathcal{B}_2). \tag{9.15}$$

Remark. The above scheme has a *deterministic* signing algorithm. But there is a variant signing algorithm that chooses b at random. As long as no message is signed twice, there is no way to distinguish the randomised signing algorithm from the correct, deterministic signing algorithm.

This phenomenon is mostly uninteresting, except in the case where we care about so-called *subliminal channels*, encoding useful information in the "noise" in other data in such a way that it cannot be detected. This is an interesting approach to leaking secrets out of secure systems.

9.2.3 Example: A Stronger RSA-FDH Proof – Maybe

Instead of making a modified RSA-FDH, we shall now investigate if there are other reasonable assumptions that we can base the security on. Recall that the exponent e must be relatively prime to $(p-1)(q-1)$ for the exponent d to be defined. It turns out that deciding if e works for a given n seems to be hard when p and q are not given.

We begin by classifying RSA moduluses according to the structure of its group of eth roots of unity. Let p, q be distinct primes, $n = pq$ and let $e < \min\{p, q\}$ be a prime. We want to understand three different cases:

- e does not divide $(p-1)$ or $(q-1)$.
- e divides either $p-1$ or $q-1$, but not both.
- e divides $p-1$ and $q-1$.

Define the two sets

$$G_{n,e} = \{x \in \mathbb{Z}_n^* \mid x^e = 1\} \text{ and } H_{n,e} = \{a + \langle n \rangle \in G_{n,e} \mid 1 < \gcd(a-1, n) < n\}.$$

The first is the eth roots of unity modulo n, and it is a subgroup of $\mathbb{Z}_n^*$. The latter set will allow us to factor n quickly. We begin by determining the group structure of $G_{n,e}$ and the size of $H_{n,e}$.

E *Exercise 9.14.* Show that $G_{n,e}$ is either trivial, or isomorphic to $\mathbb{Z}_e^+$ or $\mathbb{Z}_e^+ \times \mathbb{Z}_e^+$.

E *Exercise 9.15.* Show that

$$|H_{n,e}| = \begin{cases} 0 & G_{n,e} \text{ is trivial,} \\ e-1 & G_{n,e} \simeq \mathbb{Z}_e^+, \\ 2e-2 & G_{n,e} \simeq \mathbb{Z}_e^+ \times \mathbb{Z}_e^+. \end{cases}$$

This result tells us that if we have some algorithm that samples from the uniform distribution on a non-trivial $G_{n,e}$, then we can factor either with probability to $1 - 1/e$ or close to $2/e$. However, we shall not actually need this factoring algorithm, since for our purposes it will be sufficient to detect RSA moduluses that could be factored in this way.

The idea is that we turn a random message adversary against TB-RSA into an algorithm that finds eth roots of unity, sampled from the uniform distribution. If (n, e) was output by the RSA key generation algorithm, the only eth root of unity is 1. This allows us to distinguish the output of the RSA key generation algorithm from certain other products of two primes.

D **Definition 9.4.** Let $\mathcal{X}_0, \mathcal{X}_1$ be efficiently sampleable probability spaces on pairs of RSA moduluses and integers greater than 2, with the property that if (n, e) was sampled from $\mathcal{X}_b$, then $G_{n,e}$ is trivial if and only if $b = 0$. A τ-*distinguisher* for $(\mathcal{X}_0, \mathcal{X}_1)$ is an algorithm $\mathcal{A}$ that on input of (n, e) outputs 0 or 1, using runtime at most τ. The advantage of $\mathcal{A}$ is

$$\mathbf{Adv}_{\mathcal{X}_0, \mathcal{X}_1}^{\phi\text{-hiding}}(\mathcal{A}) = 2|\Pr[\mathcal{A}(n, e) = b \mid b \xleftarrow{r} \{0, 1\}, (n, e) \xleftarrow{r} \mathcal{X}_b] - 1/2|.$$

Remark. The above distinguishing problem is often called the ϕ-*hiding prob-lem*, where ϕ refers to *Euler's ϕ-function*. Euler's ϕ-function of a positive integer n is the number of invertible elements in $\mathbb{Z}_n$. For an RSA modulus $n = pq$, we have that $\phi(n) = (p-1)(q-1)$. If the above distinguishing prob-lem is hard for some $(\mathcal{X}_0, \mathcal{X}_1)$, it says that some information about $\phi(n)$ is hard to decide without knowing the factorisation, $\phi(n)$ is in some sense *hidden*.

An algorithm that finds eth roots when $G_{n,e}$ is trivial may not be able to find roots when the group is non-trivial. Therefore, we need to be a bit careful when we structure the proof, so that we do not squander our advantage.

Lemma 9.14. *Let $(\mathcal{X}_0, \mathcal{X}_1)$ be as in Definition 9.4, and let $\mathcal{A}$ be an algorithm that on input of of (n, e) sampled from $\mathcal{X}_0$ or $\mathcal{X}_1$ outputs either 0 or an element from $G_{n,e}$, using runtime at most τ. Suppose that if $G_{n,e}$ is non-trivial, then given that $\mathcal{A}$ does not output 0, it outputs 1 with probability at most $1/e$. Then there exists a τ_0-distinguisher $\mathcal{B}_0$ and a τ_1-distinguisher $\mathcal{B}_1$ for $(\mathcal{X}_0, \mathcal{X}_1)$, where τ_0 and τ_1 are essentially equal to τ, such that*

$$\Pr[\mathcal{A}(n, e) \neq 0 \mid (n, e) \xleftarrow{r} \mathcal{X}_0] \leq \mathbf{Adv}^{\phi\text{-hiding}}_{\mathcal{X}_0, \mathcal{X}_1}(\mathcal{B}_0) + 2\mathbf{Adv}^{\phi\text{-hiding}}_{\mathcal{X}_0, \mathcal{X}_1}(\mathcal{B}_1).$$

Proof. The algorithm $\mathcal{B}_b$ gets (n, e) as input, computes $x \leftarrow \mathcal{A}(n, e)$ and outputs 0 if $x = b$ or $x = 0$, otherwise outputs 1. Since the algorithm essentially just runs $\mathcal{A}$ with input sampled from its expected distribution, the runtime τ_b of $\mathcal{B}_b$ is essentially equal to τ. Note the difference between $\mathcal{B}_0$ and $\mathcal{B}_1$.

For $k, b \in \{0, 1\}$, let $E_{k,b}$ be the event that $\mathcal{A}$ on input sampled from $\mathcal{X}_b$ outputs k. Also, let $E_{nt,1}$ be the event that $\mathcal{A}$ on input sampled from $\mathcal{X}_1$ does not output 0 or 1, but a non-trivial element of $G_{n,e}$. Since $G_{n,e}$ is trivial when (n, e) has been sampled from $\mathcal{X}_0$, we have that $\Pr[E_{0,0}] + \Pr[E_{1,0}] = 1$. Also,

$$\Pr[\mathcal{A}(n, e) \neq 0 \mid (n, e) \xleftarrow{r} \mathcal{X}_0] = \Pr[E_{1,0}].$$

When the input has been sampled from $\mathcal{X}_1$, then if $\mathcal{A}$ does not output 0, it outputs 1 with probability at most $1/e$. Therefore,

$$\Pr[E_{nt,1}] \geq (1 - \Pr[E_{0,1}])(1 - 1/e) \geq (1 - \Pr[E_{0,1}])/2$$

since $e \geq 2$. By definition we get that

$$\mathbf{Adv}^{\phi\text{-hiding}}_{\mathcal{X}_0, \mathcal{X}_1}(\mathcal{B}_0) = |\Pr[E_{0,0}] - \Pr[E_{0,1}]|.$$

The claim follows since $\mathbf{Adv}^{\phi\text{-hiding}}_{\mathcal{X}_0, \mathcal{X}_1}(\mathcal{B}_1) = \Pr[E_{nt,1}]$ and

$$\Pr[E_{nt,1}] \geq (1 - \Pr[E_{0,1}])/2 = (\Pr[E_{1,0}] + \Pr[E_{0,0}] - \Pr[E_{0,1}])/2$$
$$\geq (\Pr[E_{1,0}] - |\Pr[E_{0,0}] - \Pr[E_{0,1}]|)/2. \qquad \square$$

We now turn an adversary against TB-RSA into an eth root of unity finder.

Lemma 9.15. *Let A be a (τ, l_h, l_s)-random message adversary against* TB-RSA $= (\mathcal{K}, \mathcal{S}, \mathcal{V})$. *Then there exists an algorithm B that on input of (n, e) outputs either 0 or an element of $G_{n,e}$, the runtime of B is essentially equal to a small multiple of τ and*

$$\Pr[B(n, e) \neq 0 \mid (n, e) \xleftarrow{r} \mathcal{X}_0] = \mathbf{Adv}^{\mathsf{suf\text{-}rma}}_{\text{TB-RSA}}(A),$$

where $\mathcal{X}_0$ is the probability space on RSA moduluses and exponents induced by the key generation algorithm. If $G_{n,e}$ is non-trivial and B does not output 0, its output is 1 with probability at most $1/e$.

Proof. The algorithm B takes as input (n, e) and runs a copy of A interacting with a simulator **Sim** for $\mathbf{Exp}^{\mathsf{rma}}_{\text{TB-RSA}}(A)$ that works as follows:

1. Set $C_1 = C_2 = \emptyset$. Send (n, e) to the adversary A.

2. When the adversary sends a query $\perp$ (*target*):

    (a) Sample $r \xleftarrow{r} \mathbb{Z}_n^*$, compute $m \leftarrow r^e$, add (r, m) to C_2 and send m to the adversary.

3. When the adversary sends a query m (*chosen message*):

    (a) Stop if $(r, m) \notin C_2$ for some r. Otherwise, add (m, r) to C_1 and send r to the adversary.

If A exceeds its bounds, then B stops and outputs 0. Otherwise, the adversary A outputs (m_0, σ_0). If $\sigma_0^e = m_0$, $(r, m_0) \in C_2$ for some r and $(m_0, \sigma_0) \notin C_1$, then B outputs σ_0/r. Otherwise, B outputs 0.

When (n, e) was output by $\mathcal{K}$, **Sim** perfectly simulates $\mathbf{Exp}^{\mathsf{rma}}_{\text{TB-RSA}}(A)$. If A outputs a valid forgery, then B does not output 0. It follows that

$$\Pr[B(n, e) \neq 0 \mid (n, e) \xleftarrow{r} \mathcal{X}_0] = \mathbf{Adv}^{\mathsf{suf\text{-}rma}}_{\text{TB-RSA}}(A).$$

For every target query, B will do one exponentiation more than $\mathbf{Exp}^{\mathsf{rma}}_{\text{SIG}}(A)$ would have done. Since the number l_h of target queries is at most a small multiple of τ, the runtime of B is at most a small multiple of τ.

We must now consider the behaviour of B when its input has been sampled from $\mathcal{X}_1$. Suppose B does not output 0. Then $(\sigma_0/r)^e = 1$, so B outputs an element of $G_{n,e}$. We must now consider two cases.

First, suppose $(m_0, r) \in C_1$ for some r distinct from σ_0. In this case, B always outputs a non-trivial element of $G_{n,e}$.

Next, suppose there is no tuple (m_0, σ) in C_1 for any σ. There will be at least one tuple $(r, m_0) \in C_2$ for some r, but since $G_{n,e}$ is the kernel of the map $x \mapsto x^e$, the adversary has no information about which preimage r of m_0 is known to B. It follows that σ_0 equals r with probability at most $1/|G_{n,e}| \leq 1/e$. The claim follows. $\qquad\square$

The result for TB-RSA follows directly from the previous two lemmas.

Proposition 9.16. *Let $\mathcal{A}$ be a (τ, l_h, l_s)-random message adversary against* TB-RSA $= (\mathcal{K}, \mathcal{S}, \mathcal{V})$. *Then there exists a τ_0-distinguisher $\mathcal{B}_0$ and a τ_1-distinguisher $\mathcal{B}_1$ for $(\mathcal{X}_0, \mathcal{X}_1)$, where τ_0 and τ_1 are essentially equal to small multiples of τ, such that*

$$\mathbf{Adv}^{\text{suf-rma}}_{\text{TB-RSA}}(\mathcal{A}) \leq \mathbf{Adv}^{\phi\text{-hiding}}_{\mathcal{X}_0, \mathcal{X}_1}(\mathcal{B}_0) + 2\mathbf{Adv}^{\phi\text{-hiding}}_{\mathcal{X}_0, \mathcal{X}_1}(\mathcal{B}_1).$$

9.2.4 Provable Security: What Does It All Mean?

We now have two security proofs for textbook RSA under random message attack, and we have a very different security proof for a variant TB-RSA' of textbook RSA under random message attack. What does this mean? Which signature scheme should we use, and which security proof should we rely on.

Practically, our goal is to choose key sizes such that our theorems provide meaningful security guarantees for the cryptosystem against adversaries we care about. In that sense, determining the most efficient scheme is easy. More theoretically, we want to better understand the differences between the schemes and the theorems.

Let us first consider the difference between textbook RSA and the variant TB-RSA', in terms of the adversary's capability. Superficially, it seems like the variant should be less secure, since the adversary now has a choice of which hash to produce an eth root for, which makes the job easier. However, this is a minor issue, compared to the signing oracle. In fact, the right way to look at the variant TB-RSA' compared to textbook RSA, is that the variant has an *unreliable* signing oracle where the adversary cannot perfectly predict which message will get signed. One plausible adversarial strategy (see Exercise 4.14) is to ask for a large set of targets, find a subset of the random targets that will assist in creating the forgery, and then ask for signatures for these targets. For textbook RSA, this is straight-forward, but for the unreliable signing oracle this is much more complicated. For some attacks, this merely results in a doubling of the number of signing queries, but for other plausible attack strategies the increase may be so large as to make the attack impractical.

The gap in the security results is justified, since the difference in adversarial capability is plausibly important and there may be a gap in security levels.

The two security results for textbook RSA differ in the underlying computational problem, either the RSA problem or the ϕ-hiding problem.

One way to compare computational problems is through the machinery of reductions. The RSA problem is clearly not easier than the ϕ-hiding problem, since if we can find eth roots, we will be able to solve the ϕ-hiding problem. The converse is unproven. Since it is hard to imagine how the ability to recognise moduluses output by the RSA key generation algorithm make solving the RSA problem easier, it seems unlikely to be provable that the ϕ-hiding problem is not easier than the RSA problem. In fact, it is perfectly plausible that it is easy to recognise moduluses, but that the RSA problem is hard to solve.

Another way to compare computational problems is to look at the best known algorithms for solving the problems. The best algorithm for solving the RSA problem is to factor the modulus, and the best algorithm for solving the ϕ-hiding problem is to factor the modulus. Note that the ϕ-hiding problem has almost certainly received less attention than the RSA problem, so we should have less confidence in this statement, and probably use a larger modulus.

If we decide to believe that the ϕ-hiding problem is as hard as the RSA problem, it follows that the variant signature scheme TB-RSA$'$ is no more secure than textbook RSA. Which implies that the unreliability of the signing oracle does not matter. The justification for this would be that factoring the modulus is the best attack, so the signing oracle would be, effectively, useless.

Against this, we should also note that the cost of using the variant is actually very small. The reason is the design of the hash function $h_{n,e} : S \to \mathbb{Z}_n \times \mathbb{Z}_n$. In practice, this hash function will be the composition of two hash functions $h_0 : S \to T$ and $h_{1,n,e} : T \to \mathbb{Z}_n \times \mathbb{Z}_n$, where the first just needs to be collision resistant and the second should be modelled as a random oracle. Technically, we would use both Proposition 9.7 and Proposition 9.8 to analyse this construction. We would use a similar construction for RSA-FDH as well. Computing the second hash function is very cheap and does not depend on the length of the message. Likewise, the PRF would be computed on the intermediate value, so this computation would also be very cheap. The extra PRF key needs to be managed, but this may very well be cheap or even free, as in some implementations one of the primes p or q could do the job (although some care is needed in this case). The length of the signature is unchanged. The verifier's cost is similarily very small, since we verify the signature by computing the eth power of the signature and compare it to the hash values.

We now have three options: Use slightly larger RSA moduluses with textbook RSA to compensate for the weaker security result or less studied underlying problem, use slightly smaller RSA moduluses with textbook RSA and a stronger faith in the underlying problem, or use a very slightly more expensive scheme TB-RSA$'$ with smaller RSA moduluses.

As the above discussion shows, provable security is not an exact science. It is not at all clear which option we should choose.

9.2.5 Example: Multivariate Quadratic Equations

A *polynomial equation* over a field $\mathbb{F}$ is of the form $f(X) = \alpha$, with $f(X) \in \mathbb{F}[X]$ and $\alpha \in \mathbb{F}$. A *multivariate polynomial equation* is of the form $f(X_1, \ldots, X_\nu) = \alpha$. We adopt the notation $\mathbf{X} = (X_1, \ldots, X_\nu)$ and $f(\mathbf{X}) = \alpha$. A *multivariate quadratic equation* is a multivariate polynomial equation $f(\mathbf{X}) = \alpha$ where f has degree 2. We denote a *system of multivariate quadratic equations* $f_i(\mathbf{X}) = \alpha_i$, $i = 1, 2, \ldots, l$, by $\mathbf{f}(\mathbf{X}) = \boldsymbol{\alpha}$.

We can modify a polynomial equation by applying a coordinate change $\mathbf{S}$ to the unknowns in the polynomial, letting $\mathbf{X} = \mathbf{S}\mathbf{Y}$, which gives us a new multivariate polynomial $\bar{g}(\mathbf{Y}) = f(\mathbf{S}\mathbf{Y})$ and corresponding equation. Such a coordinate change does not change the degree of the polynomial, and in particular maps quadratic equations to quadratic equations. This coordinate change also applies to systems of equations.

If we have a collection of multivariate polynomials $\bar{\mathbf{g}}(\mathbf{Y})$, we can create a new collection of multivariate polynomials by computing linear combinations of the old multivariate polynomials. In the usual manner, we organise the coefficients in a matrix $\mathbf{T}$ and denote the new polynomials by $\mathbf{g}(\mathbf{Y}) = \mathbf{T}\bar{\mathbf{g}}(\mathbf{Y})$.

The coordinate change and the linear combination gives us a way to map a system of polynomial equations $\mathbf{f}(\mathbf{X}) = \boldsymbol{\alpha}$ to a new system of polynomial equations $\mathbf{g}(\mathbf{Y}) = \boldsymbol{\beta}$ where $\mathbf{g}(\mathbf{Y}) = \mathbf{T}\mathbf{f}(\mathbf{S}\mathbf{Y})$ and $\boldsymbol{\beta} = \mathbf{T}\boldsymbol{\alpha}$.

A polynomial equation over a finite field is easy to solve. A multivariate quadratic equation will in general not have a unique solution. A system of multivariate quadratic equations may be very hard to solve, even over small finite fields such as $\mathbb{F}_2$. However, some systems of multivariate equations are easy to solve:

$$X_l + \bar{f}_1(X_{l+1}, \ldots, X_\nu) = \alpha_1$$
$$X_{l-1} + \bar{f}_2(X_l, \ldots, X_\nu) = \alpha_2$$
$$\vdots \tag{9.16}$$
$$X_1 + \bar{f}_l(X_2, \ldots, X_\nu) = \alpha_l$$

This system has a unique solution if $l = \nu$. The system is *underdefined* and without a unique solution if $l < \nu$, but it is still easy to find solutions: choose arbitrary values for $X_{l+1}, \ldots, X_\nu$ and solve the resulting system. If $l > \nu$, we should interpret the above system of equations as having ν equations with the special structure and $l - \nu$ arbitrary equations. In general, we cannot expect such a system to have any solutions at all, but we can find a unique solution by ignoring the arbitrary equations.

Remark. There are many other possible structures for easy-to-solve linear systems of equations. For instance, we could require that no equation contains a term of the form $X_i X_j$ for $1 \leq i, j \leq l$.

If $\mathbf{g}(\mathbf{Y}) = \boldsymbol{\beta}$ is a system of multivariate quadratic equations derived from a system of the form (9.16), then this is not a randomly chosen system of multivariate quadratic equations, so there is no a priori reason to believe it should be difficult to solve, even if randomly chosen systems are hard to solve.

Given a collection of polynomials $\mathbf{g}(\mathbf{Y})$, it is very easy to generate a system of equations that has a solution: choose arbitrary values $y_1, \ldots, y_\nu \in \mathbb{F}$ for $Y_1, \ldots, Y_\nu$ and compute $\boldsymbol{\beta} \leftarrow \mathbf{g}(\mathbf{y})$.

We can now define a signature scheme based on such systems of equations.

Example 9.8. The *multi-variate quadratic* signature scheme $\text{SIG} = (\mathcal{K}, \mathcal{S}, \mathcal{V})$ works as follows:

- The *key generation* algorithm $\mathcal{K}$ samples a collection of polynomials $\mathbf{f}(\mathbf{X})$ of the form (9.16) from the uniform distribution on collections of such quadratic polynomials, as well as two invertible matrices $\mathbf{S} \in \mathbb{F}^{\nu \times \nu}$ and $\mathbf{T} \in \mathbb{F}^{l \times l}$. It computes $\mathbf{g}(\mathbf{Y}) \leftarrow \mathbf{Tf}(\mathbf{SY})$. It outputs the verification key $vk = \mathbf{g}(\mathbf{Y})$ and the signing key $sk = (\mathbf{T}, \mathbf{f}(\mathbf{X}), \mathbf{S})$.

- The *signing* algorithm $\mathcal{S}$ takes $sk = (\mathbf{T}, \mathbf{f}(\mathbf{X}), \mathbf{S})$ and $m \in \mathbb{F}^l$ as input, computes a solution $\mathbf{x}$ to $\mathbf{f}(\mathbf{X}) = \mathbf{T}^{-1}m$ and outputs $\mathbf{S}^{-1}\mathbf{x}$.

- The *verification* algorithm $\mathcal{V}$ takes $vk = \mathbf{g}(\mathbf{Y})$, $m \in \mathbb{F}^l$ and $\sigma \in \mathbb{F}^\nu$ as input. It outputs 1 if $\mathbf{g}(\sigma) = m$, and otherwise outputs 0.

E | *Exercise 9.16.* Show that the above scheme is a signature scheme.

E | *Exercise 9.17.* Let the polynomials $\mathbf{f}(\mathbf{X})$ be of the form (9.16) with $l \leq \nu$. Let $\mathbf{g}(\mathbf{Y})$ be derived from $\mathbf{f}(\mathbf{X})$ via a coordinate change $\mathbf{S}$ and linear combiner $\mathbf{T}$ as described above.

(a) Suppose that we sample $\mathbf{x}$ from the uniform distribution on $\mathbb{F}^\nu$ and compute $\boldsymbol{\alpha} \leftarrow \mathbf{f}(\mathbf{x})$. Alternatively, we sample $\boldsymbol{\alpha} \xleftarrow{r} \mathbb{F}^l$ and solve for $\mathbf{x}$ using the above approach.

Show that the probability distributions on $\mathbb{F}^\nu \times \mathbb{F}^l$ induced by the two methods of sampling $(\mathbf{x}, \boldsymbol{\alpha})$ are identical.

(b) Suppose that we sample $\mathbf{y}$ from the uniform distribution on $\mathbb{F}^\nu$ and compute $\boldsymbol{\beta} \leftarrow \mathbf{g}(\mathbf{y})$. Alternatively, we sample $\boldsymbol{\alpha} \xleftarrow{r} \mathbb{F}^l$ and solve for $\mathbf{y}$ using $\mathbf{S}$, $\mathbf{T}$ and $\mathbf{f}$ and the above approach.

Show that the probability distributions on $\mathbb{F}^\nu \times \mathbb{F}^l$ induced by the two methods of sampling $(\mathbf{x}, \boldsymbol{\alpha})$ are identical.

We can now apply the general machinery we have developed to this signature scheme, and deduce that any random oracle adversary will be an adversary that essentially solves such a system of equations. In particular, by Exercise 9.17 it is easy to simulate signatures on random messages.

Unfortunately, the key to getting reasonable results for RSA-based signatures was the rerandomisability of RSA, which does not apply here. Which means that we get a weak result, even with $\nu < l$ or randomised hashes as in Example 9.6. Also, the strategy does not produce a solver for general systems of multivariate quadratic equations, just systems derived from our easy-to-solve systems.

E | *Exercise 9.18.* Suppose $\mathcal{A}$ is a random message adversary against the signature scheme. Show that there is a solver $\mathcal{B}$ for systems of equations derived from (9.16) with advantage equal to (i) about $1/l_h$th of $\mathcal{A}$'s advantage if $\nu = l$, or (ii) $\mathcal{A}$'s advantage if ν is sufficiently smaller than l. The runtime of $\mathcal{B}$ is at most a small multiple of the runtime of $\mathcal{A}$.

9.3 IDENTIFICATION SCHEMES

The hash-then-sign paradigm works well for primitives that are somehow similar to the RSA problem (often called a *trapdoor one-way problem*), but it does not work that well for other primitives. Therefore, we want to study another paradigm for signature scheme construction, namely applying the famous *Fiat-Shamir heuristic* an *identification protocol.*

Morally, an identification protocol is cryptographic protocol that allows a prover to convince a verifier that the prover's claimed identity is correct. Technically, the protocol allows the prover to convince a verifier that the prover has some cryptographic capability. It is only when the particular cryptographic capability is (somehow) tied to an identity that we get identification.

The cryptographic capability in question can take many forms, for instance the ability to decrypt ciphertexts or sign messages. The obvious protocol is then that the verifier makes a ciphertext or a message as a *challenge*, and the prover then *responds* with the decryption of the ciphertext or a signature on the message. However, such protocols would not make sense for our current purpose, which is designing signature schemes.

We shall therefore look at a very special class of identification protocols, built using more primitive tools. These protocols will have a very special form: the prover first makes a *commitment*. The verifier samples a challenge from the uniform distribution on some set. The prover then responds, after which the verifier accepts or rejects.

Remark. Before we go further with this discussion of identification protocols, we should emphasise that on their own these identification protocols are remarkably useless. While identification is a common task for physical humans, we rarely just want to identify someone in a cryptographic context. The identification must typically be tied to some other cryptographic context, such as a secure channel or some transaction. Typically these identification protocols must be augmented somehow to support these applications, typically by introducing associated data of some form.

This also means that the security notions we discuss probably will not be suitable for anything other than the intended application: signature scheme construction and security proofs.

Remark. The class of identification protocols and related security notions that we consider here are very similar to so-called *sigma protocols*, and we return to this class of protocols in Chapter 11.

9.3.1 Definitions

As discussed above, the identification protocols we shall use are of a special form, which means that the definition does not actually define the protocol, but instead algorithms used at various stages of the protocol.

D | **Definition 9.5.** An *identification scheme* ID $= (\mathfrak{B}, \mathcal{IK}, \mathcal{II}, \mathcal{IR}, \mathcal{IV})$ consists of a set of challenges $\mathfrak{B}$ and four algorithms:

- The *key generation* algorithm $\mathcal{IK}$ takes no input and outputs a *verification key* ivk and a *secret key* isk.

- The *commitment* algorithm $\mathcal{II}$ takes as input a secret key isk and outputs a commitment α and a *state* ρ.

- The *response* algorithm $\mathcal{IR}$ takes as input a secret key isk, a state ρ and a challenge β, and outputs a *response* γ.

- The *verify* algorithm $\mathcal{IV}$ takes as input a verification key, a commitment α, a challenge β and a response γ, and outputs either 0 or 1.

We require that for any key pair (ivk, isk) output by $\mathcal{IK}$, any (α, ρ) output by $\mathcal{II}(isk)$ and any challenge $\beta \in \mathfrak{B}$, we have that

$$\mathcal{IV}(ivk, \alpha, \beta, \gamma) = 1.$$

The triple (α, β, γ) is a *conversation*. It is *accepting* if $\mathcal{IV}(ivk, \alpha, \beta, \gamma) = 1$.

The actual identification protocol is trivial: the prover runs the commitment algorithm and sends the commitment to the verifier. The verifier samples a challenge and sends it to the prover. The prover then computes the response and sends it to the verifier, who verifies the response.

We shall now use the *Fiat-Shamir heuristic* to create a signature scheme from an identification scheme. The idea is simple: We replace the random challenge with a hash of the public key, the commitment and the message.

E | *Example* 9.9. Let ID $= (\mathfrak{B}, \mathcal{IK}, \mathcal{II}, \mathcal{IR}, \mathcal{IV})$, let $\mathfrak{P}$ be a set, and let h be a hash function from tuples of a verification key, a commitment and a message to $\mathfrak{B}$. The signature scheme SIG $= (\mathcal{K}, \mathcal{S}, \mathcal{V})$ works as follows:

- The *key generation* algorithm $\mathcal{K}$ computes $(ivk, isk) \leftarrow \mathcal{IK}$ and outputs the verification key $vk = ivk$ and the signing key $isk = (ivk, isk)$.

- The *signing* algorithm takes a signing key $sk = (ivk, isk)$ and a message $m \in \mathfrak{P}$ as input, computes $(\alpha, \rho) \leftarrow \mathcal{II}(isk)$, $\beta \leftarrow h(ivk, \alpha, m)$ and $\gamma \leftarrow \mathcal{IR}(isk, \rho, \beta)$, and outputs the signature (α, γ).

- The *verification* algorithm takes a verification key ivk, a message m and a signature (α, γ) as input and outputs $\mathcal{IV}(ivk, \alpha, h(ivk, \alpha, m), \gamma)$.

Remark. As can be seen from the proofs, including the verification key in the hash is not needed to prove the scheme secure. But including it costs very little. Including it is good *crypto hygiene*, since it ties the signature more closely to the verification key. Among other things, this prevents an obscure attack where an adversary that does not know the message is still able to create a modified signature that verifies under a specially created verification

The experiment for an adversary $\mathcal{A}$ proceeds as follows:

1. Set $C_0 = C_1 = \emptyset$.

2. Compute $(ivk, isk) \leftarrow \mathcal{IK}$ and send ivk to the adversary.

3. When the adversary sends a query α (*target*):
   (a) Sample $\beta \xleftarrow{r} \mathfrak{B}$. Set $C_0 = C_0 \cup \{(\alpha, \beta)\}$.
   (b) Send β to the adversary.

4. When the adversary sends a query $\perp$ (*random conversation*):
   (a) Compute $(\alpha, \rho) \leftarrow \mathcal{II}(isk)$, $\beta \xleftarrow{r} \mathfrak{B}$ and $\gamma \leftarrow \mathcal{IR}(isk, \rho, \beta)$.
   (b) Set $C_1 = C_1 \cup \{(\alpha, \beta, \gamma)\}$. Send (α, β, γ) to the adversary.

Eventually, the adversary outputs (α, β, γ).

FIGURE 9.4 The experiment for the security game for an identification scheme ID $= (\mathfrak{B}, \mathcal{IK}, \mathcal{II}, \mathcal{IR}, \mathcal{IV})$.

key. Signature schemes do not promise to protect against such attacks, but when protection is cheap, it seems like a good idea.

For some identification schemes, including our examples, there is a function f mapping the verification key ivk, challenge β and response γ to commitments, and the verification algorithm accepts if the commitment α is equal to $f(ivk, \beta, \gamma)$ (and possibly checking some other conditions).

If f is easy to compute, there is a variant signature scheme: The signature is (β, γ) instead of (α, γ), and the verification algorithm is modified to check if $h(ivk, f(ivk, \beta, \gamma), m)$ equals β, instead of comparing $f(ivk, \beta, \gamma)$ to α.

This is an advantage when the challenge β is shorter than the commitment α, since the signature becomes shorter, which is good.

We shall now define what security means for an identification protocol, but again we emphasise that these definitions are carefully constructed for our class of schemes and our purpose. They may not be generally applicable.

Definition 9.6. A (τ, l_h, l_s)-*adversary* against an identification scheme ID is an interactive algorithm $\mathcal{A}$ that interacts with the experiment in Figure 9.4 making at most l_h target queries and l_s random conversations queries, and where the runtime of the adversary and the experiment is at most τ.

The *advantage* of the adversary is defined to be

$$\mathbf{Adv}_{\text{ID}}^{\text{id-honest}}(\mathcal{A}) = \Pr[E],$$

where E is the event that $\mathcal{A}$ outputs a conversation (α, β, γ) that is accepting, $(\alpha, \beta) \in C_0$ and $(\alpha, \beta, \gamma) \notin C_1$.

The *collision probability* of the scheme is defined to be

$$\mathbf{Col}_{\mathrm{ID}}(\mathcal{A})\Pr[F],$$

where F is the event that a random conversation query returns a conversation (α, β, γ) and for some $\beta' \neq \beta$, $(\alpha, \beta') \in C_0$ or $(\alpha, \beta', \cdot) \in C_1$.

Remark. Collision probability is not a natural security notion for identification schemes. However, our intended proof strategy relies on the experiment lists C_0 and C_1 essentially defining a partial function from commitments to challenges. If commitment values are likely to be reused, this strategy will fail.

The schemes we will study essentially sample α values from sufficiently large sets, so collisions will be sufficiently rare.

Remark. Note that we allow multiple target queries for the same commitment, which seems incorrect since the target query in some sense should model a hash function. But for our application each commitment may be hashed with a different message, and since our identification schemes do not care about messages we must allow multiple queries.

We now show how to turn an attack on the signature scheme from Example 9.9 into an attack on the underlying identification scheme. The strategy is fairly simple. Hash queries correspond to target queries, while random conversation queries correspond to signature queries. Both types of queries are integrated into a random oracle simulation that works, because challenges are chosen uniformly at random and any failure results in a collision attack. Finally, any signature forgery becomes a forged accepting conversation.

Proposition 9.17. *Let* SIG *be the signature scheme from Example 9.9 based on an identification scheme* ID. *Let* $\mathcal{A}$ *be a* (τ, l_s)-*adversary against* SIG *in the random oracle model, making at most* l_h *queries to the random oracle* h. *Then there exists a* (τ', l_h, l_s)-*adversary against* ID, *with* τ' *essentially equal to* τ, *such that*

$$\mathbf{Adv}_{\mathrm{SIG}}^{\mathsf{suf\text{-}cma}}(\mathcal{A}) \leq \mathbf{Adv}_{\mathrm{ID}}^{\mathsf{id\text{-}honest}}(\mathcal{B}) + \mathbf{Col}_{\mathrm{ID}}(\mathcal{B}).$$

Proof. Our adversary $\mathcal{B}$ runs a copy of $\mathcal{A}$ and the following simulator **Sim**:

- If $vk' \neq vk$ for a hash query $(vk', \cdot, \cdot)$, **Sim** responds as a random oracle.
- When $\mathcal{A}$ makes a hash query (vk, α, m), if $(\alpha, m, \beta, \cdot)$ is recorded for some β, send β to $\mathcal{A}$. Otherwise, send a target query α to the experiment and get the response β. Then record $(\alpha, m, \beta, \perp)$ and send β to adversary.
- When $\mathcal{A}$ makes a signature query m, send a random conversation query to the experiment and get the response (α, β, γ) from $\mathcal{A}$. If $(\alpha, m, \cdot, \cdot)$ is already recorded, $\mathcal{B}$ stops and outputs a random conversation. Otherwise, record $(\alpha, m, \beta, \gamma)$ and send (α, γ) to $\mathcal{A}$.

When $\mathcal{B}$ receives ivk, it sends $vk = ivk$ to $\mathcal{A}$. If $\mathcal{A}$ outputs a message m and a signature (α, γ) and the simulator has recorded $(\alpha, m, \beta, \perp)$, then $\mathcal{B}$ outputs (α, β, γ). If no such record exists, or if $\mathcal{A}$ exceeds its bounds, $\mathcal{B}$ outputs a random conversation.

Let F be the event that $\mathcal{A}$ outputs a valid forgery when interacting with $\mathbf{Exp}_{\mathrm{SIG}}^{\mathrm{cma}}(\mathcal{A})$. Let F_0 be the event that $\mathcal{A}$ outputs a valid signature forgery when interacting with $\mathbf{Sim}$ and $\mathbf{Exp}_{\mathrm{ID}}^{\mathrm{id\text{-}honest}}(\mathcal{B})$. Let F_1 be the event that $\mathcal{B}$ stops and outputs a random conversation while $\mathbf{Sim}$ responds to a signature query.

By inspection, we see that unless F_1 happens $\mathbf{Sim}$ and the identification scheme experiment perfectly simulate the signature experiment. It follows that

$$|\Pr[F] - \Pr[F_0]| \le \Pr[F_1] \qquad \Rightarrow \qquad \Pr[F] \le \Pr[F_0] + \Pr[F_1].$$

A valid signature forgery (α, β, γ) would have $(\alpha, \beta) \in C_0$, but (α, β, γ) would not be in C_1. Also, F_1 is the event a random conversation query gives a collision. In other words,

$$\Pr[F_0] = \mathbf{Adv}_{\mathrm{ID}}^{\mathrm{id\text{-}honest}}(\mathcal{B}) \qquad \text{and} \qquad \Pr[F_1] = \mathbf{Col}_{\mathrm{ID}}(\mathcal{B}). \qquad \square$$

9.3.2 Simulating Random Conversations

We shall now study examples of identification schemes that we can apply the Fiat-Shamir heuristic to. Our first example is the Katz-Wang identification scheme. The security of the Katz-Wang scheme is based on a class of "impossible" verification keys, for which it is almost impossible to create signatures. It follows that if there is an an effective forger for the scheme, the ordinary verification keys can be distinguished from the "impossible" verification keys.

As we have seen, responding to the various queries without knowing the secret key is a significant obstacle to proving cryptosystems secure. For the Katz-Wang scheme, making use of the "impossible" verification keys precludes have access to a signing key. This is then an obstacle when we want to create random conversations.

The answer is to simulate random conversations, and for some schemes this can be done without access to the secret key. Therefore, we only have to consider adversaries making target queries, not random conversation queries.

Example 9.10. The *Katz-Wang* identification scheme ID $= (\mathcal{IK}, \mathcal{II}, \mathcal{IR}, \mathcal{IV})$ based on a cyclic group G of prime order n with generator g_1 works as follows:

- The *key generation* algorithm $\mathcal{IK}$ samples a secret key $a \xleftarrow{r} \{0, 1, \ldots, n-1\}$ and a generator $g_2 \xleftarrow{r} G \setminus \{1\}$. It computes $x_1 \leftarrow g_1^a$ and $x_2 \leftarrow g_2^a$. The verification key is $ivk = (g_2, x_1, x_2)$ and the secret key is $isk = (g_2, a)$.
- The *commitment* algorithm $\mathcal{II}$ samples $r \xleftarrow{r} \{0, 1, \ldots, n-1\}$, computes $\alpha_1 \leftarrow g_1^r$ and $\alpha_2 \leftarrow g_2^r$, and outputs commitment (α_1, α_2) and state r.
- The *response* algorithm $\mathcal{IR}$ takes as input a secret key $isk = (g_2, a)$, a state r and a challenge β, computes $\gamma \leftarrow r - \beta a \bmod n$, and outputs γ.

- The *verification* algorithm $\mathcal{IV}$ takes as input a verification key $ivk = (g_2, x_1, x_2)$, a commitment (α_1, α_2), a challenge β and a response γ, and outputs 1 if and only if $\alpha_1 = g_1^\gamma x_1^\beta$ and $\alpha_2 = g_2^\gamma x_2^\beta$.

The next claim is about adversaries that do not make any random conversation queries. We shall resolve this issue later in this section.

Proposition 9.18. *Consider the Katz-Wang identification scheme* ID *based on a cyclic group G of prime order n with generator g_1. Let $\mathcal{A}$ be a $(\tau, l_h, 0)$-adversary against* ID. *Then there exists a τ'-distinguisher $\mathcal{B}$ for Decision Diffie-Hellman, with τ' essentially equal to τ, such that*

$$\mathbf{Adv}_{\text{ID}}^{\text{id-honest}}(\mathcal{A}) \leq \mathbf{Adv}_{G,g_1}^{\text{DDH}}(\mathcal{B}) + 1/n + l_h/|\mathfrak{B}|.$$

Proof. The distinguisher $\mathcal{B}$ takes as input a tuple $(g_2, x_1, x_2) \in G^3$. It runs a copy of $\mathcal{A}$ and the identification scheme experiment, modified to use (x_1, x_2) as the verification key instead of doing key generation. If $\mathcal{A}$ outputs a valid forgery, $\mathcal{B}$ outputs 1. If $\mathcal{A}$ either does not output a valid forgery or exceeds its bounds, $\mathcal{B}$ outputs 0.

Let F_D be the event that $\mathcal{A}$ outputs 1 when its input is sampled from the uniform distribution on DDH tuples. Let F_R be the event that $\mathcal{A}$ outputs 1 when its input is sampled from the uniform distribution on G^3. The advantage of a Decision Diffie-Hellman distinguisher is $|\Pr[F_D] - \Pr[F_R]|$.

When (g_2, x_1, x_2) is a DDH tuple, $\mathcal{B}$ perfectly simulates the identification scheme experiment, so $\mathcal{A}$ does not exceed its bounds, and we get that

$$\Pr[F_D] = \mathbf{Adv}_{\text{ID}}^{\text{id-honest}}(\mathcal{A}).$$

Next, we analyse what happens when the input is a random tuple. Let $x_1 = g_1^a$, and let Δ be such that $x_2 = g_2^{a+\Delta}$. For the ith target query $(\alpha_{i1}, \alpha_{i2})$ with challenge β_i, there exist r_i, δ_i such that $\alpha_{i1} = g_1^{r_i}$ and $\alpha_{i2} = g_2^{r_i + \delta_i}$. If $(\alpha_{i1}, \alpha_{i2}, \beta_i, \gamma_i$ is accepting, we get that

$$r_i \equiv \gamma_i - \beta_i a \pmod{n}, \qquad r_i + \delta_i \equiv \gamma_i - \beta_i(a + \Delta) \pmod{n}.$$

However, this pair of equations can only be consistent if $\delta_i \equiv \beta_i \Delta$. Note that Δ and δ_i is determined before the experiment samples β_i from $\mathfrak{B}$, which means that if $\Delta \neq 0$ the probability that this particular system of equations can be consistent is at most $1/|\mathfrak{B}|$. If $\Delta \neq 0$, then with l_h target queries, the probability that at least one query gives $\delta_i \equiv \beta_i \Delta$ is at most $l_h/|\mathfrak{B}|$.

When (g_2, x_1, x_2) has been sampled from the uniform distribution on G^3, it will not be a DDH tuple with probability $(1 - 1/n)$. It follows that

$$\Pr[F_R] \leq \frac{1}{n} + \frac{l_h}{|\mathfrak{B}|}. \qquad \square$$

We have only proved something for adversaries that do not make random conversation queries. This is because the impossibility that gives us security also causes a different problem: We cannot efficiently respond to random conversation queries when the verification key comes from the DDH tuple. The answer to this obstacle is that it is possible to simulate such conversations without knowing the secret key. We can then respond to random conversation queries with simulated conversations. This result requires new concepts.

Definition 9.7. A *conversation sampler* SG for an identification scheme ID is an algorithm that on input of a key pair (ivk, isk) outputs an accepting conversation (α, β, γ) or $\bot$. Any identification scheme has an *honest conversation sampler* G, namely the algorithm that computes $(\alpha, \rho) \leftarrow \mathcal{II}(isk)$, $\beta \xleftarrow{r} \mathfrak{B}$ and $\gamma \leftarrow \mathcal{IR}(isk, \rho, \beta)$ and outputs (α, β, γ).

We say that a conversation sampler is *witness indistinguishable* if its output distribution is independent of isk for any key pair (ivk, isk) output by $\mathcal{IK}$, and it is *public* if its output is independent of isk.

We say that two conversation samplers are ϵ-indistinguishable if their output distributions are ϵ-close for a key pair output by $\mathcal{IK}$. We say that they are *perfectly indistinguishable* if they are 0-indistinguishable. We say that a public conversation sampler is ϵ-*indistinguishable* if it is ϵ-indistinguishable from the honest conversation sampler, and that it is *perfect* if it is perfectly indistinguishable from the honest conversation sampler.

Remark. We could define a notion of computational indistinguishability for conversation samplers, but we do not need it. We return to this in Chapter 11.

Note that a public conversation generator does not actually need a secret key as input, since any input will do. We included the secret key to simplify the description. Witness-indistinguishable is only interesting when the key generation algorithm can output key pairs that have identical verification keys and distinct secret keys, which becomes important in Section 9.3.4.

Example 9.11. A public conversation sampler SG for the Katz-Wang identification scheme is the algorithm that on input of (g_2, x_1, x_2) samples $\beta \xleftarrow{r} \mathfrak{B}$ and $\gamma \xleftarrow{r} \{0, 1, \ldots, n-1\}$, computes $\alpha_1 \leftarrow g_1^\gamma x_1^\beta$ and $\alpha_2 \leftarrow g_2^\gamma x_2^\beta$, and outputs $(\alpha_1, \alpha_2, \beta, \gamma)$.

Proposition 9.19. *The public conversation sampler from Example 9.11 for the Katz-Wang identification scheme is perfect.*

Proof. Given (x_1, x_2) and β, the map taking γ to $(g_1^\gamma x_1^\beta, g_2^\gamma x_2^\beta)$ and the map taking (α_1, α_2) to $\log_{g_1} \alpha_1 - \beta \log_{g_1} x_1 \bmod n$ are inverse bijections. Therefore, it does not matter if we sample $r = \log_{g_1} \alpha_1 = \log_{g_2} \alpha_2$ or γ. $\square$

Remark. This result only holds for verification keys that could be output by the key generation algorithm. If the key generation algorithm could not output a particular verification key, that means that there is no corresponding secret key and the honest conversation sampler does not exist. This does not matter!

We use this notion to remove the need for random conversation queries by simulating the responses using the public conversation sampler.

Proposition 9.20. *Let* ID *be an identification scheme, let* SG *be an* ϵ-*indistinguishable public conversation sampler with runtime cost* τ_0, *and let* A *be a* (τ, l_h, l_s)-*adversary against* ID. *Then there exists a* $(\tau', l_h, 0)$-*adversary* B *against* ID, *with* τ' *essentially equal to* $\tau + l_s\tau_0$, *such that*

$$\mathbf{Adv}_{\mathrm{ID}}^{\mathsf{id\text{-}honest}}(A) \leq l_s\epsilon + \mathbf{Adv}_{\mathrm{ID}}^{\mathsf{id\text{-}honest}}(B).$$

Proof. The adversary B runs a copy of A and the following simulator **Sim**:

- **Sim** gets the verification key ivk from B and sends it to A.
- **Sim** forwards any target query and its response.
- For a random conversation query, **Sim** computes $(\alpha, \beta, \gamma) \leftarrow SG(ivk, \perp)$ and sends (α, β, γ) to A.

If the adversary A exceeds its bounds, B stops and outputs a random conversation. Otherwise, B outputs the conversation output by A.

The simulator **Sim** forwards messages between A and its experiment. The only significant computational cost involved is generating l_s conversations, with total cost $l_s\tau_0$. Because bounds on A are enforced B_2 is a $(\tau', l_h, 0)$-adversary against ID, with τ' essentially equal to $\tau + l_s\tau_0$.

The only difference between A interacting with the experiment, and A interacting with **Sim** and the experiment, is how the random conversations have been generated. By Exercise 6.10, the output distributions differ in statistical distance by at most $l_s\epsilon$, and the claim follows. □

We get the Katz-Wang signature scheme by applying the construction from Example 9.9 to the Katz-Wang identification scheme from Example 9.10. The following theorem follows by applying Proposition 9.17, then Propositions 9.19 and 9.20, and finally Proposition 9.18. In addition, we need that $l_h l_s/n$ is an upper bound on the collision probability for the identification scheme.

Theorem 9.21. *Consider the Katz-Wang signature scheme* SIG *based on a cyclic group* G *of prime order* n *with generator* g_1. *Let* A *be a* (τ, l_s)-*adversary against the Katz-Wang signature scheme in the random oracle model, making at most* l_h *queries to the random oracle. Then there exists a* τ'-*distinguisher* B *for Decision Diffie-Hellman, with* τ' *essentially equal to* τ, *such that*

$$\mathbf{Adv}_{\mathrm{SIG}}^{\mathsf{suf\text{-}cma}}(A) \leq \mathbf{Adv}_{G,g_1}^{\mathsf{DDH}}(B) + l_h/|\mathfrak{B}| + (l_h l_s + 1)/n.$$

9.3.3 The Forking Lemma

We first discussed the Schnorr signature scheme in Section 4.4. The Schnorr identification scheme is seemingly simpler than the Katz-Wang identification scheme, but this simplicity introduces a new difficulty in the security argument. We cannot rely on "impossible" verification keys, but we shall introduce a new technique to prove Schnorr signatures secure.

Example 9.12. The *Schnorr* identification scheme $\text{ID} = (\mathcal{IK}, \mathcal{II}, \mathcal{IR}, \mathcal{IV})$ based on a cyclic group G of order n with generator g works as follows:

- The *key generation* algorithm $\mathcal{IK}$ samples a secret key $a \xleftarrow{r} \{0, 1, \ldots, n-1\}$ and computes $x \leftarrow g^a$. The verification key is $ivk = x$ and the secret key is $isk = a$.

- The *commitment* algorithm $\mathcal{II}$ samples $r \xleftarrow{r} \{0, 1, \ldots, n-1\}$, computes $\alpha \leftarrow g^r$ and outputs the commitment α and state r.

- The *response* algorithm $\mathcal{IR}$ takes as input a secret key $isk = a$, a state r and a challenge β. It computes $\gamma \leftarrow r - \beta a \bmod n$ and outputs γ.

- The *verification* algorithm $\mathcal{IV}$ takes as input a verification key x, a commitment α, a challenge β and a response γ and outputs 1 if and only if $\alpha = g^\gamma x^\beta$.

Like for the Katz-Wang scheme, there is a public conversation sampler for the Schnorr identification scheme that allows us to simulate random conversations and apply Proposition 9.20.

Example 9.13. A public conversation sampler for the Schnorr identification scheme is an algorithm that on input of x samples $\beta \xleftarrow{r} \mathfrak{B}$ and $\gamma \xleftarrow{r} \{0, 1, \ldots, n-1\}$, computes $\alpha \leftarrow g^\gamma x^\beta$ and outputs (α, β, γ).

Proposition 9.22. *The public conversation sampler from Example 9.13 for the Schnorr identification scheme is perfect.*

Proof. Given x and β, the map taking γ to $g^\gamma x^\beta$ and the map taking α to $\log_g \alpha - \beta \log_g x \bmod n$ are inverse bijections. Therefore, it does not matter if we sample $r = \log_g \alpha$ or γ. $\qquad\square$

The "impossibility" argument used for the Katz-Wang scheme no longer applies. However, an intuitive argument suggests that the scheme should be secure. Since the number of target queries should be small compared to the number of possible challenges, there must be some target query for which the adversary is able to create a response, not just for one particular challenge, but for many challenges. It follows that the adversary must be able to create responses for two distinct challenges. Again, we shall need some new concepts.

D | **Definition 9.8.** Let ID be an identification scheme and let ivk be a verification key. A *forked conversation pair* is a pair of accepting conversations (α, β, γ) and $(\alpha', \beta', \gamma')$, where $\alpha = \alpha'$ and $\beta \neq \beta'$.

A forked conversation pair is exactly what the above intuitive argument said the adversary should be able to produce. For the Schnorr identification scheme, such a forked conversation pair reveals the secret key, and any algorithm that finds such a pair can be used to compute discrete logarithms, as the following exercise readily shows.

E | *Exercise 9.19.* Let n be prime, and let (α, β, γ) and $(\alpha, \beta', \gamma')$ be a forked conversation pair for the verification key x. Show that

$$\log_g x \equiv -\frac{\gamma - \gamma'}{\beta - \beta'} \pmod{n}.$$

The problem is how to make the adversary produce responses for two distinct challenges. The intuitive idea is that we should simply *rewind* the adversary to the point where the target query was made and sample a different challenge. For an adversary making a single target query, this would work. (We would essentially turn the adversary into two algorithms, one algorithm creates the query and a state. The other algorithm processes the response, based on the state.) But our adversary makes many target queries, so we must work with a somewhat more general situation, for which it makes sense to prove a theorem of more general applicability.

Note that when the adversary makes at most l_h queries target queries, we can sample l_h response values at the start and use them sequentially for the query responses, instead of sampling values as we go. Therefore, the experiment and the adversary can be considered a deterministic algorithm that takes as input the key pair, random tapes for the experiment and the adversary, l_h values from $\mathfrak{B}$. (These values actually come from the experiment's random tape, but since thit is easy to separate them out.) This algorithm outputs a conversation (α, β, γ). We can augment the algorithm so that the algorithm also outputs which target query added (α, β) to C_0, or 0 if no such query exists or the conversation is not accepting or the conversation is in C_1.

The Forking lemma says that any such algorithm can be turned into an algorithm that outputs a forked conversation pair as follows. First, the algorithm is run once with output i and (α, β, γ). If i is 0, the attempt failed. Otherwise we run the algorithm again with the same key pair and the same random tapes and the same $i - 1$ first values, but the remaining values sampled fresh, subject to the value in the ith position being new. This time, the algorithm outputs i' and $(\alpha', \beta', \gamma')$. If $i = i'$, we have a forked pair.

T | **Theorem 9.23** (Forking lemma). *Let $\mathcal{A}$ be a deterministic algorithm that takes l values from a set S and outputs a pair (k, π), $k \in \{0, 1, \ldots, l\}$. If $\Pr[k \neq 0 \mid (s_1, \ldots, s_l) \xleftarrow{r} S^l, (k, \pi) \leftarrow \mathcal{A}(s_1, \ldots, s_l)] = \epsilon$, then the reduction*

The reduction works as follows:

- Sample two l-tuples $(s_1, s_2, \ldots, s_l), (s'_1, s'_2, \ldots, s'_l) \xleftarrow{r} S^l$.
- Compute $(k, \pi) \leftarrow \mathcal{A}(s_1, \ldots, s_l)$.
- If $k = 0$ or $s_k = s'_k$, output $\perp$ and stop.
- Compute $(k', \pi') \leftarrow \mathcal{A}(s_1, \ldots, s_{k-1}, s'_k, s'_{k+1}, \ldots, s'_l)$.
- If $k' \neq k$, output $\perp$ and stop.
- Otherwise, output π and π'.

FIGURE 9.5 The Forking lemma reduction for a deterministic algorithm $\mathcal{A}$ taking a tuple from S^l to $k \in \{0, 1, 2, \ldots, l\}$ and π.

from Figure 9.5 does not output $\perp$ with probability at least $\epsilon^2/(4l) - 1/|S|$. If the runtime of $\mathcal{A}$ is bounded by τ, then the runtime of the reduction is bounded by τ', where τ' is essentially equal to 2τ plus the cost of sampling $2l$ elements from S.

Proof. The runtime claim is straight-forward. The main difficulty is that the two runs of $\mathcal{A}$ in the reduction are not independent, since the first $k-1$ input elements in both runs are identical. The solution is to count very carefully.

Let $T_k = \{(s_1, \ldots, s_l) \in S^l \mid \mathcal{A}(s_1, \ldots, s_l) = (k, \cdot)\}$, $k = 1, 2, \ldots, l$. Now partition each T_i such that two tuples are in the same subset if and only if the first $i-1$ elements are identical. Let $\{V_{ij}\}$ be the partition of T_i, which contains at most $|S|^{i-1}$ distinct equivalence classes. We get that $\sum_{i,j} |V_{ij}| = \epsilon |S|^l$.

We need to consider the relative size $|V_{ij}|/|S|^{l-i+1}$, which is essentially the probability that the second run of $\mathcal{A}$ will succeed. We begin by considering the equivalence classes of small relative size. Let $W = \{(i, j) \mid |V_{ij}|/|S|^{l-i+1} < \epsilon/(2l)\}$. Then

$$\sum_{(i,j) \in W} |V_{ij}| = \sum_i \sum_{j:(i,j) \in W} |V_{ij}| < \sum_i \sum_{j:(i,j) \in W} \frac{\epsilon}{2l} |S|^{l-i+1}$$

$$\leq \frac{\epsilon}{2l} \sum_i |S|^{l-i+1} |\{j \mid (i,j) \in W\}| \leq \frac{\epsilon}{2l} \sum_i |S|^{l-i+1} |S|^{i-1} = \epsilon |S|^l / 2.$$

In other words, the equivalence classes of small relative size make up at most half of all tuples for which $\mathcal{A}$ does not output $k = 0$.

It follows that with probability $\epsilon/2$, the first tuple $(s_1, \ldots, s_l)$ chosen in the reduction lies in an equivalence class with relative size at least $\epsilon/(2l)$. The second run will abort if the kth element of the second tuple is identical to the kth element of the first tuple, which happens with probability $1/|S|$. It follows that the reduction succeeds with probability at least

$$\frac{\epsilon}{2}\left(\frac{\epsilon}{2l}-\frac{1}{|S|}\right)$$

and the claim follows. $\qquad\square$

The security proof for the Schnorr scheme now follows the above outline.

Theorem 9.24. *Consider the Schnorr signature scheme* SIG *based on a cyclic group G of prime order n with generator g. Let $\mathcal{A}$ be a (τ, l_s)-adversary against the Schnorr signature scheme in the random oracle model, making at most l_h queries to the random oracle. Then there exists a τ'-solver $\mathcal{B}$ for discrete logarithms in G, with τ' essentially equal to 2τ, such that*

$$\mathbf{Adv}_{\mathrm{SIG}}^{\mathsf{suf\text{-}cma}}(\mathcal{A}) \leq 4\sqrt{l_h}\sqrt{\mathbf{Adv}_G^{\mathsf{DLog}}(\mathcal{B}) + 1/|\mathfrak{B}|} + \frac{l_h l_s}{n}.$$

Proof. We apply Proposition 9.17 to $\mathcal{A}$ to get a (τ_1, l_h, l_s)-adversary $\mathcal{A}_1$ against the Schnorr identification scheme ID, where τ_1 is essentially equal to τ and

$$\mathbf{Adv}_{\mathrm{SIG}}^{\mathsf{suf\text{-}cma}}(\mathcal{A}) \leq \mathbf{Adv}_{\mathrm{ID}}^{\mathsf{id\text{-}honest}}(\mathcal{A}_1) + l_h l_s/n,$$

where $l_h l_s/n$ is the collision probability for the identification scheme.

Then Propositions 9.20 and 9.22 give us a $(\tau_2, l_h, 0)$-adversary $\mathcal{A}_2$ against the Schnorr identification scheme, where τ_2 is essentially equal to τ_1, since the cost of the public conversation sampler will be small compared to τ_1, and

$$\mathbf{Adv}_{\mathrm{ID}}^{\mathsf{id\text{-}honest}}(\mathcal{A}_1) \leq \mathbf{Adv}_{\mathrm{ID}}^{\mathsf{id\text{-}honest}}(\mathcal{A}_2).$$

We can consider $\mathcal{A}_2$ interacting with the identification scheme experiment as a deterministic algorithm $\mathcal{B}_1$ mapping a verification key x output by the key generation algorithm, a random tape r and values $\beta_1, \beta_2, \ldots, \beta_{l_h} \in \mathfrak{B}$ to an integer i and a conversation (α, β, γ), such that either $i = 0$ or the conversation is accepting and the ith target query was for α and $\beta = \beta_i$. The runtime of $\mathcal{B}_1$ is bounded by τ_2.

Let $\mu(x, r)$ be the probability that the Schnorr key generation scheme outputs x and the random tape r is used. Let

$$f(x, r) = \Pr[i \neq 0 \mid (\beta_1, \ldots, \beta_{l_h}) \xleftarrow{r} \mathfrak{B}^{l_h}, (i, (\alpha, \beta, \gamma)) \leftarrow \mathcal{B}_1(x, r, \beta_1, \ldots, \beta_{l_h})].$$

Then $\epsilon = \mathbf{Adv}_{\mathrm{ID}}^{\mathsf{id\text{-}honest}}(\mathcal{A}_2) = \sum_{x,r} f(x, r)\mu(x, r)$.

By a standard argument, there exists a set S of pairs (x, r) such that

$$\sum_{(x,r)\in S} f(x, r)\mu(x, r) \geq \epsilon/2 \qquad \text{and} \qquad (x, r) \in S \Rightarrow f(x, r) \geq \epsilon/2.$$

Fix a pair $(x, r) \in S$. Applying the Forking lemma to $\mathcal{B}_1(x, r, \dots)$, we get an algorithm $\mathcal{B}_2(x, r)$ that outputs a forked pair with probability at least $f(x, r)^2/(4l_h) - 1/|\mathfrak{B}|$. The runtime of $\mathcal{B}_2(x, r)$ is essentially equal to $2\tau_2$.

We get an algorithm $\mathcal{B}_3$ that on input x chooses a random tape r and runs $\mathcal{B}_2(x, r)$. The algorithm $\mathcal{B}_3$ outputs a forked pair with probability at least

$$\sum_{x,r} \left(\frac{f(x,r)^2}{4l_h} - \frac{1}{|\mathfrak{B}|} \right) \mu(x,r) \geq -\frac{1}{|\mathfrak{B}|} + \frac{\epsilon}{8l_h} \sum_{(x,r) \in S} f(x,r)\mu(x,r) \geq -\frac{1}{|\mathfrak{B}|} + \frac{\epsilon^2}{16l_h}.$$

The runtime of $\mathcal{B}_3$ is essentially equal to $2\tau_2$.

The key generation algorithm essentially samples x from the uniform distribution on G, which is the instance distribution for the discrete logarithm problem. Therefore, the τ'-solver $\mathcal{B}$ we get by applying Exercise 9.19 to $\mathcal{B}_3$, where τ' is essentially equal to $2\tau_2$, gives the advantage bound

$$\mathbf{Adv}_{\mathrm{ID}}^{\mathrm{id\text{-}honest}}(\mathcal{A}_2) \leq 4\sqrt{l_h}\sqrt{\mathbf{Adv}_G^{\mathrm{DLog}}(\mathcal{B}) + 1/|\mathfrak{B}|}. \qquad \square$$

Remark. It is often useful to "test" general conjectures using extremal values. For any identification scheme with a public conversation sampler, there is a straight-forward attack that works as follows: Run the public conversation sampler l_h times. For each conversation (α, β, γ), make a target query α. If the response is β, we have our forgery.

The probability that any single target query succeeds is $1/|\mathfrak{B}|$, so the probability that at least one query succeeds is bounded by $l_h/|\mathfrak{B}|$ and quickly approaches 1 as l_h approaches and exceeds $|\mathfrak{B}|$. If $\mathfrak{B}$ is small, this is a significant advantage for a realistic adversary.

The Forking lemma reduction would not work on this adversary, since for each target query there is a single challenge that will work, which means that rewinding cannot be successful, even when $\mathfrak{B}$ is small.

But for this particular adversary, Theorem 9.23 says that the reduction has an advantage that is at least a negative number. So all is well in the world.

Remark. The analysis of the runtime is actually not complete. The reason is that if the adversary actually makes fewer than l_h target queries, sampling l_h times may actually constitute a non-trivial increase in runtime, which would roughly result in a bound of 3τ instead of 2τ. However, a more careful reduction could avoid the extra sampling, so the requirement is not essential. We do not present this more careful reduction, since the presentation of the proof would be significantly more complicated.

Remark. Since forked conversation pairs reveal the secret key, Schnorr signatures (and related schemes such as DSA) are vulnerable to faulty random number generation. One strategy to make schemes more robust against such faults is to *derandomise* the scheme. The idea is to use a pseudo-random function to derive the randomness for the scheme from the message to be signed.

Exercise 9.20. Prove that derandomisation is safe, that is, any adversary against the derandomised scheme leads to an adversary against the pseudo-random function or the underlying signature scheme.

What does it all mean? We continue the discussion from Section 9.2.4.

Comparing the Katz-Wang signature scheme and the Schnorr signature scheme, we see that Katz-Wang requires about twice the number of exponentiations required for Schnorr. If instantiated with the same group, Schnorr seems to be much faster. However, it is instructive to compare the tightness of the bounds Theorem 9.21 and Theorem 9.24.

As a numerical example, we may want an adversary using time 2^{80} and making (say) 2^{70} hash queries and 2^{40} signature queries to have an advantage against the signature scheme of 2^{-20} (which is fairly big, as things go). Since the group order will be at least 2^{160} we can safely ignore the terms in the advantage bound with n in the denominator. We shall also assume, at least as a first approximation, that for both schemes we will choose $\mathfrak{B}$ such that any term with $|\mathfrak{B}|$ in the denominator can be ignored.

For Schnorr signatures, we must choose a group such that discrete logarithm solver using time 2^{81} has advantage at most 2^{-114}. For Katz-Wang signatures, we must choose a group such any DDH distinguisher using time 2^{80} has advantage at most 2^{-20}.

For some families of groups coming from elliptic curves it seems reasonable that the advantage ϵ of the best known discrete logarithm solvers using time at most τ roughly satisfies $\tau^2/\epsilon \approx n$. For these families, the best known DDH distinguisher is to compute a discrete logarithm. This suggests that for Schnorr signatures the group must be of size roughly 2^{276}, while for Katz-Wang a group of size 2^{180} would be sufficient. Since exponentiation algorithms are somewhere between quadratic and cubic in the length of the group order, we would expect the cost of the exponentiations for the Schnorr signatures to be between $(276/180)^2 \approx 2.4$ and $(276/180)^3 \approx 3.6$ times more expensive than the exponentiations for the Katz-Wang signatures.

In other words, because of the loose bounds in Theorem 9.24, Katz-Wang may actually be the more efficient scheme. We should immediately contradict that suggestion: The best known attack on either scheme is computing discrete logarithms in the group. Also, our analysis assumed that the best attack on DDH was computing discrete logarithms, which is not obvious. Theorems 9.21 and 9.24 merely provide upper bounds on the difference.

9.3.4 Witness Indistinguishability

Our approach so far has been to simulate the random conversation queries, and then either use indistinguishable "impossible" verification keys or extract the secret key using the Forking Lemma. But sometimes it is hard to simulate random conversations without knowing a secret key.

One approach is schemes with many possible secret keys for each verification key, but where it should be hard to find more than one secret key for a verification key. For these schemes, we could use the Forking Lemma to extract a secret key, and then hope that the extracted secret key is different from the secret key we already know. And the property we need to make this likely is *witness indistinguishability* from Definition 9.7.

The idea is that if the random conversations are independent of the secret key (that is, we need a secret key to generate the conversation efficiently, but its distribution depends only on the verification key), there will be no correlation between the extracted secret key and the already known secret key. And as long as there are at least two possible secret keys, we will have some probability of getting two distinct secret keys.

We begin with a simple scheme based on discrete logarithms.

Example 9.14. The *Okamoto* identification scheme $\text{ID} = (\mathcal{IK}, \mathcal{II}, \mathcal{IR}, \mathcal{IV})$ based on a cyclic group G of order n with generators g_1, g_2 works as follows:

- The *key generation* algorithm $\mathcal{IK}$ samples secret keys $a_1, a_2 \xleftarrow{r} \{0, 1, \ldots, n-1\}$ and computes $x \leftarrow g_1^{a_1} g_2^{a_2}$. The verification key is $ivk = x$ and the secret key is $isk = (a_1, a_2)$.

- The *commitment* algorithm $\mathcal{II}$ samples $r_1, r_2 \xleftarrow{r} \{0, 1, \ldots, n-1\}$, computes $\alpha \leftarrow g_1^{r_1} g_2^{r_2}$ and outputs the commitment α and the state (r_1, r_2).

- The *response* algorithm γ takes as input a secret key $isk = (a_1, a_2)$, a state (r_1, r_2) and a challenge β. It computes $\gamma_1 \leftarrow r_1 - \beta a_1 \bmod n$ and $\gamma_2 \leftarrow r_2 - \beta a_2 \bmod n$ and outputs (γ_1, γ_2).

- The *verification* algorithm $\mathcal{IV}$ takes as input a verification key $ivk = x$, a commitment α, a challenge β and a response (γ_1, γ_2), and outputs 1 if and only if $\alpha = g_1^{\gamma_1} g_2^{\gamma_2} x^{\beta}$.

It is easy to verify that the scheme is complete, and also that for every verification key there are many corresponding secret keys. In the following, we shall assume that n is prime.

Exercise 9.21. Show that given a forked conversation pair for the Okamoto identification scheme with verification key x, a secret key a_1, a_2 for x can be easily computed.

Exercise 9.22. Show that this scheme is perfectly witness-indistinguishable.

The Okamoto signature scheme is derived from the identification scheme in the usual manner of Example 9.9.

Exercise 9.23. Prove a theorem similar to Theorem 9.24 for the Okamoto signature scheme.

We now proceed with a somewhat more complicated example based on lattices. The main issue we have to deal with is making the responses in the scheme independent of the secret. The answer is, somewhat surprisingly, to make the identification scheme incomplete and allow the response algorithm to output $\perp$. As long as the response algorithm does not output $\perp$ too often, this causes no significant problems.

Example 9.15. Let R_p be a ring, with $\mathfrak{B} \subseteq R_p$ and $S_r, S_v, S_k, S_0 \subseteq R_p^l$ such that:

L-1. For any $\mathbf{a} \in S_k$, $\gamma \in S_v$ and $\beta \in \mathfrak{B}$, we have that $\gamma - \beta \mathbf{a} \in S_r$.

L-2. For any $\mathbf{r}, \mathbf{r}' \in S_r$, we have that $\mathbf{r} - \mathbf{r}' \in S_0$.

L-3. For any $\mathbf{a}, \mathbf{a}' \in S_k$, $\beta, \beta' \in \mathfrak{B}$, if $(\beta - \beta')(\mathbf{a} - \mathbf{a}') = 0$, then either $\beta = \beta'$ or $\mathbf{a} = \mathbf{a}'$.

L-4. If $\mathbf{g} \xleftarrow{r} R_p^l$, $\mathbf{a} \xleftarrow{r} S_k$ and $\mathbf{y} = \mathbf{g} \cdot \mathbf{a}$, then the equation $\mathbf{g} \cdot \mathbf{X} = \mathbf{y}$ has a unique solution with probability at most δ_0.

L-5. For any $\mathbf{a} \in S_k$, then if $\mathbf{r} \xleftarrow{r} S_r$ and $\beta \xleftarrow{r} \mathfrak{B}$, then $\mathbf{r} + \beta \mathbf{a} \in S_v$ with probability δ_1.

The *Lyubashevsky* identification scheme VL-ID works as follows:

- The key generation algorithm $\mathcal{IK}$ samples $\mathbf{g} \xleftarrow{r} R_p^l$ and $\mathbf{a} \xleftarrow{r} S_k$, and computes $\mathbf{y} \leftarrow \mathbf{g} \cdot \mathbf{a}$. The verification key is $ivk = (\mathbf{g}, \mathbf{y})$, the secret key is $isk = (\mathbf{g}, \mathbf{a})$.

- The commitment algorithm $\mathcal{II}$ takes as input a secret key $isk = (\mathbf{g}, \mathbf{a})$, samples $\mathbf{r} \xleftarrow{r} S_r$, computes $\alpha \leftarrow \mathbf{g} \cdot \mathbf{r}$ and outputs the commitment α and a state $\mathbf{r}$.

- The response algorithm $\mathcal{IR}$ takes as input a secret key $isk = (\mathbf{g}, \mathbf{a})$, a state $\mathbf{r}$ and a challenge $\beta \in \mathfrak{B}$. It computes $\gamma \leftarrow \mathbf{r} + \mathbf{a}\beta$. It outputs γ if $\gamma \in S_v$, otherwise it outputs $\perp$.

- The verification algorithm $\mathcal{IV}$ takes as input a verification key $ivk = (\mathbf{g}, \mathbf{y})$, a commitment α, a challenge β and a response γ. It outputs 1 if $\gamma \in S_v$ and $\mathbf{g} \cdot \gamma = \mathbf{y}\beta + \alpha$, otherwise it outputs 0.

The two probabilities δ_0 and δ_1 can to a certain extent be tuned. The first probability must be very small, since it measures the probability that our reduction will not work. The second probability must be fairly large, since it measures the probability that the prover will not reject the conversation.

E *Exercise 9.24.* Show that the honest conversation generator for VL-ID outputs $\perp$ with probability $1 - \delta_1$. Compute the expected number of attempts before a conversation is output, in terms of δ_1.

The scheme is witness indistinguishable. This result would not hold if the scheme did not sometimes output $\perp$, since then there might not be plausible randomness for every possible secret key. Note also that the below proof crucially relies on the fact that the adversary will not see how many attempts are required to generate a conversation, which may in principle depend on the secret key. (We certainly do not prove that it does not).

T **Proposition 9.25.** *The Lyubashevsky scheme from Example 9.15 is perfectly witness-indistinguishable.*

Proof. Since the scheme samples the randomness $\mathbf{r}$ from the uniform distribution on S_r and the randomness implied by either secret is in S_r by L-1, any accepting conversation is equally likely under any secret key. □

Remark. Witness indistinguishability does not require that there is more than one secret key. If there is a unique secret key for a given public key, independence of that particular secret key holds vacuously.

E *Exercise 9.25.* Suppose $\mathbf{g} \xleftarrow{r} R^l$ and $\mathbf{a} \xleftarrow{r} S_k$, and let $\mathbf{y} = \mathbf{g} \cdot \mathbf{a}$. Given a forked conversation pair (α, β, γ) and $(\alpha, \beta', \gamma')$, show that except with small probability $\gamma' - \gamma + (\beta - \beta')\mathbf{a}$ is not equal to zero and lies in S_0.

T **Proposition 9.26.** *Consider a (τ, l_h, l_s)-adversary $\mathcal{A}$ against the Lyubashevsky identification scheme from Example 9.15. Then there exists a τ'-solver $\mathcal{B}$, with τ' essentially equal to 2τ, such that*

$$\mathbf{Adv}_{\text{VL-ID}}^{\text{id-honest}}(\mathcal{A}) \leq 4\sqrt{l_h}\sqrt{\mathbf{Adv}_{R_p, l, S_0}^{\text{RSIS}}(\mathcal{B}) + 1/|\mathfrak{B}|}.$$

Proof. We can consider $\mathcal{A}$ interacting with the identification scheme experiment as a deterministic algorithm $\mathcal{B}_1$ mapping $\mathbf{g} \in R_p^l$, a random tape r and values $\beta_1, \beta_2, \ldots, \beta_{l_h} \in \mathfrak{B}$ to an integer i and a conversation (α, β, γ), such that either $i = 0$ or the conversation is accepting and the ith target query was for α and $\beta = \beta_i$. The runtime of $\mathcal{B}_1$ is bounded by τ.

Let $\mu(\mathbf{g}, r)$ be the probability that $\mathbf{g}$ is sampled and the random tape r is used. Let $f(\mathbf{g}, r)$ be the probability that $\mathcal{B}_1$ outputs $i \neq 0$ when $(\beta_1, \ldots, \beta_{l_h})$ has been sampled from the uniform distribution on $\mathfrak{B}^{l_h}$. Then

$$\epsilon = \mathbf{Adv}_{\text{VL-ID}}^{\text{id-honest}}(\mathcal{A}) = \sum_{\mathbf{g}, r} f(\mathbf{g}, r)\mu(\mathbf{g}, r).$$

By a standard argument, there exists a set S of pairs $(\mathbf{g}, r)$ such that

$$\sum_{(\mathbf{g}, r) \in S} f(\mathbf{g}, r)\mu(\mathbf{g}, r) \geq \epsilon/2 \quad \text{and} \quad (\mathbf{g}, r) \in S \Rightarrow f(\mathbf{g}, r) \geq \epsilon/2.$$

Fix a pair $(\mathbf{g}, r) \in S$. Applying the Forking lemma to $\mathcal{B}_1(\mathbf{g}, r, \dots)$, we get an algorithm $\mathcal{B}_2(\mathbf{g}, r)$ that outputs a forked pair with probability at least $f(\mathbf{g}, r)^2/(4l_h) - 1/|\mathfrak{B}|$. The runtime of $\mathcal{B}_2(\mathbf{g}, r)$ is essentially equal to 2τ.

We then get an algorithm $\mathcal{B}$ that on input of $\mathbf{g}$ chooses a random tape r and runs $\mathcal{B}_2(\mathbf{g}, r)$. If $\mathcal{B}_2$ outputs a forked pair (α, β, γ) and $(\alpha, \beta', \gamma')$, then $\mathcal{B}$ outputs $\gamma' - \gamma + (\beta - \beta')\mathbf{a}$. Otherwise, it outputs $\bot$.

When the input $\mathbf{g}$ has been sampled from the uniform distribution on R_p^l, we have that $\mathcal{B}$ outputs a forked pair with probability

$$\sum_{\mathbf{g}, r}(f(\mathbf{g}, r)^2/(4l_h) - 1/|\mathfrak{B}|)\mu(\mathbf{g}, r) \geq -\frac{1}{|\mathfrak{B}|} + \frac{\epsilon}{8l_h}\sum_{(\mathbf{g}, r) \in S} f(\mathbf{g}, r)\mu(\mathbf{g}, r)$$

$$\geq -\frac{1}{|\mathfrak{B}|} + \frac{\epsilon^2}{16l_h}.$$

By Exercise 9.25, if $\mathcal{B}$ does not output $\bot$, it outputs an element in S_0 whenever $\mathcal{B}_2$ outputs a forked pair.

By L-4, for any execution there is more than one secret key for the given public key except with probability δ_0. If there are two distinct secret keys, then for any accepting conversation, L-3 says that the randomness implied by each secret key is distinct. It follows that if $\mathcal{B}$ does not output $\bot$, it does not output 0 except with probability at most ϵ_0.

Since the key generation algorithm samples $\mathbf{g}$ from the uniform distribution on R_p^l, just as in the R-SIS problem, the claim follows. $\qquad\square$

We get the Lyubashevsky signature scheme VL-SIG by applying the construction from Example 9.9 to the identification scheme from Example 9.15.

E *Exercise* 9.26. Consider a (τ, l_h, l_s)-adversary $\mathcal{A}$ against the Lyubashevsky signature scheme VL-SIG. Show that there exists a τ'-solver $\mathcal{B}$, with τ' essentially equal to 2τ, such that

$$\mathbf{Adv}_{\text{VL-SIG}}^{\text{suf-cma}}(\mathcal{A}) \leq 4\sqrt{l_h}\sqrt{\mathbf{Adv}_{R_p, l, S_0}^{\text{RSIS}}(\mathcal{B}) + 1/|\mathfrak{B}|}.$$

Remark. As for the Schnorr signature scheme, the Lyubashevsky signature scheme has fairly loose bounds, which means that if we are to use Exercise 9.26 to decide the key sizes, we must choose fairly large key sizes. As usual, one may instead look at the best known attack on the scheme which will suggest that smaller key sizes could be acceptable.

Remark. A more serious problem for this type of argument is the fact that we are running the adversary twice. Classically, this is not a problem, but if $\mathcal{A}$ is a quantum algorithm, one run may have carefully constructed certain superpositions and looked at them in the first run. It is then impossible to restore the superpositions for the second run. Even worse, an algorithm may submit superpositions as target queries.

Submitting superpositions as target queries sounds unrealistic, and for identification protocols it probably is. But recall that our main purpose is the design of signature schemes, and in this setting the target queries actually represent hash computations (which we have first idealised as random oracle queries and then replaced with target queries). The adversary is obviously able to do hash computations on superpositions, and in order to model this we have to switch to the *quantum random oracle model*.

There are a number of alternatives. First of all, it turns out that the Fiat-Shamir approach is in many cases sound in the quantum random oracle model. Alternatively, we may switch to a completely different approach that is sound in the quantum random oracle model, some of which are costly.

A third alternative is to simply ignore these issues, and in practice this seems to be an attractive approach. The motivation is the existing situation where the random oracle model is known not to be sound, in the sense that schemes exist that can be proven secure in the random oracle model and insecure in the real world (standard model), regardless of how the random oracle is instantiated. However, these separations are very carefully constructed and in some sense unnatural, with one exception (simulatable encryption) that this author knows about. This suggests that the random oracle model is a very solid heuristic for security against classical real world adversaries.

Given the known separations between the quantum random oracle model and the random oracle model, and the range of partial results for the Fiat-Shamir heuristic in the quantum random oracle model, it seems likely that the random oracle model is a very solid heuristic, even for security against real-world quantum adversaries.

9.4 MESSAGING

We discussed how to use symmetric cryptography with a shared secret to realise a secure channel in Section 7.4. Now we shall show how to use asymmetric cryptography to realise secure messaging for users that do not share any secrets. We do require that they have the public keys of their communication partners.

The idea is fairly straight-forward. The sender encrypts each message and signs the ciphertext. This immediately gives us assurance of who sent the message, sc-1. To get assurance of message order, sc-3, we can include a counter. We can either include this as associated data when encrypting the message or sign the counter along with the data.

In general, public key encryption and signatures are slower than pure symmetric encryption, so the above protocol is actually less efficient than a channel protocol using just symmetric cryptography. The main advantage would be its flexibility. Our analysis so far only covers a single channel, but using public key encryption and signatures would allow any number of communication partners, without the need for agreeing on new symmetric keys.

But when there are more than two parties, the simple layering of encryption and signatures fails. One possible attack is that Eve can make Bob believe a message from Alice actually came from Eve, by stripping Alice' signature and adding her own signature instead. This will break confidentiality under a chosen ciphertext attack. This can be fixed by including the sender name in the associated data in the encryption.

If Alice and Bob have multiple channels active at the same time, Eve can mix up the messages, breaking integrity. This can be fixed by somehow distinguishing channels. The solution is to include some channel identity in the signatures and encryptions. One possibility is to let this be some checksum of transmitted ciphertexts and signatures, or Alice and Bob could run a simple protocol for deciding on a random channel identifier. (Alice sends a nonce to Bob, Bob responds with his own nonce and a signature on both nonces and Alice' name. Finally, Alice responds with her signature on both nonces and Bob's name. There are a few more details to work out.)

We shall return to these topics in Chapter 13.

CHAPTER 10

Key Exchange

Key exchange protocols are crucial tools for any system designer, either directly or because they are required by other tools used by the system designer.

We first saw key exchange protocols in Chapter 2, but we need a more careful definition of what a key exchange protocol is, as well as associated words. We also give some examples of important key exchange protocols.

Our next task is to define the meaning of security for key exchange protocols, in some generality. The definition is our most complex yet. There are a number of obstacles, in particular that key exchange security is intimately tied to which instances agreed on keys with each other, and what guarantees we should have about the existence and beliefs of partner instances.

Through a series of simplifying results, we describe two design strategies for key exchange schemes: The classical design is a core unauthenticated key exchange scheme augmented with digital signatures to ensure authentication. The second strategy proceeds via implicitly authentication.

We should emphasise that none of our design strategies achieve our strongest security notions. In fact, none of our examples achieve these security notions. We could achieve this by combining digital signatures as in Section 10.2.5 with the techniques from Section 10.8, but we would learn nothing new, and the construction itself would be of limited practical interest.

Our main examples are based on the Diffie-Hellman key exchange, but we prove several results on relations between key exchange and key encapsulation mechanisms, allowing us to apply our strategies to other primitives, for instance lattice-based schemes.

We also discuss some important key exchange variants, single-message key exchange, single-sided authentication and role-symmetric key exchange.

10.1   KEY EXCHANGE PROTOCOLS

A key exchange protocol is a cryptographic protocol that allows Alice and Bob to arrive at a shared secret key, usually called a *session key*. One of them will initiate the key exchange, and the other will respond. The one who initiates plays the *initiator role*, while the other party plays the *responder role*.

Definition 10.1. A *key exchange* protocol KEX is a tuple $(\mathfrak{K}, \mathcal{K}, \mathcal{I}, \mathcal{R})$, where $\mathfrak{K}$ is a set of keys and the other three are algorithms:

- The *key generation* algorithm $\mathcal{K}$ takes no input and outputs a *public key* pk and a *private key sk*.

- The interactive *initiator* algorithm $\mathcal{I}$ takes as input associated data ad, a key pair (pk_I, sk_I) and a public key pk_R. It alternates between sending and receiving messages. Eventually, it either outputs a session key k or the special symbol $\bot$ signifying failure.

- The interactive *responder* algorithm $\mathcal{R}$ takes as input associated data ad, a public key pk_I and a key pair (pk_R, sk_R). It alternates between receiving and sending messages. Eventually, it either outputs a session key k or the special symbol $\bot$ signifying failure.

Consider an *instance* of $\mathcal{I}(ad, pk_I, sk_I, pk_R)$ that sends and receives a sequence of messages $(m_1, m_2, \ldots, m_l)$. The *transcript* tr of the instance is $(ad, pk_I, pk_R, m_1, m_2, \ldots, m_l)$. The transcript of an instance of $\mathcal{R}(ad, pk_I, pk_R, sk_R)$ is defined in the same way.

We require that if (pk_I, sk_I) and (pk_R, sk_R) have been output by $\mathcal{K}$, $pk_I \neq pk_R$ and ad is some associated data, then for any two instances $\mathcal{I}(ad, pk_I, sk_I, pk_R)$ and $\mathcal{R}(ad, pk_I, pk_R, sk_R)$ that both output keys and have identical transcripts, they will output identical keys.

A key exchange protocol is *unauthenticated* if the key exchange protocol always outputs $(\bot, \bot)$, and the initiator and responder algorithms reject any keys other than $\bot$. In this case, the above requirement must also hold when $pk_I = pk_R$, since both are equal to $\bot$.

The associated data is intended to encode information that the two parties should agree on and which is arrived at outside of the protocol, such as identities, certificates or what the key exchange is used for. The latter can be used to ensure cryptographic separation between two distinct systems using the same key infrastructure.

Except for unauthenticated key exchange protocols, the definition does not require correctness when the initiator and the responder public keys are identical (self-exchange). Indeed, some reasonable key exchange protocols explicitly disallow this. Other reasonable key exchange protocols allow it, and there are cases where this is useful, so we do not forbid it, and the remaining definitions will allow either case.

Remark. It is important that a protocol explicitly forbids identical public keys if the protocol will be insecure or incorrect when run with identical public keys.

Key exchange is usually described with only two parties involved, but in practice Alice and Bob will only be two out of many parties. Also, there will be many key exchanges happening at the same time, possibly involving the same users. This means that in the real world, there will be a large collection of protocol instances running in parallel, with few or no restrictions on which role is run how many times in parallel by how many users. The different instances of a given role only differ in associated data and public and private keys.

Remark. Having more than two parties and allowing multiple key exchanges in parallel makes things more complicated, but this is unavoidable. We can design key exchange protocols that are secure with only two users or with no concurrent key exchange, but are insecure in a more realistic model.

Even though there are many parties involved, our key exchange protocol definition does not actually include any prescribed method of identifying the parties. This is deliberate. Key exchange is difficult enough without also including the public key infrastructure problem. Instead, the key exchange protocol only cares about two things: they keys involved and the *associated data* involved. Typically, the keys involved are tied to named parties somehow, and the associated data allow users to ensure agreement on these ties.

However, this simplification does not actually match how key exchange protocols are used in practice. Very often public keys are exchanged during the key exchange protocol, together with information tying the keys to some identity which will make up the associated data. This means that often a key exchange role will need to start without knowledge of the other party's public key and the associated data. For most schemes we look at, this changes nothing since the public keys and associated data is only used at the end of the protocol, but some schemes actually need to know the associated data before sending their first message.

The public and private keys are often called *long-term keys* or *static keys*. For many key exchange protocols there is an identifiable *ephemeral key* or *short-term key*, but we shall not use this term in general.

Example 10.1. The Diffie-Hellman protocol from Chapter 2, based on a group G of prime order p with generator g, works as follows:

- The *key generation* algorithm $\mathcal{K}$ outputs the empty key pair $(\perp, \perp)$.
- The *initiator* algorithm ignores its input, samples $a \xleftarrow{r} \{0, 1, \ldots, p-1\}$, computes $x \leftarrow g^a$ and sends x. When it later receives the message y, it computes $k \leftarrow y^a$ and outputs k.
- The *responder* algorithm ignores its input and samples $b \xleftarrow{r} \{0, 1, \ldots, p-1\}$. When it receives x, it computes $y \leftarrow g^b$ and $k \leftarrow x^b$. Then it sends y and outputs k.

The celebrated Diffie-Hellman protocol is the first asymmetric key exchange protocol. It has a number of useful properties, but it can only be passively secure, since there is no public key. It is, however, the basis for a wide range of key exchange protocols and other cryptographic constructions.

Example 10.2. We consider the signed Diffie-Hellman protocol from Section 4.6, based on a group G of prime order p with generator g and a signature scheme $\text{SIG} = (\mathcal{K}_0, \mathcal{S}, \mathcal{V})$. We shall assume that the set of tuples of associated data, two signature verification keys, two group elements and zero or one signatures is a subset of the signing plaintext set.

The key exchange scheme works as follows:

- The *key generation* algorithm $\mathcal{K}$ is $\mathcal{K}_0$.

- The *initiator* algorithm takes as input associated data ad, public keys vk, vk' and a private key sk. It samples $a \xleftarrow{r} \{0, 1, \ldots, p-1\}$, computes $x \leftarrow g^a$ and sends x.

  When it receives the message (y, σ'), it verifies that $\mathcal{V}(vk', tr, \sigma') = 1$, with $tr = (ad, vk, vk', x, y)$. If not, it outputs $\perp$ and stops. Otherwise it computes $\sigma \leftarrow \mathcal{S}(sk, (tr, \sigma'))$ and $k \leftarrow y^a$, sends σ and outputs k.

- The *responder* algorithm takes as input associated data ad, public keys vk, vk' and a private key sk'. It receives x, samples $b \xleftarrow{r} \{0, 1, \ldots, p-1\}$, computes $y \leftarrow g^b$ and $\sigma' \leftarrow \mathcal{S}(sk', tr)$, and sends (y, σ').

  When it receives σ, it verifies that $\mathcal{V}(vk, (tr, \sigma'), \sigma) = 1$. If not, it outputs $\perp$ and stops. Otherwise, it computes $k \leftarrow x^b$ and outputs k.

Example 10.3. We consider a Diffie-Hellman-based protocol, based on a group G of prime order p with generator g, and a key derivation function kdf with range $\mathfrak{K}$. It works as follows:

- The *key generation* algorithm $\mathcal{K}$ samples $b \xleftarrow{r} \{0, 1, \ldots, p-1\}$, computes $y \leftarrow g^b$ and outputs the public key y and private key b.

- The *initiator* algorithm takes as input associated data ad, public keys y_i, y_r and a private key b_i. It samples $r_i \xleftarrow{r} \{0, 1, \ldots, p-1\}$, computes $x_i \leftarrow g^{r_i}$ and sends x_i.

  When it receives the message x_r, it computes $k \leftarrow kdf(tr, y_r^{r_i}, x_r^{b_i})$, with $tr = (ad, y_i, y_r, x_i, x_r)$ and outputs k.

- The *responder* algorithm takes as input associated data ad, public keys y_i, y_r and a private key b_r. It receives x_i, samples $r_r \xleftarrow{r} \{0, 1, \ldots, p-1\}$, computes $x_r \leftarrow g^{x_r}$ and $k \leftarrow kdf(tr, x_i^{b_r}, y_i^{r_r})$, sends x_r and outputs k.

This is a variant of the *Key Exchange Algorithm* (KEA) with some now-standard improvements.

Remark. Naturally, the *key derivation rule*, what the *session key* is derived from, in the above protocol is extremely important. The actual example above is just one possible choice among a large family of protocols.

We divide the hashed terms into two groups. The transcript tr ensures that if the parties agree on the key, they also agree on these terms, which ensures matching conversations. Which terms to include here varies a lot among protocols. The remaining terms ensure session key confidentiality and authentication, since it is hard to compute both without either of the long-term keys.

If we add $x_i^{r_r} = x_r^{r_i}$ to the second group of hashed terms to get

$$k \leftarrow kdf(tr, y_r^{r_i}, x_r^{b_i}, x_r^{r_i}) = kdf(tr, y_i^{r_r}, x_i^{b_r}, x_i^{r_r}), \tag{10.1}$$

the key is hard to compute even if the long-term keys are later compromised.

If we add $y_i^{b_r} = y_r^{b_i}$ to the second group of hashed terms to get

$$k \leftarrow kdf(tr, y_r^{b_i}, y_r^{r_i}, x_r^{b_i}, x_r^{r_i}) = kdf(tr, y_i^{b_r}, y_i^{r_r}, x_i^{b_r}, x_i^{r_r}), \tag{10.2}$$

the key is hard to compute even if the states are later compromised.

Example 10.4. The *(Hashed) MQV* (Menezes-Qu-Vanstone) protocol can be considered a variant of the above protocol. It boils down to a clever way to compute the second group of inputs. The hashed variant is called *HMQV*.

The first idea is to compute a single group element as a linear combination of the four elements included in (10.2). That is,

$$k \leftarrow kdf(tr, g^{b_i b_r}(g^{b_i r_r})^{\alpha_1}(g^{r_i b_r})^{\alpha_2}(g^{r_i r_i})^{\alpha_3}).$$

A predictable choice for $\alpha_1, \alpha_2, \alpha_3$ will not work, but an unpredictable choice should work. We can make such a choice by hashing the transcript as $\alpha_i = h_i(tr)$, $i = 1, 2, 3$. Even very simple functions h_i, as in MQV, may be safe.

So far, the protocol has just been made more complicated and more expensive, while nothing of substance has been gained. However, the second idea is the clever part. The idea is that we can set $\alpha_3 = \alpha_1 \alpha_2$. This does not change the unpredictability of the linear combination, but the observation that

$$b_i b_r + b_i r_r \alpha_1 + r_i b_r \alpha_2 + r_i r_i \alpha_1 \alpha_2 = (b_i + \alpha_1 r_i)(b_r + \alpha_2 r_r)$$

allows us to compute the linear combination as

$$g^{b_i b_r}(g^{b_i r_r})^{\alpha_1}(g^{r_i b_r})^{\alpha_2}(g^{r_i r_i})^{\alpha_3} = (g^{b_i}(g^{r_i})^{\alpha_1})^{b_r + \alpha_2 r_r} \qquad = (g^{b_r}(g^{r}{}_r)^{\alpha_2})^{b_i + \alpha_1 r_i}$$
$$= (y_i x_i^{\alpha_1})^{b_r + \alpha_2 r_r} \qquad\qquad = (y_r x_r^{\alpha_2})^{b_i + \alpha_1 r_i}.$$

This costs just two exponentiations to compute, while computing the terms separately costs four exponentiations. The (Hashed) MQV protocol therefore achieves strong security properties in an efficient way. If clever exponentiation algorithms are used, (Hashed) MQV achieves even better performance.

The previous constructions were all based on a cyclic group. Theoretically, it is interesting to try to understand how key exchange relates to other cryptographic primitives, such as digital signatures and public key encryption. Practically, such relations give us alternative design strategies, which is useful.

Example 10.5. Let KEM $= (\mathcal{KK}, \mathcal{KE}, \mathcal{KD})$ be a key encapsulation mechanism for $\mathfrak{K}_s$. We get the following unauthenticated key exchange protocol:

- The *initiator* algorithm ignores its input, computes $(ek, dk) \leftarrow \mathcal{KK}$ and sends ek. It then receives x, computes $k \leftarrow \mathcal{KD}(dk, x)$ and outputs k.

- The *responder* algorithm ignores its input. It receives ek, computes $(x, k) \leftarrow \mathcal{KE}(ek)$, sends x and outputs k.

Compare this protocol to the plain Diffie-Hellman protocol from Example 10.1.

Example 10.6. We consider a generalisation of the signed Diffie-Hellman protocol from Example 10.2, based on a key encapsulation mechanism KEM $= (\mathcal{KK}, \mathcal{KE}, \mathcal{KD})$ and a signature scheme SIG $= (\mathcal{K}_0, \mathcal{S}, \mathcal{V})$. We shall assume that the set of tuples of associated data, two signature verification keys, a KEM encapsulation key, a KEM ciphertext and zero or one signatures is a subset of the signing plaintext set.

The key exchange scheme works as follows:

- The key generation algorithm $\mathcal{K}$ is $\mathcal{K}_0$.

- The initiator algorithm takes as input associated data ad, public keys vk, vk' and a private key sk. It computes $(ek, dk) \leftarrow \mathcal{KK}$ and sends ek.

  When it receives the message (x, σ'), it verifies that $\mathcal{V}(vk', tr, \sigma') = 1$, with $tr = (ad, vk, vk', ek, x)$, and computes $\sigma \leftarrow \mathcal{S}(sk, (tr, \sigma'))$ and $k \leftarrow \mathcal{KD}(dk, x)$. If verification or decapsulation fails, it outputs $\perp$ and stops. Otherwise it sends σ and outputs k.

- The responder algorithm takes as input associated data ad, public keys vk, vk' and a private key sk'. It receives ek, computes $(x, k) \leftarrow \mathcal{KE}(ek)$ and $\sigma' \leftarrow \mathcal{S}(sk', tr)$, and sends (x, σ').

  When it receives σ, it verifies that $\mathcal{V}(vk, (tr, \sigma'), \sigma) = 1$. If not, it outputs $\perp$ and stops. Otherwise, it outputs k.

Example 10.7. Let KEM $= (\mathcal{KK}, \mathcal{KE}, \mathcal{KD})$ be a key encapsulation mechanism for $\mathfrak{K}_s$, and let kdf be a key derivation function. Consider the implicitly authenticated key exchange protocol that works as follows:

- The key generation algorithm computes $(ek, dk) \leftarrow \mathcal{KK}$, and outputs the public key ek and the private key dk.

- The initiator algorithm takes as input associated data ad, public keys ek, ek' and a private key dk. It computes $(ek'', dk'') \leftarrow \mathcal{KK}$ and $(x', k_0') \leftarrow \mathcal{KE}(ek')$, and sends (ek'', x').

  When it receives the message (x, x''), it computes $k_0 \leftarrow \mathcal{KD}(dk, x)$ and $k_0'' \leftarrow \mathcal{KD}(dk'', x'')$. If either key is $\perp$, it outputs $\perp$ and stops. Otherwise, it computes $k \leftarrow kdf(tr, k_0, k'_0, k''_0)$, with $tr = (ad, ek, ek', ek'', x', x, x'')$, and outputs k.

- The responder algorithm takes as input associated data ad, public keys ek, ek' and a private key dk'. It receives (ek'', x'), computes $(x, k_0) \leftarrow \mathcal{KE}(ek)$, $k'_0 \leftarrow \mathcal{KD}(dk', x')$ and $(x'', k''_0) \leftarrow \mathcal{KE}(ek'')$. If $k'_0 = \bot$, it stops and outputs $\bot$. Otherwise, it sends (x, x''), computes $k \leftarrow kdf(tr, k_0, k'_0, k''_0)$ and outputs k.

This protocol is remarkably similar to Example 10.3 using the key derivation rule from (10.1). Unlike the key derivation rule from (10.2), this generic construction does not easily allow for the inclusion of a term that only depends on the long-term keys $(y_i^{b_r} = y_r^{b_i})$.

Remark. Key exchange may also happen between users who share some secret key, or with the help of some trusted third party who both parties share a key with. We shall not discuss such protocols, but their analysis is similar to the analysis of key exchange protocols.

10.2   DEFINING SECURITY

Defining security for key exchange schemes is significantly more complicated than for encryption and signature schemes. One reason is that we are considering a design space that is much larger. But we shall also have stronger security requirements for a key exchange scheme.

As usual, we begine by describing the basic definition. Then we prove several results about this definition, typically reducing the number of various queries, thereby simplifying the definition. Along the way, we shall consider several examples. We emphasise that the theorems and examples together describe two different design strategies for key exchange schemes.

As usual, one of our motivating thoughts is that we do not design key exchange protocols for particular applications, but for general usage. The protocol designer does not at design-time know what the protocol will be used for. In particular, the designer does not know how many users there will be, how many sessions there will be, how it is decided which user exchanges a key with which other user, or how the session keys will be used.

The users of a key exchange scheme have two main security goals: the session keys exchanged should be secret (*confidentiality*) and partners should be in agreement about associated data and keys used (*authentication*). Unlike for symmetric cryptography, there is little benefit to be had from two separate security notions, so we define a single notion of security for key exchange.

10.2.1   The Definition

Like our previous examples, the security definition begins with an experiment.

The real-or-random experiment proceeds as follows:

1. Sample $b, b'' \xleftarrow{r} \{0, 1\}$.

2. When the adversary sends the jth query $\perp$ (*key generation*), do:

   (a) Compute $(pk, sk) \leftarrow \mathcal{K}$, record (j, pk, sk) and send pk to $\mathcal{A}$.

3. When the adversary sends the ith query (ρ, j, pk', ad) (*execute*), do:

   (a) If (j, pk, sk) is not recorded, send $\perp$ to the adversary and stop.

   (b) If $\rho = \mathcal{I}$, start the ith instance as $\mathcal{I}(ad, (pk, sk), pk')$. When the instance sends m, send (i, m) to $\mathcal{A}$. Otherwise, start the ith instance as $\mathcal{R}(ad, pk', (pk, sk))$ and send i to $\mathcal{A}$.

4. When the adversary sends the query (i, m) (*send*), do

   (a) If (i, s_0, s_0) is recorded, send $\perp$ to adversary.

   (b) Otherwise, send m to the ith instance. If the ith instance outputs $\perp$, record $(i, \perp, \perp)$ and send $(i, \perp)$ to the adversary. If the ith instance outputs $k_0 \in \mathfrak{K}$, sample $k_1 \xleftarrow{r} \mathfrak{K}$ and record (i, k_0, k_1). If the instance sent a message m', send (i, m') to the adversary, otherwise send $(i, \top)$ to the adversary.

5. When the adversary sends the query (test, i) (*test*), do:

   (a) If (i, s_0, s_1) is recorded, send (i, s_b) to the adversary.

6. When the adversary sends the query $(\mathsf{session}, i)$ (*session key reveal*), do:

   (a) If (i, s_0, s_1) is recorded, send (i, s_0) to $\mathcal{A}$.

7. When the adversary sends the query (state, i) (*state reveal*), do:

   (a) Send the random tape of the ith player instance to $\mathcal{A}$.

8. When the adversary sends the query (ltk, j) (*long-term key*), do:

   (a) If (j, pk, sk) is recorded, send (pk, sk) to $\mathcal{A}$.

Eventually, the adversary outputs $b' \in \{0, 1\}$.

FIGURE 10.1 Experiment $\mathbf{Exp}_{\mathrm{KEX}}^{\mathsf{kex}}(\mathcal{A})$ for the real-or-random game for a key exchange protocol $\mathrm{KEX} = (\mathfrak{K}, \mathcal{K}, \mathcal{I}, \mathcal{R})$. The bit b'' is not used in the experiment but is used to simplify the calculation of advantage.

The Experiment   We shall model the security game in the usual fashion, as an interaction between an adversary and the experiment given in Figure 10.1. The experiment models the honest users and the adversary's interactions with them. It does this by generating key pairs and starting instances of the key exchange roles with specified key pairs and partner keys, and then allowing the adversary to reveal various information about these instances.

We associate a *game transcript* to a game between the experiment and an adversary, which consists of a list of queries sent to the experiment and the corresponding responses. Note that every instance transcript can be recovered from the game transcript, and that all of the information encoded in the game transcript is known by the adversary.

The first step in the experiment samples a pair of bits used to define confidentiality. The adversary should guess the first bit, unless it has been trivially revealed, in which case the adversary should guess the second bit. Steps 2–4 model honest users using the key exchange scheme, by generating key pairs, starting instances and sending messages to instances. Step 5 deals with the test query that we use to define confidentiality. The remaining steps allow the adversary to reveal various information about the instances.

An adversary will have some influence on the honest users of a key exchange scheme. We model this by giving the adversary full control. New instances of roles can be started with an *execute query*, and all of its parameters can be specified by the adversary.

The experiment models only honest instances, which will only run with a key pair generated through a *key generation query*. These are *honestly generated* key pairs. The adversary is free to choose the partner public key, either an honestly generated key or an adversary-generated key.

The adversary will typically control the network. We model this by having each instance give any messages it wants to send to the adversary. The adversary freely chooses which messages the instances should receive, through *send queries*. These messages may be messages sent by instances or arbitrary messages chosen by the adversary.

Remark. Other network models could be used. We shall model a passive network, which will not require any significant changes to the definition.

One adversarial goal is to be able to say something about session keys. We model this through a real-or-random game, where the adversary uses a *test query* to get either a session key or a random key. The adversary may inspect more than one instance in this way, and it will either always get the instance's session key, or always get a random key.

Remark. Some researchers prefer a key exchange model where some test queries may respond with the session key, while others respond with random keys. This approach makes it easier to design key exchange schemes with certain properties, but this security notion is less useful for designing higher-level protocols using key exchange schemes. As an example, the claim in Proposition 13.2 would be weaker and the proof more complicated.

The adversary may also reveal things about honest instances and their keys. Such secrets may trivially reveal session keys, so we must be careful to exclude trivial attacks when evaluating if the adversary was successful.

The bare minimum an adversary can do is to learn session keys, which we model using the *session key reveal query*. Revealing a session key will also reveal the session key of any partner. Sometimes real-world adversaries are able to steal secrets, including secret keys. We model this through the *long-term key reveal* query. Knowledge of a long-term key allows impersonation of the key owner, but it should not allow anything more.

More complicated is the ability to reveal the state of an instance, modelled through a *state reveal query*. The intention is that an implementation of a cryptographic system erases any randomness used and intermediate results computed, so that only the output is left in memory. The real-world phenomenon we are considering is a class of implementation faults including side channel leaks, failures to erase ephemeral keys or failures in (pseudo-)random generators. Depending on the real-world effect we are trying to model, the state reveal query is either too powerful, too weak or about right.

Remark. There are other options for how to model the state reveal query. One alternative is to have the state reveal query not reveal the instance's random tape, but instead protocol-specific information, which may include randomness and intermediate computations. However, while such a strategy is more flexible, the practical effect depends strongly on how well we are able to model the exact real-world situation. Practice suggests that our record in predicting new attack forms is less than perfect.

One difference between this experiment and previous examples is that we do not explicitly prevent the adversary from winning in a trivial way. For example, in the encryption experiments, we prevent the adversary from asking for decryptions of challenge ciphertexts, which would trivially reveal the challenge bit. The reason is that the accounting required to decide this would significantly complicate the description of the experiment. Instead, we shift this complexity to the partnering notion and the freshness notion.

Partnering   Key exchange is about two parties agreeing on a session key, so we need to be able to decide which of the experiment's instances have agreed on session keys with each other. Because the network is adversary-controlled, there is no explicit correspondence between a message sent by one instance and a message received by another instance. Instead, we will infer partners by looking at their transcripts, which describe their intentions (associated data and public keys involved) and actions (messages sent and received).

Definition 10.2. A *partnering notion* defines a symmetric relation $\sim$ on the set of transcripts satisfying:

P1 A transcript of an instance that outputs a session key is related to itself.

P2 Two related transcripts record the same associated data and public keys.

P3 If two instances output session keys and their transcripts are related, they output the same session key.

We also require that there exists a *bookkeeping* algorithm that receives a game transcript one event at a time and computes for each event the current graph on the set of instancs induced by $\sim$.

We will use the partnering notion to determine related instances in our games. Since instance transcripts can be recovered from the game transcript, related instances can be determined purely from the game transcript. For most partnering notions, the bookkeeping amounts to little more than maintaining a sorted list of instance transcripts, which has a trivial cost.

We shall mostly use the matching conversations partnering notion, which by the above requirements is also the strictest possible partnering notion that makes sense. We shall study less strict partnering notions in Section 10.2.6. Unless otherwise stated, our generic results on key exchange schemes are independent on the exact partnering notion used.

Definition 10.3. The *matching conversation* relation says that two transcripts $(ad_0, pk_0, pk_0', m_{01}, \ldots, m_{0,l})$ and $(ad_1, pk_1, pk_1', m_{11}', \ldots, m_{1,l'}')$ are related if and only if $ad_0 = ad_1$, $pk_0 = pk_1$, $pk_0' = pk_1'$, both transcripts contain at least one message, and the two message lists are either identical, or identical except that a final sent message in one transcript need not appear as a received message in the other transcript.

The adversary is the network in our model, so we can never guarantee that messages reach their destination. We would still like to say that an instance has a partner, even if its last sent message has not reached the partner. This is why we match conversations up to a final sent message.

Authentication   An authentication notion should describe when the authentication has been achieved, or when the adversary has broken authentication in a trivial way, for example by revealing key material.

The intuitive notion of authentication in the context of key exchange is that when Alice believes she has agreed upon a key with Bob, then Bob has agreed on that key with Alice. It turns out that this intuitive notion is quite hard to formalise and has more variants than we should expect. We must define several notions of authentication for key exchange schemes. But we need some basic rules for authentication notions in order to prove useful generic results.

D **Definition 10.4.** An *authentication predicate* is a predicate ϕ_a on instances in a game transcript satisfying:

A1 Suppose the predicate holds for a given instance: If we create a new game transcript by inserting additional long-term key reveal queries in the transcript, the predicate still holds for the instance relative to the new game transcript.

A2 Suppose the predicate does not hold for a given instance: If we create a new game transcript by removing items pertaining to instances unrelated to the given instance, the predicate still does not hold for the given instance relative to the new game transcript.

We say that the predicate holds for the game transcript if it holds for all the instances in the transcript.

The authentication predicates we consider are essentially statements about the number and types of partners in the transcript for the given instance.

D **Definition 10.5.** Let $\sim$ be a partnering relation. Consider an instance that has output a session key, and with associated data ad and public key pk, partner public key pk', where pk and pk' were both honestly generated.

The *explicit authentication predicate* ϕ_a^e holds for the instance if the set of related instances contains exactly two instances and they play different roles, or the set of related instances contains exactly one instance and there was a long-term key reveal for pk' before the instance output its session key.

The *implicit authentication predicate* ϕ_a^i holds for the instance if the set of related instances contains at most two instances that output a session key and they play different roles.

The *weak implicit authentication predicate* ϕ_a^w holds for the instance if the set of related instances contains at most one initiator instance.

The *no authentication predicate* ϕ_a^0 holds for the instance.

Explicit authentication essentially says that there is no confusion about which instance has exchanged a key with which instance, nor about who initiated the exchange. The adversary may impersonate instances, but only if the corresponding long-term key has been revealed.

Implicit authentication essentially also says that there is no confusion about which instance has exchanged a key with which instance, nor about who initiated the exchange. But in this case the adversary may impersonate instances without compromising the long-term key. Interestingly, this property allows someone to pretend that they have had a conversation with someone, which may in some cases afford a measure of *deniability*.

Weak implicit authentication say that there is no confusion about which instance initiated an exchange. Also, the adversary may impersonate instances without compromising the long-term key. This is mostly intended for single-message key exchange schemes, where there is no way to guarantee that a responder instance exists, nor any method of preventing more than one responder instance from receiving the intiator's single message.

Remark. ϕ_a^{e} logically implies ϕ_a^{i}, which implies ϕ_a^{w}, which implies ϕ_a^{0}.

Freshness The adversary breaks confidentiality by sending test queries and then deciding what the response was. However, if an adversary sends a test query to one instance and reveals the session key of the instance's partner, the adversary has won trivially. A freshness predicate describes when an adversary has trivially broken the confidentiality of some tested session key, in which case the adversary's answer should not count towards the advantage. Again, we need some basic rules in order to prove useful generic results.

Without loss of generality, we shall assume that the adversary never makes more than one test query for any single instance.

Definition 10.6. A *freshness predicate* is a predicate ϕ_f on the test queries in a game transcript satisfying:

F1 Suppose the freshness predicate holds for a given test query in a game transcript: If we create a new game transcript by replacing test queries for other instances with corresponding session key reveal queries, the freshness predicate still holds.

F2 Suppose the freshness predicate holds for a given test query in a game transcript: If we create a new game transcript by removing one or more other queries whose instances do not have public keys in common with the instance in the test query, the freshness predicate still holds.

We say that the predicate holds for the game transcript if it holds for all the test queries in the transcript.

The freshness notions we shall consider essentially describe how the adversary is allowed to use its powers to reveal things about instances. Again, freshness notions depend strongly on our notions of partnering, but it is the freshness notion that describes the security we want to achieve. Weaker partnering notions can therefore allow us to prove security for interesting protocols, as we shall see in Section 10.2.6.

Definition 10.7. Let $\sim$ be a partnering relation. Consider an instance that has output a session key, and with associated data ad and public key pk, partner public key pk', where pk and pk' were both honestly generated.

We say that an instance is *trivially compromised* if a session key reveal query has been made for that instance, or both a state reveal query has been made for the instance and a long-term key reveal has been made for the instance key pair. An instance is *exposed* if it is trivially compromised or a test query has been made for that instance.

The freshness predicate $\phi_f^{\mathsf{fe}/\mathsf{e}}$ holds for the instance if it is not trivially compromised and there is a partner instance that is not exposed.

The freshness predicate $\phi_f^{\mathsf{fe}/\mathsf{i}}$ holds for the instance if it is not trivially compromised and either there is a partner instance that is not exposed or there is no long-term key reveal query for pk'.

For $x \in \{\mathsf{e}, \mathsf{i}\}$, the freshness predicate $\phi_f^{\mathsf{f}/x}$ holds for the instance if $\phi_f^{\mathsf{fe}/x}$ holds and no state has been revealed for this instance or any partner instance.

For $x \in \{\mathsf{e}, \mathsf{i}\}$, the freshness predicate $\phi_f^{\mathsf{s}/x}$ holds for the instance if $\phi_f^{\mathsf{f}/x}$ holds and there are no long-term key reveal queries for this instance's identity or its partner identity.

The freshness predicate $\phi_f^{\mathsf{s}/\mathsf{w}}$ holds if $\phi_f^{\mathsf{s}/\mathsf{i}}$ holds and in addition there are no session key reveal queries for any partner instance.

The freshness predicate $\phi_f^{0/0}$ holds if the instance is not trivially compromised and there exists a partner instance that is not exposed.

Remark. For $x \in \{\mathsf{e}, \mathsf{i}\}$, $\phi_f^{\mathsf{fe}/x}$ logically implies $\phi_f^{\mathsf{f}/x}$, which implies $\phi_f^{\mathsf{s}/x}$.

The above freshness predicates do not allow the adversary to test both partners of a key exchange. Morally, this should be allowed. But since partnering is public knowledge, an adversary can simulate a consistent test query. And since the bookkeeping required to ensure consistent responses in the random case is significant, the sensible choice is to disallow testing both partners.

The freshness predicate $\phi_f^{0/0}$ is intended for protocols that may not satisfy any authentication properties, such as plain Diffie-Hellman. Primarily, we shall use such protocols as building blocks to build protocols with stronger security properties. But such protocols may also give us what we can call *optimistic security*, in the sense that if the adversary does not interfere, we get confidentiality. Optimistic secure seems worthwhile in some anonymity use cases, since many adversaries against anonymity are passive. When the communicating partners want to be anonymous, they cannot exchange identities or public keys before the key exchange happens. This means we cannot use authenticated key exchange, so optimistic security is a reasonable compromise.

Remark. The freshness predicate that allows for state reveal queries, but not long-term key reveal queries is missing from the above definition. This is because it seems less useful. Indeed, the above selection of freshness predicates is not intended to be exhaustive. There are many other freshness predicates that could be possible.

The freshness predicates on their own do not guarantee secrecy. But when combined with appropriate authentication we get guarantees. The authentication and freshness predicates are designed to match in the obvious way. We simplify notation for these predicates to kex-fe-e, kex-f-i, kex-s-w, etc. Non-obvious matchings result in security notions that mostly make no sense.

The notions kex-s-e, kex-f-e and kex-fe-e guarantee that there is no confusion about which instance has exchange a key with which instance, nor about who initiated the exchange. The adversary is allowed to impersonate an instance after revealing their long-term key, but if the adversary did not impersonate an instance, nor trivially compromise the session key, the session key is secret. Trivially compromised means that the adversary used a session reveal query on a party (kex-s-e), used a session reveal or a state reveal query on a party (kex-f-e), or used a session reveal query or both a state reveal and long-term key reveal query on a party (kex-fe-e).

The notions kex-s-i, kex-f-e and kex-fe-i guarantees that there is no confusion about which instance has exchange a key with which instance, nor about who initiated the exchange. The adversary is allowed to impersonate instances without revealing their long-term key. However, if the adversary does not reveal the long-term key, nor trivially compromised the session key, the session key is secret. Note that the adversary may reveal a long-term key long after it impersonated an instance, and then recover the secret key at that point. This is important, because freshness may vary with time, unlike for the explicit authentication, which is surprising.

The notion kex-s-w guarantees that there is no confusion about which instance initiated an exchange. The adversary is allowed to impersonate instances without revealing their long-term key. If the adversary did not impersonate the initiator instance, nor directly reveal the session key, the session key is secret. Again, "directly reveal" has similar meaning to above.

Finally, The Definition The actual security definition is a standard real-or-random definition. It combines the experiment from Figure 10.1, a partnering notion, an authentication predicate and a freshness predicate.

Definition 10.8. A (τ, l_t, l_u, l_s)-*adversary* against a key exchange protocol KEX is an interactive algorithm $\mathcal{A}$ that interacts with the experiment in Figure 10.1 making at most l_t test queries, l_u key generation queries and l_s execute queries, and where the runtime of the adversary, the experiment and determining the partnering relation graph is at most τ.

Let $\sim$ be a partnering notion for KEX, let ϕ_a be an authentication predicate and let ϕ_f be a freshness predicate. The *advantage* of the adversary $\mathcal{A}$ with respect to $\sim$, ϕ_a and ϕ_f is

$$\mathbf{Adv}_{\mathrm{KEX}}^{\mathsf{kex}\text{-}\phi_f\text{-}\phi_a}(\mathcal{A}) = \max\{2|\Pr[E_d] - 1/2|, \Pr[E_a]\},$$

where E_d is the event that the adversary's guess equals b if the execution is fresh, or that the adversary's guess equals b'' if the execution is not fresh; and E_a is the event that the authentication predicate ϕ_a does not hold for some instance that outputs a session key.

Remark. We count the *total number of execute queries*. Sometimes it is more convenient to count the number of execute queries *per honest key*.

Definition 10.9. A key exchange adversary is *non-invasive* if it does not reveal any states, and *non-adaptive* if it additionally reveals no long-term keys.

For any game transcript, we can define a set S of responder instances with honestly generated partner keys that received a message, the set T of initiator instances and the subset $T' \subseteq T$ of initiator instances that received at least one message. The adversary is *network-passive* if for any game transcript, there is an injective map $\iota : S \to T$ that is onto T' and such that for any responder instance mapped to an initiator instance, their transcripts agree, up to the last sent message which may not have been received.

A non-adaptive and network-passive adversary is *passive*.

Note that security against a non-invasive adversary and the freshness predicate $\phi_f^{f/x}$ is a weaker requirement than just security under $\phi_f^{f/x}$, since the adversary could use information extracted from one instance to attack other instances. The easiest example is a key exchange protocol using Schnorr signatures. Revealing an instance's random tape reveals the random value used in the Schnorr signature, which reveals the signing key. Similar remarks apply to other types of adversary and freshness predicates.

A non-adaptive adversary does not model a system where all users are honest, since the adversary may use any public key in an execute query.

Exercise 10.1. Let $\mathcal{A}$ be a (τ, l_t, l_u, l_s)-adversary against KEX. Then for any $x \in \{\mathsf{fe}, \mathsf{f}, \mathsf{s}\}$ and $y \in \{\mathsf{e}, \mathsf{i}\}$ we have that

$$\mathbf{Adv}_{\mathrm{KEX}}^{\mathsf{kex}\text{-}\mathsf{fe}\text{-}y}(\mathcal{A}) \geq \mathbf{Adv}_{\mathrm{KEX}}^{\mathsf{kex}\text{-}\mathsf{f}\text{-}y}(\mathcal{A}) \geq \mathbf{Adv}_{\mathrm{KEX}}^{\mathsf{kex}\text{-}\mathsf{s}\text{-}y}(\mathcal{A})$$

and

$$\mathbf{Adv}_{\mathrm{KEX}}^{\mathsf{kex}\text{-}x\text{-}\mathsf{e}}(\mathcal{A}) \geq \mathbf{Adv}_{\mathrm{KEX}}^{\mathsf{kex}\text{-}x\text{-}\mathsf{i}}(\mathcal{A}) \geq \mathbf{Adv}_{\mathrm{KEX}}^{\mathsf{kex}\text{-}x\text{-}\mathsf{w}}(\mathcal{A}).$$

10.2.1.1 Example: Diffie-Hellman

The famous Diffie-Hellman protocol was proposed only for passive adversaries (an eavesdropper). As long as the group has been carefully chosen so that the Decision Diffie-Hellman problem is hard, the protocol really is hard to attack for a passive adversary.

Proposition 10.1. *Let $\mathcal{A}$ be a passive (τ, l_t, l_u, l_s)-adversary against the Diffie-Hellman key exchange protocol from Example 10.1 using a group G with generator g. Then there exists a τ'-adversary $\mathcal{B}$ against Decision Diffie-*

Hellman, where τ' is essentially equal to τ, and

$$\mathbf{Adv}_{\mathrm{DH}}^{\mathrm{kex}\text{-}0\text{-}0}(\mathcal{A}) \leq 2\mathbf{Adv}_{G,g}^{\mathrm{DDH}}(\mathcal{B}) + 1/p.$$

It is easiest to understand the proof if we consider a single pair of instances. The messages sent by the instances and the session key they agree make up a Diffie-Hellman tuple. The two messages and a random group element make up a random tuple. In other words, we can pretend that the group elements are the messages and the test query result. Then we use the adversary to decide if the test query returned the session key or a random group element. If the adversary decides on the former, we guess Diffie-Hellman tuple. If the adversary decides on the latter, we guess random tuple.

Remark. The above argument is actually sufficient to prove a variant of the claim, when combined with theorems we shall prove. But we want a *tighter* result, that is, the bound on the adversary's advantage in terms of some other advantage should be tight, for which we must use random self-reducibility.

E | *Exercise* 10.2. Prove Proposition 10.1.
Hint: See Example 8.3. It is possible to remove the factor 2.

Remark. It is tempting to change the session key derivation rule in the plain Diffie-Hellman protocol from g^{ab} to $kdf(g^{ab})$. This means that it is no longer sufficient for the adversary recognise Diffie-Hellman tuples, but it must actually solve the Computational Diffie-Hellman problem.

Indeed, the adversary must solve the CDH problem to win in the random oracle model, since it must query the hash oracle with the correct answer to the CDH problem instance to have any advantage in distinguishing the key. But we do not know how to detect *exactly when* the adversary has solved the problem. In fact, detecting the correct solution is the Decision Diffie-Hellman problem. When we cannot detect which hash query contains the right answer, we are little closer to finding the answer, since the adversary may make a large number of hash queries, and the adversary may choose not to make the correct answer its final hash query (for no other reason than to be difficult).

The answer is to use the Gap Diffie-Hellman problem instead. The Gap Diffie-Hellman problem is somewhat similar to the Strong Diffie-Hellman problem, except that the adversary sends queries (x', y', z') to which the experiment responds with 1 if it is a DDH tuple, and 0 if not. This allows us to recognise when we get the correct answer, and more importantly, to remove the requirement that the adversary is passive.

If we derive the session key as $kdf(g^a, g^b, g^{ab})$, that simplifies the reduction.

The Gap Diffie-Hellman problem is very different from other problems, in the sense that the Decision Diffie-Hellman problem must be solved to respond to queries, which we only know how to do for elliptic curve groups with feasible pairings. For most groups, the Gap Diffie-Hellman problem is not a traditional problem, but the problem of reducing Computational Diffie-Hellman to Decision Diffie-Hellman. This still seems to be hard, though.

E *Exercise* 10.3. Prove a result similar to Proposition 10.1 for the hashed variant of plain Diffie-Hellman discussed in the above remark.

10.2.2   Implicit Authentication Suffices – Maybe

Some applications get by well with implicit authentication. Typically, the exchanged key is used in a manner that makes authentication explicit once the application proceeds. For instance, if the key exchange protocol is composed with a secure channel protocol, which is a common application, the party that finishes first sends some message through the channel. If the other party then sends some message through the channel, explicit authentication follows.

However, many applications find it easier to deal with explicit authentication. Fortunately, it is easy to add explicit authentication to a key exchange protocol. Proving that this works is a highly technical exercise. The main complicating factor is that the protocol's authentication guarantees crucially depend on the secrecy of the exchanged key. This is also the main limiting factor of the technique, because explicit authentication should hold also in situations where the session key is compromised.

The idea is to let the exchanged key contain two additional tags. Each player then sends one of the tags to the other player. When the player receives a tag it has not sent, this proves that someone else must know that tag. But since only a partner could know the tag, it follows that receiving the tag proves the existence of a partner.

E *Example* 10.8. Let $\text{KEX}_0 = (\mathfrak{K}_0, \mathcal{K}_0, \mathcal{I}_0, \mathcal{R}_0)$ be a key exchange scheme with $\mathfrak{K}_0 = \mathfrak{K} \times \mathfrak{T}^2$. The key exchange scheme $\text{KEX} = (\mathfrak{K}, \mathcal{K}_0, \mathcal{I}, \mathcal{R})$ works as follows:

- The interactive algorithms of KEX are based on the corresponding interactive algorithms of KEX_0. Let $\mathcal{P}$ denote either $\mathcal{I}$ or $\mathcal{R}$, and let $\mathcal{P}_0$ denote the corresponding choice of $\mathcal{I}_0$ or $\mathcal{R}_0$.

  The interactive algorithm $\mathcal{P}$ runs the $\mathcal{P}_0$ algorithm with its own input. If $\mathcal{P}_0$ ever outputs $\perp$, $\mathcal{P}$ immediately outputs $\perp$ and stops. Any messages $\mathcal{P}_0$ sends $\mathcal{P}$ sends, and $\mathcal{P}$ forwards any received messages to $\mathcal{P}_0$, with the following modifications:

  - If $\mathcal{P}_0$ sends a message m and outputs a session key $(k, t_0, t_1) \in \mathfrak{K}_0$, $\mathcal{P}$ remembers (k, t_1) and sends the pair (m, t_0).
  - If $\mathcal{P}$ receives a pair (m, t_0'), it sends m to $\mathcal{P}_0$. If $\mathcal{P}_0$ sends any message, $\mathcal{P}$ outputs $\perp$. If $\mathcal{P}_0$ does not send a message and outputs a session key (k, t_0, t_1) with $t_0' = t$, then $\mathcal{P}$ sends t_1 and outputs k. Otherwise, $\mathcal{P}$ outputs $\perp$.

      – If $\mathcal{P}_0$ output (k, t_0, t_1) and $\mathcal{P}$ receives a tag t_1', then it outputs k if $t_1 = t_1'$, otherwise it outputs $\perp$.

Remark. Key exchange schemes typically exchange keys in some smaller set. To get a key exchange scheme with the convenient key set we use in this construction, we would use a key derivation function (or just a pseudo-random generator) to extend a small key into a larger key.

Proposition 10.2. *Let $\mathcal{A}$ be a (τ, l_t, l_u, l_s)-adversary against* KEX. *Then there exists a $(\tau_1', 0, l_u, l_s)$-adversary $\mathcal{B}_1$, a $(\tau_2', 1, l_u, l_s)$-adversary $\mathcal{B}_2$ and a $(\tau_3', 1, l_u, l_s)$-adversary $\mathcal{B}_3$, all against* KEX$_0$, *with τ_1', τ_2' and τ_3' essentially equal to τ, such that*

$$\mathbf{Adv}_{\mathrm{KEX}}^{\mathsf{kex\text{-}x\text{-}e}}(\mathcal{A}) \leq 2\mathbf{Adv}_{\mathrm{KEX}_0}^{\mathsf{kex\text{-}x\text{-}i}}(\mathcal{B}_1) + 2l_s\mathbf{Adv}_{\mathrm{KEX}_0}^{\mathsf{kex\text{-}x\text{-}i}}(\mathcal{B}_2) + 2l_s\mathbf{Adv}_{\mathrm{KEX}_0}^{\mathsf{kex\text{-}x\text{-}i}}(\mathcal{B}_3) + \frac{2l_s}{|\mathfrak{T}|},$$

for $\mathsf{x} \in \{\mathsf{f}, \mathsf{s}\}$. If $\mathsf{x} = \mathsf{f}$, all the adversaries should be non-invasive. If $\mathsf{x} = \mathsf{s}$, all the adversaries should be non-adaptive.

    The proof is structured as a sequence of games. The initial game is the key exchange experiment interacting with the adversary $\mathcal{A}$. The next game is a simple administrative game, where we replace the experiment for KEX with an experiment for KEX$_0$, inserting a simulator to handle the additional messages. We also introduce some bookkeeping to track partners among KEX$_0$ instances.

    We then modify the game so that it terminates if implicit authentication ever fails for the underlying protocol, that is, if there is ever more than one partner for some instance. Any change in behaviour would result in an adversary against implicit authentication. Finally, we modify the game so that it rejects the additional messages unless there is a partner. Any change in behaviour means that the adversary is capable of predicting the additional messages, breaking the confidentiality of the underlying key.

    In the final game it is now possible to prove that any guessing advantage that remains can be turned into guessing advantage for KEX$_0$.

Proof of Proposition 10.2 In the following, let $E_{d,i}$ and $E_{a,i}$ be events in Game i, initially corresponding to E_d and E_a in Definition 10.8 with the intended freshness notion and explicit authentication, and later reinterpreted.

Game 0 The initial game is the key exchange experiment $\mathbf{Exp}_{\mathrm{KEX}}^{\mathsf{kex}}$ interacting with the adversary $\mathcal{A}$. Let $E_{d,0}$ and $E_{a,0}$ be the events in this game corresponding to the events E_d and E_a in Definition 10.8 for KEX. Then

$$\mathbf{Adv}_{\mathrm{KEX}}^{\mathsf{kex\text{-}x\text{-}e}}(\mathcal{A}) = \max\{2|\Pr[E_{d,0}] - 1/2|, \Pr[E_{a,0}]\}. \tag{10.3}$$

Game 1 In this game, we replace the experiment $\mathbf{Exp}_{\mathrm{KEX}}^{\mathsf{kex}}$ with the experiment $\mathbf{Exp}_{\mathrm{KEX}_0}^{\mathsf{kex}}$, and introduce a simulator **Sim** that works as follows:

- It samples $b_{sim} \xleftarrow{r} \{0, 1\}$.

- When the ith instance of KEX$_0$ finishes, possibly sending a final message, **Sim** makes a session key reveal query to get (k_0, t_0, t_1) or $\bot$. If it is $\bot$, **Sim** records $(i, \bot, \bot)$. Otherwise, **Sim** samples $k_1 \xleftarrow{r} \mathfrak{K}$ and uses t_0 and t_1 as the final two tags, as in the KEX specification. If the tag verification fails **Sim** records $(i, \bot, \bot)$, otherwise it records (i, k_0, k_1).

- The simulator forwards any key generation, state reveal, long-term key reveal and execute queries, as well as send queries not handled above, to $\mathbf{Exp}_{\text{KEX}_0}^{\text{kex}}$, and forwards the responses to $\mathcal{A}$.

- The simulator keeps track of which instances of KEX$_0$ are partners.

- When $\mathcal{A}$ makes a test query for the ith instance and (i, s_0, s_1) is recorded, **Sim** responds with $s_{b_{sim}}$.

By inspection, we verify that the experiment $\mathbf{Exp}_{\text{KEX}_0}^{\text{kex}}$ together with **Sim** is functionally identical to the experiment $\mathbf{Exp}_{\text{KEX}}^{\text{kex}}$. If we reinterpret $E_{d,1}$ with respect to b_{sim}, and $E_{a,1}$ with respect to the composite of instances of KEX$_0$ and messages simulated by **Sim**, we get that

$$\Pr[E_{d,1}] = \Pr[E_{d,0}], \qquad \Pr[E_{a,1}] = \Pr[E_{a,0}].$$

Compared to the experiment in Game 0, the simulator **Sim** in this game introduces some overhead by forwarding messages, and also keeping track of partners. However, for any interesting adversary $\mathcal{A}$, this overhead is small, so the runtime bound τ_1 is essentially equal to τ_0.

Game 2 We modify **Sim** so that the game stops if any instance ever has more than one partner with respect to KEX$_0$. Denote this event by F_2.

Unless F_2 happens, this game proceeds exactly as the previous game. If we denote by $E_{d,2}$ and $E_{a,2}$ the events corresponding to $E_{d,1}$ and $E_{a,1}$, we get

$$|\Pr[E_{d,2}] - \Pr[E_{d,1}]| \leq \Pr[F_2], \quad |\Pr[E_{a,2}] - \Pr[E_{a,1}]| \leq \Pr[F_2]. \quad (10.4)$$

Since we keep track of partners, checking if the authentication predicate holds has trivial cost. So the runtime bound τ_2 is essentially equal to τ_1.

The event F_2 is an authentication failure, and we can build an adversary that uses this to break the key exchange scheme. This adversary makes no attempt at guessing the challenge bit. Indeed, it makes no test queries.

Lemma 10.3. *There exists a* $(\tau_1', 0, l_u, l_s)$*-adversary* $\mathcal{B}_1$ *against* KEX$_0$*, with* τ_1' *essentially equal to* τ_2*, such that*

$$\mathbf{Adv}_{\text{KEX}}^{\text{kex-x-i}}(\mathcal{B}_1) = \Pr[F_2].$$

Proof. The adversary $\mathcal{B}_1$ runs a copy of $\mathcal{A}$ and the simulator **Sim**. If $\mathcal{A}$ stops or **Sim** detects too many partners for some instance, $\mathcal{B}_1$ stops with output 0.

The key exchange experiment for KEX_0 and $\mathcal{B}_1$ together behave exactly as Game 2, except when $\mathcal{B}_1$ stops early. It follows that τ_1' is essentially equal to τ_2. Also, an instance will have too many partners (breaking implicit authentication) with probability equal to $\Pr[F_2]$. The claim follows. □

Ⓖ **Game 3** In this game any KEX instance that receives a final message matching its key material, rejects unless it has a partner or there is a long-term key reveal for its partner public key. Let F_3 denote the event that an instance rejects, even though the received message matches its key material.

Game 2 ruled out any instance having more than one partner. Therefore, by design the authentication predicate ϕ_a^e holds for every KEX instance that outputs a key in this game, and

$$\Pr[E_{a,3}] = 0. \tag{10.5}$$

If F_3 never occurs, this and the previous games proceed the same. We get

$$|\Pr[E_{d,3}] - \Pr[E_{d,2}]| \leq \Pr[F_3], \quad |\Pr[E_{a,3}] - \Pr[E_{a,2}]| \leq \Pr[F_3]. \tag{10.6}$$

We need to bound the difference in behaviour between this game and the previous game, and then bound the guessing advantage in this game, which is done by the next two lemmas.

Note that while F_3 is an authentication failure, it is only an explicit authentication failure for KEX, not an implicit authentication failure for KEX_0.

Ⓣ **Lemma 10.4.** *There exists a $(\tau_2', 1, l_u, l_s)$-adversary $\mathcal{B}_2$ against KEX_0, with τ_2' essentially equal to τ_3, such that*

$$\mathbf{Adv}_{\text{KEX}_0}^{\text{kex-x-i}}(\mathcal{B}_2) \geq |\Pr[F_3]/l_s - 1/|\mathfrak{T}||.$$

Proof. The adversary $\mathcal{B}_2$ runs a copy of **Sim** from Game 2, modified as follows:

- It samples $j \xleftarrow{r} \{1, 2, \ldots, l_s\}$.
- When the jth instance of KEX_0 finishes (or some instance whose partner is the jth instance), **Sim** makes a test query for the appropriate instance to get (k_0, t_0, t_1) or $\perp$. It uses the same value for any partner instance, and does not make a session reveal query. It then continues as before.
- If the jth instance receives a final message matching its key material, but it does not have a partner (as a KEX instance) and there is no long-term key reveal for the partner public key, $\mathcal{B}_2$ stops with output 0.

If $\mathcal{A}$ stops or exceeds its bounds, $\mathcal{B}_2$ stops and outputs 1.

The enforced bounds on $\mathcal{A}$ ensures that τ_2' is essentially equal to τ_3, and also that $\mathcal{B}_2$ does not make too many queries.

If $b = 0$, **Sim** and the experiment for KEX_0 perfectly simulate Game 2 until $\mathcal{B}_2$ stops. The probability that $\mathcal{B}_2$ outputs 0 is at least $\Pr[F_3]/l_s$.

If $b = 1$, then the adversary has no information about the jth instance's key material. The probability that $\mathcal{B}_2$ outputs 0 is at most $1/|\mathfrak{T}|$.

Note that if F_3 happens and it involves the jth instance, then the jth instance does not have a partner and its partner's long-term key is not compromised, so it is fresh with respect to KEX_0. □

Lemma 10.5. *There exists a $(\tau_3', 1, l_u, l_s)$-adversary $\mathcal{B}_3$ against KEX_0, with τ_3' essentially equal to τ_3, such that*

$$\mathbf{Adv}_{\text{KEX}_0}^{\text{kex-x-i}}(\mathcal{B}_3) \geq |\Pr[E_{d,3}] - 1/2|/l_s.$$

Proof. The adversary $\mathcal{B}_3$ runs a copy of **Sim** from Game 3, modified as follows:

- It samples $j \xleftarrow{r} \{1, 2, \ldots, l_s\}$.
- When the jth instance of KEX_0 finishes (or some instance whose partner is the jth instance), **Sim** makes a test query for the appropriate instance to get (k_0, t_0, t_1) or $\perp$. It uses the same value for any partner instance, and does not make a session reveal query. It then continues as before.

If $\mathcal{A}$ exceeds its bounds, $\mathcal{B}_3$ stops and outputs a random bit. Otherwise, $\mathcal{A}$ eventually stops with output b'. If $\mathcal{A}$ did not make a test query for the jth instance, or if the jth instance is not fresh at the end, $\mathcal{B}_3$ outputs a random bit. Otherwise, $\mathcal{B}_3$ outputs 0 if $b' = b_{sim}$ and 1 if $b' \neq b_{sim}$.

The enforced bounds on $\mathcal{A}$ ensures that τ_3' is essentially equal to τ_3, and also that $\mathcal{B}_3$ does not make too many queries.

If $b = 1$, then k_0 and k_1 for the jth instance are both independent of anything else, so $\mathcal{A}$ has no information about b_{sim} if the test query is made for the jth instance. It follows that $\mathcal{B}_3$ outputs 0 with probability $1/2$.

If $b = 0$, then the key exchange experiment for KEX_0 and **Sim** perfectly simulate Game 3. Also, $\mathcal{A}$ has no information about j. The claim now follows from a tedious and technical calculation. □

Proposition 10.2 follows from equations (10.3)–(10.6) and the above lemmas.

Remark. Some key exchange schemes achieve slightly stronger security properties than we claim in the proposition, in the sense that even in the presence of state reveal and long-term key reveal queries, authentication should still hold for instances that have not been trivially compromised. In principle, we could have defined further variants of explicit authentication that more precisely capture these properties. The goal would be to ensure that any instance for which we expect authentication to hold should also be fresh with respect to KEX_0. But we have already introduced sufficient complexity, so we leave this as an exercise for the interested reader.

10.2.3 A Single Test Suffices – Maybe

As we shall see, multiple test queries makes key exchange much easier to handle in application security proofs. However, this also makes our proof obligation much more complicated. In particular, we cannot guess with any reasonable chance of success which instances an adversary will make test queries for. For a single test query, we can guess with reasonable success probability. It will turn out that this is something we shall need to do, so we need this theorem.

T **Proposition 10.6.** *Let $\mathcal{A}$ be a (τ, l_t, l_u, l_s)-adversary against* KEX. *Then there exists a $(\tau', 1, l_u, l_s)$-adversary $\mathcal{B}$ against* KEX, *where τ' is essentially equal to τ, and*

$$\mathbf{Adv}_{\mathrm{KEX}}^{\mathsf{kex}}(\mathcal{A}) \leq l_t \mathbf{Adv}_{\mathrm{KEX}}^{\mathsf{kex}}(\mathcal{B}).$$

The proof is a standard hybrid argument, where we replace all but one test query with session reveal queries, and then deliberately randomise the session key for some of these queries. This approach relies crucially on the technical fact that for any freshness notion, it does not actually matter if we replace some of the test queries with session reveal queries.

E *Exercise* 10.4. Prove Proposition 10.6.

10.2.4 A Single Pair of Users Suffices – Maybe

The previous result showed us that we can limit ourselves to a single test query, if we accept the loss in adversary power. The next question is if we can simplify the situation further, by reducing the number of honestly generated public keys. In fact, we can reduce this to two, which is obviously the minimum.

The idea is that for a single test query, we can guess the two public keys involved, and simulate all the other keys. We can also guess which public keys will be involved in the first break of authentication. It follows that we can have two honest key pairs, and we can simulate the other key pairs.

Technically, we simulate the other key pairs by running a second copy of the key exchange experiment, forwarding to it all queries not involving the two main keys. Queries involving one of the two main keys and keys from our second experiment can be dealt with, since our experiment happily accepts public keys that it has not generated. And then we end up hoping that the adversary chooses to attack an instance using the two main keys.

T **Proposition 10.7.** *Let $\mathcal{A}$ be a $(\tau, 1, l_u, l_s)$-adversary against* KEX. *Then there exists a $(\tau, 1, 2, l_s)$-adversary $\mathcal{B}$ against* KEX, *where τ' is essentially equal to τ, and*

$$\mathbf{Adv}_{\mathrm{KEX}}^{\mathsf{kex}}(\mathcal{A}) \leq l_u^2 \mathbf{Adv}_{\mathrm{KEX}}^{\mathsf{kex}}(\mathcal{B}).$$

Proof. The adversary $\mathcal{B}$ runs a copy of $\mathcal{A}$ and a copy **GExp** of the key exchange experiment, and does as follows:

- It samples $i, j \xleftarrow{r} \{1, 2, \ldots, l_u\}$.

- The ith and the jth key generation queries are forwarded to $\mathbf{Exp}_{\text{KEX}}^{\text{kex}}$. Every other key generation query is forwarded to **GExp**.

- Any query involving the ith or jth key pair or an instances using the key pairs is forwarded to $\mathbf{Exp}_{\text{KEX}}^{\text{kex}}$. Any other query is forwarded to **GExp**.

If $\mathcal{A}$ outputs b', $\mathcal{B}$ outputs b'.

First, observe that the key exchange experiment treats every instance independently. Therefore, the fact that we have distributed our instances across two key exchange experiments cannot be noticed by the adversary. Note that the experiments have different challenge bits, but since there will only be one test query, this is not observable by the adversary.

The adversary will now either break authentication or confidentiality. In the authentication case, there may be more than one instance involved, in which case we choose the first one. Let F_d be the event that the tested instance uses no other public keys than the ith and the jth, and let F_a be the corresponding event for the first instance that breaks authentication. Also, let E_y^x be the event that E_y holds when the adversary x interacts with $\mathbf{Exp}_{\text{KEX}}^{\text{kex}}$.

Since the simulation is perfect, the public keys used by these instances are independent of i and j. We have that for $x \in \{d, a\}$

$$\Pr[F_x] \geq \frac{1}{l_u^2}.$$

We first compute the adversary's probability of breaking authentication. If the adversary $\mathcal{A}$ breaks authentication and F_a happens, then $\mathcal{B}$ also breaks authentication. It follows that

$$\Pr[E_a^{\mathcal{B}}] \geq \Pr[E_a^{\mathcal{A}} \wedge F_a] = \Pr[E_a^{\mathcal{A}}]\Pr[F_a] \geq \frac{1}{l_u^2}\Pr[E_a^{\mathcal{A}}].$$

Next, we compute the adversary's advantage in guessing the challenge bit correctly. If the adversary $\mathcal{A}$ guesses correctly and F_d happens, then $\mathcal{B}$ also guesses correctly. However, if F_d does not happen, then $\mathcal{B}$ has no information about the challenge bit of its experiment, which means that its probability of guessing correctly is $1/2$. The claim follows from

$$\left|\Pr[E_d^{\mathcal{B}}] - \frac{1}{2}\right| = \left|\Pr[E_d^{\mathcal{A}} \wedge F_d] + \frac{1}{2}\Pr[\neg F_d] - \frac{1}{2}\right| = \left|\Pr[E_d^{\mathcal{A}}]\Pr[F_d] - \Pr[F_d]/2\right|$$

$$= \Pr[F_d]\left|\Pr[E_d^{\mathcal{A}}] - \frac{1}{2}\right| \geq \frac{1}{l_u^2}\left|\Pr[E_d^{\mathcal{A}}] - \frac{1}{2}\right|. \qquad \square$$

We remark on two crucial properties of the key exchange experiment. First, instances are independent, because the experiment does not handle partnering notions, authentication predicates and freshness predicates.

Second, the experiment allows arbitrary public keys as partner public keys. This models adversarially controlled users who may not generate keys in the same way honest users do, so the experiment could not be restricted to keys output by a key generation query, even if these keys could be revealed.

This also illustrates why it is a good idea to separate the cryptography from support structures such as public key infrastructures.

The previous simplification of only using one test query is essential. If we had more than one test query, we could not isolate a single pair of keys. While our copy of the key exchange simulator could handle these test queries, its responses would not be consistent with the real simulator. In order to get the proof to work, it seems like we would have to do them one query at a time, just like the proof of Proposition 10.6.

Can we make the security game even simpler? Do we need to have many sessions in parallel? The answer is that we almost certainly do. The reason is that it is possible to design protocols that seem secure if they ever only run one session, but if they run multiple sessions in parallel, there are attacks. We also see this in the obvious approaches to proving such a statement, where we quickly run into problems when we must simulate the other sessions, which is hard without the secret keys.

It is possible to simplify the security game if we specify that the players should have no long-term keys. In this case, any damage caused by multiple sessions could be simulated by an adversary. On the other hand, such protocols cannot be sufficiently secure. This means that any simplification must be part of a more general proof strategy. We return to this question in Section 10.2.5.

Example: A Diffie-Hellman-based protocol Consider one of the Diffie-Hellman-based family of protocols from Example 10.3, with the session key derived as in (10.2)

$$k \leftarrow kdf(ad, g^{b_i}, g^{b_r}, g^{r_i}, g^{r_r}, g^{b_i b_r}, g^{r_i b_r}, g^{b_i r_r}, g^{r_i r_r}). \qquad (10.7)$$

The idea is that as long as the adversary does not trivially compromise the test instance or a partner, we can embed a Computational Diffie-Hellman problem in the test instance. The difficult part is simulating the other instances. We shall rely on the random oracle model, and we shall use a technique called lazy evaluation. Sometimes we do not know exactly where to evaluate the hash function, but in the random oracle model, we can still choose the hash value, as long as we can recognise when someone else queries the hash at the correct value and thereby maintain consistency. We can do this through the Strong Diffie-Hellman oracle.

We need to consider some variants of the Strong Diffie-Hellman problem. The first is the Double Strong Diffie-Hellman problem, where the Strong Diffie-Hellman oracle is made available for both terms in the Diffie-Hellman problem. Second, we need the Square Diffie-Hellman problem, which is when the two terms in the Diffie-Hellman problem are the same. For this variant, we also need to Strong Diffie-Hellman oracle for this problem.

T **Proposition 10.8.** *Let* KEX *be the Diffie-Hellman-based variant* KEX *from Example 10.3 based on a group G with generator g, deriving the session key as in* (10.7). *Let $\mathcal{A}$ be a $(\tau, 1, 2, l_s)$-adversary against* KEX *in the random oracle model, making at most l_h hash queries to* kdf. *Then there exists a (τ'_1, l_h)-adversary $\mathcal{B}_1$ against Strong Diffie-Hellman, a $(\tau'_2, 2l_h)$-adversary $\mathcal{B}_2$ against Strong Diffie-Hellman, a $(\tau'_3, 2l_s)$-adversary $\mathcal{B}_3$ against Double Strong Diffie-Hellman and a $(\tau'_4, 2l_h)$-adversary $\mathcal{B}_4$ against Strong Square Diffie-Hellman, with $\tau'_1, \tau'_2, \tau'_3, \tau'_4$ all essentially equal to a small multiple of τ, such that*

$$\mathbf{Adv}^{\text{kex-fe-i}}_{\text{KEX}}(\mathcal{A}) \leq l_s^2 \mathbf{Adv}^{\text{SDH}}_{G,g}(\mathcal{B}_1) + 2l_s \mathbf{Adv}^{\text{SDH}}_{G,g}(\mathcal{B}_2) + l_s \mathbf{Adv}^{\text{DSDH}}_{G,g}(\mathcal{B}_3)$$
$$+ 2l_s \mathbf{Adv}^{\text{SSqDH}}_{G,g}(\mathcal{B}_4) + \frac{(7l_h + 1)l_s^2}{|G|}.$$

We structure the proof as a series of lemmas.

T **Lemma 10.9.** *The authentication predicate ϕ^i_a holds except with probability $l_s^2/|G|$.*

Proof. Consider first an initiator instance that outputs a session key. It has then sent one message and received one message. If there are any partnered instances, they must have sent the receiving messages. It follows that if there is more than one responder instance partnered to the initiator instance, those responder instances sent the same message.

Next, consider a responder instance that outputs a session key. If there is more than one initiator instance partnered to the responder instance, the initiator instances must have sent the same message.

By the above arguments, if the authentication predicate does not hold, there must be either two responder instances or two initiator instances that sent exactly the same message. Since the messages sent are sampled from the uniform distribution on the group G, the birthday paradox gives us that

$$\Pr[E_a] \leq l_s^2/|G|. \qquad \square$$

Next, we begin to prove that unless the adversary makes a certain query to the hash oracle, it will not have any advantage at all. Suppose the tested instance has transcript $(ad, g^{b_i}, g^{b_r}, g^{r_i}, g^{r_r})$. Let F be the event that the adversary queries the hash oracle at the point given by (10.7).

T **Lemma 10.10.** *We have that*

$$2|\Pr[E_d] - 1/2| \leq \Pr[F].$$

Proof. Because we model the hash value as a random oracle, the hash value at the specific point given by (10.7) is independent of any other hash value. It follows that for someone who has not queried the hash oracle at the specific point, the hash value is indistinguishable from a random value. It follows that

$\Pr[E_d \mid \neg F] = 1/2$, and we get

$$2|\Pr[E_d] - 1/2| = 2|\Pr[E_d \mid F]\Pr[F] + \Pr[E_d \mid \neg F](1 - \Pr[F]) - 1/2| \le \Pr[F].$$
□

We will consider three distinct cases in order to bound $\Pr[F]$. Either:

1. the tested instance has a partner instance and there is no state reveal query for either instance; or

2. the tested instance has a partner instance and there is a state reveal query for exactly one of the instances; or

3. there is a state reveal query for the tested instance, and either there is no partner instance or the partner instance state has been revealed.

For the third case, there are two subcases: either the tested instance's initiator and responder public keys are distinct, or the public keys are the same.

Denote the events that the adversary makes the specific hash query in each case by F_1, F_2, F_{3a} and F_{3b}, respectively. Since the events are mutually exclusive, we get that

$$\Pr[F] = \Pr[F_1] + \Pr[F_2] + \Pr[F_{3a}] + \Pr[F_{3b}].$$

We bound each probability separately.

T **Lemma 10.11.** *There exists a (τ_1', l_h)-adversary $\mathcal{B}_1$ against Strong Diffie-Hellman, with τ_1' essentially equal to τ plus a small multiple of the runtime required for up to l_h Strong Diffie-Hellman oracle queries, such that*

$$\Pr[F_1] \le l_s^2 \mathbf{Adv}_{G,g}^{\mathsf{SDH}}(\mathcal{B}_1) + 2l_s^2 l_h / |G|^2.$$

Proof. The adversary $\mathcal{B}_1$ gets (x, y) as input, samples $i, j \xleftarrow{r} \{1, 2, \dots, l_s\}$ (guesses for the tested instance and its partner), and runs a copy of $\mathcal{A}$ and the key exchange experiment, modified as follows:

- Let the common input of the ith and the jth instances be the associated data ad, the initiator public key pk_I and the responder public key pk_R. The ith instance uses x as its message, and the jth instance uses y.

- If the tested instance is not the ith instance, the ith instance's partner key is not honest, the partner of the ith instance is not the jth instance, or $\mathcal{A}$ reveals the state or the session key of the ith or the jth instance, then $\mathcal{B}_1$ stops immediately with output 1.

- Before the ith and the jth instance have been started: If the adversary queries the hash oracle with (x, y) or (y, x) in the fourth and fifth coordinates, then $\mathcal{B}_1$ outputs 1 and stops.

- After the ith or jth instance have been started: Whenever the adversary queries the hash oracle at $(ad, pk_I, pk_R, x, y, pk_I^{sk_R}, x^{sk_R}, y^{sk_I}, z)$, or at

$(ad, pk_I, pk_R, y, x, pk_I^{sk_R}, y^{sk_R}, x^{sk_I}, z)$ if the ith instance is a responder instance, $\mathcal{B}_1$ uses the Strong Diffie-Hellman oracle to decide if z is the correct answer to the Diffie-Hellman problem (x, y). If z is the correct answer, $\mathcal{B}_1$ outputs z.

When $\mathcal{A}$ stops, $\mathcal{B}_1$ outputs 1 and stops.

Until $\mathcal{B}_1$ stops, the simulation is perfect. The only case where F_1 could have happened, but $\mathcal{B}_1$ stopped would be if a hash query involving x and y happens before the ith instance is started. However, since no information about x or y is available to $\mathcal{A}$ before the ith instance is started, this happens with probability at most $2l_h/|G|^2$.

If F_1 happens with probability $\Pr[F_1]$ when $\mathcal{A}$ interacts with the key exchange experiment, the corresponding event when $\mathcal{A}$ interacts with $\mathcal{B}_1$ happens with probability at least $\Pr[F_1]/l_s^2 - 2l_h/|G|^2$, and when the correct hash query happens $\mathcal{B}_1$ outputs the correct answer. It follows that

$$\mathbf{Adv}_{G,g}^{\mathsf{SDH}}(\mathcal{B}_1) \geq \frac{1}{l_s^2}\Pr[F_1] - \frac{2l_h}{|G|^2}.$$

There is a small change in the cost of answering the ith and jth execute queries. Also, there is a small amount of extra work involved in each hash query, and every hash query may result in a Strong Diffie-Hellman query. The runtime of this extra work is bounded by a small multiple of the runtime required for l_h Strong Diffie-Hellman queries. $\qquad\square$

We now need to embed the challenge into a long-term key, which means that we have to handle instances using this long-term key. We do so by choosing a random session key for these instances and then recognising when the adversary makes hash queries with the correct Diffie-Hellman secrets.

Lemma 10.12. *There exists a $(\tau_2', 2l_h)$-adversary $\mathcal{B}_2$ against Strong Diffie-Hellman, with τ_2' essentially equal to τ plus a small multiple of the runtime required for up to l_h Strong Diffie-Hellman oracle queries, such that*

$$\Pr[F_2] \leq 2l_s\mathbf{Adv}_{G,g}^{\mathsf{SDH}}(\mathcal{B}_2) + \frac{2l_s^2 l_h}{|G|}.$$

Proof. We shall assume without loss of generality that the adversary $\mathcal{A}$ does not send the test query and a state reveal query to the same instance. Since we actually guess which instance *can* receive the test query, in the event that the adversary sends the test query to the partner instance, we can simply redirect the test query, with no loss of advantage.

The adversary $\mathcal{B}_2$ gets (x, y) as input, samples $i \xleftarrow{r} \{1, 2, \ldots, l_s\}$ and $j \xleftarrow{r} \{1, 2\}$ (guesses for the tested instance and the partner key), and runs a copy of $\mathcal{A}$ and the key exchange experiment, modified as follows:

- It samples $k_0, k_1 \xleftarrow{r} \mathfrak{K}$. The test query response will be k_b.

- For the jth key generation query, $\mathcal{B}_2$ returns y as the public key.

- When the ith execute query happens with associated data ad, partner public key y and key pair (pk, sk), $\mathcal{B}_2$ uses x as its message.

- If the tested instance is not the ith instance, its partner key is not the jth key, the ith instance's state or session key is revealed, or the jth long-term key is revealed, then $\mathcal{B}_2$ stops immediately with output 1.

- Before the ith instance has started: If the adversary queries the hash oracle with x, then $\mathcal{B}_2$ outputs 1 and stops.

- After the ith instance has started: Whenever the adversary queries the hash oracle at $(ad, pk, y, x, m, z', z, z'', z''')$, or $(ad, y, pk, m, x, z', z'', z, z''')$ if the ith instance is a responder instance, $\mathcal{B}_2$ uses the Strong Diffie-Hellman oracle to decide if z is the correct answer to the Diffie-Hellman problem (x, y). If z is the correct answer, $\mathcal{B}_2$ outputs z.

- For any execute query (other than the ith) using the jth key pair, $\mathcal{B}_2$ does not query the hash oracle, but samples a session key $k \xleftarrow{r} \mathfrak{K}$.

  For such an instance with associated data ad, initiator public key pk_I, responder public key pk_R, initiator message m_1 and responder message m_2, the adversary $\mathcal{B}_2$ stores $(ad, pk_I, pk_R, m_1, m_2, \rho, r, k)$, where ρ is the instance's role and r is the instance's random value.

  For any subsequent hash oracle query $(ad, pk_I, pk_R, m_1, m_2, z_1, z_2, z_3, z_4)$ that matches a record for an instance, the adversary $\mathcal{B}_2$ uses the value r and two queries to the Strong Diffie-Hellman oracle to check if this is the hash query that the instance would have made. If so, $\mathcal{B}_2$ returns k as the response to the hash query. Otherwise, it forwards the query to the hash oracle and returns its response.

If $\mathcal{A}$ stops or exceeds its bounds, $\mathcal{B}_2$ outputs 1 and stops.

If some hash oracle query includes a group element that an instance later samples as its message (the tested instance's x value included), the test query response or the hash oracle modifications in the final bullet may cause inconsistent hash oracle responses. Since $\mathcal{A}$ would have to guess one of at most l_s values later independently sampled from the uniform distribution on G by $\mathcal{B}_2$, we can bound the probability of this happening by $l_s l_h / |G|$.

If the hash oracle responses are consistent, the simulation is perfect until $\mathcal{B}_2$ stops. If F_2 happens with probability $\Pr[F_2]$ when $\mathcal{A}$ interacts with the key exchange experiment, the corresponding event happens when $\mathcal{A}$ interacts with $\mathcal{B}_2$ with probability at least $\Pr[F_2]/(2l_s) - l_s l_h / |G|$, and when the correct hash query happens $\mathcal{B}_2$ outputs the correct answer. It follows that

$$\mathbf{Adv}_{G,g}^{\mathsf{SDH}}(\mathcal{B}_2) \geq \frac{1}{2l_s} \Pr[F_2] - \frac{l_s l_h}{|G|}.$$

There is a small change in the cost of answering the jth key generation query and the ith execute query. Also, each execute and hash query requires a small amount of extra work, and every hash query may result in two Strong Diffie-Hellman queries. The runtime of this extra work is bounded by a small multiple of the runtime required for l_h Strong Diffie-Hellman queries. □

Since the states are revealed in this case, we must embed the challenge in the public keys, which requires two Strong Diffie-Hellman oracles. Alternatively, if the two public keys happen to be identical, we have to rely on a different problem, the Strong Sqare Diffie-Hellman problem.

Lemma 10.13. *There exists a $(\tau_3', 2l_h)$-adversary $\mathcal{B}_3$ against Double Strong Diffie-Hellman, with τ_3' essentially equal to τ plus a small multiple of the runtime required for up to l_h Strong Diffie-Hellman oracle queries, such that*

$$\Pr[F_{3a}] \leq l_s \mathbf{Adv}_{G,g}^{\mathsf{DSDH}}(\mathcal{B}_3) + \frac{l_s^2 l_h}{|G|}.$$

Lemma 10.14. *There exists a $(\tau_4', 2l_h)$-adversary $\mathcal{B}_4$ against Strong Square Diffie-Hellman, with τ_4' essentially equal to τ plus a small multiple of the runtime required for up to l_h Strong Diffie-Hellman oracle queries, such that*

$$\Pr[F_{3b}] \leq 2l_s \mathbf{Adv}_{G,g}^{\mathsf{SSqDH}}(\mathcal{B}_4) + \frac{2l_s^2 l_h}{|G|}.$$

Exercise 10.5. Prove the above two lemmas.
   Hint: Model the proofs on the proof of the lemma for $\mathcal{B}_2$.

Proposition 10.8 follows from the above lemmas.

Remark. In the first confidentiality case (where the tested instance has a partner), we could have avoided the use of the Strong Diffie-Hellman problem, and instead relied on the Decision Diffie-Hellman problem. If the input (x, y, z) is not a Diffie-Hellman tuple, the probability that the adversary ever queries the hash oracle at z is negligible, so if we ever observe a query including z, we know that (x, y, z) is a Diffie-Hellman tuple.

Remark. The reliance on Double Strong Diffie-Hellman in Proposition 10.8 is probably acceptable, since it seems unlikely that the second Strong Diffie-Hellman oracle will help the adversary, and the problem is weaker than the Gap Diffie-Hellman problem. The reliance on Strong Square Diffie-Hellman is also probably acceptable. A solver for Square Diffie-Hellman can be used to solve Computational Diffie-Hellman. The obstacle is the Strong Diffie-Hellman oracle. However, if Strong Square Diffie-Hellman is seen as problematic, it can be removed if we modify the protocol to reject identical initiator and responder public keys. That is, the protocol would not allow Alice to agree on a key with herself. This is usually not desirable.

Alternatively, we can get a proof that relies only on Strong Diffie-Hellman if we restrict attention to non-invasive adversaries. In this case, however, it is possible to do a more careful analysis that does not rely on Propositions 10.6 and 10.7 to get a much stronger result.

E *Exercise 10.6.* Show that the protocol from Example 10.7 cannot be secure against an adversary that makes state reveal queries.

Prove a result similar to Proposition 10.8 for a non-invasive adversary.

E *Exercise 10.7.* A *key compromise impersonation attack* is an attack where Eve compromises the long-term key of Alice and uses this to impersonate Bob to Alice. In this case, Bob's long-term key is not compromised, which means that any Alice instance whose state has not been revealed should expect security.

Consider the variant of the Diffie-Hellman-based protocol from Example 10.3 that derives the session key using

$$k \leftarrow kdf(ad, g^{b_i}, g^{b_r}, g^{r_i}, g^{r_r}, g^{b_i b_r}, g^{r_i r_r}).$$

Show that this protocol is insecure when long-term keys are compromised.

Example: The NAXOS Trick When deriving the session key using (10.7), the term $g^{r_i r_r}$ provides security against long-term key reveals, $g^{b_i b_r}$ protects against state reveals, and the terms $g^{r_i b_r}$ and $g^{b_i r_r}$ provide authentication.

There is a trick that provides protection against state reveals using only three terms, which is faster. The so-called *NAXOS trick* is to hash the per-session randomness r_i and r_r with the corresponding long-term keys to get the per-session exponents $\tilde{r}_i = kdf_1(ad, b_i, r_i)$ and $\tilde{r}_r = kdf_2(ad, b_r, r_r)$, sending the messages $x_i \leftarrow g^{\tilde{r}_i}$ and $x_r \leftarrow g^{\tilde{r}_r}$ and deriving the key as

$$k \leftarrow kdf_0(ad, g^{b_i}, g^{b_r}, g^{\tilde{r}_i}, g^{\tilde{r}_r}, g^{b_i \tilde{r}_r}, g^{\tilde{r}_i b_r}, g^{\tilde{r}_i \tilde{r}_r}).$$

An adversary that gets either of b_i and r_i, but not both, knows nothing about $\tilde{r}_i$, which means essentially that its security reduces to that of the scheme from Proposition 10.8, but without session reveal queries.

E *Exercise 10.8.* Let $\mathcal{A}$ be a $(\tau, 1, 2, l_s)$-adversary against the above key exchange protocol NAXOS. Specify one or more adversaries against Gap Diffie-Hellman (depending on $\mathcal{A}$), explain the relationship their runtime to the runtime of $\mathcal{A}$, and bound $\mathbf{Adv}^{\text{kex-fe-e}}_{\text{NAXOS}}(\mathcal{A})$ using the advantage of the specified adversaries.

E *Exercise 10.9.* Adapt the NAXOS trick to the KEM-based key exchange protocol from Example 10.7. Prove it secure.

The NAXOS trick is somewhat problematic, in the sense that it exploits features of our cryptographic model that may not be valid in the real world. Specifically, our model allows the adversary to learn an instance's random tape, but it does not allow an adversary to learn the initiator's intermediate value $\tilde{r}_i = kdf_2(ad, r_i, b_i)$. It may very well be realistic that r_i can leak without b_i leaking, for example if the latter only exists in protected storage. But suppose the real world adversary learns r_i because it was not erased. It is then not obvious why $\tilde{r}_i$ would be erased.

One plausible case where the cryptographic model is realistic, is if the pseudo-random generator fails, but it fails in a somewhat specific way, namely that it produces non-repeating outputs that are predictable (they have low entropy). This has happened in the real world, with the likely compromise of the Dual-EC-DRBG pseudo-random generator standard as the best example. In this somewhat special case, NAXOS would provide real security.

Remark. The above discussion shows that it is important to validate the cryptographic model against the real world, and in particular be aware of cryptographic schemes trying to "cheat" by exploiting features of the cryptographic model in an unrealistic way.

10.2.5 Unauthenticated Security Suffices – Maybe

Our goal now is to add authentication to an unauthenticated protocol by signing the protocol transcript, thereby achieving stronger security. We will achieve this as follows: The party that sends the last message of the underlying unauthenticated key exchange protocol will attach a signature to it. And then it will wait for the other party to reply with a signature of its own.

This design paradigm for key exchange schemes is useful, as shown by the discussion at the end of the section. The scheme in this section essentially generalises the Signed Diffie-Hellman scheme from Example 10.2. Its message structure is also very similar to the construction from Example 10.8.

Example 10.9. Let $\text{KEX}_0 = (\mathfrak{K}, \mathcal{K}_0, \mathcal{I}_0, \mathcal{R}_0)$ be an unauthenticated key exchange protocol, and let $\text{SIG} = (\mathcal{K}, \mathcal{S}, \mathcal{V})$ be a signature scheme whose plaintext set contains the set of possible transcripts for KEX. We construct a new key exchange protocol $\text{KEX} = (\mathfrak{K}, \mathcal{K}, \mathcal{I}, \mathcal{R})$ as follows:

- The *key generation* algorithm $\mathcal{K}$ is just the signature key generation algorithm, treating the verification key vk as the public key and the signing key sk as the private key.

- The interactive algorithms of KEX are based on the corresponding interactive algorithms of KEX_0. Let $\mathcal{P}$ denote either $\mathcal{I}$ or $\mathcal{R}$, and let $\mathcal{P}_0$ denote the corresponding choice of $\mathcal{I}_0$ and $\mathcal{R}_0$. We denote the partner verification key by vk'' and the transcript of the KEX_0 instance by tr_0.

The interactive algorithm $\mathcal{P}$ takes as input associated data ad, initiator public keys vk, responder public key vk' and a private key sk. It runs the $\mathcal{P}_0$ algorithm with input $(ad, \bot, \bot, \bot)$. If $\mathcal{P}_0$ ever outputs $\bot$, $\mathcal{P}$ immediately outputs $\bot$ and stops. Any messages $\mathcal{P}_0$ sends $\mathcal{P}$ sends, and $\mathcal{P}$ forwards any received messages to $\mathcal{P}_0$, with the following modifications:

- If $\mathcal{P}_0$ sends a message m and outputs a session key k, $\mathcal{P}$ computes $\sigma \leftarrow \mathcal{S}(sk, tr)$, where $tr = (ad, vk, vk', tr_0)$, and sends (m, σ).
- If $\mathcal{P}$ receives a message m with a signature σ such that $\mathcal{V}(vk'', tr) = 1$, it sends m to $\mathcal{P}_0$. If $\mathcal{P}_0$ outputs $\bot$ or sends any message, $\mathcal{P}$ outputs $\bot$. Otherwise, $\mathcal{P}_0$ does not send a message and outputs a session key k, and $\mathcal{P}$ computes $\sigma' \leftarrow \mathcal{S}(sk, (tr, \sigma))$, sends σ' and outputs k.
- If $\mathcal{P}$ sent $(\cdot, \sigma)$ after $\mathcal{P}_0$ output k, and $\mathcal{P}$ then receives σ', $\mathcal{P}$ outputs k if $\mathcal{V}(vk'', (tr, \sigma), \sigma') = 1$, otherwise it outputs $\bot$.

Theorem 10.15. *Let* KEX$_0$ *be an unauthenticated key exchange protocol and let* KEX *be the above construction using the signature scheme* SIG. *Let $\mathcal{A}$ be a non-invasive (τ, l_t, l_u, l_s)-adversary against* KEX. *Then there exists a non-invasive $(\tau_1', l_t, 1, l_s)$-adversary $\mathcal{B}_1$ against* KEX$_0$ *and a (τ_2', l_u, l_s)-adversary $\mathcal{B}_2$ against* SIG, *with τ_1' and τ_2' essentially equal to τ, such that*

$$\mathbf{Adv}_{\text{KEX}}^{\text{kex-f-e}}(\mathcal{A}) \leq \mathbf{Adv}_{\text{KEX}_0}^{\text{kex-0-0}}(\mathcal{B}_1) + 2\mathbf{Adv}_{\text{SIG}}^{\text{suf-mu-cma}}(\mathcal{B}_2).$$

We first modify the game between the adversary and the experiment by replacing signature forgeries with invalid signatures. This either ensures that every instance that outputs a key will have a partner, or gives us an adversary against the signature scheme. Next, we show that the signatures can be simulated and create an adversary against the unauthenticated key exchange protocol where any instance that outputs a key has a partner.

This adversary need not be network-passive. It can tamper with any message, and even make up messages. This will eventually be discovered, however.

Proof of Theorem 10.15   In the following, let $E_{d,i}$ and $E_{a,i}$ be the events in Game i corresponding to E_d and E_a in Definition 10.8.

Game 0   The initial game is the key exchange experiment interacting with the adversary $\mathcal{A}$. Then

$$\mathbf{Adv}_{\text{KEX}}^{\text{kex-f-e}}(\mathcal{A}) = \max\{2|\Pr[E_{d,0}] - 1/2|, \Pr[E_{a,0}]\}. \tag{10.8}$$

Let $\tau_0 = \tau$ be the runtime of the experiment and the adversary.

Game 1   We modify the experiment to invalidate forged signatures:

- When the experiment sends a signature σ on behalf of some instance to the adversary, it recreates the instance's transcript and records the transcript and the signature on the public keys and the transcript. Then the signature is forwarded to the adversary.

- When the adversary sends a signature σ to an instance whose long-term key has not been revealed, the experiment recreates the instance's transcript and checks if this transcript and signature has been recorded for these public keys. If it has, then the signature is processed as usual. Otherwise, the experiment samples $b' \xleftarrow{r} \{0,1\}$ and stops, pretending the adversary output b'.

Unless a forgery happens, this game proceeds exactly as the previous game. If a forgery happens, the game immediately stops. So the runtime bound τ_1 for this game is essentially equal to τ_0.

By inspection, we see that $\Pr[E_{a,1}] = 0$. This observation together with (10.8) and the following two lemmas prove the theorem.

The first lemma shows that if the adversary's advantage has changed from Game 0 to Game 1, then we have an adversary against the signature scheme. The second lemma shows that there is an adversary against the underlying key exchange protocol with respect to $\phi_f^{0/0}$ with the same guessing advantage as our adversary in Game 1.

Lemma 10.16. *There exists a (τ_2', l_u, l_s)-adversary $\mathcal{B}_2$ against* SIG, *with τ_2' essentially equal to τ_1, such that*

$$\mathbf{Adv}_{\mathrm{SIG}}^{\mathsf{suf\text{-}mu\text{-}cma}}(\mathcal{B}_2) \geq \max\{|\Pr[E_{a,1}] - \Pr[E_{a,0}]|, |\Pr[E_{d,1}] - \Pr[E_{d,0}]|\}.$$

Proof. The adversary $\mathcal{B}_2$ runs a copy of $\mathcal{A}$ and the experiment from this game, modified as follows:

- For the ith key generation query, $\mathcal{B}_2$ makes a key generation query and receives a verification key vk_i, which it sends to $\mathcal{A}$.
- When $\mathcal{A}$ reveals the long-term key for the ith public key, $\mathcal{B}_2$ makes a key reveal query i and receives a signing key sk_i, which it sends to $\mathcal{A}$.
- When the experiment needs to sign a message m with the ith signing key, $\mathcal{B}_2$ makes a sign query for (i, m) and receives a signature σ, and the experiment uses σ as the signature.
- If the experiment receives a forgery σ for a transcript m for the ith public key, $\mathcal{B}_2$ outputs (i, m, σ) and stops.

If $\mathcal{A}$ stops for any other reason, $\mathcal{B}_2$ stops without any output.

The adversary $\mathcal{B}_2$ perfectly simulates Game 1. Since $\mathcal{B}_2$ just forwards some messages to the signature experiment and does some extra accounting compared to Game 1, it follows that τ_2' is essentially equal to τ_1.

Let F_i be the event that there is a signature forgery in Game i. By inspection, if $\mathcal{A}$ forges a signature, $\mathcal{B}_2$ will output a valid forgery, so

$$\mathbf{Adv}_{\mathrm{SIG}}^{\mathsf{suf\text{-}mu\text{-}cma}}(\mathcal{B}_2) = \Pr[F_1].$$

Game 0 and 1 behave identically we get a forgery. By Lemma 7.7, we get

$$|\Pr[E_{a,1}] - \Pr[E_{a,0}]| \leq \Pr[F_1] \quad \text{and} \quad |\Pr[E_{d,1}] - \Pr[E_{d,0}]| \leq \Pr[F_1]. \quad \square$$

T **Lemma 10.17.** *There exists a non-invasive* $(\tau_1', l_t, 1, l_s)$-*adversary* $\mathcal{B}_1$ *against* KEX$_0$, *with* τ_1' *essentially equal to* τ_1, *such that*

$$\mathbf{Adv}_{\mathrm{KEX}_0}^{\mathrm{kex\text{-}0\text{-}0}}(\mathcal{B}_1) = 2|\Pr[E_{d,1}] - 1/2|.$$

Proof. The adversary $\mathcal{B}_1$ runs a copy of $\mathcal{A}$ and the experiment from this game, modified as follows:

- When $\mathcal{A}$ makes its first key generation query, $\mathcal{B}_1$ makes a key generation query to the key exchange experiment for KEX$_0$.

- When the key exchange experiment for KEX would start an instance of KEX$_0$, $\mathcal{B}_1$ instead sends an execute query to the key exchange simulator for KEX$_0$. Any message that would be sent to an instance of KEX$_0$ is instead sent as a send query to the key exchange experiment for KEX$_0$. Any message returned by the key exchange simulator for KEX$_0$ is treated as if it was output by the corresponding KEX$_0$ instance.

- When $\mathcal{A}$ makes a session key reveal query or a test query, $\mathcal{B}_1$ makes a corresponding session key reveal query or test query to the key exchange experiment for KEX$_0$ and returns the result.

By inspection we see that $\mathcal{B}_1$ and the key exchange experiment for KEX$_0$ perfectly simulate the key exchange experiment for KEX. Since $\mathcal{B}_1$ just forwards messages and does some accounting, τ_1' is essentially equal to τ_1.

The simulator **Sim** ensures that for every instance that outputs a session key, either there is a partner instance or there is a long-term key reveal for its partner. In other words, for the test query instances in Game 1, $\phi_f^{\mathrm{f}/\mathrm{e}}$ holds if and only if $\phi_f^{0/0}$ holds. The claim follows. $\qquad\square$

If SIG is deterministic, we could remove the requirement that $\mathcal{A}$ is non-invasive. The reason we can only deal with a non-invasive adversary is that revealing the randomness used to generate a signature may reveal the signing key, which means that a state reveal query may be functionally equivalent to a long-term key reveal, which would allow the adversary to win trivially.

The proof fails because the signature adversary we create does not know the randomness used in signing queries, which means that it could not respond correctly to state reveal queries. For a deterministic signature scheme, this obstacle is removed. A derandomised signature scheme is deterministic.

E *Exercise* 10.10. State and prove a variant of Theorem 10.15 for a general adversary and deterministic signature scheme.

E *Exercise* 10.11. Let KEX$_0$ be an unauthenticated key exchange scheme. Let $\mathcal{A}$ be a passive $(\tau, 1, 1, l_s)$ adversary against KEX$_0$. Show that there exists a passive $(\tau', 1, 1, 2)$-adversary $\mathcal{B}$ against KEX$_0$, with τ' essentially equal to τ,

such that
$$\mathbf{Adv}_{\text{KEX}_0}^{\text{kex-0-0}}(\mathcal{A}) \leq l_s^2 \mathbf{Adv}_{\text{KEX}_0}^{\text{kex-0-0}}(\mathcal{B}).$$

This exercise concludes the classical key exchange scheme design paradigm of adding signatures to an unauthenticated key exchange scheme. The idea is that we design an extremely simple unauthenticated key exchange scheme and prove it secure when there is a single initiator instance and a single responder instance. Then we apply Exercise 10.11 to show that this protocol is secure also when there are many instances in addition to the tested instance. Then we apply Proposition 10.6 to prove that the protocol is secure when there are many test queries. Next, we apply the construction in this section together with Theorem 10.15 to get a signature scheme with strong security. One problem is that the resulting security result is not very tight.

It is tempting to combine a result like Proposition 10.1 with the above design strategy, instead of using Exercise 10.11 and Proposition 10.6. This does not work, since Proposition 10.1 applies to a passive adversary, but Theorem 10.15 gives a non-invasive adversary.

There are a alternative strategies for proving tighter results about concrete key exchange schemes, but these are beyond the scope of this book.

10.2.6 Benignly Malleable Conversations

The matching conversation partnering notion from Definition 10.3 is somewhat inconvenient, because it is very strong. In particular, the design paradigm using digital signatures from Section 10.2.5 requires very strong security from the digital signature schemes used. If a signature on a message can be manipulated to create a new signature on the same message, an adversary can break partnering, allowing a test query for one instance and a session reveal for the other instance. In some sense, this is not a real attack, so we want to investigate if we can prove something less.

Before we define our partnering notion, we need to define some structures. Let $\text{KEX} = (\mathfrak{K}, \mathcal{K}, \mathcal{I}, \mathcal{R})$ be a key exchange scheme with associated data set $\mathfrak{F}$. Define S_1, S_2 and S_{12} be the set of public keys, secret keys and key pairs, respectively, that $\mathcal{K}$ can output. Let S_3 be the set of random tapes for instances of $\mathcal{I}$ or $\mathcal{R}$, and let S_4 be the set of transcripts for instances of $\mathcal{I}$ and $\mathcal{R}$.

Let $T_{0,\mathcal{I}}$ be the subset of $S_1 \times S_1 \times S_2 \times \mathfrak{F} \times S_3 \times S_4 \times \mathfrak{K}$ of tuples $(pk_I, pk_R, sk, ad, r, tr, k)$ such that $(pk_I, sk) \in S_{12}$ and an instance of $\mathcal{I}$ with random tape r and input ad, (pk_I, sk) and pk_R could have transcript tr and output k. Let $T_{0,\mathcal{R}}$ be the corresponding set for $\mathcal{R}$. Let T be the subset of $S_1 \times S_1 \times \mathfrak{F} \times S_4 \times \mathfrak{K}$ of tuples (pk_I, pk_R, ad, tr, k) for which r and sk exist such that $(pk_I, pk_R, sk, ad, r, tr, k) \in T_{0,\mathcal{I}}$ and r' and sk' exist such that $(pk_I, pk_R, sk', ad, r', tr, k) \in T_{0,\mathcal{R}}$.

Definition 10.10. Let KEX be a key exchange scheme, and let T be as defined above. The *benignly malleable conversations* relation $\sim_b$ on transcripts for KEX says that transcripts tr and tr' are related if and only if they contain the same public keys pk_I, pk_R and the same associated data ad, and for some k we have that $(pk_I, pk_R, ad, tr, k) \in T$ and $(pk_I, pk_R, ad, tr', k) \in T$, where one of the transcripts may be augmented by the final message of the other transcript.

We say that the relation $\sim_b$ is *admissible* for a key exchange scheme if there exists a bookkeeping algorithm for deciding if two transcripts are related with runtime cost roughly the same as running the key exchange protocol, and there exists an algorithm for determining the relation graph for a collection of instance transcripts that has roughly the same computational cost as sorting the list of transcripts with a comparison operation that has roughly the same computational cost as running the key exchange protocol.

The relation $\sim_b$ is symmetric, and any transcript of an instance that outputs a key is related to itself. It follows that as long as $\sim_b$ is admissible for a key exchange scheme, it is also a valid partnering notion for that key exchange scheme.

Exercise 10.12. Show that $\sim_b$ is admissible for the Signed Diffie-Hellman protocol from Example 10.2.

Remark. Our generic theorems are independent of the particular partnering notion used, so we do not have to reprove every theorem for a new partnering notion. The proofs for concrete constructions would have to be redone, but the constructions would likely be different, so this is not unexpected.

10.3   KEY EXCHANGE FROM KEY ENCAPSULATION

We shall now use the two design paradigms we studied in Section 10.2 to understand the security of the three key exchange schemes from Examples 10.5–10.7 based on key encapsulation mechanisms from Section 8.2.

10.3.1   Key Exchange with Signatures

We begin with the unauthenticated scheme from Example 10.5. If we just have one initiator instance and one responder instance, unauthenticated security follows from the chosen plaintext security of the key encapsulation mechanism. The idea is simple: We use the encapsulation key as the initator instance's message and the single challenge ciphertext as the responder instance's message. If the adversary's behaviour ruins this plan, the corresponding behaviour would result in a non-fresh execution when the adversary interacts with the key exchange experiment, which we can safely ignore.

Proposition 10.18. *Let* KEX *be the key exchange scheme from Example 10.5 based on a KEM* KEM. *Let* $\mathcal{A}$ *be a non-invasive* $(\tau, 1, 1, 2)$-*adversary against* KEX. *Then there exists a* $(\tau', 1, 0)$-*adversary* $\mathcal{B}$ *against* KEM, *where* τ' *is* *essentially equal to τ, such that*

$$\mathbf{Adv}_{\text{KEX}}^{\text{kex-0-0}}(\mathcal{A}) \leq \mathbf{Adv}_{\text{KEM}}^{\text{ror-cpa}}(\mathcal{B}).$$

Proof. The adversary $\mathcal{B}$ gets an encapsulation key ek as input, makes a challenge query to get an encapsulation x and a key k, and then runs a copy of $\mathcal{A}$ together with a simulator **Sim** that works as follows:

- When $\mathcal{A}$ makes its key generation query, **Sim** responds with $(\perp, \perp)$.
- When $\mathcal{A}$ makes the first execute query for an initiator instance and ith execute query overall, **Sim** responds with (i, ek).
- When $\mathcal{A}$ makes the first execute query for a responder instance and jth execute query overall, **Sim** responds with j.
- When $\mathcal{A}$ makes the send query (j, ek), **Sim** responds with (j, x).
- When $\mathcal{A}$ makes the send query (i, x), **Sim** does nothing.
- When $\mathcal{A}$, after making the above send query to an instance, makes a test query to that instance, **Sim** responds with k.
- If $\mathcal{A}$ makes any other queries than described above, $\mathcal{B}$ samples $b' \xleftarrow{r} \{0,1\}$ outputs b' and stops.

When $\mathcal{A}$ outputs b', $\mathcal{B}$ outputs b' and stops.

Until $\mathcal{B}$ outputs a random bit and stops, **Sim** and the key encapsulation experiment with challenge bit b together perfectly simulate the key exchange experiment with challenge bit b.

When $\mathcal{B}$ outputs a random bit, that bit equals the challenge bit of the key encapsulation mechanism with probability $1/2$. Also, the corresponding interaction with the key exchange experiment would either result in a non-fresh execution or no test query, in which case the bit output by $\mathcal{A}$ would equal, respectively, the random bit b'' or the unused challenge bit b of the key exchange experiment with probability $1/2$.

Because of the perfect simulation, when interacting with **Sim** and the key encapsulation experiment, $\mathcal{A}$ outputs a bit which equals the key encapsulation experiment's challenge bit with the same probability that $\mathcal{A}$ interacting with the key exchange experiment results in a fresh execution and the $\mathcal{A}$'s output bit equals the key exchange experiment's challenge bit. The claim follows. $\square$

The signed version of the unauthenticated protocol from Example 10.6 is (except for minor differences in message layout, which we can ignore) equal to the construction from Section 10.2.5 applied to Example 10.5. By applying Theorem 10.15 to deal with the signatures, Proposition 10.6 to deal with the many test queries, Exercise 10.11 to deal with the many sessions, and finally Proposition 10.18 we have proved the following result.

Theorem 10.19. *Let* KEX *be the key exchange protocol from Example 10.6 based on a key encapsulation mechanism* KEM *and a digital signature scheme* SIG. *Let* $\mathcal{A}$ *be a non-invasive* (τ, l_t, l_u, l_s)-*adversary against* KEX. *Then there* *exists a* (τ_1, l_s, l_u)-*adversary* $\mathcal{B}_1$ *against* SIG *and a* $(\tau_2, 1, 0)$-*adversary* $\mathcal{B}_2$ *against* KEM, *with* τ_1 *and* τ_2 *essentially equal to* τ, *such that*

$$\mathbf{Adv}_{\text{KEX}}^{\text{kex-f-e}}(\mathcal{A}) \le 2\mathbf{Adv}_{\text{SIG}}^{\text{suf-mu-cma}}(\mathcal{B}_1) + l_t l_s^2 \mathbf{Adv}_{\text{KEM}}^{\text{ror-cpa}}(\mathcal{B}_2).$$

10.3.2 Implicitly Authenticated Key Exchange

We now study the implicitly authenticated scheme from Example 10.7. The proof is in many ways similar to the proof of the Diffie-Hellman based protocol from Proposition 10.8. Note that this protocol is not secure when an adversary reveals the state of the partners, which simplifies the proof somewhat. But the general approach is the same: Either the test instance has a partner, or the test instance is an initiator and the responder's key pair is never compromised, or the test instance is a responder and the initiator's key pair is never compromised.

Proposition 10.20. *Let* KEX *be the key exchange protocol from Example 10.7 based on a key encapsulation mechanism* KEM *and a key derivation function* kdf. *Let* $\mathcal{A}$ *be a* (τ, l_t, l_u, l_s)-*adversary against* KEX *with the random oracle* kdf. *Then there exists a* $(\tau_1, l_s, 0)$-*adversary* $\mathcal{B}_1$ *against* KEM *and a* (τ_2, l_s, l_s)-*adversary* $\mathcal{B}_2$, *with* τ_1 *and* τ_2 *essentially equal to* τ, *such that*

$$\mathbf{Adv}_{\text{KEX}}^{\text{kex-f-i}}(\mathcal{A}) \le 2\mathbf{Col}_{\text{KEM}}(l_s) + l_s^2 \mathbf{Adv}_{\text{KEM}}^{\text{ror-cpa}}(\mathcal{B}_1) + l_s \mathbf{Adv}_{\text{KEM}}^{\text{ror-cca}}(\mathcal{B}_2).$$

We structure the proof as a sequence of smaller results.

The first result shows that implicit authentication holds. The only way it can not hold is if there is a collision among keys or ciphertexts.

Exercise 10.13. Show that the authentication predicate ϕ_a^i fails with probability at most $2\mathbf{Col}_{\text{KEM}}(l_s)$.

The next result shows that the adversary has no advantage unless the adversary queries the random oracle at the right point. Let the test instance have transcript $(ad, ek, ek', ek'', x', x, x'')$, where k, k' and k'' are the decapsulations of x, x' and x'', respectively. Let F be the event that the adversary queries the random oracle at $(ad, ek, ek', ek'', x, x', x'', k, k', k'')$.

Exercise 10.14. Show that $2|E_d - 1/2| \le \Pr[F]$.

When we want to bound $\Pr[F]$ we can consider two disjoint cases: Either the tested instance has a partner instance, or it does not. We denote the event that the adversary makes the specific hash query in each case by F_1 and F_2, respectively. Since the two events are disjoint, we get that

$$\Pr[F] = \Pr[F_1] + \Pr[F_2].$$

The strategy is now to embed the challenge in the test instance. In the former case we embed it in ek'' and x''. In the latter case, we embed it in the partner's long-term key and the corresponding encapsulation. In the latter case, we need to be able to decapsulate ciphertexts in order to correctly simulate other instances using the partner's long-term key.

Exercise 10.15. Show that there exists a $(\tau_1, l_s, 0)$-adversary $\mathcal{B}_1$, with τ_1 essentially equal to τ, such that

$$\Pr[F_1] \leq l_s^2 \mathbf{Adv}_{\mathrm{KEM}}^{\mathrm{ror\text{-}cpa}}(\mathcal{B}_1).$$

Exercise 10.16. Show that there exists a (τ_2, l_s, l_s)-adversary $\mathcal{B}_2$, with τ_2 essentially equal to τ, such that

$$\Pr[F_2] \leq 2l_s \mathbf{Adv}_{\mathrm{KEM}}^{\mathrm{ror\text{-}cca}}(\mathcal{B}_2).$$

Exercises 10.13–10.16 and the arguments above prove Proposition 10.20.

We use the construction from Section 10.2.2 to the scheme from Example 10.7 to get a key exchange scheme with explicit authentication. Then we apply Proposition 10.2 to deal with the tags, Proposition 10.6 to deal with the many test queries, Proposition 10.7 to deal with the many users, and finally Proposition 10.20 applies, which proves the following theorem.

Theorem 10.21. *Let* KEX_0 *be the scheme from Example 10.7 based on a KEM* KEM, *and let* KEX *be the construction from Section 10.2.2 applied to* KEX_0. *Let* $\mathcal{A}$ *be a non-invasive* (τ, l_t, l_u, l_s)-*adversary against* KEX. *Then there exists a* (τ', l_s, l_s)-*adversary* $\mathcal{B}$ *against* KEM, *with* τ' *essentially equal to* τ, *such that*

$$\mathbf{Adv}_{\mathrm{KEX}}^{\mathrm{kex\text{-}f\text{-}e}}(\mathcal{A}) \leq l_t l_s^3 l_u^2 \mathbf{Adv}_{\mathrm{KEM}}^{\mathrm{ror\text{-}cca}}(\mathcal{B}) + 2l_t l_s l_u^2 \mathbf{Col}_{\mathrm{KEM}}(l_s) + \frac{l_s}{|\mathfrak{T}|}.$$

10.3.3 Key Encapsulation from Key Exchange

The two previous sections studied how to build key exchange schemes based on key encapsulation mechanisms. This section studies how to build a key encapsulation mechanism from a particular class of key exchange schemes.

Remark. This section is included because it is neat, not because it is useful.

We begin with a two-round unauthenticated key exchange scheme $\mathrm{KEX} = (\mathfrak{K}, -, \mathcal{I}, \mathcal{R})$. The initiator $\mathcal{I}$ is an interactive algorithm that takes no input, sends a single initiator message, waits for a single responder message and then outputs $\perp$ or a session key.

The responder $\mathcal{R}$ is an interactive algorithm that takes no input, waits for a single initiator message and then either outputs $\perp$ or sends a single responder message and outputs a session key. Note that while we say that the responder is interactive, it does not actually have any interaction, so we may consider it to be an algorithm that takes as input an initiator message and outputs either $\perp$ or a responder message and a session key. By the correctness of the key exchange scheme, we know that as long as the initiator message was output by $\mathcal{I}$, the responder will never output $\perp$.

The initiator $\mathcal{I}$ is probabilistic, which essentially just means that it is a deterministic algorithm that has access to a random tape. We may also allow ourselves to assume that the random tape is bounded in length. (While it is certainly possible to design key exchange algorithms that use an unbounded amount of randomness, in practice we never use such schemes since they also have unbounded runtime. This is impractical.) In other words, we may assume that the initiator algorithm is a deterministic algorithm that on input of a random tape, first sends a message and then outputs $\perp$ or a session key after receiving a responder message. It is then clear that we can construct two distinct deterministic algorithms: One that on input of a random tape outputs an initiator message, and one that on input of a random tape and a responder message outputs either $\perp$ or a session key.

And at this point, we have our key encapsulation mechanism.

Example 10.10. Let $\text{KEX} = (\mathfrak{K}, -, \mathcal{I}, \mathcal{R})$ be a two-round unauthenticated key exchange scheme, where the random tape used by $\mathcal{I}$ is sampled uniformly at random from a set $\mathfrak{N}$. The key encapsulation mechanism $\text{KEM} = (\mathfrak{K}, \mathcal{KK}, \mathcal{KE}, \mathcal{KD})$ works as follows:

- The *key generation* algorithm $\mathcal{KK}$ samples $r \xleftarrow{r} \mathfrak{N}$ and runs $\mathcal{I}$ with random tape r until it outputs an initiator message m_0. Then it outputs the encapsulation key $ek = m_0$ and the decapsulation key $dk = r$.

- The *encapsulation* algorithm $\mathcal{KE}$ takes as input an encapsulation key $ek = m_0$. It runs $\mathcal{R}$ and gives it the message m_0, after which it either outputs $\perp$ or sends m_1 and outputs k. In the former case, $\mathcal{KE}$ outputs $\perp$. In the latter case, $\mathcal{KE}$ outputs $x = m_1$ and k.

- The *decapsulation* algorithm $\mathcal{KD}$ takes as input a decapsulation key $dk = r$ and an encapsulation $x = m_1$. The decapsulation algorithm runs $\mathcal{I}$ with random tape dk and ignores the message it sends. It then sends $\mathcal{I}$ the message m_1, after which it outputs either $\perp$ or k. In the former case, $\mathcal{KD}$ outputs $\perp$. In the latter case, $\mathcal{KD}$ outputs k.

Exercise 10.17. Show that the above is a key encapsulation mechanism.

The chosen plaintext security of the key encapsulation method follows from the unauthenticated security of the key exchange scheme. The idea is that in the chosen plaintext game, the adversary gets one key pair and is then allowed to get many encapsulations and either the encapsulated keys or independent random keys. And this exactly corresponds to starting one initiator instance, many responder instances, giving each of the responder instances the message output by the initiator instance and then sending test queries for each of the responder instances to get either the session key or random keys. (We are careful never to send any messages to the initiator instance, so the responders all have the initiator as a partner. Therefore, they will all be fresh.)

T **Proposition 10.22.** *Let $\mathcal{A}$ be a $(\tau, l_c, 0)$-adversary against* KEM *from Example 10.10. Then there exists a non-invasive, $(\tau', l_c, 1, l_c + 1)$-adversary $\mathcal{B}$ against* KEX, *with τ' essentially equal to τ, and*

$$\mathbf{Adv}_{\mathrm{KEM}}^{\mathrm{ror\text{-}cpa}}(\mathcal{A}) = \mathbf{Adv}_{\mathrm{KEX}}^{\mathrm{kex\text{-}0\text{-}0}}(\mathcal{B}).$$

E *Exercise* 10.18. Prove the above proposition.

10.4 SINGLE-MESSAGE KEY EXCHANGE

In many communications networks the latency is non-trivial. It is therefore important to minimise the number of messages sent between parties in a protocol. The example protocols we have looked at so far are two- and three-message protocols. It is worthwhile investigating if we can do key exchange in a single message. Unfortunately, it can be proven that we cannot achieve strong authentication with a single message, nor can we achieve security against long-term key reveals, unless we allow long-term keys to change.

E *Example* 10.11. Let KEM $= (\mathcal{KK}, \mathcal{KE}, \mathcal{KD})$ be a key encapsulation mechanism for $\mathfrak{K}$ with associated data from $\mathfrak{F}$ and let SIG $= (\mathcal{K}, \mathcal{S}, \mathcal{V})$ be a signature scheme. We construct the following key exchange protocol KEX $= (\mathfrak{K}, \mathcal{K}, \mathcal{I}, \mathcal{R})$:

- The *key generation* protocol $\mathcal{K}$ computes $(ek, dk) \leftarrow \mathcal{KK}$, $(vk, sk) \leftarrow \mathcal{K}$ and outputs the public key (ek, vk) and the private key (dk, sk).

- The *initiator* algorithm takes as input ad, (ek, vk), (ek', vk') and (dk, sk), computes $(c, k) \leftarrow \mathcal{KE}(ek', (ad, ek, vk, ek', vk'))$ and $\sigma \leftarrow \mathcal{S}(sk, c)$, sends (c, σ) and outputs k.

- The *responder* algorithm takes as input ad, (ek, vk), (ek', vk') and (dk', sk'). When it receives $m = (c, \sigma)$, it checks that $\mathcal{V}(vk, c) = 1$ and computes $k \leftarrow \mathcal{KD}(dk, (ad, ek, vk, ek', vk'), c)$. If the signature verification or the decryption fails, it outputs $\perp$. Otherwise it outputs k.

E *Exercise* 10.19. Show that the above protocol is a key exchange protocol.

T **Proposition 10.23.** *Let $\mathcal{A}$ be a non-invasive $(\tau, 1, l_u, l_s)$-adversary against the above key exchange protocol* KEX. *Then there exists a (τ'_1, l_s, l_s)-adversary $\mathcal{B}_1$ against* KEM *and a (τ'_2, l_s)-adversary $\mathcal{B}_2$ against* SIG, *where τ'_1 and τ'_2 are essentially equal to τ, such that*

$$\mathbf{Adv}_{\mathrm{KEM}}^{\mathrm{kex\text{-}s\text{-}w}}(\mathcal{A}) \leq 2\mathbf{Col}_{\mathrm{KEM}}(l_s) + 2l_u\mathbf{Adv}_{\mathrm{KEM}}^{\mathrm{ror\text{-}cca}}(\mathcal{B}_1) + 2\mathbf{Adv}_{\mathrm{SIG}}^{\mathrm{suf\text{-}mu\text{-}cma}}(\mathcal{B}_2).$$

The idea is that the key encapsulation method keeps the session key secret, while the signatures ensures authentication. There are some subtleties, which is why we include everything in the encapsulation associated data.

E | *Exercise* 10.20. Suppose we use just *ad* as associated data when creating the encapsulation. Find an attack that then succeeds with probability close to 1.

E | *Exercise* 10.21. Suppose we let the associated data be just (ad, ek, vk). Find an attack that succeeds with probability close to 1. (This is more obscure, reusing parts of someone else's public key as your own.)

The proof goes as follows. The first change is technical, making sure that all the ciphertexts for any encapsulation key are distinct. This prevents trivial (but unlikely) attacks against authentication. The next step ignores any forgeries. If such forgeries are likely, the key exchange adversary is essentially an adversary against the signature scheme. At this point, distinct ciphertexts and no forgeries implies that the authentication predicate holds for every instance.

Next, we make another technical step, rejecting any test query except when the instance tested has a specific responder public key, chosen at random. This reduces the adversary's advantage, but simplifies the further analysis. In the final game, we do not use the encapsulated keys, but instead use random values when deriving the session keys. This exactly corresponds to the real-or-random game for the key encapsulation mechanism. The result is a game where the adversary has no distinguishing advantage.

E | *Exercise* 10.22. A significant complication in the proof is long-term key compromise. Prove Proposition 10.23 for non-adaptive adversaries.

E | *Exercise* 10.23. Prove Proposition 10.23.

E | *Exercise* 10.24. As usual, the signature scheme prevents the proof from working for invasive adversaries. Assume a deterministic signature scheme and prove Proposition 10.23 for general adversaries.

10.5 SINGLE-SIDED AUTHENTICATION

Defining single-sided security is in some sense easy. The key generation algorithm is run once and the resulting key pair is made public. The party that does not authenticate will use this key pair. That is, the non-authenticating users are equivalent to a single user that has had its long-term key revealed.

However, it is better to design a dedicated protocol for single-sided authentication. We say that a protocol is a *single-sided authentication key exchange protocol* if there is a special key pair $(pk_\perp, sk_\perp)$, the initiator role only accepts $(pk_\perp, sk_\perp)$ as its key pair, the responder role only accepts $pk_\perp$ as its initiator public key, and $pk_\perp$ is never accepted as a responder public key. In practice, the initiator role simply does not take an initiator key pair as input, and the responder role does not take an initiator public key.

With respect to security notions, we simply consider the special key pair $(pk_\perp, sk_\perp)$ to be revealed, which means that we can reuse partnering notions, authentication predicates and freshness predicates.

| E | *Exercise* 10.25. A single user suffices. Let $\mathcal{A}$ be a $(\tau, 1, l_u, l_s)$-adversary against a single-sided authentication key exchange protocol KEX. Show that there exists a $(\tau', 1, 1, l_s)$-adversary $\mathcal{B}$ against KEX, where τ' is essentially equal to τ, and

$$\mathbf{Adv}^{\text{kex}}_{\text{KEX}}(\mathcal{A}) \leq l_u \mathbf{Adv}^{\text{kex}}_{\text{KEX}}(\mathcal{B}).$$

Hint: Compare with Proposition 10.7 from Section 10.2.4.

The best known single-sided authentication protocol is derived from a KEM. The initiator creates an encapsulation and the responder decapsulates.

| E | *Example* 10.12. Given a key encapsulation mechanism KEM $= (\mathcal{KK}, \mathcal{KE}, \mathcal{KD})$ with key set $\mathfrak{K}_s$ and associated data set $\mathfrak{F}$, we define the following key exchange protocol KEX $= (\mathfrak{K}_s, \mathcal{K}, \mathcal{I}, \mathcal{R})$:

- The *key generation* protocol $\mathcal{K}$ is identical to the key encapsulation scheme's key generation protocol, where the public key is the encapsulation key and the secret key is the decapsulation key.

- The *initiator* algorithm takes as input associated data ad and a public key pk. It computes $(c, k) \leftarrow \mathcal{KE}(pk, ad)$. It sends c and outputs k.

- The *responder* algorithm takes as input associated data ad and a key pair (pk, sk). When it receives c, it computes $k \leftarrow \mathcal{KD}(sk, ad, c)$. If the decapsulation fails, it outputs $\perp$. Otherwise it ouputs k.

| E | *Exercise* 10.26. Show that the above protocol is a key exchange protocol.

| E | *Exercise* 10.27. Let $\mathcal{A}$ be a non-invasive $(\tau, 1, 1, l_s)$-adversary against the above key exchange protocol KEX. Show that there exists a (τ', l_s, l_s)-adversary $\mathcal{B}$ against KEM, where τ' is essentially equal to τ, such that

$$\mathbf{Adv}^{\text{kex-s-w}}_{\text{KEX}}(\mathcal{A}) \leq \mathbf{Adv}^{\text{ror-cca}}_{\text{KEM}}(\mathcal{B}).$$

Hint: Proposition 10.23 is a similar case, though more complicated.

While the above protocol is quite efficient, we would still like security against future long-term key reveals, and perhaps also security against state reveal. Of course, since we consider the initiator key pair to be revealed, the initiator's state cannot also be revealed. But we can achieve some security against responder state reveal.

E *Exercise* 10.28. Design a variant of Signed Diffie-Hellman from Example 10.2 for single-side authentication key exchange, and state and prove a result similar to Theorem 10.15.

E *Exercise* 10.29. Design a variant of the implicitly authenticated Diffie-Hellman-based protocol from Proposition 10.8 for single-side authentication key exchange, and state and prove a result similar to the proposition.

10.6 CONTINUOUS KEY EXCHANGE

It is sometimes desirable to refresh keys often, to reduce the consequences of session key compromise. This also speeds up recovery time if the key compromise was a one-time event, in particular if the compromise was not detected.

Remark. This is a somewhat obscure threat model, but it is not unrealistic. One example is an adversary using an unknown implementation flaw to compromise a terminal without detection. A subsequent update to the terminal fixes the flaw and as a side effect removes the terminal from the adversary's control. If the key material is not then refreshed, the adversary retains access to the terminal's communications.

The natural way to refresh keys is to exchange keys continously. But at the same time, it is also desirable to have as few network exchanges as possible. It would therefore be good if the continuous key exchange could piggy-back on the regular network exchanges. (The amount of data sent over the network still increases, but there is sometimes a non-trivial per-message cost involved in network communications, and avoiding it is good.)

The natural thing to do now is to use a two-round key exchange, but this can only achieve implicit authentication. That is in some sense acceptable, because we anyway intend to use the session key immediately, thereby conferring explicit authentication on the session key.

In a conversation between Alice and Bob, the idea would be that Alice sends her initiator message to Alice. Bob responds with a responder message and a key confirmation token but also sends his own intiator message. Alice responds with her own key confirmation token, a message encrypted under the session key, a responder message for Bob's initiator message, and her own second initiator message. And so it goes.

In principle, Bob could have sent an encrypted message along with his first responder message, but at that point in time he does not actually know that Alice is present. For the subsequent key exchanges, it makes sense to be a bit more aggressive, sending messages encrypted under a session key together with the responder message that establishes the session key, provided we tie the session key to previous session keys, perhaps by including a complete or partial network history in the key exchange associated data.

Some key exchange protocols have a form of symmetry in terms of computation. The prime example is the Diffie-Hellman key exchange from Example 10.1. The initiator and the responder both generate their messages in the same way. They also generate the session key in the same way.

When we design a key encapsulation mechanism from a two-round key exchange scheme, we essentially consider a situation where we run the initiator many times with the same randomness. As we have shown, this is safe. For a two-round scheme with the symmetry we have just observed, we may also ask if it is safe to run an initiator and a responder instance with the same randomness. If so, we can improve the above general approach by having Alice and Bob send only one key exchange message, but letting this message play the role of both responder message and initiator message.

Definition 10.11. A *(2-round) role-symmetric key exchange scheme* KEX_{rs} is a tuple $(\mathfrak{K}, \mathfrak{N}_{rs}, \mathcal{K}, \mathcal{MG}, \mathcal{SK})$, where $\mathfrak{K}$ is a set of keys, $\mathfrak{N}_{rs}$ is a finite set of *ephemeral keys*, and the other three are algorithms:

- The *key generation* algorithm $\mathcal{K}$ outputs a key pair (pk, sk).

- The deterministic *message generator* algorithm $\mathcal{MG}$ takes a secret key sk and a *ephemeral key* r as input and outputs a *message* m.

- The deterministic *session key* algorithm $\mathcal{SK}$ takes as input a role $\rho \in \{0,1\}$, associated data ad, a key pair (pk, sk), a public key pk', a message m and an ephemeral key r. It then outputs a key $k \in \mathfrak{K}$.

We require that for any associated data ad, any distinct key pairs (pk_0, sk_0) and (pk_1, sk_1) that have been output by $\mathcal{K}$ and any $r_0, r_1 \in \mathfrak{N}_{rs}$, we have that

$$\mathcal{SK}(0, ad, (pk_0, sk_0), pk_1, \mathcal{MG}(sk_1, r_1), r_0)$$
$$= \mathcal{SK}(1, ad, (pk_1, sk_1), pk_0, \mathcal{MG}(sk_0, r_0), m_0, r_1).$$

The scheme KEX_{rs} is *unauthenticated* if the key exchange protocol always outputs $(\perp, \perp)$ and the above requirement also holds when $pk_0 = pk_1$.

The scheme KEX_{rs} is *simple* if the output of $\mathcal{MG}$ is independent of sk.

For convenience we shall simply write $(r, m) \leftarrow \mathcal{MG}(sk)$ instead of writing that we sample $r \xleftarrow{r} \mathfrak{N}_{rs}$ and compute $m \leftarrow \mathcal{MG}(sk, r)$.

We could have defined a more general notion of role-symmetric that allowed for more than one message in each direction, but this does not seem so useful. Therefore, we drop the words "2-round" in the discussion.

A role-symmetric key exchange scheme is not technically a key exchange scheme, but it can be trivially turned into a key exchange scheme. In general, when we talk about a concrete role-symmetric key exchange scheme as a key exchange scheme, it will be the following construction we are talking about.

Example 10.13. Let $\text{KEX}_{rs} = (\mathfrak{K}, \mathcal{K}, \mathcal{MG}, \mathcal{SK})$ be a role-symmetric key exchange scheme. We construct the key exchange scheme $(\mathfrak{K}, \mathcal{K}, \mathcal{I}, \mathcal{R})$ as follows:

- The initiator algorithm $\mathcal{I}$ takes as input associated data ad, a key pair (pk, sk) and a public key pk. It samples $r_0 \xleftarrow{r} \mathfrak{N}_{rs}$, computes $m_0 \leftarrow \mathcal{MG}(sk, r_0)$ and sends m_0. When it receives a message m_1, it computes $k \leftarrow \mathcal{SK}(0, ad, (pk, sk), pk', m_1, r_0)$ and outputs k.

- The responder algorithm $\mathcal{R}$ takes as input associated data ad, a public key pk' and a key pair (pk, sk). It receives a message m_0, samples $r_1 \xleftarrow{r} \mathfrak{N}_{rs}$, computes $m_1 \leftarrow \mathcal{MG}(sk, r_1)$ and $k \leftarrow \mathcal{SK}(1, ad, (pk, sk), pk', m_0, r_1)$, sends m_1 and outputs k.

Note that neither algorithm will ever output $\perp$.

Remark. Note that the only randomness in both the initiator and responder algorithms is the ephemeral key the initiator samples.

We have already mentioned that Diffie-Hellman is role-symmetric. The schemes derived from key encapsulation mechanisms are in general not role-symmetric. But there are other schemes as well.

Exercise 10.30. The HMQV scheme from Example 10.4 is role-symmetric and simple. Write out the algorithms.

Exercise 10.31. The key exchange scheme from Proposition 10.8 is role-symmetric and simple. Write out the algorithms.

Exercise 10.32. The key exchange scheme from Section 10.2.4 using the NAXOS trick is role-symmetric, but not simple. Write out the algorithms.

We could define security of a role-symmetric key exchange scheme in terms of the key exchange scheme defined in Example 10.13, but as suggested before the definition, we want more from a role-symmetric scheme.

Definition 10.12. Let $\mathbf{Exp}_{\text{KEX}_{rs}}^{\text{rskex}}$ be the following modification of the key exchange experiment $\mathbf{Exp}_{\text{KEX}_{rs}}^{\text{kex}}$ from Figure 10.1, for a role-symmetric key exchange scheme KEX_{rs}:

- A *mirror execute* query (i, ad') for which the ith execute query was (ρ, j, pk', ad) and started an instance with key pair (pk, sk) and ephemeral key r. If no earlier mirror execute query has been made for i, then the experiment starts an instance of the opposite role with associated data ad', key pair (pk, sk), partner public key pk' and *with ephemeral key r*.

We number the new instance as $-i$ and refer to the ith and $-i$th instances as *mirrors*.

- When the experiment processes a state reveal query, the experiment also does a state reveal query for the mirror instance.

The *advantage* of a (τ, l_t, l_u, l_s)-adversary against a role-symmetric key exchange scheme is defined exactly as for an ordinary key exchange scheme, except with respect to $\mathbf{Exp}_{\mathrm{KEX}_{rs}}^{\mathrm{rskex}}$ instead of $\mathbf{Exp}_{\mathrm{KEX}_{rs}}^{\mathrm{kex}}$.

The idea is that an adversary is allowed to use a given ephemeral key both in the initiator role and in the responder role. The experiment must also double the state reveal query, otherwise we would allow trivial attacks where the adversary reveals the state of one instance and uses it to attack the mirror instance, which would be fresh.

Remark. We could have allowed the mirror execute query to specify a different partner public key. Instead of role symmetry as we shall need it later, this would model more general randomness reuse, which is not what we want.

E | *Exercise* 10.33. Explain why Diffie-Hellman as defined in Example 10.1 is not secure as a role-symmetric key exchange scheme, not even against passive adversaries.

E | *Exercise* 10.34. Prove analogues of Propositions 10.6 and 10.7 for role-symmetric key exchange schemes.

E | *Exercise* 10.35. Prove an analogue of Proposition 10.8 when we consider the key exchange scheme as a role-symmetric key exchange scheme.

CHAPTER 11

Arguments

We shall now study the class of two-party protocols called arguments. These protocols are arguments, in the sense that one party is trying to convince the other party of something, and the other party will either become convinced and accept the argument, or reject the argument.

In more detail, both parties have been given a statement. One party will attempt to convince the other party that the statement is in some sense true, and in our telling, this party has also been given some evidence proving that the statement is true. We see that revealing the evidence is a trivial solution to this task, but for our intended applications this is not an acceptable answer.

We encountered one motivating example in Section 8.5.1, when discussing the problem of election integrity. The problem is that whoever decrypts ciphertexts can simply lie about the decryptions. In this case, the statement would consist of an encryption key, a ciphertext and a message, with the intended interpretation that the given message is the decryption of the ciphertext. The evidence would be the decryption key. The decryption key must be kept private in order to preserve confidentiality, so it should not be revealed.

Another problem related to public key encryption is that someone who claims to own an encryption key should also be able to decrypt ciphertexts under that key. We havel already seen this problem in Section 9.3 where the idea was to define identification in terms of possession of some unique cryptographic capability. We now want to define a pure version of this problem, where a party does not just claim to have a cryptographic capability, but claims to have actual *knowledge* of a particular secret.

There are very generic approaches to the construction of these arguments, using ideas from computational complexity theory. While these generic approaches are surprisingly useful in many cases, we will often want to study specialised arguments for some particular problems.

DOI: 10.1201/9781003149422-11

11.1   ARGUMENTS

Our motivating goal in this chapter is to explain how a prover can convince a verifier that a given statement is true. In order to formalise this, we must first explain precisely what we mean by statement and what we mean by statements being true.

11.1.1   Witness Relations

Our *statements* shall simply be set elements belonging to some set X. The true statements are defined to be the statements belonging to some subset $L \subseteq X$. The subset L is usually called a *language*.

Remark. In mathematical logic the set X is usually the set of strings of symbols from some alphabet. The set L is the subset of strings generated from a set of axioms and some interesting rules, with the pious hope that the strings of the language can be interpreted as true mathematical statements. While we shall consider other sets, we retain this use of language.

When we have a statement in a language, we will often need a piece of evidence for this fact. This piece of evidence will be called a *witness*.

We now define these structures precisely. Consider a set X of *statements*, a subset (*language*) $L \subseteq X$, a set W (*witnesses*) and a relation $R \subseteq X \times W$. We say that R is a *witness relation* for L if for any $x \in X$, $x \in L$ if and only if there exists a witness $w \in W$ such that $x \, R \, w$.

Given this, proving that a witness exists proves that a statement $x \in X$ belongs to L. Proving that we *know* a witness is a stronger statement.

Remark. Sometimes, we shall assume that the tuple (X, L, W, R) comes equipped with probability distributions on X, L, $X \setminus L$ and W, along with suitable algorithms for sampling according to these distributions.

Example 11.1. Let l be an integer and let X be the set of pairs of graphs with l vertices $\{1, 2, \ldots, l\}$. Let $L \subseteq X$ be the pairs with isomorphic graphs. Let W be the set of permutations on $\{1, 2, \ldots, l\}$. Let R be the relation defined by $(G_0, G_1) \, R \, \pi$ if and only if π is a isomorphism from G_0 to G_1.

Example 11.2. Let l be an integer and let X be the set of graphs with l vertices $\{1, 2, \ldots, l\}$. Let L be the set of Hamiltonian graphs with l vertices. Let W be the set of cycles for graphs with l vertices. Let R be the relation defined by $x \, R \, w$ if and only if w is an Hamiltonian cycle on x.

Example 11.3. Let G be a cyclic group of prime order n, and let g, h be generators. Let $X = G \times G$, and let $L = \{(g^w, h^w) \mid w \in \mathbb{Z}\}$. With $W = \mathbb{Z}$, let

R be the relation defined by (x, y) R w if and only if $x = g^w$ and $y = h^w$. (Sometimes we consider a variant where g and h are part of the statement.)

Example 11.4. Let G be a cyclic group of prime order n, and let g be a generator. Let $X = G$, and let $L = X$. With $W = \mathbb{Z}$, let R be the relation defined by x R w if and only if $x = g^w$.

Example 11.5. Let G be a cyclic group of prime order n, and let g, h be generators. Let $X = G$, and let $L = X$. With $W = \mathbb{Z} \times \mathbb{Z}$, let R be the relation defined by x R (r, a) if and only if $x = g^r h^a$.

Exercise 11.1. Let n be a product of two distinct large primes. Let $X = \{0, 1, 2, \ldots, n - 1\}$, and let $L = \{x \in X \mid \exists w \in X : w^2 \equiv x \pmod{n}\}$. Let $W = X$ and let R be the relation defined by x R w if and only if $x \equiv w^2 \pmod{n}$. Show that R is a witness relation for L.

These structures are somewhat different. Examples 11.1–11.3 and Exercise 11.1 all concern languages that are proper subsets of X, but in Example 11.4 and 11.5 the language L is equal to the set X. In these cases, it is not the language itself we care about, nor the existence of a witness, but rather knowledge of the witness. We will return to this point in Section 11.1.3.

Composition of witness relations   Let (X_1, L_1, W_1, R_1) and (X_2, L_2, W_2, R_2) be two languages with corresponding witness relations.

We can create a new relation $R_3 \subseteq (X_1 \times X_2) \times (W_1 \times W_2)$ given by (x_1, x_2) R (w_1, w_2) if and only if x_1 R_1 w_1 and x_2 R_2 w_2. This gives us a a composed witness relation $(X_1 \times X_2, L_1 \times L_2, W_1 \times W_2, R_3)$.

We can create another new relation $R_4 \subseteq (X_1 \times X_2) \times (W_1 \times W_2)$ given by (x_1, x_2) R (w_1, w_2) if and only if x_1 R_1 w_1 or x_2 R_2 w_2 or both (*inclusive* or). This gives us a a composed witness relation $(X_1 \times X_2, L_1 \times L_2, W_1 \times W_2, R_4)$.

We will denote R_3 by $R_1 \wedge R_2$ and R_4 by $R_1 \vee R_2$.

Remark. Given a language, a witness relation and an efficient way to verify that relation, there is no obvious witness set and relation for the complement of L in X. For this reason, an *exclusive* or composition of relations seems to be hard in general. It may be possible for special cases, of course.

11.1.2   Interactive Arguments

Let $\mathcal{P}$ (*prover*) and $\mathcal{V}$ (*verifier*) be interactive algorithms, where $\mathcal{P}$ takes as input $x \in X$ and $w \in W$, and $\mathcal{V}$ takes as input $x \in X$. The two algorithms then alternate sending messages to each other. We say that $(\mathcal{P}, \mathcal{V})$ is an *interactive argument* if, regardless of input, $\mathcal{V}$ will eventually terminate and either accept or reject. We call x the *public input* and w the *private input*. These algorithms may also take additional *auxillary data aux* as input, which may depend on the other input. We shall sometimes also include *associated data*.

We say that the interactive argument $(\mathcal{P}, \mathcal{V})$ is *complete with respect to* (X, L, W, R) if whenever $x \, R \, w$, running $\mathcal{P}$ and $\mathcal{V}$ together, with private input w and public input x, results in $\mathcal{V}$ accepting.

Remark. This theory is usually developed in more generality, considering provers that may require unreasonably amounts of resources instead of witnesses. In this case, the interactive argument is complete if the verifier always accepts when $x \in L$. Witnesses then appear naturally later. We have opted to use witnesses from the start, thereby simplifying the exposition, even though generality and applicability is lost.

Also, a more powerful notion of *interactive proof* is usually defined first. Again, we forgo generality and power in order to simplify our presentation.

We now present several interactive arguments. The first argument is mostly of educational value and is not so useful in practice. The next two interactive arguments are not useful, but are examples of how not to achieve various goals. The final three interactive arguments are more useful, the first of which is the celebrated Schnorr argument from Section 9.3.3.

Example 11.6. Consider the following interactive argument for the witness relation from Example 11.1:

- The public input is (G_0, G_1) and the private input is π.

- The prover chooses a random permutation ρ on $\{1, 2, \ldots, l\}$, computes the graph $H = \rho(G_0)$, and then sends H to the verifier.

- The verifier samples $\beta \xleftarrow{r} \{0, 1\}$ and sends β to the prover.

- If $\beta = 0$, the prover sets $\gamma = \rho$. If $\beta = 1$, the prover sets $\gamma = \rho \circ \pi^{-1}$. Then the prover sends γ to the verifier.

- The verifier accepts if and only if $\gamma : G_\beta \to H$ is a graph isomorphism.

If the challenger chooses $\beta = 0$, the prover sends $\gamma = \rho$, which is an isomorphism from G_0 to H, so the verifier accepts. If the challenger chooses $\beta = 1$, the prover sends $\gamma = \rho \circ \pi^{-1}$. Since π^{-1} and ρ are a graph isomorphisms, $\rho \circ \pi^{-1}$ is a graph isomorphism of G_1 with H. Therefore, the verifier accepts. This shows that the interactive argument is complete.

We shall return to this interactive argument in Sections 11.1.3 and 11.1.4.

Exercise 11.2. Consider the following interactive argument for the witness relation from Example 11.3:

- The public input is $(x_0, x_1) \in X$ and the private input is $w \in \mathbb{Z}$.

- The prover sends w to the verifier.
- The verifier accepts if and only if $x_0 = g^w$ and $x_1 = h^w$.

Show that this interactive argument is complete.

E *Exercise* 11.3. Consider the following interactive argument for the witness relation from Example 11.3:

- The public input is $(x_0, x_1) \in X$ and the private input is $w \in \mathbb{Z}$.
- The prover sends 1 to the verifier.
- The verifier accepts.

Show that this interactive argument is complete.

E *Exercise* 11.4. Consider the following interactive argument for the witness relation from Example 11.3:

- The public input is (x_0, x_1) and the private input is w. (Sometimes we also include g and h in the public input.)
- The prover samples $\rho \xleftarrow{r} \{0, 1, 2, \dots, n-1\}$, computes $\alpha_0 = g^\rho$ and $\alpha_1 = h^\rho$, and sends (α_0, α_1) to the verifier.
- The verifier samples $\beta \xleftarrow{r} \{0, 1, 2, \dots, n-1\}$ and sends β to the prover.
- The prover computes $\gamma = \rho - \beta w \bmod n$ and sends γ to the verifier.
- The verifier $\mathcal{V}$ accepts if and only if $g^\gamma x_0^\beta = \alpha_0$ and $h^\gamma x_1^\beta = \alpha_1$.

Show that this interactive argument is complete.

Compare the following interactive argument to the work done in Sections 4.4 and 9.3.3, and to Exercise 11.4.

E *Exercise* 11.5. Let $\mathcal{B} \subseteq \{0, 1, 2, \dots, n-1\}$. Consider the following interactive argument for the witness relation from Example 11.4:

- The public input is x and the private input is w.
- The prover samples $\rho \xleftarrow{r} \{0, 1, 2, \dots, n-1\}$, computes $\alpha = g^\rho$, and sends α to the verifier.
- The verifier samples $\beta \xleftarrow{r} \mathcal{B}$ and sends β to the prover.
- The prover computes $\gamma = \rho - \beta w \bmod n$ and sends γ to the verifier.
- The verifier $\mathcal{V}$ accepts if and only if $g^\gamma x^\beta = \alpha$.

Show that this interactive argument is complete.

E *Exercise* 11.6. Consider the following interactive argument for the witness relation from Exercise 11.1:

- The modulus is n, the public input is x and the private input is w.
- The prover $\mathcal{P}$ samples $\rho \xleftarrow{r} X$, computes $\alpha = \rho^2 \bmod n$ and sends α.
- The verifier $\mathcal{V}$ samples $\beta \xleftarrow{r} \{0,1\}$ and sends β to the prover.
- The prover $\mathcal{P}$ computes $\gamma = w^\beta \rho \bmod n$ and sends γ to $\mathcal{V}$.
- The verifier $\mathcal{V}$ accepts if and only if $\gamma^2 \equiv \alpha x^\beta \pmod{n}$.

Show that this interactive argument is complete.

For an argument, we say that the *conversation* (or transcript) is the list of all messages sent in the protocol. Compare this to the notions developed in Section 9.3 and Chapter 10.

Remark. As we shall see, for general protocols there may not be agreement on the (temporal) order of messages. But for interactive arguments where the players alternate sending and receiving, the order is given and agreed on.

The verifier's *view* consists of the conversation as well as the random choices made by the verifier during the execution of the interactive argument. Note that unlike the conversation, the view always contains enough information to decide whether or not the verifier accepted.

We say that an interactive argument is *public coin* if, given the conversation, it is easy to find random choices for the verifier such that, upon receiving the prover's messages from the conversation, the exact same conversation would be reproduced, and for any such random choices the verifier's conclusion would be the same. For a public coin interactive argument, the conversation and the view are essentially equivalent. (We say *public* coin, because all the coin tosses made by the verifier that matter are all essentially recoverable from the conversation, which in a sense is public.)

E *Exercise* 11.7. Show that the interactive arguments in Example 11.6 and Exercises 11.4–11.6 are all public coin.

Furthermore, for a public coin protocol, we can decide based on the conversation alone if the verifier accepted or not. We say that a conversation is *accepting* if the verifier accepts at the end of the conversation.

One important class of public coin interactive arguments are *sigma protocols*, which are 3-move (3-message) interactive arguments where the prover sends the first and last message, and the verifier's message is sampled from the uniform distribution on some set (hence public coin). We actually studied some properties of these sigma protocols in Section 9.3.

E *Exercise* 11.8. Show that the interactive arguments from Example 11.6 and Exercises 11.4–11.6 are all sigma protocols.

The prover's first message is called the *commitment*, the verifier's message is called the *challenge* and the prover's final message is called the *response*.

As the exercise suggests, sigma protocols are very common, and they are important for practical cryptography because they can be composed in very useful ways, giving us a rich toolbox when designing cryptographic protocols.

Composition of arguments Composition of protocols involves using protocols as *subprotocols* in a larger protocol. In this section we shall consider two simple forms of composition of interactive arguments. Let (X_1, L_1, W_1, R_1) and (X_2, L_2, W_2, R_2) be two languages with corresponding witness relations, and consider the witness relation $R_1 \wedge R_2$.

The *sequential* composition of the two interactive arguments $(\mathcal{P}_1, \mathcal{V}_1)$ and $(\mathcal{P}_2, \mathcal{V}_2)$ is an interactive argument $(\mathcal{P}_s, \mathcal{V}_s)$ where $\mathcal{P}_s$ first runs $\mathcal{P}_1$ with the first part of its input and then runs $\mathcal{P}_2$ with the second part of its input, while $\mathcal{V}_s$ first runs $\mathcal{V}_1$ with the first part of its input, and if $\mathcal{V}_1$ accepts it runs $\mathcal{V}_2$ with the second part of its input. If $\mathcal{V}_2$ also accepts, $\mathcal{V}_s$ accepts. Otherwise, $\mathcal{V}_s$ rejects. If one party sends both the last message of the first protocol and the first message of the second protocol, we combine these messages into one so that the prover and verifier still alternate sending and receiving messages.

> ⓔ *Exercise* 11.9. Prove that if $(\mathcal{P}_1, \mathcal{V}_1)$ is complete for (X_1, L_1, W_1, R_1) and $(\mathcal{P}_2, \mathcal{V}_2)$ is complete for (X_2, L_2, W_2, R_2), then $(\mathcal{P}_s, \mathcal{V}_s)$ is complete for $(X_1 \times X_2, L_1 \times L_2, W_1 \times W_2, R_1 \wedge R_2)$.

The *parallel* composition of the two interactive arguments $(\mathcal{P}_1, \mathcal{V}_1)$ and $(\mathcal{P}_2, \mathcal{V}_2)$ is an interactive argument $(\mathcal{P}_p, \mathcal{V}_p)$ where $\mathcal{P}_p$ simultaneously runs $\mathcal{P}_1$ with the first part of its input and $\mathcal{P}_2$ with the second part of its input. Likewise, $\mathcal{V}_p$ simultaneously runs $\mathcal{V}_1$ with the first part of its input and $\mathcal{V}_2$ with the second part of its input. We let $\mathcal{V}_p$ accept only if both $\mathcal{V}_1$ and $\mathcal{V}_2$ accept. Otherwise $\mathcal{V}_p$ rejects. When $\mathcal{P}_p$ and $\mathcal{V}_p$ sends messages, they need to mark them appropriately, so that the recipient knows who the message is for. We synchronise the two protocols so that each message sent by $\mathcal{P}_p$ and $\mathcal{V}_p$ contains a message from both of the subprotocols, as far as possible.

> ⓔ *Exercise* 11.10. Prove that if $(\mathcal{P}_1, \mathcal{V}_1)$ is complete for (X_1, L_1, W_1, R_1) and $(\mathcal{P}_2, \mathcal{V}_2)$ is complete for (X_2, L_2, W_2, R_2), then $(\mathcal{P}_p, \mathcal{V}_p)$ is complete for $(X_1 \times X_2, L_1 \times L_2, W_1 \times W_2, R_1 \wedge R_2)$.

Sigma protocols The parallel composition of two sigma protocols is also a sigma protocol, but sequential composition would not be a sigma protocol.

11.1.3 Soundness

Our motivation for an interactive argument was for the verifier to be convinced that the public input x is in the language L. The goal of a *cheating prover* is to convince the verifier to accept for some $x \notin L$.

D **Definition 11.1.** Consider an interactive argument $(\mathcal{P}, \mathcal{V})$ for (X, L, W, R). A *cheating prover* is an interactive algorithm that takes as input $x \in X$, and interacts with $\mathcal{V}$ such that $\mathcal{V}$ eventually terminates. A cheating prover $\mathcal{P}^*$ has *(soundness) advantage* ϵ if for some $x \in X \setminus L$, the probability that $\mathcal{P}^*$ can make $\mathcal{V}$ accept is ϵ.

Informally, we say that $(\mathcal{P}, \mathcal{V})$ has *soundness error* ϵ if there is an "interesting" cheating prover with advantage ϵ, but no "interesting" cheating prover has advantage much greater than ϵ. In many cases, the soundness error remains an informal concept, but in some cases we can determine a precise value. (We leave the meaning of "interesting" undefined, but sometimes we only care about cheating provers whose running time is not too large.)

An argument with soundness error meaningfully smaller than 1 is *sound*.

E *Exercise* 11.11. Consider the interactive arguments from Exercises 11.2 and 11.3. Show that one argument has soundness error 0, while the other has soundness error 1 (exhibit a cheating prover with advantage 1).

T **Proposition 11.1.** *Consider the interactive argument $(\mathcal{P}, \mathcal{V})$ from Example 11.6. It has soundness error $1/2$.*

Proof. In the interactive argument, the cheating prover commits to a graph H. The verifier then asks for an isomorphism from H to one of G_0 and G_1.

Unless H is isomorphic to at least one of them, there is no way that the cheating prover can provide such an isomorphism. If the two graphs G_0 and G_1 are not isomorphic, the graph H cannot be isomorphic to both of them. Therefore, the verifier will ask for an isomorphism between two non-isomorphic graphs with probability at least $1/2$. Since $\mathcal{P}^*$ cannot provide such an isomorphism, the verifier rejects with probability at least $1/2$.

Also, there is a trivial adversary that has advantage $1/2$. It presents G_0 as its commitment. If the verifier asks to see an isomorphism to G_0, it responds with the identity map, after which the verifier will accept. If the verifier asks to see an isomorphism to G_1, our cheating prover responds with a random map, which will not be an isomorphism, so $\mathcal{V}$ will reject. □

A soundness error of $1/2$ does not sound useful. Indeed, compared to results from previous chapters, a cheating prover winning with probability $1/2$ is absymal. However, repeating the argument in sequence many times will quickly reduce the overall soundness error. Parallel repetition may also work.

This simple strategy is expensive, so we need other techniques.

Proposition 11.2. *Consider the interactive argument* $(\mathcal{P}, \mathcal{V})$ *from Example 11.4. It has soundness error* $1/n$.

Proof. We shall calculate the probability that the verifier chooses a challenge β for which there is a response that will make $\mathcal{V}$ accept.

Suppose the public input is (x_0, x_1) with $x_0 = g^w$ and $x_1 = h^{w+\delta}$, for some $w, \delta \in \{0, 1, 2, \ldots, n-1\}$. Since $(x_0, x_1) \notin L$, we know that $\delta \neq 0$.

A cheating prover must first commit to (α_0, α_1), and we know that $\rho, \Delta \in \{0, 1, \ldots, n-1\}$ exist such that $\alpha_0 = g^\rho$ and $\alpha_1 = h^{\rho+\Delta}$.

Note that only when $\mathcal{P}^*$ has committed to δ and Δ will $\mathcal{V}$ choose its challenge β. This means that β is independent of δ and Δ.

The verifier accepts only if the conversation $((\alpha_0, \alpha_1), \beta, \gamma)$ satisfies

$$g^\gamma x_0^\beta = \alpha_0 \qquad \text{and} \qquad h^\gamma x_1^\beta = \alpha_1.$$

Inserting w, ρ, δ and Δ in the equations, we get that

$$g^\gamma g^{w\beta} = g^\rho \qquad \text{and} \qquad h^\gamma h^{w\beta+\delta\beta} = h^{\rho+\Delta}.$$

Equivalently, we have that

$$\gamma + w\beta \equiv \rho \pmod{n} \qquad \text{and} \qquad \gamma + w\beta + \delta\beta \equiv \rho + \Delta \pmod{n}.$$

These equations are only consistent if $\delta\beta \equiv \Delta \pmod{n}$. Recall that $\mathcal{P}^*$ committed to δ and Δ before $\mathcal{V}$ chose β. Also, since $\delta \neq 0$, there is exactly one possible choice of β that will give a consistent system of equations. The probability that $\mathcal{V}$ chooses exactly this value is $1/n$, since n is prime.

For any fixed (x_0, x_1, γ), the verifier will accept with probability $1/n$. $\square$

Arguments of knowledge Sometimes, it is not enough to know that a statement is in the language, which in our case implies that a witness exists. For example, if a user is supposed to know a password in order to be authenticated, it is not enough for that user to show that the password exists. Rather, the user must show that he *knows* the password.

What does it mean to know a witness? One answer is that you know something if you are able to tell someone. However, we are now studying algorithms, not people. We cannot ask a cheating prover to reveal what we think it should know. Instead, we have to somehow extract the information from the cheating prover. This cannot be done by inspecting the description of the algorithm. Instead we must use a *knowledge extractor*, an algorithm that uses the cheating prover, possibly repeatedly, to find a witness.

Previously, we have defined a cheating prover's advantage in terms of convincing the verifier to accept an element not in the language. We shall now consider a cheating prover's success probability in terms of convincing the verifier to accept, even though the prover has not been given a witness.

Soundness advantage is defined only for statements not in the language. For arguments of knowledge, statements outside the language do not have witnesses, so we must consider only statements in the language.

D **Definition 11.2.** Let $(\mathcal{P}, \mathcal{V})$ be an interactive argument for (X, L, W, R). A cheating prover $\mathcal{P}^*$ has *(knowledge) advantage* ϵ if when x has been sampled from L, the probability that $\mathcal{P}^*$ can make $\mathcal{V}$ accept is ϵ.

D **Definition 11.3.** Let $(\mathcal{P}, \mathcal{V})$ be an interactive argument for (X, L, W, R). An *(knowledge) extractor* for this witness relation is an algorithm that on input of x sampled from L and given access to a cheating prover $\mathcal{P}^*$ outputs a witness w. The *advantage* of the extractor is the probability that the extractor outputs a witness such that $x \ R \ w$.

Note of course that the advantage of a given extractor will depend on the success probability of the cheating prover it is used with. If the cheating prover never succeeds, we should not expect the extractor to succeed very often.

Informally, we say that an interactive argument has *knowledge error* δ if there is an extractor such that for any $\mathcal{P}^*$ with success probability $\epsilon + \delta$, then the extractor will find a witness using not too much more time than $\mathcal{P}^*$ with a probability that is not too much smaller than ϵ. If we have a good knowledge extractor, we say that the interactive argument is a *argument of knowledge*.

Compare the soundness error and the knowledge error: one measures how certain an accepting verifier is that the public input x is in L, while the other measures how certain an accepting verifier is that the prover really knows a witness to the fact that x is in L.

Remark. It is tempting to argue that if the verifier is certain that the prover knows a witness, the verifier must be equally certain that $x \in L$. However, knowledge advantage is only defined with respect to statements sampled from the language, so if we can recognise the language, we can define arguments that have knowledge error 0 and soundness error 1.

If an argument of knowledge is not sound, we can use a cheating prover and the knowledge extractor to distinguish L from $X \setminus L$. But if we can distinguish L from $X \setminus L$, soundness is not really relevant.

T **Proposition 11.3.** *Consider the interactive argument from Example 11.6. Let $\mathcal{P}^*$ be a cheating prover that has (knowledge) advantage $1/2 + \epsilon$ for some $\epsilon > 0$. The extractor given in Figure 11.1 running with $\mathcal{P}^*$ has advantage at least ϵ. Its running time is at most twice that of $\mathcal{P}^*$ plus some trivial amount of computation.*

The idea is that a cheating prover that wins with probability greater than $1/2$ must on average be able to respond to both challenges, at least to some degree. This means that if we can get the cheating prover to respond to both challenges, we should have a reasonable probability of getting isomorphisms from the cheating prover's commitment to both of our graphs, from which we can easily deduce a graph isomorphism. Getting the cheating prover to respond to both challenges is done using rewinding.

The extractor gets as input a pair of isomorphic graphs (G_0, G_1).

- The extractor runs the cheating prover $\mathcal{P}^*$ with (G_0, G_1) as input.
- When the cheating prover commits to the graph H, the extractor duplicates the state of the cheating prover, getting two identical instances of the cheating prover that both believe they just committed to H.
- The extractor sends $\beta = 0$ to the first instance and $\beta = 1$ to the second instance. The instances respond with γ_0 and γ_1, respectively.

The extractor outputs the map $(\gamma_1)^{-1} \circ \gamma_0$.

FIGURE 11.1 Extractor for the interactive argument from Example 11.6.

Proof. We first note that the extractor from Figure 11.1 essentially runs (part of) the cheating prover's computation twice, and then does some trivial computations, so the run time of the extractor is at most twice that of the cheating prover plus some trivial amount of computation.

The extractor produces two conversations (H, β, γ) and (H, β', γ'). If both are accepting, the map $(\gamma')^{-1} \circ \gamma$ is an isomorphism of G_0 and G_1.

It now remains to compute the extractor's advantage. Let E be the event that the conversation that $\mathcal{P}^*$ and $\mathcal{V}$ have is accepting. Let E_β be the corresponding event, given that the challenge in the conversation is β. Let $F_{G_0, G_1, r}$ be the event that the public input is the pair (G_0, G_1) and that the cheating prover has random tape r. The advantage of our extractor is

$$\sum_{G_0, G_1, r} \Pr[E_0 \wedge E_1 \mid F_{G_0, G_1, r}] \Pr[F_{G_0, G_1, r}].$$

Given the graphs G_0, G_1, the random tape r and the challenge β, the outcome is fully determined, so $\Pr[E_\beta \mid F_{G_0, G_1, r}]$ is either 0 or 1.

Since our cheating prover has advantage $1/2 + \epsilon$, we must have that

$$1/2 + \epsilon = \Pr[E] = \sum_{G_0, G_1, r} \Pr[E \mid F_{G_0, G_1, r}] \Pr[F_{G_0, G_1, r}]$$

$$= \sum_{G_0, G_1, r} \frac{1}{2} (\Pr[E_0 \mid F_{G_0, G_1, r}] + \Pr[E_1 \mid F_{G_0, G_1, r}]) \Pr[F_{G_0, G_1, r}].$$

There must be a set S of triples (G_0, G_1, r) such that $\Pr[E_0 \mid F_{G_0, G_1, r}] = \Pr[E_1 \mid F_{G_0, G_1, r}] = 1$ and

$$\sum_{(G_0, G_1, r) \in S} \Pr[F_{G_0, G_1, r}] \geq \epsilon.$$

Note that for any triple (G_0, G_1, r) we must have that $(G_0, G_1) \in L$, since otherwise the cheating prover cannot respond correctly to both challenges.

Since the outcome is fully determined given G_0, G_1, r and β, we have that

$$\sum_{G_0, G_1, r} \Pr[E_0 \wedge E_1 \mid F_{G_0, G_1, r}] \Pr[F_{G_0, G_1, r}]$$

$$= \sum_{(G_0, G_1, r) \in S} \Pr[E_0 \wedge E_1 \mid F_{G_0, G_1, r}] \Pr[F_{G_0, G_1, r}]$$

$$= \sum_{(G_0, G_1, r) \in S} \Pr[F_{G_0, G_1, \sigma}] \geq \epsilon.$$

It follows that if the cheating prover has advantage $1/2 + \epsilon$, then our extractor has advantage ϵ in producing a witness (graph isomorphism). □

Remark. It is not known how hard it is to find graph isomorphisms. We know that it is easy to find isomorphisms for many classes of graphs, but it is not known to be easy for all graphs, nor is it known to be as hard as other classes of problems strongly suspected to be hard. The interactive argument and the extractor is therefore not very useful. The main value is pedagogical.

Sigma protocols   Sigma protocols have a natural notion of security closely related to arguments of knowledge. Recall that sigma protocols are 3-move public coin protocols: the prover commits, the verifier challenges and the prover responds. Also, two accepting conversations (α, β, γ) and $(\alpha, \beta', \gamma')$ for the same statement with $\beta \neq \beta'$ is called a *forked conversation pair*.

Definition 11.4. Let $(\mathcal{P}, \mathcal{V})$ be a sigma protocol for a witness relation (X, L, W, R). A τ-*(special soundness-)extractor* $\mathcal{ZE}$ is an algorithm that takes as input a forked conversation pair for $x \in L$ and outputs a witness w such that $x \, R \, w$ using time at most time τ.

Informally, we say that $(\mathcal{P}, \mathcal{V})$ has τ-*special soundness* if there exists a τ-special soundness extractor. If the time bound τ is trivial, we ignore the bound and simply say *(special soundness) extractor* and *special soundness*.

Remark. The notion of extractor used in special soundness generalises to more than two conversations, though we shall not develop this further.

Remark. Extractor is a word in common use in cryptography. Compare Definitions 11.3 and 11.4. When which notion of extractor is not clear from context, we shall prefix the word with an explanatory word or phrase.

We have essentially already seen the first special-soundness extractor already, in the knowledge extractor given in Figure 11.1. We also encountered this notion while studying identification protocols in Section 9.3.

The extractor gets as input $x \in L$:

- The extractor runs the cheating prover $\mathcal{P}^*$ with x as input.

- When the cheating prover commits to α, the extractor duplicates the state of the cheating prover, getting two identical instances of the cheating prover that both believe they just committed to α.

- The extractor samples two distinct challenges β and β' from the uniform distribution on the challenge space. The extractor sends β to the first instance and β' to the second instance. The instances respond with γ and γ', respectively.

- If both conversations are accepting, the extractor computes $w \leftarrow \mathcal{ZE}(x, (\alpha, \beta, \gamma), (\alpha, \beta, \gamma))$. Otherwise, the extractor samples w from W.

The extractor outputs the witness w.

FIGURE 11.2 Knowledge extractor from a special-soundness extractor.

Example 11.7. Consider the interactive argument from Example 11.6. Our extractor gets two accepting conversations $(H, 0, \gamma_0)$ and $(H, 1, \gamma_1)$ for the public input (G_0, G_1). We then know that γ_0 is an isomorphism of G_0 and H, while γ_1 is an isomorphism of G_1 and H. Then $(\gamma_1)^{-1} \circ \gamma_0$ is a graph isomorphism from G_0 to G_1, and therefore the witness our extractor outputs. The computations are trivial, so the interactive argument has special soundness.

Exercise 11.12. Show that the interactive arguments from Exercises 11.4, 11.5 and 11.6 all have special soundness.

Theorem 11.4. *Let $(\mathcal{P}, \mathcal{V})$ be a sigma protocol with τ-special soundness, where the verifier's response is chosen from a set with l elements. Let $\mathcal{P}^*$ be a cheating prover with advantage $1/l + \epsilon$, $\epsilon > 0$, using time at most τ'. Then the knowledge extractor given in Figure 11.2 has advantage at least ϵ^2 and uses time $2\tau' + \tau$ plus some trivial amount of computation.*

Proof. We first note that the extractor from Figure 11.2 essentially runs the cheating prover twice, and then does some trivial computations plus the special-soundness extraction, so the run time of the extractor is twice that of the cheating prover plus the cost of the special soundness extraction plus some trivial amount of computation. We need to compute the probability that the two runs of the cheating prover result in two accepting conversations.

Let E be the event that the cheating prover convinces the verifier on input sampled from L. Let E_β denote the event E when the challenge is β. Let $E_{\neq \beta}$ denote the event E when then the challenge is not β. Let $F_{x,r}$ be the event that x is the input and the cheating prover's random tape is r, so that

$$1/l + \epsilon = \Pr[E] = \sum_{x,r} \Pr[E \mid F_{x,r}]\Pr[F_r].$$

The outcome of E_β is fully determined given x and r, while $E_{\neq\beta}$ may depend on the actual challenge chosen. Note that these events are independent.

The event $E_{\neq\beta}$ amounts to ignoring one particular challenge value. Ignoring a single challenge value will decrease the cheating prover's advantage by at most $1/l$, which would happen only when the cheating prover would succeed with absolute certainty for the ignored challenge. This means that $\Pr[E_{\neq\beta} \mid E_\beta \wedge F_{x,r}] \geq \Pr[E \mid F_{x,r}] - 1/l$, so we get

$$\Pr[E_\beta \wedge E_{\neq\beta} \mid F_{x,r}] \geq \Pr[E_\beta \mid F_{x,r}](\Pr[E \mid F_{x,r}] - 1/l).$$

Computing the extractor's advantage, we get

$$\sum_{x,r}\sum_{\beta}\frac{1}{l}\Pr[E_\beta \wedge E_{\neq\beta} \mid F_{x,r}]\Pr[F_{x,r}]$$

$$\geq \sum_{x,r}(\Pr[E \mid F_{x,r}] - 1/l)\left(\sum_{\beta}\Pr[E_\beta \mid F_{x,r}]/l\right)\Pr[F_{x,r}]$$

$$= \left(\sum_{x,r}\Pr[E \mid F_{x,r}]^2\Pr[F_{x,r}]\right) - 1/l\left(\sum_{x,r}\Pr[E \mid F_{x,r}]\Pr[F_{x,r}]\right)$$

$$\geq \Pr[E]^2 - \Pr[E]/l = (1/l + \epsilon)\Pr[E] - \Pr[E]/l = \epsilon\Pr[E] \geq \epsilon^2. \quad \square$$

Remark. The above proof is a simple variant of the Forking lemma from Section 9.3.3, which we will revisit in Section 11.3.3.

A result similar to Proposition 11.3 follows from the above result and Example 11.7. Note that the Proposition 11.3 is a stronger result. This is because the number of challenges l. The above proof could also be improved for $l = 2$, but when l is large we cannot expect to do much better than ϵ^2.

E| *Exercise* 11.13. As suggested in the above remark, the proof of Proposition 11.3 can be adapted to the interactive argument from Exercise 11.5 when the set of challenges is $\{0,1\}$. Do so. Then apply Theorem 11.4 and the results of Exercise 11.12 to the interactive argument with challenge set $\{0, 1, 2, \ldots, 2^{80} - 1\}$. Compare the knowledge errors.

Special soundness is preserved by parallel composition, since given two conversations for the composed protocol with identical commitments, we trivially get two conversations for each of the subprotocols with identical commitments.

The sequential composition of sigma protocols is no longer a sigma protocol. However, many of the techniques used for sigma protocols can be used to analyse the sequential compositions of sigma protocols. Except for special cases, we will not develop this further.

11.1.4 Zero Knowledge

Interactive arguments (proofs) have been studied for their own sake, but in cryptograhy we care about a very special property of interactive arguments, namely that the verifier must learn nothing from the conversation with the prover. Again, just as we had to give precise meaning to the informal concept of knowledge, we have to define precisely what it means to learn nothing.

The premise is that we have a *cheating verifier* that follows the protocol and attempts to learn the witness, or at least some information about the witness. It would have been good to follow an approach similar to that of semantic security for encryption, where the adversary specifies two witness distributions, and must decide which distribution the witness comes from. However, unlike for encryption the adversary typically has much less influence over the witness distribution, so this approach would not model the real world well. Also, this approach would not work for a number of reasonable arguments, in particular because it is possible to verify if a guess is correct.

We shall instead use an approach based on *simulation*. The idea is that we should be able to simulate the cheating verifier's interaction with the prover, but without knowing a witness. If the simulation cannot be distinguished from the real thing, it follows that anything a cheating verifier learned by interacting with the prover could be learned without interaction with a prover.

Remark. While simulation makes sense for the purpose of defining security, its beauty is best seen when we use arguments as components of a larger scheme.

Morally, a cheating verifier is an algorithm that interacts with the prover and then outputs some information about the witness. In order to simplify our definition, we *shall not care* about this output. The reason is that this output can be recreated given the conversation and the cheating verifier's random tape, called the cheating verifier's *view*. For a cheating verifier, the view and the conversation are *not* equivalent, even if the protocol is public coin.

Our goal is to create a view that looks like the view a cheating verifier would have, but without interaction with the prover and its witness. The algorithm that creates this view is called a *simulator*.

If the simulator's output really looks like the cheating verifier's view, then since the simulator does not know a witness, anything interesting that the cheating verifier learned about the witness, it learned without the help of the prover and its witness. In other words, the cheating verifier learned nothing from the interaction, which was our overall objective.

For reasons that will become apparent, we shall allow the simulator to give up, signalled by outputting the special symbol $\perp$.

Definition 11.5. Let $(\mathcal{P}, \mathcal{V})$ be an interactive argument for (X, L, W, R).

A *cheating verifier* $\mathcal{V}^*$ is an interactive algorithm that takes as input $x \in X$, interacts with the prover $\mathcal{P}$ (that gets x and a witness w for x as input),

---

The simulator works as follows:

- Start the cheating verifier V^* with input (G_0, G_1) and random tape r.
- Sample $\beta' \xleftarrow{r} \{0,1\}$ and a random permutation γ on $\{1, 2, \ldots, l\}$, compute $H = \gamma(G_{\beta'})$ and send H to V^*.
- The cheating verifier responds with β. If $\beta \neq \beta'$, output $\perp$.
- Otherwise, output the view (H, β, γ, r).

FIGURE 11.3 Zero knowledge simulator for the interactive argument from Example 11.6 and a cheating verifier V^*.

and eventually terminates. The *view* of the cheating verifier is its conversation and its random tape.

A *simulator* for V^* is an algorithm that on input of x eventually outputs a view (a conversation and a random tape) or $\perp$. The probability that the simulator outputs $\perp$ should be independent of x and not too large.

We say that a protocol is *zero knowledge* if the simulator's output (when it does not output $\perp$) is indistinguishable from the cheating verifier's output.

We have the usual variants of indistinguishable: perfectly indistinguishable if the two distributions are identical, statistically indistinguishable if the two distributions are ϵ-close for some small ϵ, and computational indistinguishability otherwise. We get *perfect zero knowledge*, *statistical zero knowledge* and *computational zero knowledge*. The two former are much easier to work with.

Remark. We can avoid letting the simulator output $\perp$. One option is to allow the simulator to have unbounded running time, but bounded *expected running time*. Using the above definition, suppose the simulator requires running time at most τ and the probability that $\perp$ is output is ϵ. Running the simulator until it outputs something other than $\perp$ is a process that never outputs $\perp$. Its running time is unbounded, but its expected running time is τ/ϵ.

Another option is to run the simulator a fixed number of times. If $\perp$ is output by all the simulator runs, our simulator will output something random. The resulting simulator will not have perfect zero knowledge, but it might be statistical zero knowledge, which for most practical purposes is good enough.

E *Exercise* 11.14. Consider the interactive arguments from Exercises 11.2 and 11.3. Show that there is a perfect simulator for one argument, but not for the other, since a simulator would be an algorithm that finds a witness.

Proposition 11.5. *Consider the interactive argument* $(\mathcal{P}, \mathcal{V})$ *from Example 11.6. Let* $\mathcal{V}^*$ *be a cheating verifier. Then the simulator* $\mathcal{S}$ *for* $\mathcal{V}^*$ *given in Figure 11.3 is perfect, has essentially the same running time as* $\mathcal{V}^*$ *and outputs* $\perp$ *with probability* $1/2$.

The simulator pretends to be the prover. It does not have a witness, so to be able to reply, it guesses the cheating verifier's challenge and chooses its commitment accordingly. If the guess is correct, the simulator has a correctly distributed view. If the guess is incorrect, the simulator gives up.

Remark. This idea works for any sigma protocol with a small challenge set.

Proof. The first thing we observe is that in the simulator's conversation with $\mathcal{V}^*$, the commitment is a graph that is isomorphic to both G_0 and G_1 and is chosen uniformly at random from this isomorphism class. This is exactly the commitment distribution that $\mathcal{P}$ would generate. Furthermore, because of this the actual choice of H is independent of β'. Which means that the cheating verifier's view after seeing the commitment is exactly as if it was interacting with $\mathcal{P}$, and its choice of β is independent of β'. It follows that the probability that $\beta' \neq \beta$ is $1/2$, so the probability that $\perp$ is output is $1/2$, as claimed.

When $\beta' = \beta$, the simulator gives $\mathcal{V}^*$ a random graph isomorphism from G_β to H, which is exactly what $\mathcal{V}^*$ would get if it was interacting with $\mathcal{P}$.

We see that if the simulator does not output $\perp$, the cheating verifier sees a simulated conversation that is identical to what it would have seen if it was interacting with $\mathcal{P}$. It follows that the view of $\mathcal{V}^*$ when interacting with the simulator is identical to when it is interacting with $\mathcal{P}$. $\square$

Exercise 11.15. Show that the interactive argument from Exercise 11.6 has a perfect simulator that aborts with probability $1/2$.

Exercise 11.16. Show that the interactive argument from Exercise 11.5 with challenge set $\mathfrak{B} = \{0, 1\}$ has a perfect simulator that aborts with probability $1/2$. What if $\mathfrak{B} = \{0, 1, 2, 3\}$? What if $\mathfrak{B} = \{0, 1, 2, \ldots, 2^{80} - 1\}$?

With respect to composition, zero knowledge is more complicated than soundness. Parallel composition does not in general preserve zero knowledge. Likewise, sequential composition of interactive arguments does not in general preserve zero knowledge. The problem for sequential composition is that simulators for the subprotocols are not sufficient to create a simulator for the composed protocol, since the cheating verifier's interaction with the second subprotocol may depend on the interaction with the first subprotocol. The usual strategy is to augment our zero knowledge definition to include some auxillary input, essentially to pass the history of previous interactions to future simulators. We do not explore this further.

Honest verifiers It is often difficult to get both decent soundness or knowledge error and useful levels of zero knowledge at the same time.

E *Exercise* 11.17. Consider the interactive argument from Exercise 11.5, where we can vary the size of the set challenges are chosen from. Based on the results of Exercises 11.13 and 11.16, compare the effect of the size of the challenge set on the knowledge error and the efficiency of the zero knowledge simulator.

As we shall see in Section 11.3, a good building block is a protocol that does not have full zero knowledge, but rather zero knowledge against a verifier that follows the protocol, but wants to know as much about the witness as possible. Such a verifier is often called *honest-but-curious*.

For a public coin protocol, where the conversation and the verifier's view is essentially equivalent, anything a curious verifier can compute after the interaction can also be computed by someone who has just seen the conversation.

D **Definition 11.6.** Let $(\mathcal{P}, \mathcal{V})$ be a public coin interactive argument for (X, L, W, R). A *(honest verifier) simulator* is an algorithm that on input of x eventually terminates and outputs a conversation, or the special symbol $\perp$. The probability that the simulator outputs $\perp$ should be independent of x.

Remark. We could have defined an honest verifier simulator for a general interactive argument (not public coin), in which case we would have to consider the *view* of the verifier. We will not explore honest verifier zero knowledge for general interactive arguments, so we avoid this extra difficulty.

Again, the idea is that if the simulator's conversation looks like a conversation between the prover and an honest-but-curious verifier, then since the simulator did not have access to a witness, anything the curious verifier could learn with the help of $\mathcal{P}$ and a witness could be computed without the witness.

An interactive argument is *honest verifier zero knowledge* (HVZK) if a conversation output by the simulator is indistinguishable from a conversation between $\mathcal{P}$ and $\mathcal{V}$. As for ordinary zero knowledge, we have three flavours of indistinguishability for HVZK: computational, statistical and perfect.

T **Proposition 11.6.** *Consider the interactive argument $(\mathcal{P}, \mathcal{V})$ from Example 11.6. There is a perfect honest verifier simulator that never outputs $\perp$ that requires trivial amounts of computation.*

Proof. The simulator works as follows:

- Sample $\beta \xleftarrow{r} \{0, 1\}$ and a random permutation γ on $\{1, 2, \ldots, l\}$. Compute $H = \gamma(G_\beta)$. Output the conversation (H, β, γ).

The simulator outputs an accepting conversation. Furthermore, the commitment is chosen uniformly at random from the isomorphism class of G_0 and G_1. This is exactly the commitment distribution that $\mathcal{P}$ would generate. Furthermore, because of this the actual choice of H is independent of β. Finally, γ is a random graph isomorphism from G_β to H. □

E| *Exercise* 11.18. Show that the argument from Exercise 11.4 has a perfect honest verifier simulator.

E| *Exercise* 11.19. Show that the interactive argument from Exercise 11.5 with challenge set $\mathfrak{B} = \{0,1\}$ has a perfect honest verifier simulator that never outputs $\perp$. What if $\mathfrak{B} = \{0,1,2,3\}$? What if $\mathfrak{B} = \{0,1,2,\ldots,2^{80}-1\}$? Compare with the results of Exercise 11.16.

Unlike for zero knowledge, sequential and parallel composition preserve honest verifier zero knowledge, since we can simulate a conversation for each protocol separately and construct a conversation for the composed protocol by either concatenating transcripts or combining them.

Sigma protocols Again, sigma protocols have a natural notion of security that is closely related to honest verifier zero knowledge. Recall that sigma protocols are 3-move public coin protocols: the prover makes a commitment, the verifier issues a challenge and the prover responds.

D| **Definition 11.7.** Let $(\mathcal{P}, \mathcal{V})$ be a sigma protocol. A *(special honest verifier)* *simulator* is an algorithm that on input of x and a challenge β outputs an accepting conversation with β as its challenge.

Remark. Note that for sigma protocols, we have given up on the option to abort by outputting $\perp$. We now expect the simulator to always work.

The argument is *special honest verifier zero knowledge* (SHVZK) if the output of the simulator is indistinguishable from conversations between $\mathcal{P}$ and $\mathcal{V}$. Again, we have the usual flavours.

A special honest verifier simulator gives an honest verifier simulator by sampling a challenge and then running the special honest verifier simulator.

An honest verifier simulator cannot be turned into a special honest verifier simulator, because the honest verifier simulator is allowed to choose its own challenge, and may use this power to avoid difficult (but unlikely) challenges or coordinate the sampling of the challenge with the sampling of other things, a coordination that may be difficult to achieve for a given challenge.

E| *Exercise* 11.20. Consider the interactive argument $(\mathcal{P}, \mathcal{V})$ from Example 11.6. Show that there is a perfect special honest verifier simulator for this argument that requires trivial amounts of computation.

E| *Exercise* 11.21. Show that the interactive argument from Exercise 11.5 with challenge set $\mathfrak{B} = \{0,1\}$ has a perfect special honest verifier simulator. What if $\mathfrak{B} = \{0,1,2,3\}$? What if $\mathfrak{B} = \{0,1,2,\ldots,2^{80}-1\}$? Compare with the results of Exercise 11.16.

Witness indistinguishability The identification protocols we studied in Section 9.3 are very similar to sigma protocols, though we did simplify some things. We defined *witness indistinguishability* for identification schemes. It can be defined for interactive arguments, but we shall not do so.

11.2 NON-INTERACTIVE ARGUMENTS

In certain situations it is inconvenient or impossible for the prover and the verifier to have a conversation. The canonical example is when Alice needs to prove something to Bob, but Alice comes to town only on Mondays and Tuesdays, while Bob is in town only on Thursdays and Fridays. Running a sigma protocol will then require 10 days. Alice therefore wants to generate something that will convince Bob directly, without a conversation.

Unfortunately, it is impossible to use our simulation ideas to prove that such a protocol is zero knowledge, because any simulator should be able to generate an argument that looks like what the real prover would generate. But if it looks like a real argument, it will also fool the verifier.

This means that we need something else. One technique is a *common reference string* (CRS), a limited source of public randomness. The key observation is that the verifier's role in many interactive arguments is to issue random challenges to the prover. There is no real intelligence involved, and the important contribution is unpredictability.

The idea is to find some other source of unpredictability that can replace the verifier. Of course, as we have seen, if the verifier can be predicted, everything usually falls apart. And any unpredictability found lying around in a common reference string must be known to a cheating prover. Which means that we cannot directly replace the verifier by public randomness.

Remark. There is a lot of essentially unpredictable (and hence random-looking) data lying around, from weather patterns and sunspot activity to results of human activities such as birth counts and stock market data. But everyone must agree on how the common reference string is generated and from which random data and at which points in time. This makes it quite inconvenient in practical terms. Many cryptographers therefore tend to avoid common reference strings or use non-random, but somehow random-looking, data such as the binary expansion of the real number π.

Remark. We consider a completely different way to achieve non-interactive zero knowledge in Section 11.3.3, which avoids the CRS by exploiting the properties of the random oracle model.

Definition 11.8. A *non-interactive argument* for a witness relation (X, L, W, R) consists of three algorithms $(\mathcal{G}, \mathcal{P}, \mathcal{V})$:

- The *CRS generator* algorithm $\mathcal{G}$ outputs a common reference string.

- The *prover* algorithm $\mathcal{P}$ takes a common reference string, $x \in L$, $w \in W$ and outputs an argument.

- The *verifier* algorithm $\mathcal{V}$ takes as input a common reference string, $x \in L$ and an argument and either accepts or rejects.

The scheme is *complete* if when $x \in L$, the common reference string has been generated by the CRS generator, the argument has been created by the honest prover with the witness as input, and the verifier runs on input of the same common reference string, x and the generated argument, the verifier accepts except with small probability.

We shall consider an example that is related to Exercise 11.1, but for this specific example, the witness would not be sufficient to run the prover part of the protocol efficiently. We extend the witness relation in a suitable way.

Let p and q be two distinct primes that are congruent to 3 modulo 4, and let $n = pq$. Let G be the elements of $\mathbb{Z}_n^*$ that have Jacobi symbol 1, and let H be the squares in $\mathbb{Z}_n^*$. Then $|G| = (p-1)(q-1)/2$ and $|H| = (p-1)(p-1)/4$.

E | *Exercise* 11.22. Prove that the Jacobi symbol of -1 is 1, but $1 \notin H$. Then prove that for any $x \in G \setminus H$, there exists $w \in H$ such that $x = -w^2$.

E | *Exercise* 11.23. Let $X = G$, $L = G \setminus H$, $W = H \times \mathbb{Z} \times \mathbb{Z}$ and define R as $x \, R \, (w, p, q)$ if and only if $x = -w^2$ and $pq = n$. Show that (X, L, W, R) is a witness relation, and how to sample from the uniform distribution on X, H and L.

E | *Exercise* 11.24. Let $\mathcal{A}$ be an algorithm that on input of G and $x \in G$ outputs a square root of x with probability ϵ and with runtime τ. Give an algorithm $\mathcal{B}$ that on input of G outputs (p, q) with probability $\epsilon/2$ and with runtime essentially equal to τ.

Hint: Use Proposition 3.13 and the related discussion.

E | *Example* 11.8. Consider the following non-interactive argument for witness relation from Exercise 11.23.

- The CRS generator $\mathcal{G}$ samples α from the uniform distribution on X. The common reference string output is α.

- On input of α, x and (w, p, q), the prover algorithm $\mathcal{P}$ checks if $\alpha \in H$. If it is, it computes $\gamma \in H$ such that $\gamma^2 = \alpha$. If $\alpha \notin H$, it computes $\gamma \in H$ such that $\gamma^2 x = \alpha$. The argument is γ.

- On input of α, x and the argument γ, the verifier $\mathcal{V}$ computes γ^2 and $\gamma^2 x$. The verifier accepts if either value is α, otherwise it rejects.

Unless the prover knows p and q, it cannot compute the square roots quickly.

This argument would have better soundness if the common reference string was -1. But that would make simulation impossible.

11.2.1 Soundness

Soundness for a non-interactive argument is much the same as soundness for an interactive argument, in that if our verifier accepts a argument, we should be able to strictly limit the probability that the statement is not in the language.

Definition 11.9. Let $(\mathcal{G}, \mathcal{P}, \mathcal{V})$ be a non-interactive argument for (X, L, W, R). A *cheating prover* $\mathcal{P}^*$ is an algorithm that on input of a CRS string generated by $\mathcal{G}$ and $x \in X$ outputs an argument. We say that a cheating prover $\mathcal{P}^*$ has *(soundness) advantage* ϵ if for some $x \in X \setminus L$, the probability that $\mathcal{V}$ accepts an argument output by $\mathcal{P}^*$ is ϵ.

Informally, again, we say that $(\mathcal{G}, \mathcal{P}, \mathcal{V})$ has *soundness error* ϵ if there is an "interesting" cheating prover with advantage ϵ, but no "interesting" cheating prover has advantage greater than ϵ. (Again, we leave the exact meaning of "interesting" undefined.)

Proposition 11.7. *Consider the non-interactive argument* $(\mathcal{G}, \mathcal{P}, \mathcal{V})$ *for the witness relation from Exercise 11.23 described in Example 11.8. It has soundness error* $1/2$.

Proof. We known that the probability that a random element from $\mathbb{Z}_n^*$ is a non-quadratic residue is $1/2$. If x is not a non-quadratic residue, half the time the CRS string will be a non-quadratic residue, and it will be impossible for the cheating prover to produce an argument γ that is accepted, since both $x\gamma^2$ and γ^2 will be quadratic residues. □

There are two strategies to improve this soundness bound. First, we can generate many arguments with independent common reference strings. Since each argument is independent, the probabilities multiply and quickly grow small. The other strategy is to have multiple elements from $\mathbb{Z}_n^*$ in the CRS, with the prover doing its computation for each of them. Essentially, this is exactly the same as repeating the argument many times.

Note also that the same CRS can be reused for many different elements of $X \setminus L$ without affecting soundness.

11.2.2 Arguments of Knowledge

Again, a non-interactive argument may prove not only that the statement is in the language but also that the prover actually knows a witness. As for interactive arguments, we define an extractor to be an algorithm that extracts a witness from the cheating prover.

Definition 11.10. Consider a non-interactive argument $(\mathcal{G}, \mathcal{P}, \mathcal{V})$ for (X, L, W, R). A *(non-interactive) (knowledge) extractor* for this witness relation is an algorithm that on input of x sampled from L and given access to a cheating prover $\mathcal{P}^*$ outputs a witness w. Note that the extractor is allowed to decide the common reference string used by the cheating prover.

The *advantage* of the extractor is the probability that the extractor outputs a witness such that $x\ R\ w$.

As usual, the advantage of a given extractor will depend on the success probability of the cheating prover it is used with. Informally, we say that a non-interactive argument has *knowledge error* δ if there is an extractor such that for any $\mathcal{P}^*$ with advantage $\delta+\epsilon$, the extractor will find a witness using not too much more time than $\mathcal{P}^*$ with a probability that is not too much smaller than ϵ. If we have a good non-interactive knowledge extractor, we say that the non-interactive argument is an *non-interactive argument of knowledge*.

Proposition 11.8. *Consider the non-interactive argument from Example 11.8 for the witness relation from Exercise 11.23. Let $\mathcal{P}^*$ be a cheating prover that for $x \in X$ has knowledge advantage $1/2 + \epsilon$ for some $\epsilon > 0$. Then there exists an extractor running with $\mathcal{P}^*$ that has advantage at least ϵ for $x \in X$. Its running time is essentially equal that of $\mathcal{P}^*$.*

Proof. We first build an algorithm that works as follows:

- It samples $r \xleftarrow{r} H$, computes $\alpha \leftarrow -r^2$ and runs $\mathcal{P}^*$ on α and x.
- If $\mathcal{P}^*$ outputs an argument γ such that $\alpha = x\gamma^2$, then the algorithm outputs r/γ. Otherwise the algorithm outputs $\perp$.

The algorithms's running time is the same as the running time of the cheating prover plus some trivial amount of computation. Likewise, if the cheating prover outputs a correct argument, the algorithm will not output $\perp$.

The algorithm samples the CRS from the uniform distribution on L. The probability that the algorithm outputs a witness is therefore equal to the probability that the cheating prover outputs a correct argument, given that the common reference string was sampled from the uniform distribution on L.

Let E be the event that the cheating prover runs on an input $x \in L$ with a CRS α generated by $\mathcal{G}$ and outputs an argument γ that $\mathcal{V}$ accepts. Let F be the event that a CRS α generated by $\mathcal{G}$ is in H.

The probability $\Pr[E \mid \neg F]$ is exactly the probability that the extractor outputs a witness. Then, we compute

$$1/2 + \epsilon = \Pr[E] = \Pr[E \mid F]\Pr[F] + \Pr[E \mid \neg F]\Pr[\neg F]$$
$$= \frac{1}{2}(\Pr[E \mid F] + \Pr[E \mid \neg F])$$
$$\Rightarrow\ \Pr[E \mid \neg F] = 1 + 2\epsilon - \Pr[E \mid F] \geq 2\epsilon.$$

We have not built an extractor for a witness, but an algorithm that finds square roots in $\mathbb{Z}_n$. By Exercise 11.24, such an algorithm can be used to find p and q with probability at least ϵ, which can then be used to find a witness. $\quad\square$

11.2.3 Zero Knowledge

Next, we shall prove that the argument itself does not reveal anything. As explained above, the simulator must in this case depend on more than the statement to be proved, and the extra information is the common reference string. We shall therefore need a simulator that produces both the CRS and the argument at the same time. Contrary to zero knowledge for interactive arguments, the verifier is not active during the argument generation, so we do not need to take a verifier into account. This simplifies our work.

D **Definition 11.11.** Let $(\mathcal{G}, \mathcal{P}, \mathcal{V})$ be a non-interactive argument. A *simulator* for the argument is an algorithm that on input of x outputs a common reference string and an argument.

We say that a non-interactive argument is *non-interactive zero knowledge* (NIZK) if the simulator's output is indistinguishable from the output of the CRS generation algorithm and the prover. As usual, we distinguish between *computational*, *statistical* and *perfect* non-interactive zero knowledge.

T **Proposition 11.9.** *Consider the non-interactive argument $(\mathcal{G}, \mathcal{P}, \mathcal{V})$ for the witness relation from Exercise 11.23 described in Example 11.8. There exists a perfect non-interactive zero knowledge simulator.*

Proof. The simulator gets x as input and does as follows:

- It samples $\gamma \xleftarrow{r} H$ and $b \xleftarrow{r} \{0, 1\}$, and computes $\alpha \leftarrow x^b \gamma^2$.

Since γ^2 is uniformly distributed over H and $x \notin H$, $x^b r^2$ will be uniformly distributed over G. Our simulator therefore simulates the CRS perfectly. Next, given the statement x and CRS, the correct response is uniquely determined, which means that the argument is also correctly distributed. □

Unlike for soundness, we cannot easily say that non-interactive zero knowledge is preserved if we reuse the CRS, because the simulator stops working.

11.3 USING HVZK

If we look at the argument discussed in Exercises 11.5, 11.13, 11.16, 11.19 and 11.21, we see that there is a trade-off between zero knowledge and soundness error. Also, we seem to have a large number of honest verifier zero knowledge arguments. It is interesting to see if we can use these protocols to build an argument that is zero knowledge. We shall now study a few methods where we can strengthen arguments so that we get zero knowledge instead of honest verifier zero knowledge. These methods will rely on different tools and give us different results.

11.3.1 From HVZK to ZK

We begin with a simple method to convert a suitable sigma protocol into an interactive argument (which will no longer be a sigma protocol) that is zero knowledge and still an argument of knowledge.

The general idea is to have the prover and verifier cooperate in choosing two possible challenges for the underlying sigma protocol, and then have the verifier choose which of the two challenges to use.

This is an interactive argument between $\mathcal{P}$ and $\mathcal{V}$, where $\mathcal{P}$ takes x and w as input, while $\mathcal{V}$ takes x as input.

- The prover runs $\mathcal{P}_0$ with input x and w, and sends its message α to $\mathcal{V}$.
- The verifier samples $\Delta \xleftarrow{r} \mathcal{B} \setminus \{0\}$ and sends Δ to $\mathcal{P}$.
- The prover samples $\beta_0 \xleftarrow{r} \mathcal{B}$ and sends β_0. Both compute $\beta_1 = \Delta - \beta_0$.
- The verifier samples $b \xleftarrow{r} \{0, 1\}$ and sends b to $\mathcal{P}$.
- The prover sends β_b to $\mathcal{P}_0$ and sends its response γ to $\mathcal{V}$.
- The verifier accepts if $(\alpha, \beta_b, \gamma)$ is an accepting conversation for $(\mathcal{P}_0, \mathcal{V}_0)$.

FIGURE 11.4 Interactive argument $(\mathcal{P}, \mathcal{V})$ based on a sigma protocol $(\mathcal{P}_0, \mathcal{V}_0)$ whose challenge set $\mathcal{B}$ has a group structure.

The protocol is given in Figure 11.4. Before we prove that this protocol is zero knowledge, we need to prove that we still have some measure of soundness. We do this by showing that any cheating prover can be used to create two accepting conversations with the same commitment, which can then be used by a special soundness extractor to find a witness. In other words, the protocol is an argument of knowledge.

Proposition 11.10. *Let $(\mathcal{P}, \mathcal{V})$ be the interactive argument from Figure 11.4 for the witness relation (X, L, W, R), based on a sigma protocol $(\mathcal{P}_0, \mathcal{V}_0)$ for the same relation. Suppose the sigma protocol has τ-special soundness.*

If $\mathcal{P}^*$ is a cheating prover with advantage $1/2 + \epsilon$, $\epsilon \geq 0$, then there exists a knowledge extractor that when running with $\mathcal{P}^*$ has advantage at least ϵ. If the cheating prover $\mathcal{P}^*$ uses time at most τ', then the extractor uses time at most $\tau + 2\tau'$ plus some trivial administrative overhead.

Remark. This proof is essentially identical to the proof of Proposition 11.3.

Our next task is to prove that the protocol is zero knowledge, and we shall do this by exhibiting a perfect zero knowledge simulator.

The simulator S takes x as input.

- It samples $b' \xleftarrow{r} \{0,1\}$, computes $(\alpha, \beta_{b'}, \gamma) \leftarrow S_0$ and starts V^* with random tape r.
- It sends α to V^*, to which V^* responds with Δ.
- The simulator computes $\beta_{1-b'} = \Delta - \beta_{b'}$ and sends β_0 to V^*.
- Then V^* responds with b. If $b \neq b'$, the simulator outputs $\perp$ and stops.
- Otherwise, the simulator outputs the view $(\alpha, \Delta, \beta_0, b, \gamma, r)$.

FIGURE 11.5 Zero knowledge simulator for the interactive argument $(\mathcal{P}, \mathcal{V})$ from Figure 11.4, using a cheating verifier V^* and an honest verifier simulator S_0 for the underlying sigma protocol $(\mathcal{P}_0, \mathcal{V}_0)$.

T **Proposition 11.11.** *Consider the interactive argument $(\mathcal{P}, \mathcal{V})$ given in Figure 11.4, based on a sigma protocol $(\mathcal{P}_0, \mathcal{V}_0)$ with a perfect honest verifier simulator S_0. The simulator S given in Figure 11.5 is a perfect zero knowledge simulator for $(\mathcal{P}, \mathcal{V})$ that outputs $\perp$ with probability $1/2$.*

E *Exercise* 11.25. Prove Propositions 11.10 and 11.11.

11.3.2 From SHVZK to ZK Using Commitments

There are two drawbacks with the construction from the previous section. First, there is an additional message. And second, the construction introduces a significant knowledge error. In this section, we shall explain how to keep the knowledge error small. We will still need an extra message.

The idea is for the verifier to commit to his challenge before the prover makes his commitment (see Exercise 4.18). This is the extra message, and it forces a cheating verifier to make his choice before seeing the prover's commitment, which in effect forces him to behave semi-honestly, in the sense that he will not look at the prover's commitment when choosing his challenge. This is only semi-honest behaviour, since the verifier does not have to choose a random challenge. However, if our protocol is special honest verifier zero knowledge, semi-honest behaviour is sufficient. Finally, the hides the challenge from the prover, so a cheating prover does not get any extra advantage.

Given a sigma protocol, we can construct the interactive argument given in Figure 11.6. If we use a carefully constructed commitment scheme, we can prove the new protocol zero knowledge. We will also need to show that our modification does not ruin soundness.

We will use the commitment scheme from Section 8.4.1. It is both equivocable and extractable, which means that there is a way to generate "trapdoored" commitment keys that either allow us to open commitments to any value, or

This is an interactive argument between $\mathcal{P}$ and $\mathcal{V}$, where $\mathcal{P}$ takes x and w as input, while $\mathcal{V}$ takes x as input.

- The verifier samples $\beta \xleftarrow{r} \mathfrak{B}$, computes $(u, o) \leftarrow \mathcal{CC}(\beta)$ and sends u.
- The prover runs $\mathcal{P}_0$ with input x and w, and sends its message α to $\mathcal{V}$.
- The verifier sends (β, o) to $\mathcal{P}$.
- If $\mathcal{CO}(u, \beta, o) = 1$, the prover sends β to $\mathcal{P}_0$ and its response γ to $\mathcal{V}$.
- The verifier accepts if (α, β, γ) is an accepting conversation for $(\mathcal{P}_0, \mathcal{V}_0)$.

FIGURE 11.6 An interactive argument $(\mathcal{P}, \mathcal{V})$ for a witness relation (X, L, W, R) based on a sigma protocol $(\mathcal{P}_0, \mathcal{V}_0)$ for the same relation and a commitment scheme whose message space contains the sigma protocol's challenge space $\mathfrak{B}$.

extract the message from a commitment. We shall use the former property to create adversaries against soundness and argument of knowledge, and the latter against zero knowledge.

Soundness Let $\mathcal{P}^*$ be a cheating prover against the interactive argument from Figure 11.6. We shall create a cheating prover $\mathcal{P}_0^*$ against $(\mathcal{P}_0, \mathcal{V}_0)$.

We begin with the observation that if we use the equivocable commitment key generator to create the commitment key in the common reference string, the cheating prover's advantage should not change.

Next, we change the honest verifier so that it first commits to 0, then uses the equivocation property to open the commitment to its challenge. Again, the cheating prover's advantage should not change. Clearly, it does not now matter if the honest verifier delays deciding on the challenge until after the cheating prover has sent its first message.

Our cheating prover $\mathcal{P}_0^*$ for $(\mathcal{P}_0, \mathcal{V}_0)$ gets x as input and does:

- Compute $(ck, a) \leftarrow \mathcal{CK}_{eq}$ and $(u, o) \leftarrow \mathcal{CC}(ck, 0)$.
- Run $\mathcal{P}^*$ with common reference string ck and input x. Send u to $\mathcal{P}^*$.
- When $\mathcal{P}^*$ sends α, send α to $\mathcal{V}$ and receive the challenge β.
- Compute $o' \leftarrow \mathcal{CQ}(ck, a, u, o, 0, \beta)$ and send (β, o') to $\mathcal{P}^*$.
- When $\mathcal{P}^*$ outputs γ, output γ.

By the above argument, this cheating prover $\mathcal{P}_0^*$ should work about as well against the underlying sigma protocol as $\mathcal{P}^*$ does against the interactive argument from Figure 11.6.

It follows that we have not lost soundness. If we ever had it.

E| *Exercise* 11.26. Verify the details of the above argument.

Argument of Knowledge We shall now show that any cheating prover for the interactive argument from Figure 11.6 can be used in a slightly modified version of the extractor from Figure 11.2. This shows that the proof of Theorem 11.4 still works, so that our protocol is an argument of knowledge.

By Exercises 8.43 and 8.44, we know that the cheating prover running with a commitment key generated by $\mathcal{CK}_{eq}$ and seeing a commitment to the challenge that comes from $\mathcal{CQ}$ will behave in essentially the same way.

Next, we describe the extractor. The only modification we need to make to the extractor from Figure 11.2 is to generate an equivocable commitment key in the common reference string using $\mathcal{CK}_{eq}$. We also need to give the cheating prover a commitment to 0 before it sends β. When the cheating prover outputs its first message α, we create openings of our commitment to two distinct challenge values using $\mathcal{CQ}$. When we give the cheating prover the challenges, we also need to give the corresponding opening. Otherwise, the extractor is unchanged.

All that remains is to show that the proof of Theorem 11.4 still works with the modified extractor, which means that we have an argument of knowledge.

E| *Exercise* 11.27. Verify the details of the above argument.

Zero knowledge This time, we have a cheating verifier. We now generate an extractable commitment key in the common reference string using $\mathcal{CK}_{ex}$. By the arguments in Section 8.4.1 the cheating verifier should not notice. Furthermore, using the "trapdoor" to the commitment key, we can extract the challenge from the cheating verifier's commitment, and we can then use the sigma protocol's special honest verifier simulator to generate a suitable conversation for that challenge. By Exercise 8.45, the honest prover would never accept any other challenge than the one extracted.

E| *Exercise* 11.28. Verify the details of the above argument.

11.3.3 From HVZK to NIZK in the ROM

There is a technique that turns any sigma protocol into a non-interactive argument. If the sigma protocol has soundness, special soundness and honest verifier zero knowledge, then the non-interactive argument will be sound, an argument of knowledge and zero knowledge in the random oracle model.

For these non-interactive arguments, we add the ability to tie associated data to the argument. This is useful when using non-interactive arguments in protocols, for example to ensure that an argument generated by one party cannot be reused by a different party. Note that associated data is different from the auxillary information that is needed for proof-technical reasons.

The non-interactive argument for the witness relation (X, L, W, R) works as follows:

- The prover $\mathcal{P}$ gets as input $x \in L$, $w \in W$ and associated data ad as input. It runs $\mathcal{P}_0$ with x and w as input, and $\mathcal{P}_0$ outputs α.
- The prover $\mathcal{P}$ computes $\beta \leftarrow h(x, \alpha, ad)$ and sends β to $\mathcal{P}_0$.
- When $\mathcal{P}_0$ outputs γ, $\mathcal{P}$ outputs the argument (α, γ).
- The verifier $\mathcal{V}$ gets x, (α, γ) and ad as input. It computes $\beta \leftarrow h(x, \alpha, ad)$. If (α, β, γ) is an accepting conversation for the sigma protocol, then $\mathcal{V}$ accepts, otherwise it rejects.

FIGURE 11.7 Non-interactive argument $(\mathcal{P}, \mathcal{V})$ for a witness relation (X, L, W, R) using a sigma protocol $(\mathcal{P}_0, \mathcal{V}_0)$ and a hash function h.

The non-interactive argument is given in Figure 11.7 and is known as the Fiat-Shamir heuristic. We have already studied this heuristic in connection with signature schemes and identification protocols in Section 9.3.

In some sense, we replace the honest verifier's random choice by a hash. If the hash function is random-looking, the cheating prover cannot know what the challenge will be until it has decided on α and computed the hash of the element. From a cheating prover's point of view, this is very similar to the interactive setting: the verifier's choice is unknown until the prover commits.

The main difference is that in the interactive setting, the number of times a cheating prover can try is typically small, while in the non-interactive setting, the only limit is the cheating prover's available computing resources.

Zero Knowledge The zero knowledge property essentially follows from the honest verifier property of the sigma protocol. This is achieved by *tampering* with the random oracle. For this tampering to work, the probability of guessing the prover's commitment should be sufficiently small. If the sigma protocol does not already have this property, it is easy to achieve, e.g. simply by adding random bits to the prover's commitment that are subsequently ignored. This means that the probability bound in the following result can be made arbitrarily small, at the expense of a slightly longer argument.

Proposition 11.12. *Suppose there is a perfect honest verifier simulator $\mathcal{S}$ for the sigma protocol $(\mathcal{P}_0, \mathcal{V}_0)$ for the witness relation (X, L, W, R). Suppose also that the probability of $\mathcal{P}_0$ choosing any given commitment is at most δ. Then, there is a perfect simulator for the non-interactive argument from Figure 11.7 that outputs $\perp$ with probability at most $l\delta$, where l is the number of queries to the random oracle.*

Proof. The simulator computes $(\alpha, \beta, \gamma) \leftarrow \mathbf{Sim}(x, ad)$ and reprograms the random oracle so that $h(x, \alpha, ad) = \beta$, returning $\perp$ if reprogramming fails.

The honest verifier simulator is perfect, so the only way this can be noticed is if the random oracle has already been queried at (x, α, ad). If there have been at most l random oracle queries, the probability that the random oracle has already been queried at (x, α, ad) is at most $l\delta$. □

It is worth noting that this simulator in this theorem manipulates the random oracle. This means that it would be non-trivial to model two different constructions using the same random oracle. Fortunately, it is easy to cryptographically separate hash functions using unique prefixes.

Related to this, the result should also be compared to Proposition 9.20 and the overall strategy used in that section.

The requirement that the honest verifier simulator is perfect is not essential. A standard hybrid argument proves that we can run the underlying honest verifier simulator l_s times to generate l_s non-interactive arguments, and any adversary's distinguishing advantage increases by at most l_s.

E | *Exercise* 11.29. State carefully and prove the claim in the previous paragraph.

Soundness   Again, soundness essentially follows from the soundness of the sigma protocol. The proof is straight-forward. We simulate the random oracle and guess which query will be used for the cheating prover's output. We then use the hash query as our cheating prover's commitment and the verifier's challenge as the random oracle response. If we guessed the query correctly, the interactive verifier accepts. If we guess wrong or the cheating adversary fails to produce a valid argument, we fail to convince the interactive verifier.

T | **Proposition 11.13.** *Let $\mathcal{P}^*$ be a cheating prover for the non-interactive argument from Figure 11.7 with soundness advantage ϵ, making l queries to the random oracle. Then, there exists a cheating prover $\mathcal{P}_0^*$ against $(\mathcal{P}_0, \mathcal{V}_0)$ with soundness advantage at least ϵ/l. The runtime of $\mathcal{P}_0^*$ is essentially the same as the runtime of $\mathcal{P}^*$.*

As for zero knowledge, l can be very large, so the soundness error of the underlying sigma protocol must be very small. Fortunately, for sigma protocols this is usually quite easy to achieve.

Argument of Knowledge   Showing that the non-interactive argument is an argument of knowledge is somewhat more difficult, and while superficially similar to the rewinding argument used in Section 11.1.3, the actual result is much weaker. The problem is that since the argument is non-interactive, we cannot know exactly which hash value the cheating prover will use as its challenge. This gives the cheating prover more freedom, and makes our extractor's job more difficult.

▏T▕ **Proposition 11.14.** *Consider the non-interactive argument* $(\mathcal{G}, \mathcal{P}, \mathcal{V})$ *from Figure 11.7 based on the sigma protocol* $(\mathcal{P}_0, \mathcal{V}_0)$ *with* τ-*special soundness.*

Let $\mathcal{P}^*$ *be a cheating prover in the random oracle model making at most* l *oracle queries and having soundness advantage* ϵ. *Then, there exists a knowledge extractor with knowledge advantage* $\epsilon^2/(16l) - 1/|\mathfrak{B}|$. *The runtime of the knowledge extractor is essentially at most twice the runtime of the cheating prover plus* τ *plus the cost of sampling* $2l$ *elements from* $\mathfrak{B}$.

The proof is essentially a slightly simplified variant of the proof of Theorem 9.24, first creating a derandomised algorithm and applying the Forking lemma (Theorem 9.23), after which we apply the special soundness extractor.

▏E▕ *Exercise* 11.30. Write out the proof of Proposition 11.14.

Remark. While ϵ^2 may be very small, l is the number of hash queries and the only reasonable bound on l is the cheating prover's runtime, which means that l will be very large. The power of this result is therefore fairly weak.

Remark. The above result relies on rewinding, which is difficult to apply in some contexts. In particular, trying to use the extracted witness in the execution from which it was extracted can, unless great care is exercised, cause an excessive number of rewindings, increasing the extractor runtime exponentially. Soundness and zero knowledge does not suffer from this problem.

Public Coin Arguments The above approach works for any public coin interactive argument with respect to zero knowledge. It does not work in general for soundness. Consider the example of the sequential composition of l instances of Example 11.4 with challenge set $\mathfrak{B} = \{0, 1\}$. As an interactive argument, it has soundness error $1/2^l$. Made non-interactive in the obvious fashion, an adversary making an expected $3l/2$ random oracle queries can forge a proof. The problem is that our forger can essentially rewind the verifier.

The approach works for many protocols, but we do not develop this theory.

▏E▕ *Exercise* 11.31. Verify the above claim with respect to zero knowledge.

11.4 FURTHER USEFUL ARGUMENTS

We shall now study a collection of useful arguments. In particular, we shall study how to prove properties of encryptions and commitments. We shall also describe how to combine sigma protocols in a very useful way. Finally, we shall give a variant of an argument that is theoretically very important.

11.4.1 Encryption Arguments

Example 11.3 and Exercise 11.4 show how to argue that two discrete logarithms are the same. This argument has immediate applications.

Example 11.9. Consider the ElGamal cryptosystem based on a cyclic group G of prime order n with a generator g. We would like to argue that the decryption of a given ElGamal ciphertext is a specific message.

The ElGamal encryption key is $y \in G$ and the decryption key is $b \in \mathbb{Z}$ such that $y = g^b$. A ciphertext $(x, w) \in G \times G$ decrypts to $m = wx^{-b}$. But this means that $w/m = x^b$, so the decryption of (x, w) is m if and only if

$$\log_x w/m = \log_g y.$$

The interactive argument from Exercise 11.4 can be used to prove that the discrete logarithms are the same. Which means that the argument can be used to prove that (x, w) decrypts to m.

The above example can only be used by the owner of the decryption key. Others may need to prove that they have encrypted a given message.

Exercise 11.32. Based on the previous example, recall that an ElGamal ciphertext is $(x, w) = (g^r, y^r m)$. Explain how the sigma protocol from Exercise 11.4 can be used to prove that (x, w) is an encryption of m.

We shall make this argument more useful in the next section.

Note that the same technique as in the above exercise can be used with the ElGamal-based commitment scheme from Section 8.4.2, to prove that a commitment can be opened to a given message.

Exercise 11.33. Consider the Pedersen commitment scheme from Section 8.4.2. Note that if o is an opening of a commitment u to a message m, then $uh^{-m} = g^o$. Use this fact together with the sigma protocol from Example 11.5 to give an argument of knowledge of an opening of u to m.

11.4.2 Or Argument

Consider two statements x_0 and x_1. Alice would like to convince Bob that at least statement x_b is true, for some $b \in \{0, 1\}$. We already have sigma protocols such that Alice can convince Bob that x_b is true if she is willing to reveal b, but she does not want to reveal b.

The idea is to run the two sigma protocols in parallel. However, Alice will use the honest verifier simulator to simulate a conversation for one of the statements. Bob makes a challenge, but this will not be a challenge for either conversation. Instead, Bob expects Alice to reply with a challenge for each conversation and an accepting response. And the two challenges Alice sent should sum to Bob's challenge, which means that Alice is free to choose the challenge for one of the conversations, but not both.

We augment the Or composition of two witness relations (X_0, L_0, W_0, R_0) and (X_1, L_1, W_1, R_1) with a witness bit indicating a relation that holds. An interactive argument for this witness relation is given in Figure 11.8.

The sigma protocol for the witness relation $(X_0 \times X_1, (L_0 \times X_1) \cup (X_0 \times L_1), W_0 \times W_1, R_0 \vee R_1)$ works as follows.

- The public input is (x_0, x_1). The private input is (w_0, w_1) and $b \in \{0, 1\}$ such that $x_b \, R_b \, w_b$.

- The prover runs S_{1-b} with input x_{1-b} until it gets a conversation $(\alpha_{1-b}, \beta_{1-b}, \gamma_{1-b})$.

- The prover runs P_b with input x_b and w_b, and gets the message α_b. The prover sends (α_0, α_1) to the verifier.

- The verifier chooses a random challenge β and sends it to the prover.

- The prover computes $\beta_b = \beta - \beta_{1-b}$ and gives the message β_b to P_b, which outputs γ_b. The prover sends (β_0, β_1) and (γ_0, γ_1) to the verifier.

- The verifier accepts if $\beta_0 + \beta_1 = \beta$ and the two conversations $(\alpha_0, \beta_0, \gamma_0)$ and $(\alpha_1, \beta_1, \gamma_1)$ are accepting.

FIGURE 11.8 Or argument (P_{or}, V_{or}) based on two sigma protocols (P_0, V_0) and (P_1, V_1) and honest verifier simulators S_0 and S_1. The two sigma protocols must have the same challenge set, and this challenge set must have some group structure (written additively).

Remark. If one of the honest verifier simulators sometimes outputs $\perp$, the prover in Figure 11.8 is unbounded. If we cared, we could either allow provers to fail sometimes like in Section 9.3.4, or go for statistical security and choose a random argument after many attempts at using the simulator.

If the underlying sigma protocols have special soundness, our composed protocol will have special soundness. Let ZE_0 and ZE_1 be extractors for the underlying sigma protocols. Suppose we have two accepting conversations $((\alpha_0, \alpha_1), \beta, (\beta_0, \beta_1, \gamma_0, \gamma_1))$ and $((\alpha_0, \alpha_1), \beta', (\beta_0', \beta_1', \gamma_0', \gamma_1'))$ for some $x \in X_0 \times X_1$ with $\beta \neq \beta'$. Then, either $\beta_0 \neq \beta_0'$ or $\beta_1 \neq \beta_1'$, so for some $b \in \{0, 1\}$ we have two accepting conversations $(\alpha_b, \beta_b, \gamma_b)$ and $(\alpha_b, \beta_b', \gamma_b')$ with $\beta_b \neq \beta_b'$. We can now run ZE_b on these two conversations to recover a witness w_b. Now choose an arbitrary witness w_{1-b}, and (w_0, w_1) will be a witness for x. We have proved the following claim.

T **Proposition 11.15.** *If the underlying sigma protocols have special soundness, then the interactive argument from Figure 11.8 has special soundness.*

E *Exercise 11.34.* Give an honest verifier zero knowledge simulator for the interactive argument from Figure 11.8. What do you need to construct a special honest verifier zero knowledge simulator?

E *Exercise 11.35.* Explain how the Or argument can be extended to l sigma protocols $(\mathcal{P}_1, \mathcal{V}_1), (\mathcal{P}_2, \mathcal{V}_2), \ldots, (\mathcal{P}_l, \mathcal{V}_l)$.

E *Exercise 11.36.* Combine the use of the sigma protocol from Exercise 11.32 with the Or argument to give a sigma protocol showing that a given ElGamal encryption (x, w) is an encryption of one of the messages $m_1, m_2, \ldots, m_l$.

E *Exercise 11.37.* Combine the use of the sigma protocol from Exercise 11.33 with the Or argument to give a sigma protocol showing that a given Pedersen commitment u is a commitment to one of the messages $m_1, m_2, \ldots, m_l$.

11.4.3   A Multiplication Argument

Suppose Alice has committed to three values, and that the product of the first two values equals the third value. In this section we shall see how Alice can convince Bob of this fact, without revealing the three values. We begin by giving a simple interactive argument that does not treat the values as secret. Then, we add commitments to this simple argument, which will give us a sigma protocol that is sound and has (special) honest verifier zero knowledge.

Consider three values $w_1, w_2, w_3 \in \mathbb{F}_p$ known to a prover, who would like to convince a verifier of the fact that $w_1 w_2 = w_3$. The prover samples $\rho_1 \xleftarrow{r} \mathbb{F}_p$ and computes $\rho_2 = w_2 \rho_1$. The verifier samples a challenge β. The prover then computes $\gamma = \beta w_1 + \rho_1$, and the verifier verifies that

$$\beta w_1 + \rho_1 = \gamma \qquad\qquad w_2 \gamma - \rho_2 - \beta w_3 = 0.$$

A simple computation shows that this is a complete argument.

Now suppose that the verifier accepts, but that the prover is cheating and the claim is untrue, which means that $w_3 = w_1 w_2 + \delta$ and $\rho_2 = w_2 \rho_1 + \Delta$ for some $\delta \neq 0$. Then $\beta w_1 = \gamma - \rho_1$ and

$$\begin{aligned}
0 &= w_2 \gamma - w_2 \rho_1 - \Delta - \beta w_1 w_2 - \beta \delta \\
&= w_2 \gamma - w_2 \rho_1 - \Delta - (\gamma - \rho_1) w_2 - \beta \delta = \Delta - \beta \delta.
\end{aligned}$$

Note that δ is determined by the three values, and the cheating prover decides on Δ before the verifier chooses β, which means that if $\delta \neq 0$ a cheating prover has a $1/p$ chance of convincing a verifier that the claim holds.

The above is thus a sound argument. We want an argument that is (special) honest verifier zero knowledge. Our approach will be the usual one, where we get a challenge β, sample γ at random and then compute $\rho_1 = \gamma - \beta w_1$ and $\rho_2 = w_2\gamma - \beta w_3$. This simulates an accepting conversation perfectly.

Of course, the protocol makes no attempt to keep the values w_1, w_2, w_3 secret. We need to hide the values and still give the verifier some means to verify relations on the hidden values. For this we shall use the group homomorphic commitments from Section 8.4.2. These schemes are such that if one knows any two of message, commitment and opening, the third is fully determined.

A group homomorphic commitment scheme with unconditional binding gives us unconditional soundness. If the commitment scheme is unconditionally hiding, we get computational soundness. Regardless, we have perfect honest verifier zero knowledge.

The commitment verification algorithm $\mathcal{CO}$ defines a natural relation between commitments and opening-message pairs. We extend this relation to a relation R on the set X of triples of commitments and the set of triples of opening-message pairs. Let W be the subset of triples of opening-message pairs $W = \{((o_1, m_1), (o_2, m_2), (o_3, m_3)) \mid m_1 m_2 = m_3\}$, and let L be the set of triples of commitments which are related to some triple in W.

The interactive argument for the witness relation (X, L, W, R) is:

- The public input is three commitments u_1, u_2, u_3. The private input is w_1, w_2, w_3 and openings o_1, o_2, o_3.
- The prover samples ρ_1, computes $(u_4, o_4) \leftarrow \mathcal{CC}(\rho_1)$, $(u_5, o_5) \leftarrow \mathcal{CC}(\rho_1 w_2)$, and sends (u_4, u_5) to the verifier.
- The verifier samples a challenge β and sends it to the prover.
- The prover computes $\gamma \leftarrow \beta w_1 + \rho_1$, $o_6 \leftarrow \beta o_1 + o_4$ and $o_7 \leftarrow \gamma o_2 - o_5 - \beta o_3$, and sends (γ, o_6, o_7) to the verifier.
- The verifier accepts if $\mathcal{CO}(u_1^\beta u_4, \gamma, o_6) = 1 = \mathcal{CO}(u_2^\gamma (u_5 u_3^\beta)^{-1}, 0, o_7)$.

FIGURE 11.9 Argument for commitment multiplication from a group homomorphic commitment scheme $(\mathcal{CK}, \mathcal{CC}, \mathcal{CO})$. We have omitted ck.

The interactive argument is given in Figure 11.9.

Proposition 11.16. *There exists a perfect honest verifier simulator for the argument in Figure 11.9.*

Since the commitment scheme is group homomorphic, the following result is immediate.

T **Proposition 11.17.** *Let u_1, u_2, u_3 be commitments and $(u_4, u_5, \beta, \gamma, o_6, o_7)$, $(u_4, u_5, \beta', \gamma', o_6', o_7')$ be two accepting conversations, with $\beta' \neq \beta$. Then u_1 can be opened to $(\gamma - \gamma')(\beta - \beta')^{-1}$ and if u_2 can be opened to t, then u_3 can be opened to $(\gamma - \gamma')(\beta - \beta')^{-1}t$.*

In the unconditionally binding case, a commitment can only be opened to a unique value, which means that unless the claim about u_1, u_2, u_3 is true, a cheating prover can only make the verifier accept for at most one challenge, which means that we have the following theorem.

T **Proposition 11.18.** *If the commitment scheme is unconditionally binding and the challenge in the above protocol is sampled from a set of ν elements, the protocol has soundness error at most $1/\nu$.*

In the perfect hiding case, we can bound the soundness error by observing that the conversations give us openings for the commitments satisfying the claim. Which means that if openings of the commitments that do not satisfy the claim are known, we have openings of these commitments to different values, which breaks binding. We have proven the following theorem.

11.4.4   Shuffle Arguments

We shall sometimes need to shuffle things and hide how they were shuffled. We shall consider three cases: Shuffles of commitments to known values, shuffles of (commitments) to unknown, random values, and shuffles of ciphertexts.

Remark. In the interest of compact notation, we shall use our notation for vectors in $\mathbb{F}^n$ also for general tuples, and extend it. We denote a tuple $(\lambda_1, \ldots, \lambda_l)$ by $\boldsymbol{\lambda}$. We denote the element-wise product of two tuples $(\lambda_1 \mu_1, \ldots, \lambda_l \mu_l$ by $\boldsymbol{\lambda\mu}$, with similar notation for $\boldsymbol{\lambda} + \boldsymbol{\mu}$. We do the same for exponentiation, denoting the tuple $(\alpha_1^{\lambda_1}, \ldots, \alpha_l^{\lambda_l})$ by $\boldsymbol{\alpha^\lambda}$. Also, $\alpha^{\boldsymbol{\lambda}}$ denotes $(\alpha^{\lambda_1}, \ldots, \alpha^{\lambda_l})$ and $\boldsymbol{\alpha}^\lambda$ denotes $(\alpha_1^\lambda, \ldots, \alpha_l^\lambda)$. If π is a permutation on the indexes $\{1, 2, \ldots, l\}$, then $\boldsymbol{\pi\lambda}$ denotes $(\lambda_{\pi(1)}, \ldots, \lambda_{\pi(l)})$. Also, $\boldsymbol{\alpha}^\pi$ denotes $(\alpha_{\pi(1)}, \ldots, \alpha_{\pi(l)})$. (If the permutation π is thought of as a permutation matrix, this notation is consistent with a natural extension to matrices, which we do not develop.)

Shuffle of Known Values   Let $m_1, m_2, \ldots, m_l$ be (non-secret) messages and let $u_1, u_2, \ldots, u_l$ be commitments under some group homomorphic commitment scheme from Section 8.4.2. Alice would like to convince Bob that there is (she knows) a permutation π on $\{1, 2, \ldots, l\}$ and openings $o_1, o_2, \ldots, o_l$ such that $\mathcal{CO}(ck, u_i, m_{\pi(i)}, o_i) = 1$ for $i = 1, 2, \ldots, l$.

E *Exercise* 11.38. Write out the witness relation for this shuffle carefully.

We begin with some linear algebra and an interactive argument about the rank of a matrix. Let $\tilde{m}_1, \tilde{m}_2, \ldots, \tilde{m}_l, \tilde{\mu}_1, \tilde{\mu}_2, \ldots, \tilde{\mu}_l$ be non-zero values. We shall relate them to Alice's messages later. Consider a matrix

$$
\mathbf{M} =
\begin{pmatrix}
\tilde{m}_1 & \tilde{\mu}_1 & & & \\
 & \tilde{m}_2 & \tilde{\mu}_2 & & \\
 & & \ddots & & \\
 & & & \tilde{m}_{l-1} & \tilde{\mu}_{l-1} \\
(-1)^l \tilde{\mu}_l & & & & \tilde{m}_l
\end{pmatrix}.
\tag{11.1}
$$

Except for the single entry in the lower left corner, this matrix has non-zero entries only on the diagonal and above the diagonal.

E │ *Exercise* 11.39. Show that $\det(\mathbf{M}) = \prod_i \tilde{m}_i - \prod_i \tilde{\mu}_i$, and that the $l-1$ right-most columns of $\mathbf{M}$ are linearly independent, so $\mathbf{M}$ has rank at least $l-1$.

Alice wants to convince Bob that $\det(\mathbf{M}) = 0$, and will do that by showing that the first column of $\mathbf{M}$ can be expressed as a linear combination of the remaining columns, which are linearly independent by Exercise 11.39. Let

$$
\mathbf{1} =
\begin{pmatrix}
-\tilde{m}_1 \\
0 \\
\vdots \\
0 \\
(-1)^{l+1}(\tilde{\mu}_l)
\end{pmatrix}
\qquad
\tilde{\mathbf{M}} =
\begin{pmatrix}
\tilde{\mu}_1 & & & \\
\tilde{m}_2 & \tilde{\mu}_2 & & \\
 & \ddots & & \\
 & & \tilde{m}_{l-1} & \tilde{\mu}_{l-1} \\
 & & & \tilde{m}_l
\end{pmatrix}.
$$

E │ *Exercise* 11.40. Let $\rho_1' = -\tilde{m}_1/\tilde{\mu}_1$ and define $\rho_i' = -\tilde{m}_i \rho_{i-1}'/\tilde{\mu}_i$, $i = 2, 3, \ldots, l-1$. Show that ρ' thus defined satisfies

$$
\tilde{\mathbf{M}}\rho' = \mathbf{1}.
\tag{11.2}
$$

The interactive argument works as follows. Alice samples ρ and computes $\alpha = \tilde{\mathbf{M}}\rho$. Bob samples a challenge β_1. Alice computes ρ' as in Exercise 11.40 and $\gamma \leftarrow \rho + \beta\rho'$, which she sends to Bob. Bob then checks if

$$
\tilde{\mathbf{M}}\gamma = \alpha + \beta_1 \mathbf{1}
\tag{11.3}
$$

and accepts if and only if this relationship holds.

If the products $\prod_i \tilde{m}_i$ and $\prod_i \tilde{\mu}_i$ are equal, then by Exercise 11.39 Alice can always find a response that will make Bob accept.

If the products are inequal, the first column of $\mathbf{M}$ is linearly independent of the other $l-1$ columns, and for some $\delta, \mathbf{\Delta}$ we have

$$
\alpha = \delta \mathbf{1} + \tilde{\mathbf{M}}\mathbf{\Delta}.
$$

If Bob accepts, then $\delta + \beta_1 = 0$, which since δ is determined before β_1 is chosen will happen only if Bob chooses $\beta_1 = -\delta$.

We can simulate such conversations. For any β_1, sample γ and set

$$\alpha = \tilde{\mathbf{M}}\gamma - \beta_1 \mathbf{1}. \tag{11.4}$$

Since $\tilde{\mathbf{M}}$ has rank $l - 1$, it does not matter if we first choose ρ and then compute α and γ, or if we choose γ and compute α (and implicitly also ρ).

We shall now extend this linear algebra to a shuffle. As for the multiplication argument, we begin by considering two sets of messages $m_1, m_2, \ldots, m_l$ and $\mu_1, \mu_2, \ldots, \mu_l$. Alice would like to convince Bob that there is a permutation π such that $m_{\pi(i)} = \mu_i$ for $i = 1, 2, \ldots, l$.

We define two polynomials

$$f(X) = \prod_{i=1}^{l}(X - m_i) \qquad g(X) = \prod_{i=1}^{l}(X - \mu_i).$$

E **Exercise 11.41.** Show that if Alice' claim is correct, then for any β_0 we will have that $f(\beta_0) = g(\beta_0)$, but if no such permutation exists then equality will hold for at most l distinct β_0 values.

Define $\tilde{m}_i = \beta_0 - m_i$ and $\tilde{\mu}_i = \beta_0 - \mu_i$ for all i. We shall assume that β_0 is distinct from $m_1, \ldots, m_l, \mu_1, \ldots, \mu_l$. Observe that with the matrix $\mathbf{M}$ defined as in (11.1) we have that $\det(\mathbf{M}) = f(\beta_0) - g(\beta_0)$.

Now we observe that if there is a permutation π such that $m_{\pi(i)} = \mu_i$ for $i = 1, 2, \ldots, l$, then $\det(\mathbf{M}) = 0$. Otherwise, $\det(\mathbf{M}) \neq 0$. This means that Alice can use the above linear algebra argument to convince Bob that $\det(\mathbf{M}) = 0$. If Bob chose β_0 at random, then if he accepts that $\det(\mathbf{M}) = 0$, he must also accept that a permutation π exists.

The argument so far has not been attempting to hide anything. Now we note that $\mu_1, \ldots, \mu_l$ are not part of the public input, but instead everyone gets commitments $u_1, \ldots, u_l$ to the messages as part of the public input. The prover also gets openings of these commitments.

The idea is that the verifier first samples β_0, thereby implicitly defining the matrix $\mathbf{M}$. The prover commits to the vector in the column space of $\tilde{\mathbf{M}}$. After the verifier sends the challenge β_1, the prover responds with γ and openings of certain linear combinations of commitments to γ.

It is convenient to rearrange (11.3) so that all the public terms are on one side, while the terms that have been committed to are on the other side:

$$\alpha_1 + \gamma_1 \tilde{\mu}_1 = \beta_1 \tilde{m}_1,$$
$$\alpha_i + \gamma_i \tilde{\mu}_i = \gamma_{i-1} \tilde{m}_i, \quad i = 2, 3, \ldots, l - 1, \text{ and}$$
$$\alpha_l + (-1)^{l+1} \beta_1 \tilde{\mu}_l = \gamma_{l-1} \tilde{m}_l.$$

The protocol is given in Figure 11.10. In addition to the above values, it must also keep track of the correct openings of the product commitments.

The interactive argument works as follows:

- The public input is u and m. The private input is a permutation π and openings o. Let $\mu = \pi m$.

- The verifier samples $\beta_0 \xleftarrow{r} \mathfrak{B} \setminus \{m_1, \ldots, m_l,\}$ and sends β_0 to $\mathcal{P}$.

- The prover samples $\rho_1, \ldots, \rho_{l-1}$ and computes $(v_1, o_1') \leftarrow \mathcal{CC}(ck, \rho_1(\beta_0 - \mu_1))$, $(v_l, o_l') \leftarrow \mathcal{CC}(ck, \rho_{l-1}(\beta_0 - m_l))$ and $(v_i, o_i') \leftarrow \mathcal{CC}(ck, \rho_{i-1}(\beta_0 - m_{i-1}) + \rho_i(\beta_0 - \mu_i))$, $i = 2, 3, \ldots, l - 1$. The prover sends v to $\mathcal{V}$.

- The verifier samples $\beta_1 \xleftarrow{r} \mathfrak{B}$ and sends β_1 to $\mathcal{P}$.

- The prover computes $\rho_1' = -(\beta_0 - m_1)/(\beta_0 - \mu_1)$, $\rho_i' = -\rho_{i-1}'(\beta_0 - m_i)/(\beta_0 - \mu_i)$ and $\gamma_i = \rho_i + \beta_1 \rho_i'$ for $i = 1, 2, \ldots, l - 1$. Then, it computes $o_1'' = o_1' - \gamma_1 o_1$, $o_i'' = o_i' - \gamma_i o_i$ for $i = 2, 3, \ldots, l-1$ and $o_l'' = o_l' - (-1)^{l+1}\beta_1 o_l$, and sends $(\gamma_1, \ldots, \gamma_{l-1}, o_1'', \ldots, o_l'')$ to the verifier.

- The verifier accepts if and only if

$$\mathcal{CO}(ck, v_i u_i^{-\gamma_i}, \gamma_{i-1}(\beta_0 - m_i) - \gamma_i \beta_0, o_i'') = 1$$

for $i = 1, 2, \ldots, l$, with $\gamma_0 = \beta_1$ and $\gamma_l = (-1)^{l+1}\beta_1$.

FIGURE 11.10 Interactive argument $(\mathcal{P}, \mathcal{V})$ for shuffle of known values, based on a group homomorphic commitment scheme $(\mathcal{CK}, \mathcal{CC}, \mathcal{CO})$. The CRS contains $ck = (g, h)$ generated by $\mathcal{CK}$.

E *Exercise 11.42.* Verify that the interactive argument from Figure 11.10 is public coin and complete.

As usual, showing that there is an honest verifier simulator is fairly straight-forward, relying on the argument from the linear algebra simulation argument combined with the properties of the commitment schemes.

E *Exercise 11.43.* Show that there is a perfect honest verifier simulator for the argument from Figure 11.10.

Finally, we prove soundness when the commitment scheme is unconditionally binding, in which case the linear algebra argument applies.

T **Proposition 11.19.** *If the commitment scheme $(\mathcal{CK}, \mathcal{CC}, \mathcal{CO})$ is unconditionally binding, then the interactive argument from Figure 11.10 has soundness error at most $l/(|\mathfrak{B}| - l) + 1/|\mathfrak{B}|$.*

Proof. Since the commitment scheme is unconditionally binding, the commitments $u_1, \ldots, u_l$ define a set of messages $\mu_1, \mu_2, \ldots, \mu_l$. Furthermore, the prover and verifier essentialy run pure linear algebra argument.

Suppose that no permutation π exists such that $\pi\mathbf{m} = \boldsymbol{\mu}$. By Exercise 11.41 there are at most l choices of β_0 such that the matrix $\mathbf{M}$ has rank less than l, so the matrix $\mathbf{M}$ has rank l except with probability $l/(|\mathfrak{B}| - l)$.

If $\mathbf{M}$ has full rank, the discussion above shows that the verifier accepts with probability at most $1/|\mathfrak{B}|$. ☐

If the commitment scheme is not unconditionally binding, things become trickier. One idea is to use several conversations with identical commitments to either recover openings to a permutation, or break binding for the commitment scheme. The result is fairly weak. We leave this for the interested reader.

11.4.4.1 Shuffle of Random Values

We want to create a variant of this argument where we use weak commitments *on both sets of messages*. By weak commitments, we mean commitments that are perfectly binding, but only hiding for random messages. That is sufficient for the application we have in mind, as we shall see.

In fact, the commitment scheme we use is extremely simple, and we shall not treat it as a commitment scheme. It will be simply exponentiation in a group. Let G be a cyclic group of prime order n with generators g and ξ, with $\xi = g^a$. The commitment to $\tilde{m}_i$ will be $u_i = g^{\tilde{m}_i}$, while the commitment to $\tilde{\mu}_i$ will be $v_i = \xi^{\tilde{\mu}_i}$. It is essential that we use distinct generators and that a is secret, otherwise it is trivial to deduce the permutation.

In this context, Alice would like to convince Bob that there is a permutation π on $\{1, 2, \ldots, l\}$ such that $\pi\mathbf{m} = \boldsymbol{\mu}$, when Bob knows $g, \xi, \mathbf{u}, \mathbf{v}$.

E| *Exercise* 11.44. Write out the witness relation for this shuffle carefully.

We want to reuse the previous argument, but evaluate (11.3) in the exponent. For example, the second equation becomes $u_2^{\gamma_1} v_2^{\gamma_2} = \alpha_2$. Since we must use distinct generators, every v_i term is multiplied by a in the exponent. Then, the determinant is $\prod_i \tilde{m}_i - \prod_i a\tilde{\mu}_i$ and our argument fails. To compensate, we augment the matrix with l rows $(\ldots \, a \, 1 \ldots)$, multiplying the first term of the determinant by a suitable power of a:

$$\mathbf{M} = \begin{pmatrix} \tilde{m}_1 & \tilde{\mu}_1 a \\ & \tilde{m}_2 & \tilde{\mu}_2 a \\ & & \ddots \\ & & & \tilde{m}_l & \tilde{\mu}_l a \\ & & & a & 1 \\ & & & & & \ddots \\ & & & & & & a & 1 \\ & 1 & & & & & & & a \end{pmatrix}. \tag{11.5}$$

Note that $\tilde{m}_i \neq 0 \neq \tilde{\mu}_i$ for any i, otherwise the argument fails. But this is easy to verify in our case, since it is easy to compute g^{β_0} and ξ^{β_0}.

Again, the idea is for the verifier to choose β_0 which implicitly defines $\mathbf{M}$. Let $\mathbf{1}$ and $\tilde{\mathbf{M}}$ be as before. The prover commits to $\tilde{\mathbf{M}}\rho$ in the exponent and finds ρ' such that $\tilde{\mathbf{M}}\rho' = \mathbf{1}$. When the verifier chooses β_1, the prover reveals $\gamma = \rho + \beta_1\rho'$. The verifier can check that $\tilde{\mathbf{M}}\gamma = \tilde{\mathbf{M}}\rho + \beta_1\mathbf{1}$ in the exponent since it can use u_i instead of m_i and v_i instead of $\mu_i a$.

E *Exercise* 11.45. Redo Exercises 11.39 and 11.40 for $\mathbf{M}$ as given in (11.5).

The interactive argument works as follows:

- The public input is $\xi \in G$, commitments $\mathbf{u}$ and $\mathbf{v}$. The private input is messages $\mathbf{m}$, a and a permutation π with $\xi = g^a$, $\mathbf{u} = g^{\mathbf{m}}$ and $\mathbf{v} = \xi^{\pi\mathbf{m}}$.

- The verifier samples $\beta_0 \xleftarrow{r} \mathfrak{B} \setminus \{m_1, \dots, m_l\}$ and sends β_0 to $\mathcal{P}$.

- The prover samples $\rho_1, \dots, \rho_{2l-1}$ and computes $\alpha_1 \leftarrow g^{\rho_1 a(\beta_0 - m_{\pi(1)})}$, $\alpha_i \leftarrow g^{\rho_{i-1}(\beta_0 - m_i) + \rho_i a(\beta_0 - m_{\pi(i)})}$, $i = 2, 3, \dots, l$, $\alpha_i \leftarrow g^{\rho_{i-1}a + \rho_i}$, $i = l+1, l+2, \dots, 2l-1$, and $\alpha_{2l} \leftarrow g^{\rho_{2l-1}a}$. The prover sends $(\alpha_1, \alpha_2, \dots, \alpha_{2l})$ to the verifier.

- The verifier samples $\beta_1 \xleftarrow{r} \mathfrak{B}$ and sends β_1 to $\mathcal{P}$.

- The prover computes $\rho'_1 = -(\beta_0 - m_1)/(a(\beta_0 - m_{\pi(1)}))$, $\rho'_i = -\rho'_{i-1}(\beta_0 - m_i)/(a(\beta_0 - m_{\pi(i)}))$ for $i = 2, 3, \dots, l$, $\rho'_i = -\rho'_{i-1}a$ for $i = l+1, \dots, 2l-1$ and finally $\gamma = \rho + \beta_1\rho'$, and sends γ to $\mathcal{V}$.

- The verifier accepts if and only if for $i = 1, 2, \dots, 2l$ it holds that

$$
\alpha_i = \begin{cases} (g^{\beta_0}u_i^{-1})^{\gamma_{i-1}}(\xi^{\beta_0}v_i^{-1})^{\gamma_i} & \text{for } i = 1, 2, \dots, l, \\ \xi^{\gamma_{i-1}}g^{\gamma_i} & \text{for } i = l+1, l+2, \dots, 2l, \end{cases}
$$

with $\gamma_0 = \gamma_{2l} = \beta_1$.

FIGURE 11.11 Interactive argument $(\mathcal{P}, \mathcal{V})$ for shuffle of secret random values, based on a group G of prime order n with generator g.

The resulting argument is given in Figure 11.11. A perfect honest verifier simulator is very similar to the simulator given in Exercise 11.43. Soundness follows since arithmetic in the exponent is the same as arithmetic modulo n.

E *Exercise* 11.46. Show that the argument in Figure 11.11 is complete.

E *Exercise* 11.47. Give a perfect honest verifier simulator for the argument in Figure 11.11.

E| *Exercise* 11.48. Show that the argument in Figure 11.11 has soundness error at most $l/(|\mathfrak{B}| - l) + 1/|\mathfrak{B}|$.

11.4.4.2 Shuffle of Ciphertexts

One important technique that we shall need later is to rerandomise (or reencrypt) and shuffle (reorder) a collection of ciphertexts. The purpose may be for Alice to break the link between the origin of a particular ciphertext and its eventual decryption, but still convince Bob that the collection of decryptions reflect the original collection of ciphertexts. Alice will do this by convincing Bob that she has done the rerandomisation and shuffling correctly, without revealing how the ciphertexts were shuffled.

Let G be a cyclic group of prime order p with generator g, and let $y \in G$ be an ElGamal encryption key. Alice has a collection of ElGamal ciphertexts $(x_1, w_1), \ldots, (x_l, w_l)$ and the shuffled rerandomisations $(\tilde{x}_1, \tilde{w}_1), \ldots, (\tilde{x}_l, \tilde{w}_l)$. Alice wants to convince Bob that randomness $r_1, r_2, \ldots, r_l$ and a permutation π exist such that $\tilde{x}_i = g^{r_i} x_{\pi(i)}$ and $\tilde{w}_i = y^{r_i} w_{\pi(i)}$.

E| *Exercise* 11.49. Write out the witness relation for this shuffle carefully.

Remark. Note that this argument is somewhat complicated by the fact that either Alice or Bob may have been involved in creating (some of) the ciphertexts. Alice may also know the decryption key, but Bob should not.

The argument is based on the unpredictability of random linear combinations. For any sequence of group elements, a random linear combination of the elements in the sequence will only rarely give the identity, unless every element in the sequence equals identity.

E| *Exercise* 11.50. Let $s_1, \ldots, s_l$ be a sequence of group elements, not all equal to the identity. Suppose we sample $\lambda_1, \ldots, \lambda_l \xleftarrow{r} \{0, 1, \ldots, p-1\}$. Show that

$$\Pr\left[\prod_i s_i^{\lambda_i} = 1\right] = \frac{1}{p}.$$

If the ciphertexts had not been shuffled, Alice could use the homomorphic property of ElGamal to compute the product ciphertext $(\prod_i (x_i/\tilde{x}_i)^{\lambda_i}, \prod_i (w_i/\tilde{w}_i)^{\lambda_i})$ and then use the argument from Section 11.32 to convince Bob that this is an encryption of 1. But Alice has shuffled the ciphertexts, so Bob does not (and must not) know how to match shuffled ciphertexts with linear coefficients.

Our idea will be that Alice commits to the coefficients and a permutation of the coefficients by raising two distinct generators to the coefficients. She uses the argument from Figure 11.11 to convince Bob that she has done this correctly. Then, she uses the equality of discrete logarithm argument from Example 11.4 to convince Bob that she has raised each ciphertext to the correct value. Finally, everything is multiplied together, and Alice proves that the difference is an encryption of zero.

The main obstacle is that Alice cannot choose the linear coefficients, and Bob cannot know them. Therefore, Alice and Bob must create them jointly. The obvious idea is for Alice and Bob each to contribute a share to the co-efficients. Again, Bob cannot know the permutation, so Alice must compute the reconstruction of the permuted coefficients from the shares. However, if Alice is allowed to choose the permutation *after* the coefficients have been chosen, Exercise 11.50 no longer applies, since the number of permutations would allow Alice to try many different sequences of group elements.

The end result is that Alice first commits to the permutation by committing to shares in the coefficients and a permutation of the shares. Then, Alice and Bob jointly create a shift (which simplifies hiding the permutation), which Alice then applies to the permutation. To ensure that Alice does not change the permutation, Bob chooses a random multiple of the shift to apply.

The proof for the above argument relies on the ciphertexts being random-looking. Since Bob may have been involved in creating the ciphertexts, they may not be random-looking. Therefore, Alice must first rerandomise the ci-phertexts before the above shuffle. For simplicity when using the argument in larger constructions, we do the same for the shuffled ciphertexts.

We will need to be able to extract the permutation from the conversa-tion. In principle, we could treat the argument as an argument of knowledge and build an extractor based on rewinding, but doing so introduces its own set of problems. Instead, the prover shal commit to the permutation under a commitment key chosen by the verifier, which allows us to extract the permu-tation. The argument also trivially reveals everything to a cheating verifier, but this is acceptable since we only aim for security for an honest verifier.

The argument is summarised in Figure 11.12.

Remark. The design goal for the argument is simplicity, both with respect to understanding, proofs and later use. There are a number of simple improve-ments to this argument with respect to size and computational cost. There is also a large body of research on improved or generalised arguments for shuffles.

E | *Exercise* 11.51. Show that the argument in Figure 11.12 is public coin and complete.

The tuples u_0 and v_0 commit the prover to a permutation. This should be the same permutation established by the argument from Figure 11.11.

E | *Exercise* 11.52. Consider an honest verifier's view of the first four steps:

- ξ, u_2, v_0 and distinct $u_{0,1}, \ldots, u_{0,l}$ were chosen;
- λ_3 was sampled from $\{0, 1, \ldots, p-1\}^l$;

The interactive argument works as follows:

- The public input is an ElGamal encryption key y and two sequences of ciphertexts $(\mathbf{x}, \mathbf{w})$ and $(\tilde{\mathbf{x}}, \tilde{\mathbf{w}})$. The private input is a permutation π on $\{1, 2, \ldots, l\}$ and randomness $\mathbf{r}$ such that $\tilde{\mathbf{x}} = g^{\pi \mathbf{r}} \mathbf{x}^{\pi}$ and $\tilde{\mathbf{w}} = y^{\pi \mathbf{r}} \mathbf{w}^{\pi}$.

- The prover samples $a \xleftarrow{r} \{0, 1, \ldots, p-1\}$ and $\mathbf{r}', \mathbf{r}'', \boldsymbol{\lambda}_0, \boldsymbol{\lambda}_2 \xleftarrow{r} \{0, 1, \ldots, p-1\}^l$, subject to $\lambda_{0,1}, \ldots, \lambda_{0,l}$ being all distinct, and computes rerandomisations $\bar{\mathbf{x}} \leftarrow g^{\mathbf{r}'} \mathbf{x}$, $\bar{\mathbf{w}} \leftarrow y^{\mathbf{r}'} \mathbf{w}$, $\dot{\mathbf{x}} \leftarrow g^{\mathbf{r}''} \tilde{\mathbf{x}}$, $\dot{\mathbf{w}} \leftarrow y^{\mathbf{r}''} \tilde{\mathbf{w}}$, $\xi \leftarrow g^a$, and permutation commitments $\mathbf{u}_0 \leftarrow g^{\boldsymbol{\lambda}_0}$, $\mathbf{u}_2 \leftarrow g^{\boldsymbol{\lambda}_2}$ and $\mathbf{v}_0 \leftarrow \xi^{\pi \boldsymbol{\lambda}_0}$. The prover sends ξ, $(\bar{\mathbf{x}}, \bar{\mathbf{w}})$, $(\dot{\mathbf{x}}, \dot{\mathbf{w}})$, $\mathbf{u}_0, \mathbf{u}_2, \mathbf{v}_0$ to $\mathcal{V}$.

- The verifier checks that $u_{0,1}, \ldots, u_{0,l}$ are all distinct, samples $\boldsymbol{\lambda}_3 \xleftarrow{r} \{0, 1, \ldots, p-1\}^l$, and sends $\boldsymbol{\lambda}_3$ to $\mathcal{P}$. Both parties compute $\mathbf{u}_1 \leftarrow g^{\boldsymbol{\lambda}_3} \mathbf{u}_2$.

- The prover computes $\boldsymbol{\lambda}_1 \leftarrow \boldsymbol{\lambda}_2 + \boldsymbol{\lambda}_3$ and $\mathbf{v}_1 \leftarrow \xi^{\pi \boldsymbol{\lambda}_1}$. It sends $\mathbf{v}_1$ to $\mathcal{V}$.

- The verifier samples $\zeta \xleftarrow{r} G$ and $\beta \xleftarrow{r} \mathfrak{B}$ and sends (ζ, β) to $\mathcal{P}$. The prover computes $\boldsymbol{\lambda} \leftarrow \boldsymbol{\lambda}_0 + \beta \boldsymbol{\lambda}_1$, and both parties compute $\mathbf{u} \leftarrow \mathbf{u}_0 \mathbf{u}_1^{\beta}$ and $\mathbf{v} \leftarrow \mathbf{v}_0 \mathbf{v}_1^{\beta}$.

- The prover computes $\hat{\mathbf{v}} \leftarrow \zeta^{\pi \boldsymbol{\lambda}}$, $\check{\mathbf{x}} \leftarrow \bar{\mathbf{x}}^{\boldsymbol{\lambda}}$, $\check{\mathbf{w}} \leftarrow \bar{\mathbf{w}}^{\boldsymbol{\lambda}}$, $\hat{\mathbf{x}} \leftarrow \dot{\mathbf{x}}^{\pi \boldsymbol{\lambda}}$ and $\hat{\mathbf{w}} \leftarrow \dot{\mathbf{w}}^{\pi \boldsymbol{\lambda}}$, and sends $\hat{\mathbf{v}}$, $(\check{\mathbf{x}}, \check{\mathbf{w}})$, $(\hat{\mathbf{x}}, \hat{\mathbf{w}})$ to $\mathcal{V}$.

- The prover and verifier run the argument from Figure 11.11 with public input ξ, $\mathbf{u}$, $\mathbf{v}$, and private input $\boldsymbol{\lambda}$, a and π.

- For $i = 1, 2, \ldots, l$, the prover and verifier run the argument from Exercise 11.4 (equality of discrete logarithms) with input as in the table:

public input	private input
$(g, \bar{x}_i, u_i, \check{x}_i)$	λ_i
$(g, \bar{w}_i, u_i, \check{w}_i)$	λ_i
$(\xi, \dot{x}_i, v_i, \hat{x}_i)$	$\lambda_{\pi(i)}$
$(\xi, \dot{w}_i, v_i, \hat{w}_i)$	$\lambda_{\pi(i)}$
$(\xi, \zeta, v_i, \hat{v}_i)$	$\lambda_{\pi(i)}$
$(g, y, \bar{x}_i/x_i, \bar{w}_i/w_i)$	r_i'
$(g, y, \dot{x}_i/\tilde{x}_i, \dot{w}_i/\tilde{w}_i)$	r_i''
$(g, y, \prod_i \check{x}_i/\hat{x}_i, \prod_i \check{w}_i/\hat{w}_i)$	$(\sum_i \lambda_i r_i') - (\sum_i \lambda_{\pi(i)}(r_{\pi(i)} + r_i'')) \bmod p$

FIGURE 11.12 Interactive argument $(\mathcal{P}, \mathcal{V})$ for a shuffle of ElGamal ciphertexts, based on ElGamal over a group G of prime order p with generator g. The operations on tuples are coordinate-wise addition, multiplication and exponentiation, and permutations on indexes are reinterpreted as permutation matrices applied to tuples.

- $v_{1,1}, \ldots, v_{1,l}$ were chosen; and
- β was sampled from $\{0, 1, \ldots, p-1\}$.

Define $u_i = u_0(u_2 g^{\lambda_3})^\beta$ and $v = v_0 v_1^\beta$. Suppose $u_1, \ldots, u_l$ are distinct and that there exists a permutation π such that $\log_g u_i = \log_\xi v_{\pi(i)}$ for $i = 1, 2, \ldots, l$. Show that $\log_g u_{0,i} = \log_\xi v_{0,\pi(i)}$, except with probability $1/p$.

E *Exercise* 11.53. Show that if no witness exists, then the verifier in the argument from Figure 11.12 accepts with probability at most $l/(|\mathfrak{B}| - l) + 1/|\mathfrak{B}|$. Conclude that the argument has soundness error at most $l/(|\mathfrak{B}| - l) + 1/|\mathfrak{B}|$.

Hint: The verification fails if any one error is detected. Since we need a lower bound, we need only consider what is most likely to go undetected.

The simulator gets ciphertexts $(x_1, w_1), \ldots, (x_l, w_l)$ and $(\tilde{x}_1, \tilde{w}_1), \ldots, (\tilde{x}_l, \tilde{w}_l)$ and does as follows:

- Sample $\beta \xleftarrow{r} \{0, 1, \ldots, p\}$, $\lambda_3 \xleftarrow{r} \{0, 1, \ldots, p\}^l$, $\xi, \zeta \xleftarrow{r} G$, $\rho, \rho_1, \rho_2, \rho_3, \rho_4 \xleftarrow{r} \{0, 1, \ldots, p-1\}^l$ and $u, u_2 \xleftarrow{r} G^l$.

- Compute $v \leftarrow \xi^\rho$, $\hat{v} \leftarrow \zeta^\rho$, $u_0 \leftarrow u(u_2 g^{\lambda_3})^{-\beta}$, $v_0 \leftarrow v(v_2 g^{\lambda_3})^{-\beta}$, $\bar{x} \leftarrow g^{\rho_1}$, $\check{x} \leftarrow u^{\rho_1}$, $\bar{w} \leftarrow g^{\rho_2}$, $\check{w} \leftarrow u^{\rho_1}$, $\dot{x} \leftarrow \xi^{\rho_3}$, $\hat{x} \leftarrow v^{\rho_3}$, $\tilde{w} \leftarrow \xi^{\rho_4}$ and $\hat{w} \leftarrow v^{\rho_4}$.

- Output $\perp$ if $u_{0,1}, \ldots, u_{0,l}$ or $v_{0,1}, \ldots, v_{0,l}$ are not all distinct.

- Simulate conversations for the argument from Figure 11.11 with public input ξ, u and v, using the honest verifier simulator from Exercise 11.47.

- For $i = 1, 2, \ldots, l$, use the honest verifier simulator from Exercise 11.18 to simulate conversations with public input $(g, \bar{x}_i, u_i, \check{x}_i)$, $(g, \bar{w}_i, u_i, \check{w}_i)$, $(\xi, \tilde{x}_i, v_i, \hat{x}_i)$, $(\xi, \tilde{w}_i, v_i, \hat{w}_i)$, $(\xi, \zeta, v_i, \hat{v}_i)$, $(g, y, \bar{x}_i/x_i, \bar{w}_i/w_i)$, $(g, y, \dot{x}_i/\tilde{x}_i, \dot{w}_i/\tilde{w}_i)$ and $(g, y, \prod_i \tilde{x}_i/\hat{x}_i, \prod_i \tilde{w}_i/\hat{w}_i)$.

FIGURE 11.13 Honest verifier simulator for the Figure 11.12 argument.

This shows that the argument is unconditionally sound, which is good. We must now show that it is honest verifier zero knowledge. The simulator given in Figure 11.13 and essentially samples a lot of random group elements and the challenges, then uses appropriate simulators for arguments from Figure 11.12 and Example 11.4. This simulator is not perfect. Showing that the simulated conversations look like real conversations is therefore somewhat more involved than our previous examples.

T **Proposition 11.20.** *Let $\mathcal{A}$ be a τ-distinguisher for the simulator from Figure 11.13. Then there exists a τ_1'-distinguisher $\mathcal{B}_1$ and a τ_2'-distinguisher $\mathcal{B}_2$ against Decision Diffie-Hellman, with τ_1' and τ_2' essentially equal to τ, such that*

$$\mathbf{Adv}(\mathcal{A}) \leq 2\mathbf{Adv}_{G,g}^{\mathsf{DDH}}(\mathcal{B}_1) + 2\mathbf{Adv}_{G,g}^{\mathsf{DDH}}(\mathcal{B}_2).$$

The proof proceeds by stepwise changes to the prover and verifier, ending up at the given simulator. We phrase this in terms of a sequence of games, which we sketch below. By inspection we see that the runtime of each new game is essentially equal to the runtime of the previous game.

The initial game is the prover and verifier generating a conversation, with the statement and the witness as input. The first modification is to replace all the subarguments by the corresponding honest verifier simulations. Since all the simulators for the subarguments are perfect, this changes nothing. The next modification is to replace the ciphertexts $(\bar{x}_i, \bar{w}_i)$ and $(\dot{x}_i, \dot{w}_i)$ by random group elements for $i = 1, 2, \ldots, l$.

E *Exercise 11.54.* Show that there exists a τ_1'-distinguisher $\mathcal{B}_1$ for Decision Diffie-Hellman, where τ_1' is essentially equal to the runtime of this game, such that the change in behaviour caused by the above modification is bounded by $2\mathbf{Adv}_{G,g}^{\mathsf{DDH}}(\mathcal{B}_1)$.

Next, we change how we sample ζ by sampling $b_0 \xleftarrow{r} \{0, 1, \ldots, p-1\}$ and computing $\zeta \leftarrow \xi^{b_0}$. We use the embedded trapdoor b_0 to compute $\hat{v} \leftarrow v^{b_0}$.

Instead of sampling λ_0 and computing λ, we sample λ and compute $\lambda_0 \leftarrow \lambda - \beta\lambda_1$, subject to $u_{0,1}, \ldots, u_{0,l}$ being unique.

We sample $\rho_1, \rho_2, \rho_3, \rho_4 \xleftarrow{r} \{0, 1, \ldots, p-1\}^l$ and compute $\bar{x} \leftarrow g^{\rho_1}$, $\bar{w} \leftarrow g^{\rho_2}$, $\check{x} \leftarrow u^{\rho_1}$, $\check{w} \leftarrow u^{\rho_2}$, $\dot{x} \leftarrow \xi^{\rho_3}$, $\dot{w} \leftarrow \xi^{\rho_4}$, $\hat{x} \leftarrow v^{\rho_3}$ and $\hat{w} \leftarrow v^{\rho_4}$.

Then, we sample $\rho \xleftarrow{r} \{0, 1, \ldots, p-1\}^l$ and compute $v \leftarrow \xi^{\rho}$.

E *Exercise 11.55.* Show that there exists a τ_2'-distinguisher $\mathcal{B}_2$ for Decision Diffie-Hellman, where τ_2' is essentially equal to τ, such that the change in behaviour caused by the above modification is bounded by $2\mathbf{Adv}_{G,g}^{\mathsf{DDH}}(\mathcal{B}_2)$.

Finally, we compute $\hat{v} \leftarrow \zeta^{\rho}$. We no longer use b_0, so we can sample $\zeta \xleftarrow{r} G$.

The result follows by observing that we began with the real argument generation process and ended with exactly the simulator from Figure 11.13.

E *Exercise 11.56.* Write out the proof of Proposition 11.20 in detail.

The argument from Figure 11.12 is not a sigma protocol, so we cannot apply the construction from Figure 11.7 (Fiat-Shamir) directly. However, we can make a similar construction and prove it secure.

E *Exercise 11.57.* Use the Fiat-Shamir heuristic to design a non-interactive argument for a shuffle of ElGamal ciphertexts, based on the argument from Figure 11.12. Show that it is sound and computational zero knowledge, and

The interactive argument for the witness relation (X, L, w, R) from Example 11.2 works as follows:

- The CRS contains ck output by $\mathcal{CK}$. The public input is a graph G. The private input is a cycle π (a permutation on G's vertices).
- The prover $\mathcal{P}$ samples a permutation ρ on the vertices of G and sends commitments to the edges of $\rho(G)$ to the verifier, in random order.
- The verifier $\mathcal{V}$ samples $\beta \xleftarrow{r} \{0, 1\}$ and sends β to $\mathcal{P}$.
- If $\beta = 0$, the prover's response is ρ together with openings of all the commitments. If $\beta = 1$, the response is the cycle $\rho \circ \pi$ and openings of the edges corresponding to the cycle $\rho \circ \pi$.
- If $\beta = 0$, the verifier accepts if the commitments open to the edges of the graph $\rho(G)$. If $\beta = 1$, the verifier accepts if the commitments open to edges along the given cycle.

FIGURE 11.14 Interactive argument $(\mathcal{P}, \mathcal{V})$ for the witness relation from Example 11.2 using the commitment scheme from Section 8.4.1.

give bounds on the failure probability of the random oracle reprogramming and on the soundness error.

E *Exercise* 11.58. Explain how to choose ζ such that the permutation can be extracted from an accepting conversation, using some information generated when ζ was chosen. (We *embed a trapdoor*.) Note that this does not extract a witness for the relation, which would include the randomness used. But for the intended application, the ability to extract the permutation is all we need.

11.4.5 An Argument for Hamiltonian Cycles

We want to prove that a graph has an Hamiltonian cycle (the witness relation from Example 11.2). If we can commit to a second graph and both open it to show that it is isomorphic, and reveal an Hamiltonian cycle in the second graph, the original graph must have an Hamiltonian cycle.

We shall use the commitment scheme from Section 8.4.1, which is both equivocable and extractable. We shall use the first property to show that our protocol is zero knowledge, while the second property will give us soundness. We do not actually need these properties, but they simplify the proof.

The interactive argument is given in Figure 11.14.

We begin by proving that the scheme is sound. Note that the argument uses a common reference string. The idea is that we can use our commitment scheme with an extractable commitment key. This turns the commitment scheme into an unconditionally binding scheme. With unconditionally binding commitments, we get unconditional soundness.

Theorem 11.21. *Consider the interactive argument* $(\mathcal{P}, \mathcal{V})$ *from Figure 11.14. Suppose there is a cheating prover* $\mathcal{P}^*$ *that has soundness advantage* $1/2 + \epsilon$ *for some* $\epsilon \geq 0$. *Then there exists an adversary against Paillier encryption with advantage* 2ϵ *and running time essentially the same as* $\mathcal{P}^*$.

Proof. Suppose we run the interactive argument with the commitment key ck in the CRS generated by $\mathcal{CK}_{ex}$. In this case, the commitment scheme is unconditionally binding. This means that if G is not Hamiltonian, the cheating prover cannot both open the commitments to a graph that is isomorphic to G and open an Hamiltonian cycle in the graph. It follows that in the modified interactive argument, the advantage of $\mathcal{P}^*$ is at most $1/2$.

We now create an adversary against Paillier encryption. Our adversary receives the public key n, asks for an encryption of 0 or an encryption of a random message, and receives a ciphertext c. Our adversary computes the commitment key as $(n, -c^2)$ and runs $\mathcal{V}$ with $\mathcal{P}^*$. If $\mathcal{V}$ rejects, our adversary guesses 0, otherwise our adversary guesses random.

If c is an encryption of a random message, then $-c^2$ is distributed exactly as the commitment key is supposed to be distributed. Which means that the probability that $\mathcal{P}^*$ makes $\mathcal{V}$ accept is $1/2 + \epsilon$.

If c is an encryption of 0, then $-c^2$ is distributed exactly as the commitment key output by $\mathcal{CK}_{ex}$, and the probability that $\mathcal{P}^*$ makes $\mathcal{V}$ accept is $1/2$.

It follows that the advantage of our Paillier adversary is 2ϵ. □

Remark. There is one difficulty with the above argument, which is that we need a non-Hamiltonian graph that $\mathcal{P}^*$ has soundness advantage $1/2 + \epsilon$ against. Even if such graphs exist, they may be hard to find. Essentially, our adversary against Paillier might need some "advice" to be able to do its job.

However, if we believe that Paillier is secure, we must also believe that this "advice" does not exist. Again, we remind the reader that our goal is not to ensure that we get a consolation prize if our cryptography is broken. Our goal is to build cryptography that will not be broken.

It is tempting to argue that the scheme has special soundness. Since we do not know how the conversations arise, the obvious extractor might not be able to find an Hamiltonian cycle, nor can we use the conversations to break Pallier encryption in some way.

The scheme is zero knowledge. As usual, we guess the adversary's challenge and make commitments to an appropriate graph which can then be revealed.

Theorem 11.22. *Consider the interactive argument* $(\mathcal{P}, \mathcal{V})$ *from Figure 11.14. Then the simulator* $\mathcal{S}$ *for* $\mathcal{V}^*$ *given in Figure 11.15 is statistical,*

The simulator takes G as input and works as follows:

- Start the cheating verifier $\mathcal{V}^*$ with input G and random tape r.

- Sample $b \xleftarrow{r} \{0,1\}$. If $b = 0$, sample a random permutation ρ on the vertices and create commitments to the graph $\rho(G)$. If $b = 1$, sample a random graph with an Hamiltonian cycle and the same number of edges as G, and create commitments to this graph.

- Send the commitments to $\mathcal{V}^*$ and get β back from $\mathcal{V}^*$.

- If $\beta \neq b$, output $\perp$.

- If $b = 0$, send openings of all the commitments and ρ to $\mathcal{V}^*$. If $b = 1$, send openings of the edges on the Hamiltonian cycle to $\mathcal{V}^*$.

- When $\mathcal{V}^*$ stops, output the view.

FIGURE 11.15 Simulator for the interactive argument from Figure 11.14 and a cheating verifier $\mathcal{V}^*$.

has essentially the same running time as $\mathcal{V}^*$ and outputs $\perp$ with probability (very close to) $1/2$.

Proof. By Exercises 8.43 and 8.44 and the arguments after it the commitment scheme we use is statistically hiding. This means that the commitments made by the simulator reveals (almost) no information about b. This means that the probability that $b \neq \beta$ is (very close to) $1/2$.

When $b = \beta = 0$, the commitments open to the graph $\rho(G)$, which is what would happen in the real interactive argument. When $b = \beta = 1$, the commitments open to an Hamiltonian cycle, which is what would happen in the real interactive argument. Since the commitment scheme is statistically hiding, the unopened commitments do not matter. Therefore, the simulated view is statistically close to the view of a conversation with the prover. □

E | *Exercise* 11.59. The above theorem and its proof do not compute the statistical distances. Use Exercise 8.43 to give an upper bound on these distances.

The equivocable and extractable commitment scheme is not essential. We need a scheme that is statistically hiding and an alternative commitment key generation algorithm for which the scheme is unconditionally binding. One candidate is a combination of ElGamal and Pedersen commitments.

We could use the equivocable and extractable properties in a different proof. Instead of Theorem 11.21, we could construct an extractor: it would extract the isomorphic graph directly from the commitments, and then send $\beta = 1$ to get the Hamiltonian cycle. To get this to work, the prover would also have to commit to the isomorphism ρ. The advantage of this extractor is that it has knowledge advantage equal to the soundness advantage minus the advantage of an adversary against Paillier, and it runs the cheating prover exactly once. Avoiding rewinding is sometimes useful.

The simulator would make commitments to zero. When the cheating verifier issues its challenge, we use equivocation to either open all the commitments to an isomorphic graph, or some of the commitments to an Hamiltonian cycle. The main advantage is that we get statistical zero knowledge, and the simulator never outputs $\perp$.

Remark. This interactive argument is theoretically important. Deciding if a graph is Hamiltonian is an NP-complete problem, so any language in NP can be reduced to deciding if graphs are Hamiltonian. We can use these techniques to turn the interactive argument into an argument for any NP language.

One would expect such an interactive argument to be inefficient in most cases. Finding practical arguments is an interesting research question for many real-world problems.

CHAPTER 12

Multi-party Computation

Alice, Bob, Carol, Dave and Eve are engaged in conversation, seated around a table. Eve asks what their average income is. Alice and Carol do not want to reveal their individual income. We shall now explain they can compute their average income (actually, the sum of their incomes) without revealing their individual incomes. When doing this computation, they shall only whisper privately to their neighbours and do some private computations. How to compute the average income without revealing any further information about individual incomes is a simple example of a multi-party computation problem.

There is a natural correspondence between the elements of a finite prime field $\mathbb{F}_q$ and the integers $\{0, 1, \ldots, q - 1\}$, so we can sum the incomes in $\mathbb{F}_q$ instead of in $\mathbb{Z}$ if q is sufficiently large.

Alice samples $r \xleftarrow{r} \mathbb{F}_q$ and computes $y_A = r + s_A$. Then she whispers y_A to her neighbour Bob. Bob computes $y_B = s_B + y_A$ and whispers the result to Carol, and this continuous around the table. When Eve tells Alice y_E, Alice computes $u \leftarrow y_E - r$ and announces the result to everyone. If everyone followed the protocol, Alice announces the correct result.

We first consider what Bob learns about the other's income. Bob hears u, but this is what we want to achieve. He also hears $r + s_1$, which from his point of view is uniformly distributed, so any value for s_1 is equally likely. Bob hears nothing else. Therefore, Bob learned nothing that we did not want him to learn. The same argument applies to all the other players, including Alice.

However, if Bob and Dave cooperate, they learn Carol's income, as well as the sum of Eve and Alice's incomes. The protocol does not protect against such coalitions pooling knowledge.

Also, Alice may cheat and announce any result she cares to, and only if her result is inconsistent with the income of one of the other players will this be detected. Note that Alice cheating in this way is distinct from any other player whispering $y_i + s_i + \Delta$, since Alice gets to see the correct result first. It might seem like the others would simply be lying about their income, but they could cheat by whispering $y_i + \Delta$, $\Delta < 0$, causing an effect on the result which no amount of lying could duplicate.)

When Bob and Dave cooperate by pooling knowledge, but otherwise follow the protocol, we say that Bob and Dave are *passive*, or *honest-but-curious*. When Alice deliberately deviates from the protocol, she is an *active* attacker. In both cases, coalitions try to achieve some effect.

We shall encounter several vector spaces in this chapter. We shall assume that they are all finite-dimensional and that any such k-dimensional space has a natural and easy-to-compute isomorphism with $\mathbb{F}^k$.

12.1   SECRET SHARING

We begin by defining new concepts and notation. We want to share a secret among a certain set of players $\mathfrak{U}$. Each player will get a *share* of the secret, and a *sharing* of s is a function f that assigns a share to each player P, and we say that $f(P)$ is the share belonging to the player P. For a given set of players $\mathfrak{U}_0$, we denote the shares of these players by $f|_{\mathfrak{U}_0}$. We denote the set of sharings by V and the set of values for $f|_{\mathfrak{U}_0}$ by $V_{\mathfrak{U}_0}$.

We need to specify which players should be able to reconstruct a secret from their given shares, which we do by specifying all the subsets of players that should be able to reconstruct the secret. This is called an *access structure* $\mathfrak{A}$, which is a subset of the power set of the player set $2^{\mathfrak{U}}$. Any set of players can pretend to be a smaller set of players, so for an any access structure $\mathfrak{A}$ and any player set $\mathfrak{U}_0 \in \mathfrak{A}$, if the player set $\mathfrak{U}_1$ contains $\mathfrak{U}_0$, then $\mathfrak{U}_1 \in \mathfrak{A}$.

D **Definition 12.1.** A *secret sharing* scheme for a set S and a set of players $\mathfrak{U}$ is a pair of algorithms $(\mathcal{D}, \mathcal{R})$:

- The *dealer* algorithm $\mathcal{D}$ takes as input $s \in S$ and outputs a sharing f.
- The deterministic *reconstruction* algorithm $\mathcal{R}$ takes as input $\mathfrak{U}_0 \subseteq \mathfrak{U}$ and $f|_{\mathfrak{U}_0}$, and outputs either $s \in S$ or the special symbol $\bot$.

The scheme $(\mathcal{D}, \mathcal{R})$ is *correct* for an *access structure* $\mathfrak{A} \subseteq 2^{\mathfrak{U}}$ if for any set of players $\mathfrak{U}_0 \in \mathfrak{A}$ and sharing f of s output by $\mathcal{D}$ we have that $\mathcal{R}(\mathfrak{U}_0, f|_{\mathfrak{U}_0}) = s$.

We could have required that for any $\mathfrak{U}_1 \notin \mathfrak{A}$, we have that $\mathcal{R}(\mathfrak{U}_1, f|_{\mathfrak{U}_1}) = \bot$. The reconstruction algorithm should not be able to compute the correct answer in this case, which means that any response other than $\bot$ is somehow incorrect. If this property is desired, the reconstruction algorithm can check if $\mathfrak{U}_0 \in \mathfrak{A}$.

E *Exercise 12.1.* Show that for any two access structures $\mathfrak{A}_1$ and $\mathfrak{A}_2$ on the player set $\mathfrak{U}$, $\mathfrak{A}_1 \cup \mathfrak{A}_2$ and $\mathfrak{A}_1 \cap \mathfrak{A}_2$ are also access structures. Explain their meaning.

While access structures can in principle be very complicated, most of the schemes we construct will usually support only quite simple access structures. In particular, one common access structure is the threshold access structure where any set containing t out of l players is allowed to reconstruct. This access structure is often called *t-out-of-l*.

Example 12.1. Let S be the set of bit strings of length $3l$. Let $\mathfrak{U} = \{1, 2, \ldots, l\}$, $l \geq 3$, and define $\mathfrak{U}_j = \{i \in \mathfrak{U} \mid i \equiv j \pmod{3}\}$ for $j = 1, 2, 3$. Let the access structure be the subsets of $\mathfrak{U}$ that non-trivially intersect with $\mathfrak{U}_1$, $\mathfrak{U}_2$ and $\mathfrak{U}_3$.

The dealer divides the secret $s \in S$ into three equal-length parts s_1, s_2, s_3. The share of a player $P \in \mathfrak{U}_j$ is s_j.

The reconstruction algorithm with input $\mathfrak{U}_0$ and $f|_{\mathfrak{U}_0}$ samples $P_j \xleftarrow{r} \mathfrak{U}_0 \cap \mathfrak{U}_j$, $j = 1, 2, 3$, and outputs the concatenation of $f(P_1)$, $f(P_2)$ and $f(P_3)$.

Example 12.2. Let G be any finite group and let $S = G$. We construct a secret sharing scheme $(\mathcal{D}, \mathcal{R})$ for the access structure $A = \{\mathfrak{U}\}$, that is, where every player's share is needed for reconstruction.

To share a secret $s \in G$ among players $\{1, 2, \ldots, l\}$, the dealer $\mathcal{D}$ samples $l - 1$ group elements $c_1, c_2, \ldots, c_{l-1}$ from the uniform distribution on G, and computes $c_l = s(c_1 c_2 \cdots c_{l-1})^{-1}$.

To reconstruct a secret from the shares $c_1, c_2, \ldots, c_l$, the reconstruction algorithm $\mathcal{R}$ computes $s = c_1 c_2 \cdots c_l$.

Exercise 12.2. Prove that the scheme from Example 12.2 is correct for the given access structure.

Exercise 12.3. Suppose we have secret sharing schemes $(\mathcal{D}_1, \mathcal{R}_1)$ and $(\mathcal{D}_2, \mathcal{R}_2)$ for access structures $\mathfrak{A}_1$ and $\mathfrak{A}_2$, respectively. Use the two secret sharing schemes to construct a secret sharing scheme for the access structures $\mathfrak{A}_1 \cup \mathfrak{A}_2$ and $\mathfrak{A}_1 \cap \mathfrak{A}_2$. For the intersection case you should also use the secret sharing scheme from Example 12.2 for two players.

The second part of this exercise illustrates the point that we will create sharings where the shares do not correspond to players but is instead used in some other fashion. Likewise, not every sharing that we will encounter will be output by $\mathcal{D}$. We say that a sharing c is *consistent* with respect to some access structure if there exists a value s such that for any $\mathfrak{U}_0$ in the access structure, $\mathcal{R}(\mathfrak{U}_0, c) = s$. Correctness implies that any share output by $\mathcal{D}$ is consistent.

12.1.1 Defining Security

With the secret sharing scheme from Example 12.1 no set of players not in the access structure can reconstruct the entire secret. However, the goal of a secret sharing scheme cannot be just to prevent reconstruction of the secret by any subset of players not in the access structure. We will need to prevent them from learning anything at all about the secret.

Semantic Security We could follow the usual approach used for encryption, and define security as a game between an adversary and an experiment. The adversary first chooses two messages. The experiment creates a sharing of one of them. The adversary adaptively queries the experiment for various players' shares, under the restriction that the set of players queried does not lie in the access structure. Finally, the adversary guesses which message was shared.

Example 12.3. Let $(\mathfrak{K}, S, \mathfrak{C}, \mathcal{E}, \mathcal{D})$ be a symmetric cryptosystem, and consider the player set $\mathfrak{U} = \{1, 2, 3\}$ and the access structure $\{\{1, 2\}, \{1, 3\}, \{2, 3\}, \mathfrak{U}\}$.

The dealer $\mathcal{D}$ samples $k_1, k_2, k_3 \xleftarrow{r} \mathfrak{K}$, computes $c_i = \mathcal{E}(k_i, s)$, $i = 1, 2, 3$, and outputs the three shares (k_1, c_2), (k_2, c_3) and (k_3, c_1).

The reconstruction algorithm will get one share with $(k_i, \ldots)$ and one share with $(\ldots, c_i)$, and outputs $\mathcal{D}(k_i, c_i) = s$.

Exercise 12.4. Give a security definition based on the above sketch, and prove that the scheme from Example 12.3 is secure if the cryptosystem is secure. (That is, provide a reduction from breaking the secret sharing scheme to breaking the symmetric cryptosystem.)

The above scheme uses two ideas: Symmetric encryption is in a certain sense a secret sharing for two players, with the key and the ciphertext being the shares. If you know just the key, you have no information about the message. If you know just the ciphertext, the encryption hides the shared secret.

Also, the above scheme creates many two-player sharings (using the first idea) and distributes these among the three players such that no one player has enough shares to recover the secret, but any two players has enough shares.

Such schemes can be extended to a larger number of players, but the individual shares quickly grow very large, so this becomes impractical.

Remark. The security definition from Exercise 12.4 is *adaptive*, in the sense that the adversary is allowed to look at some shares before it decides which other shares to look at. We could make a non-adaptive definition, where the adversary makes a single query to the simulator specifying which shares it wants to look at. Adaptive security implies non-adaptive security.

Information-Theoretic Security We shall now consider a simpler and stronger definition that is easier to work with. We want the distribution of the shares of any unauthorised set of players to be independent of the secret.

This is information-theoretic security. It implies security under the definition from Exercise 12.4. If we use the one-time pad as our cryptosystem in Example 12.3, we get that individual shares are independent of the shared secret. But if $|S| > |\mathfrak{K}|$ the shares cannot be independent of the shared secret.

Definition 12.2. A secret sharing scheme $(\mathcal{D}, \mathcal{R})$ is *information-theoretically secure* for access structure $\mathfrak{A}$ if for any set of players $\mathfrak{U}_0 \notin \mathfrak{A}$, the distribution of these parties' shares is independent of the secret shared.

E *Example* 12.4. Consider the secret sharing scheme from Example 12.2 for l players. We shall prove that for any proper subset $\mathfrak{U}_0 \subseteq \mathfrak{U}$, the distribution of the sharing $f|_{\mathfrak{U}_0}$ is independent of the value s shared. We need only consider subsets containing $l-1$ players.

This holds for the subset $\{1, 2, \ldots, l\}$, since the shares are sampled from the uniform distribution on the group. This is independent of the secret s.

If $i \notin \mathfrak{U}_0$, we need only consider the lth share, whose value is

$$c_l = sc_1^{-1} \cdots c_i^{-1} \cdots c_{l-1}^{-1}.$$

Multiplication by a group element is a bijection, so it does not matter if we sample c_i and compute c_l, or sample c_l and compute c_i. Therefore, the secret is independent of the shares.

E *Exercise* 12.5. Suppose $(\mathcal{D}, \mathcal{R})$ is information-theoretically secure. Show that $(\mathcal{D}, \mathcal{R})$ also satisfies the security notion from Exercise 12.4.

E *Exercise* 12.6. Suppose we use the one-time pad in the construction from Example 12.3. Prove that the resulting secret sharing scheme is information-theoretically secure for the given access structure.

Simulation   Another approach to defining security is the simulation approach, which we used to define zero knowledge for arguments in Chapter 11. In order to prove that a set of shares contains no information about the shared secret, it is sufficient to prove that there exists a simulator that does not know the shared secret, but can still sample simulated shares whose distribution is the same as the distribution of the real shares.

This definition seems very similar to the information theoretic definition, but it is not exactly the same, which can best be seen by comparing with the definition from Exercise 12.4. The problem is that the adversary in that definition is allowed to adaptively decide which shares to get. That is, the adversary may look at one share and then decide on which share to look at next based on the value of the first share.

If we only have a simulator that simulates shares for a given set of participants, we cannot use this to simulate the shares in the definition, because we cannot predict which shares the adversary will ask for.

Instead, we must require a simulator that can, given a set of players, their shares and a set of additional players, generate simulated shares for the new players. We allow the simulator to maintain a state.

The simulation-based security definition has many advantages compared to other definitions, in particular once we have defined one more notion. The idea so far is that we will want to simulate shares of an unknown secret. Eventually, however, we want to be able to create a sharing of this unknown secret that is consistent with the simulated shares.

Definition 12.3. A *simulator* $\mathcal{S} = (\mathcal{S}_1, \mathcal{S}_2)$ for a secret sharing scheme $(\mathcal{D}, \mathcal{R})$ and access structure $\mathfrak{A}$ is a pair of algorithms that work as follows:

- The algorithm $\mathcal{S}_1$ takes as input two sets of players $\mathfrak{U}_0, \mathfrak{U}_1 \notin \mathfrak{A}, \mathfrak{U}_0 \subseteq \mathfrak{U}_1$, a sharing $f|_{\mathfrak{U}_0}$ and a state σ_0, and outputs a sharing $h|_{\mathfrak{U}_1}$ and a state σ_1 such that $f|_{\mathfrak{U}_0}(P) = h|_{\mathfrak{U}_1}(P)$ for all $P \in \mathfrak{U}_0$.

- The algorithm $\mathcal{S}_2$ takes as input a secret s, a set of players $\mathfrak{U}_0 \notin \mathfrak{A}$, a sharing $f|_{\mathfrak{U}_0}$ and a state σ_0, and outputs a consistent sharing h of s such that $h(P) = f|_{\mathfrak{U}_0}(P)$ for all $P \in \mathfrak{U}_0$.

A secret sharing scheme $(\mathcal{D}, \mathcal{R})$ is *simulatable* if there exists a simulator $\mathcal{S}$ such that the distribution of any shares output by the simulator is indistinguishable from the distribution of the shares output by $\mathcal{D}$.

If we only consider the shares output by $\mathcal{S}_1$ on input of $\mathfrak{U}_0 = \emptyset$, the scheme is *non-adaptively simulatable*. If the state is always $\perp$, the simulator is *stateless* and the scheme is *statelessly simulatable*.

The definition comes in two flavours. Either *indistinguishable* means that the distributions are identical or sufficiently statistically close, which we call *information-theoretical* security. We use *perfect* when the distributions are identical, or *statistical* when they are usefully statistically close. The other flavour is *computationally* indistinguishable. In the latter case, we are actually considering a game between an experiment and an adversary, where the adversary chooses a secret and the experiment either

- uses $\mathcal{D}$ to create a sharing of the secret; or
- uses $\mathcal{S}_1$ to simulate shares and then $\mathcal{S}_2$ to simulate the complete sharing.

The adversary's job is to decide what the experiment is doing.

Unless we explicitly say that we are interested in computational simulatability, we will be thinking about perfect or statistical simulatability.

Remark. The statelessly simulatable schemes are important, since these allow us to apply the $\mathcal{S}_2$ algorithm to shares that have not been simulated.

We will now compare the notion of simulatability to the definition of security from Exercise 12.4 and information theoretic security. If a secret sharing scheme is simulatable, it is trivially also non-adaptively simulatable. For the other direction, there is no practical way to generate new consistent shares.

Proposition 12.1. *If a secret sharing scheme is non-adaptively information-theoretically simulatable, then it is information-theoretically secure.*

Proof. We need to prove that for any set of players $\mathfrak{U}_0 \notin \mathfrak{A}$, the distribution of the shares output by $\mathcal{D}$ is independent of the secret. However, this distribution is identical to the distribution of the shares output by the simulator on input of $\mathfrak{U}_0$. This distribution is independent of the secret, since the simulator samples from that distribution without information about the secret. $\qquad\square$

We consider the converse case. For the first part we share some random secret using the dealer. The distribution of the shares of any player set not in the access structure is independent of the secret, hence indistinguishable.

The problem is the second part of the simulator, which must output a sharing that is both consistent with a given secret and the shares that were simulated without knowledge of the secret. There is no practical way to construct a consistent sharing.

To summarise, information-theoretic security and simulatability are in some sense similar notions, but different since we want the simulator algorithms to be practical. All that remains is to connect simulatability to the security notion from Exercise 12.4. Again, one direction is straight-forward.

E | *Exercise 12.7.* Show that if a secret sharing scheme is computationally simulatable, then it is secure under the security definition from Exercise 12.4.

The other direction causes an inherent conflict between the adversary's goal of deciding which secret was shared and revealing a sharing of that secret. However, the former notion implies that the first simulator algorithm exists.

E | *Exercise 12.8.* Show that if a scheme is secure under the definition from Exercise 12.4, there exists an algorithm S_1 satisfying the requirements of computational simulatability.

Before we discuss an example, it will be convenient to prove a result about certain secret sharing schemes. For any secret sharing scheme, if V_P is the set of possible values for the share given to P, we encode the shares given to a set of players $\mathfrak{U}_0$ as a function

$$\mathfrak{U}_0 \to \prod_{P \in \mathfrak{U}_0} V_P.$$

We shall now be interested in secret sharing schemes where for any $\mathfrak{U}_0 \notin \mathfrak{A}$, any such function is not only a possible outcome of running $\mathcal{D}$ but is as likely to happen as any other outcome.

T | **Proposition 12.2.** *Let $(\mathcal{D}, \mathcal{R})$ be a secret sharing scheme with secret set S, player set $\mathfrak{U}$ and access structure $\mathfrak{A}$. Suppose that $\mathcal{D}$ works by sampling a value from the uniform distribution on a set T and computing a bijection $D : S \times T \to V$ to get the sharing. Suppose further that for any set of players $\mathfrak{U}_0 \notin \mathfrak{A}$ and secret $s \in S$ there exists a set $T_{\mathfrak{U}_0}$ and a bijection $\sigma_{s,\mathfrak{U}_0} : T_{\mathfrak{U}_0} \times V_{\mathfrak{U}_0} \to T$ such that for any $t \in T_{\mathfrak{U}_0}$ and shares $f|_{\mathfrak{U}_0}$ then*

$$D(s, \sigma_{s,\mathfrak{U}_0}(t, f|_{\mathfrak{U}_0}))|_{\mathfrak{U}_0} = f|_{\mathfrak{U}_0}.$$

If we have algorithms for computing $\sigma_{s,\mathfrak{U}_0}$ and sampling from the uniform distribution on the sets $T_{\mathfrak{U}_0}$ and V_P for any set of players $\mathfrak{U}_0 \notin \mathfrak{A}$ and any player $P \in \mathfrak{U}$, then $(\mathcal{D}, \mathcal{R})$ is simulatable.

Proof. We first describe the simulator. The $\mathcal{S}_1$ algorithm simply samples a share from the uniform distribution on V_P when the adversary requests P's share. The $\mathcal{S}_2$ algorithm completes the simulated sharing $f|_{\mathfrak{U}_0}$ by sampling t from the uniform distribution on $T_{\mathfrak{U}_0}$, computes $r = \sigma_{s,\mathfrak{U}_0}(t, f|_{\mathfrak{U}_0})$ and then runs $\mathcal{D}$ with input s and randomness r, outputting the result h.

Since the map $\sigma_{s,\mathfrak{U}_0}$ is a bijection, it does not matter if we sample $r \xleftarrow{r} T$ and compute h using $\mathcal{D}$ and r, or if we sample $f|_{\mathfrak{U}_0}$ share by share, sample $t \xleftarrow{r} T_{\mathfrak{U}_0}$, compute r using $\sigma_{s,\mathfrak{U}_0}$ and compute h using $\mathcal{D}$ and r. □

Example 12.5. Consider the secret sharing scheme from Example 12.2. To use Proposition 12.2, we need to explain how to compute the bijection $\sigma_{s,\mathfrak{U}_0}$.

The dealer algorithm samples $l-1$ group elements from the uniform distribution, which is the shares belonging to the first $l-1$ players. The last share is computed. Therefore, $T = G^{l-1}$.

If $\mathfrak{U}_0 = \{1, 2, \ldots, l-1\}$, we let $\sigma_{s,\mathfrak{U}_0}$ be the identity map. Otherwise, suppose $i \notin \mathfrak{U}_0$. We can use the above map if we can compute c_i, which we can because

$$c_i = s \left(\prod_{j \neq i} f|_{\mathfrak{U}_0}(P_j) \right)^{-1}.$$

The maps $\sigma_{s,\mathfrak{U}_0}$ we have described are trivially bijections, and compatible with $\mathcal{D}$ by inspection. It follows that we can apply Proposition 12.2 and that the secret sharing scheme is simulatable.

12.1.2 Geometric Secret Sharing

The example schemes we discussed in the previous exercise are interesting and useful, but we want more flexible schemes. In principle, we can use the results of Exercise 12.3 to construct many new access structures, but this is not convenient. Instead, we shall study a construction based on geometry.

Two dimensions We begin with an example for a 2-out-of-l access structure. We know that any two distinct, non-parallell lines in the plane intersect in exactly one point. The immediate idea is that we encode our secret as a point in the plane and choose as our shares lines through this plane. As long as no two lines are identical, any two shares can be used to reconstruct the secret point, simply by computing the intersection.

We do the computation in excrutiating detail, since it is instructive for the general construction. A line in the plane is described by a linear equation

$$\alpha X + \beta Y = \gamma.$$

Unlike in Chapter 2, we will restrict attention to rational points.

E *Exercise* 12.9. Consider the affine plane over a finite field with q elements. Show that there are $q(q+1)$ distinct lines in the plane.

To choose a random line through a secret point (x, y), we choose α, β at random such that both are not zero, and then compute $\gamma = \alpha x + \beta y$. Over a finite field with q elements, there are $q + 1$ distinct such lines.

Recall that two lines in the plane will satisfy two equations of the form

$$\alpha_1 X + \beta_1 Y = \gamma_1 \qquad \alpha_2 X + \beta_2 Y = \gamma_2.$$

To find the point of intersection, we solve the linear system

$$\begin{pmatrix} \alpha_1 & \beta_1 \\ \alpha_2 & \beta_2 \end{pmatrix} \begin{pmatrix} x \\ y \end{pmatrix} = \begin{pmatrix} \gamma_1 \\ \gamma_2 \end{pmatrix}.$$

As long as the lines are not parallel, this system will have a unique solution.

There remains one difficulty, which is that a line through (x, y) reveals information about the pair (x, y), since a line contains only q points.

The solution is to encode just a single field element s in the secret point (x, y), by choosing (x, y) as a random point on a particular public *encoding line* $\alpha_0 X + \beta_0 Y = s$. We must then be careful about not distributing this line as a share. One nice choice is $X = s$.

In other words, the dealer algorithm chooses a random point on the encoding line $X = s$, then chooses random lines through that point as shares, taking care that none are equal to the encoding line.

E *Exercise* 12.10. Prove that the above scheme is an information-theoretically secure secret sharing scheme for the 2-out-of-l access structure.

E *Example* 12.6. The above scheme is simulatable. The dealer algorithm first chooses a random point on the encoding line, and then a random line through that point. There are q points on the line and q distinct lines through the point, when we exclude the encoding line. There are then q^2 choices.

Our simulator chooses a random line through space as the single share, taking care that it is not parallel to any encoding line. There are exactly q^2 such lines. The share is then correctly distributed.

Given the secret s, the simulator computes the intersection point of the single share and the encoding line, and then samples random lines through the intersection point as the remaining shares, excluding the encoding line.

Since the single share determines a unique point on the encoding line, the distribution of the remaining shares will be identical to what the dealer would generate, given the single share.

Higher Dimensions We now generalise the above idea to higher dimension. Recall that a hyperplane in t-space is described by a equations of the form

$$\alpha_1 X_1 + \alpha_2 X_2 + \cdots + \alpha_t X_t = \gamma.$$

Note that the vector $(\alpha_1, \alpha_2, \ldots, \alpha_t)$ is the normal vector of the plane.

If we fix the normal vector $(\alpha_{0,1}, \ldots, \alpha_{0,t})$ of a public *encoding hyperplane*, we can encode our secret s by choosing normal vectors $(\alpha_{i1}, \alpha_{i2}, \ldots, \alpha_{i,t})$, $i = 1, 2, \ldots, l$ and a random point $(x_1, x_2, \ldots, x_t)$ on the hyperplane $\alpha_{0,1}X_1 + \cdots + \alpha_{l+1,t}X_t = s$, and then computing $\gamma_i = \alpha_{i1}x_1 + \cdots + \alpha_{i,t}x_t$, for $i = 1, 2, \ldots, l$. The shares are the hyperplanes.

If we have t such hyperplanes, finding the point of intersection amounts to solving a linear system of equations to recover the point $(x_1, \ldots, x_t)$ and then recompute the secret as $\alpha_{0,1}x_1 + \cdots + \alpha_{0,t}x_t = s$. Unlike the two-dimensional case, for the system to have a unique solution it is not sufficient that the hyperplanes are all distinct and non-parallel. Their normal vectors must be linearly independent, and this must be true *regardless of which shares we use.*

Consider the matrix where each row is the normal vector of one share,

$$
M = \begin{pmatrix}
\alpha_{11} & \alpha_{12} & \cdots & \alpha_{1,t} \\
\alpha_{21} & \ddots & & \vdots \\
\vdots & & \ddots & \alpha_{l-1,t} \\
\alpha_{l,1} & \cdots & \alpha_{l,t-1} & \alpha_{l,t}
\end{pmatrix}.
$$

In order for the secret to be reconstructable from every set of t shares, this matrix has to have a very special property, namely that any t rows of the matrix must be linearly independent.

Remark. This property is actually the defining property of so-called Maximum Distance Separable (MDS) codes. In fact, there is a strong connection between error correcting codes and secret sharing schemes.

For security to hold, this property must also hold if we add the normal vector $(\alpha_{0,1}, \ldots, \alpha_{0,t})$ to the matrix, otherwise fewer than t shares could reveal the secret by revealing *some* point on the encoding hyperplane.

It seems intuitive that for q large compared to l and t, we would expect most matrices M to be suitable for secret sharing. First, most choices of t vectors will be linearly independent. Further, with j vectors, we can span at most $\binom{j}{t-1}$ distinct $t-1$-dimensional subspaces. As long as $\binom{j}{t-1}$ is very small compared to q, the odds that the $j+1$th vector will be in a subspace spanned by $t-1$ of the jth first vectors is small.

However, we want to use secret sharing even for q small, and proving that a given matrix is suitable for secret sharing is cumbersome.

Fixed Normal Vectors   For a moment, we return to the two-dimensional case. Instead of choosing random lines through the point encoding our secret, we choose lines with fixed slopes, that is, we have a sequence of distinct field elements $\alpha_1, \alpha_2, \ldots, \alpha_l$, and choose lines of the form $\alpha_i X + Y = \gamma_i$ for share i, making sure that none of these have the same slope as the encoding line.

This will work, in the sense that we can create shares and recover the secret. But will it be secure?

Example 12.7. We redo the argument from Example 12.6, however, this time using Proposition 12.2.

First, observe that the simulator chooses a single point on the encoding line, perhaps by sampling r from the uniform distribution on $\mathbb{F}_q$, setting $y = r$ and solving for x. Given (x, y), the shares are now uniquely determined.

Next, a share γ_i and the secret s uniquely determines (x, y) as the intersection point of the encoding line and $\alpha_i X + Y = \gamma_i$. If (x, y) is the point of intersection, we then get that $\sigma_{s,\{P_i\}} : \mathbb{F}_q \to \mathbb{F}_q$ is given by $\sigma_{s,\{P_i\}}(\gamma_i) = y$.

The $\sigma_{s,\{P_i\}}$ maps are all bijections, which means that Proposition 12.2 applies. The scheme is simulatable.

We are now ready to describe our first practical secret sharing scheme.

Example 12.8. Let $M = (\alpha_{ij})$ be a $(l+1) \times t$ MDS matrix.

- The dealer samples a point $(x_1, x_2, \ldots, x_t)$ on $\alpha_{0,1}X_1 + \cdots + \alpha_{0,t}X_t = s$ and computes the ith share as $\gamma_i \leftarrow \sum_j \alpha_{ij}X_j$, $i = 1, 2, \ldots, l$. Note that

$$\begin{pmatrix} s \\ \gamma_1 \\ \gamma_2 \\ \vdots \\ \gamma_l \end{pmatrix} = M \begin{pmatrix} x_1 \\ x_2 \\ \vdots \\ x_t \end{pmatrix}.$$

- The reconstruction algorithm takes as input a set of at least t players $\mathfrak{U}$ and their shares. It chooses t shares and reconstructs a matrix M_t (using the normal vectors as rows) and a corresponding vector γ and computes

$$\begin{pmatrix} x_1 \\ x_2 \\ \vdots \\ x_t \end{pmatrix} = M_t^{-1}\gamma \quad \text{and} \quad s = \sum_{i=1}^{t} \alpha_{0,i}x_i.$$

Theorem 12.3. *The above secret sharing scheme is a simulatable t-out-of-l secret sharing scheme.*

Exercise 12.11. Prove Theorem 12.3.

How do we find a suitable matrix M? So-called *Vandermonde* matrices

$$M = \begin{pmatrix} 1 & \alpha_0 & \alpha_0^2 & \cdots & \alpha_0^{t-1} \\ 1 & \alpha_1 & \alpha_1^2 & \cdots & \alpha_1^{t-1} \\ \vdots & & & & \\ 1 & \alpha_l & \alpha_l^2 & \cdots & \alpha_l^{t-1} \end{pmatrix}$$

satisfy exactly this property when $\alpha_0, \alpha_1, \ldots, \alpha_l$ are distinct field elements. If $\alpha_0 = 0$, the first row in the matrix will be $(1, 0, \ldots, 0)$, which is convenient.

At this point, consider a polynomial $f(X) = \sum_{i=0}^{t} f_i X^i$, and suppose

$$\begin{pmatrix} s \\ \gamma_1 \\ \gamma_2 \\ \vdots \\ \gamma_l \end{pmatrix} = M \begin{pmatrix} f_0 \\ f_1 \\ \vdots \\ f_{t-1} \end{pmatrix}.$$

Then we see that $\gamma_i = f(\alpha_i)$. This observation gives us a much more convenient description of these secret sharing schemes.

12.1.3 Shamir's Secret Sharing

Before we begin, we shall first cover some elementary material on Lagrange interpolation. Finding a polynomial that has prescribed values at prescribed points is called *polynomial interpolation*. Suppose we have t distinct field elements $\alpha_1, \alpha_2, \ldots, \alpha_t$, values $\gamma_1, \gamma_2, \ldots, \gamma_t$, and we want to find a polynomial such that $f(\alpha_i) = \gamma_i$ for $i = 1, 2, \ldots, t$.

If we can find polynomials $\phi_1(X), \phi_2(X), \ldots, \phi_t(X)$ satisfying

$$\phi_i(\alpha_j) = \begin{cases} 1 & i = j, \\ 0 & i = 1, 2, \ldots, j-1, j+1, \ldots, t, \end{cases}$$

then

$$f(X) = \sum_{i=1}^{t} \gamma_i \phi_i(X) \tag{12.1}$$

is a suitable polynomial. The polynomial $\prod_{j \neq i}(X - \alpha_j)$ has the prescribed zeros for $\phi_i(X)$, but its value at α_i is not 1, but rather $\prod_{j \neq i}(\alpha_i - \alpha_j)$. It follows that

$$\phi_i(X) = \prod_{j \neq i} \frac{X - \alpha_j}{\alpha_i - \alpha_j} \tag{12.2}$$

is exactly the polynomial we are looking for. This is a *Lagrange polynomial* and the method is called *Lagrange interpolation*.

Proposition 12.4. Let $\alpha_1, \alpha_2, \ldots, \alpha_t$ be distinct field elements, and let $\gamma_1, \gamma_2, \ldots, \gamma_t$ be arbitrary values. Then the polynomial $f(X)$ defined by (12.1) and (12.2) satisfies $f(\alpha_i) = \gamma_i$ for $i = 1, 2, \ldots, t$. If $g(X)$ is a polynomial satisfying the same property and $\deg g(X) \leq \deg f(X)$, then $g(X) = f(X)$.

Proof. The above argument shows that the given polynomial satisfies the required properties. All that remains is to show uniqueness. The polynomial

$$f(X) - g(X)$$

has zeros at $\alpha_1, \alpha_2, \ldots, \alpha_t$. Furthermore, if $\deg f(X) \geq \deg g(X)$, then the difference has degree less than t, from which it follows that $f(X) = g(X)$. □

Example 12.9. *Shamir's secret sharing* scheme uses a finite field $\mathbb{F}_q$ and $\mathfrak{U} \subseteq \mathbb{F}_q^*$.

- The *dealer* $\mathcal{D}$ takes $s \in \mathbb{F}_q$ as input, samples a random polynomial $f(X)$ of degree less than t such that $f(0) = s$, and outputs the sharing that maps $\alpha \in \mathfrak{U}$ to $f(\alpha)$.

- The *reconstruction* algorithm $\mathcal{R}$ takes as input a set of at least t players $\mathfrak{U}_0$ and their shares $f|_{\mathfrak{U}_0}$. It uses Proposition 12.4 to find a polynomial $g(X)$ such that $g(\alpha) = f|_{\mathfrak{U}}(\alpha)$ for $\alpha \in \mathfrak{U}_0$. Then it outputs $g(0)$.

The scheme is correct for a t-out-of-l access structure by Proposition 12.4.

Theorem 12.5. *Shamir's scheme is a simulatable t-out-of-l secret sharing scheme.*

Proof. We could reduce Shamir's scheme to the scheme from Section 12.1.2, but instead we shall use Proposition 12.4.

We may assume that $\mathfrak{U}_0 = \{\alpha_1, \alpha_i, \ldots, \alpha_{t-1}\}$ with shares $\gamma_1, \gamma_2, \ldots, \gamma_{t-1}$. Proposition 12.4 applied to $0, \alpha_1, \alpha_i, \ldots, \alpha_{t-1}$ and $s, \gamma_1, \gamma_2, \ldots, \gamma_{t-1}$ gives us a polynomial $f(X) = \sum_{i=0}^{t-1} f_i X^i$. Then

$$\sigma_{s,\mathfrak{U}_0}(\alpha_1, \alpha_2, \ldots, \alpha_{t-1}) = (f_1, f_2, \ldots, f_{t-1}).$$

Since Proposition 12.4 gives us a unique polynomial, the maps are bijections and Proposition 12.2 applies. The claim follows. □

One particularily nice result for Shamir's secret sharing is a convenient formula for reconstruction. For $\alpha \in \mathfrak{U}_0$, define

$$\lambda_{\alpha,\mathfrak{U}_0} = \prod_{\beta \in \mathfrak{U}_0 \beta \neq \alpha} \frac{-\beta}{\alpha - \beta}.$$

Note that $\lambda_{\alpha,\mathfrak{U}_0}$ depends only on α and $\mathfrak{U}_0$.

Then for any sharing f and any $\mathfrak{U}_0 = \{\alpha_1, \alpha_2, \ldots, \alpha_t\}$, we have that

$$\mathcal{R}(\mathfrak{U}_0, f|_{\mathfrak{U}_0}) = \lambda_{\alpha_1,\mathfrak{U}_0} f(\alpha_1) + \lambda_{\alpha_2,\mathfrak{U}_0} f(\alpha_2) + \cdots + \lambda_{\alpha_t,\mathfrak{U}_0} f(\alpha_t). \quad (12.3)$$

This reconstruction formula is also useful in other situations. For instance, we could do simulation using the formula, since given $t - 1$ shares and the secret, the other $l - t + 1$ shares can be computed using this formula.

12.1.4   Linear Secret Sharing

One important property of secret sharing schemes is linearity. The idea is that if we have secret sharings of two secrets s_1 and s_2, we can construct a sharing of $\lambda_1 s_1 + \lambda_2 s_2$ by computing the same linear combination of each share.

D **Definition 12.4.** Consider a secret sharing scheme $(\mathcal{D}, \mathcal{R})$ with access structure $\mathfrak{A}$ for secrets from a field $\mathbb{F}$, and suppose that the share of any player lies in an $\mathbb{F}$-vector space. For a set of ν players $\mathfrak{U}_0 \in \mathfrak{A}$, we denote the function from a ν-tuple of shares to $\mathbb{F}$ computed by $\mathcal{R}$ by $\mathcal{R}_{\mathfrak{U}_0}$. The secret sharing scheme is *($\mathbb{F}$-)linear* if $\mathcal{R}_{\mathfrak{U}_0}$ is $\mathbb{F}$-linear for any $\mathfrak{U}_0 \in \mathfrak{A}$.

T **Proposition 12.6.** *Let $(\mathcal{D}, \mathcal{R})$ be any linear secret sharing scheme, let f_1, f_2 be sharings of s_1, s_2, and let $\lambda_1, \lambda_2 \in \mathbb{F}$. Define h by $h(P) = \lambda_1 f_1(P) + \lambda_2 f_2(P)$. Then h is a sharing of $\lambda_1 s_1 + \lambda_2 s_2$.*

Proof. By linearity we have that

$$\mathcal{R}_{\mathfrak{U}_0}(h|_{\mathfrak{U}_0}) = \mathcal{R}_{\mathfrak{U}_0}(\lambda_1 f_1|_{\mathfrak{U}_0} + \lambda_2 f_2|_{\mathfrak{U}_0}) =$$
$$\lambda_1 \mathcal{R}_{\mathfrak{U}_0}(f_1|_{\mathfrak{U}_0}) + \lambda_2 \mathcal{R}_{\mathfrak{U}_0}(f_2|_{\mathfrak{U}_0}) = \lambda_1 s_1 + \lambda_2 s_2. \quad \square$$

The scheme from Example 12.3 is not linear for most cryptosystems. Of particular interest is Shamir's secret sharing scheme . It is linear by the discussion at the end of Section 12.1.3. In fact, this linearity allows us to explain one particularily important application of linear secret sharing schemes.

E *Example* 12.10. Consider the ElGamal cryptosystem for a group $G = \langle g \rangle$ of prime order q, where the decryption key is b, the encryption key is $y = g^b$, an encryption of a message $m \in G$ is $(x, w) \in G \times G$, and it satisfies $w = x^b m$. Then the decryption of the ElGamal ciphertext is

$$m = wx^{-b}.$$

Now let f be a sharing of b under (t, l)-Shamir secret sharing. Define

$$x_i = x^{f(P_i)}$$

for $i = 1, 2, \ldots, l$. For any player set $\mathfrak{U}_0$ with at least t players, we have that

$$m = w \prod_{P_i \in \mathfrak{U}_0} x_i^{-\lambda_{i, \mathfrak{U}_0}}.$$

This is the basis for *distributed decryption*. We share the decryption key among l players with a threshold t. Any t players can decrypt a ciphertext by computing $x^{f(P_i)}$ and revealing x_i. We shall see that if (x, w) is honestly generated, then a *decryption share* x_i reveals no information about $f(P_i)$.

E *Example* 12.11. Another application of linearity is shifting a sharing under one scheme to a sharing under another scheme. We explain the idea.

Consider two $\mathbb{F}$-linear secret sharing schemes, $(\mathcal{D}, \mathcal{R})$ and $(\mathcal{D}', \mathcal{R}')$ with access structures $\mathfrak{A}, \mathfrak{A}'$, respectively. For simplicity, we shall assume that all shares are field elements. In this case, the reconstruction functions are simple linear combinations of shares, and we fix the notation

$$\mathcal{R}_{\mathfrak{U}_0}(f|_{\mathfrak{U}_0}) = \sum_{P \in \mathfrak{U}_0} \rho_{P,\mathfrak{U}_0} f(P) \quad \text{and} \quad \mathcal{R}'_{\mathfrak{U}_1}(h|_{\mathfrak{U}_1}) = \sum_{P' \in \mathfrak{U}_1} \rho'_{P',\mathfrak{U}_1} h(P').$$

Suppose h is a sharing under $(\mathcal{D}', \mathcal{R}')$. Let f'_P be a sharing of $h(P)$ under $(\mathcal{D}, \mathcal{R})$. For any $\mathfrak{U}_1 \in \mathfrak{A}'$, we define a sharing f under $(\mathcal{D}, \mathcal{R})$ by

$$f(P) = \sum_{P' \in \mathfrak{U}_1} \rho'_{P',\mathfrak{U}_1} f'_{P'}(P)$$

It is a sharing of the same value as h. For any $\mathfrak{U}_0 \in \mathfrak{A}$ we compute

$$\mathcal{R}(\mathfrak{U}_0, f|_{\mathfrak{U}_0}) = \mathcal{R}_{\mathfrak{U}_0}(f) = \sum_{P \in \mathfrak{U}_0} \rho_{P,\mathfrak{U}_0} f(P) = \sum_{P \in \mathfrak{U}_0} \rho_{P,\mathfrak{U}_0} \sum_{P' \in \mathfrak{U}_1} \rho'_{P',\mathfrak{U}_1} f'_{P'}(P)$$

$$= \sum_{P' \in \mathfrak{U}_1} \rho'_{P',\mathfrak{U}_1} \sum_{P \in \mathfrak{U}_0} \rho_{P,\mathfrak{U}_0} f'_{P'}(P) = \sum_{P' \in \mathfrak{U}_1} \rho'_{P',\mathfrak{U}_1} h(P')$$

$$= \mathcal{R}'_{\mathfrak{U}_1}(h) = \mathcal{R}'(\mathfrak{U}_1, h|_{\mathfrak{U}_1}),$$

which completes the argument.

This example illustrates one issue that will become important. We will sometimes create shares not by using the dealer algorithm, but by computing linear combinations of other shares. In this case, simulatability can fail because the distribution of a linear combination of shares could be very different from the distribution output by the dealer.

Definition 12.5. Let $(\mathcal{D}, \mathcal{R})$ be an $\mathbb{F}$-linear secret sharing scheme. Let $s_1, s_2, \ldots, s_l \in \mathbb{F}$, and let $\alpha_1, \alpha_2, \ldots, \alpha_l \in \mathbb{F}$ not all be zero. Suppose $f_i \leftarrow \mathcal{D}(s_i)$, $h \leftarrow \sum \alpha_i f_i$ and $h' \leftarrow \mathcal{D}(\sum \alpha_i s_i)$. The linear secret sharing scheme is *linearly simulatable* if it is statelessly simulatable, and the distribution of a h and h' is identical.

Exercise 12.12. Show that Shamir's secret sharing is linearly simulatable.

12.1.5 Multiplicative Secret Sharing

Some secret sharing schemes have stronger properties than just linearity. They may also have a multiplicative property, which more or less allows us to get shares of a product by multiplying shares of the factors.

Definition 12.6. Let $(\mathcal{D}, \mathcal{R})$ be a linear secret sharing scheme with access structure $\mathfrak{A}$. We say that the scheme *admits a multiplication operation* if there exists a binary operation $\boxdot$, an access structure $\mathfrak{A}^{\boxdot}$ and a linear secret sharing scheme $(\mathcal{D}^{\boxdot}, \mathcal{R}^{\boxdot})$ such that for any sharings f_1, f_2 of values s_1, s_2, if we let h

be the sharing defined by $h(P) = f_1(P) \boxdot f_2(P)$, then h is a sharing of $s_1 s_2$ under $(\mathcal{D}^{\boxdot}, \mathcal{R}^{\boxdot})$ with the access structure $\mathfrak{A}^{\boxdot}$.

Remark. If $(\mathcal{D}, \mathcal{R})$ is simulatable, then $\mathfrak{A}^{\boxdot} \subseteq \mathfrak{A}$, since otherwise we could reconstruct the secret using fewer than required shares by creating a product share and reconstructing the product.

By itself, this multiplication operation is less than useful. However, because of linearity, we can use the techniques from Example 12.11 to shift the sharing under $(\mathcal{D}^{\boxdot}, \mathcal{R}^{\boxdot})$ back to a sharing under $(\mathcal{D}, \mathcal{R})$. This idea turns out to be extremely useful, especially if the secret sharing scheme is linearly simulatable.

Example 12.12. Consider Shamir's secret sharing scheme from Section 12.1.3 for a t-out-of-l access structure, and suppose $2t - 1 \leq l$.

If we have two polynomials $f_1(X)$ and $f_2(X)$, each of degree at most $t - 1$, then $g(X) = f_1(X)f_2(X)$ is a polynomial of degree at most $2t - 2$. Furthermore, if α_i is any field element, then $g(\alpha_i) = f_1(\alpha_i)f_2(\alpha_i)$.

In other words, t-out-of-l Shamir's secret sharing admits a multiplication operation if we let $\boxdot$ be ordinary multiplication in the field and let $(\mathcal{D}^{\boxdot}, \mathcal{R}^{\boxdot})$ be Shamir's secret sharing for a $(2t - 1)$-out-of-l access structure.

Note also that the techniques from Example 12.11 apply, so the sharing under the $(2t - 1)$-out-of-l access structure can be turned back into a sharing under the t-out-of-l access structure.

12.2   MULTI-PARTY COMPUTATION

At the start of the chapter, we discussed a protocol that allowed people seated around a table to compute a sum. We now show how to do the same computation using linear secret sharing.

Example 12.13. We give another example of a protocol that allows our group to compute their average salary, using a linear secret sharing scheme and the ability to whisper to any one of the other players. Let $(\mathcal{D}, \mathcal{R})$ be any linear secret sharing scheme with a 5-out-of-5 access structure.

Alice runs $\mathcal{D}$ on input s_A and gets a sharing f_A. She whispers $f_A(Bob)$ to Bob, $f_A(Carol)$ to Carol, and so on. The other four do the same, in parallel.

We define a sharing h by

$$h(P) = f_A(P) + \cdots + f_E(P).$$

Observe that since the scheme is linear, Alice can compute $h(Alice)$, Bob can compute $h(Bob)$, and so on. Once this is done, everyone announces their share $h(P)$, after which everyone can use $\mathcal{R}$ to recover u.

By linearity, u will equal $s_A + \cdots + s_E$.

If the scheme is information-theoretically secure, no coalition learns anything from their shares, so the protocol is secure against any passive coalition.

However, the reconstruction formula is known. The player that outputs their share $h(P)$ last can use the reconstruction formula to choose a share that will give a desired outcome. Since the other four have already output their shares, the laggard may reconstruct the outcome privately and act upon that value. The scheme is therefore not secure against an active attacker.

In both of the two protocols we have seen, the computational effort involved is trivial. However, there are other resources of interest for protocols, and one of them is the bandwidth required for communication, private (whispering) and broadcast (speaking out loud).

We see that the first protocol requires 5 whispered messages and 1 announcement. The second protocol requires $5 \cdot 4$ whispered messages and 5 announcements. Security has a cost.

Example 12.14. Consider the situation where Alice and Bob each have a secret input s_A and s_B, and together with Carol and Dave they would like to create a sharing of the product $s_A s_B$ of their secrets. They will use a secret sharing scheme $(\mathcal{D}, \mathcal{R})$ that admits a multiplication operation, such that $\{Alice, Bob, Carol, Dave\} \in \mathfrak{A}^{\square}$.

Alice runs $\mathcal{D}$ to construct a sharing f_A of s_A and whisper the shares to the other three, retaining one share. Bob likewise creates a sharing f_B of s_B and distributes the shares.

Separately, each of them computes $h'(P) = f_A(P) \boxdot f_B(P)$.

Each of the players use $\mathcal{D}$ to share $h'(P)$ among the others.

Following Example 12.11, each player then separately uses the linearity of $(\mathcal{D}^{\square}, \mathcal{R}^{\square})$ to recover a share of $s_A s_B$ under $(\mathcal{D}, \mathcal{R})$, as desired.

12.2.1 Discussing Security

The primary goal of multi-party computation is to evaluate some function on some inputs, without revealing anything about the inputs that is not implicitly revealed by the function value. This is *confidentiality*. As usual, we also want *integrity*, namely that the correct result is computed. This is somewhat complicated by the fact that the adversarially controlled players may participate in the computation and choose their input, but the idea is that the adversary must somehow commit to its input so that it cannot adaptively change it. We also want a notion of *robustness*, so that once committed the adversary cannot prevent the honest players from computing the result. We shall not develop a security definition in detail.

When defining security, we must precisely describe the adversary's capabilities. In multiparty computation, we assume that we have secure channels, which simplifies private communication and reduces the adversary's potential benefit from controlling the network. Thus the adversary's main tool is corrupting parties in the computation. We need to decide on how and when player corruption happens and how corrupt players should behave.

How corrupt players behave   Consider first how corrupt players should behave. It seems natural to assume that an adversary that corrupts a player controls that player's every action. In this case, we can no longer predict that player's actions. We say that such an adversary is *active*.

However, some adversaries will be more circumspect. One interesting class is summed up by the name *honest-but-curious*, where the corrupted players will continue to follow the protocol, but the adversary will learn everything they do. We say that such an adversary is *passive*.

One can obviously imagine intermediate classes of adversaries. However, these two classes seem most realistic. The first is obviously plausible. The other can be enforced by a number of scenarios, for example when a computer system is compromised so that all its secret information leaks, but monitoring systems make it too risky for the adversary to change its behaviour. Another example is when a player is curious, but since a change in behaviour could be detected, legal consequences deter the adversary from any detectable misbehaviour.

How and when players are corrupted   Next, we must discuss how and when the adversary corrupts players. We inherit the idea of access structures from secret sharing, so our main restriction must be that the adversary never corrupts a set of players that like in the access structure. This is unavoidable.

The more interesting question is how the adversary corrupts players. The adversary's corruption must be invisible to the remaining honest players. This means that honest players actions cannot depend on whether another player is corrupt or not, at least until that player's actions reveal its corruption.

We model this by allowing the adversary to signal its intention to corrupt a player. Once that has been signaled, we reveal the player's secrets to the adversary, and the player's actions are thereafter controlled by the adversary.

The natural model for cryptographic protocol analysis is to allow the adversary to corrupt players whenever he wants to. However, for many applications of multiparty computations, this seems unlikely. Also, allowing the adversary to corrupt players at arbitrary times makes the job of securing protocols and proving them secure much more difficult. It therefore makes sense to look at what we can do against a less powerful adversary.

The usual solution is to force the adversary to do all of his corruption before anything interesting happens. Essentially, we will be considering a situation where a fixed set of the players are corrupted, so this is often known as *static* corruption, and we say that the adversary is *non-adaptive* or *static*. An adversary that is allowed to corrupt at any time is said to be *adaptive*, capable of adapting its corruption strategy.

For many of the protocols we shall study, it is somewhat hard to see what advantage an adaptive adversary would have over a non-adaptive adversary. The next example shows how an adaptive adversary can be stronger than a non-adaptive adversary. Note that the scheme does not quite fit our setting.

Example 12.15. Consider the following secret sharing scheme, with $2l \geq 256$ players. It is based on the $l + 1$-out-of-$l + 1$ secret sharing scheme from Example 12.2. The dealer chooses l players at random, uses the $l + 1$-out-of-$l + 1$ secret sharing scheme to create $l + 1$ shares. The first l shares are distributed among the chosen l players, while the remaining l players get the final share and a list of the chosen l players.

While reconstruction is possible for many subsets of players with at least $l + 1$ players, most such subsets of players will not be able to reconstruct the secret. In fact, the honest players must ensure that every one of the $2l$ players participate in order to ensure reconstruction.

For a static (non-adaptive) adversary, the same holds true. If t players are honest, the probability that none of them are among the l players chosen by the dealer is less than 2^{-t}.

An adaptive adversary, however, does not corrupt any players until the dealer is done. Afterwards, the adversary corrupts players until a player with the $l+1$th share and a list of players is corrupted. The adversary then corrupts the remaining players on the list and reconstructs the secret.

We see that any static adversary that corrupts $l + 1$ players has negligible chance of reconstructing the secret, while an adaptive adversary that corrupts $l + 1$ players can reconstruct the secret with probability 1.

12.2.2 A First Multiparty Computation Protocol

Each party in a multiparty protocol has its own inputs. These are either the player's input to the computation or the player's shares of a common value. These shares will be the result either of other players sharing their input to the computation or the output of previous computations. Second, each party will produce certain outputs, either results from the computation or shares of a common value to be used in further computations. We shall now discuss two simple protocols that more or less formalise Examples 12.13 and 12.14.

Every player P gets as input their share of sharings $f_1, f_2, \ldots, f_l$. The coefficients $\alpha_1, \alpha_2, \ldots, \alpha_l$ are public values.

- Player P computes $h(P) \leftarrow \sum_{i=1}^{l} \alpha_i f_i(P)$.

The output of player P is its share of the sharing h.

FIGURE 12.1 Protocol for computing a linear combination, based on a linear secret sharing scheme.

The protocols in this section will be based on a secret sharing scheme $(\mathcal{D}, \mathcal{R})$ that admits a multiplication, with secret sharing scheme $(\mathcal{D}^\square, \mathcal{R}^\square)$.

Every player P gets as input their share of sharings f_1, f_2.

- Player P computes $f_3(P) \leftarrow f_1(P) \boxdot f_2(P)$.
- Player P runs $\mathcal{D}$ on $f_3(P)$, obtaining a sharing h'_P, and sends the share $h'_P(P')$ privately to P' for all $P' \in \mathfrak{U}$.
- For some set of players $\mathfrak{U}_0$, player P receives a share $h'_{P'}(P)$ from every player $P' \in \mathfrak{U}_0$, and no shares from any player $P' \notin \mathfrak{U}_0$. If $\mathfrak{U}_0 \notin \mathfrak{A}^\boxdot$, P stops with output $\perp$.
- Player P computes

$$h(P) = \sum_{P' \in \mathfrak{U}_0} \rho^\boxdot_{P', \mathfrak{U}_0} h'_{P'}(P),$$

where $\rho^\boxdot$ are the coefficients in the linear combination that $\mathcal{R}^\boxdot_{\mathfrak{U}_0}$ evaluates.

The output of player P is its share of the sharing h.

FIGURE 12.2 Protocol for computing a product, based on a secret sharing scheme that admits a multiplication.

We shall assume that coefficients $\rho^\boxdot_{P, \mathfrak{U}_0}$ exist for all $P, \mathfrak{U}_0 \in \mathfrak{A}^\boxdot$ such that

$$\mathcal{R}^\boxdot_{\mathfrak{U}_0}(f|_{\mathfrak{U}_0}) = \sum_{P \in \mathfrak{U}_0} \rho^\boxdot_{P, \mathfrak{U}_0} f|_{\mathfrak{U}_0}(P).$$

The first protocol takes as input sharings of l common values and computes a sharing of a public linear combination of the values. It is given in Figure 12.1.

The second protocol takes as input sharings of two common values and computes a sharing of the product of the two values. It is given in Figure 12.2.

We begin our work by proving that these protocols are complete, in the sense that when everyone is honest the players end up with the desired output.

Proposition 12.7. *On input of sharings $f_1, \ldots, f_l$ and coefficients $\alpha_1, \ldots, \alpha_l$, the protocol in Figure 12.1 run by honest players outputs a sharing h satisfying*

$$\mathcal{R}(\mathfrak{U}_0, h) = \sum_{i=1}^{l} \alpha_i \mathcal{R}(\mathfrak{U}_0, f_i)$$

for any $\mathfrak{U}_0 \in \mathfrak{A}$. It does not require any communication.

Proof. We compute

$$\mathcal{R}_{\mathfrak{U}_0}(h) = \sum_{P \in \mathfrak{U}_0} \rho_{P,\mathfrak{U}_0} h(P) = \sum_{P \in \mathfrak{U}_0} \rho_{P,\mathfrak{U}_0} \sum_{i=1}^{l} \alpha_i f_i(P)$$

$$= \sum_{i=1}^{l} \alpha_i \sum_{P \in \mathfrak{U}_0} \rho_{P,\mathfrak{U}_0} f_i(P) = \sum_{i=1}^{l} \alpha_i \mathcal{R}_{\mathfrak{U}_0}(f_i). \qquad \square$$

Proposition 12.8. *On input of sharings f_1, f_2, the protocol in Figure 12.2 run by honest players outputs a sharing h satisfying*

$$\mathcal{R}(\mathfrak{U}_0, h) = \mathcal{R}(\mathfrak{U}_0, f_1)\mathcal{R}(\mathfrak{U}_0, f_2)$$

for any $\mathfrak{U}_0 \in \mathfrak{A}$. The protocol requires sending privately $l(l-1)$ shares.

Proof. We must first note that regardless of the actual set of players $\mathfrak{U}_0''$ the player P receives shares from, the computed share $h(P)$ will satisfy

$$h(P) = \sum_{P' \in \mathfrak{U}_0'} \rho^{\boxdot}_{P',\mathfrak{U}_0'} h'_{P'}(P)$$

for any set $\mathfrak{U}_0' \in \mathfrak{A}^{\boxdot}$. Now we compute

$$\mathcal{R}_{\mathfrak{U}_0}(h) = \sum_{P \in \mathfrak{U}_0} \rho_{P,\mathfrak{U}_0} h(P) = \sum_{P \in \mathfrak{U}_0} \rho_{P,\mathfrak{U}_0} \sum_{P' \in \mathfrak{U}_0'} \rho^{\boxdot}_{P',\mathfrak{U}_0'} h'_{P'}(P)$$

$$= \sum_{P' \in \mathfrak{U}_0'} \rho^{\boxdot}_{P',\mathfrak{U}_0'} \sum_{P \in \mathfrak{U}_0} \rho_{P,\mathfrak{U}_0} h'_{P'}(P) = \sum_{P' \in \mathfrak{U}_0'} \rho^{\boxdot}_{P',\mathfrak{U}_0'} (f_1(P) \boxdot f_2(P))$$

$$= \mathcal{R}_{\mathfrak{U}_0}(f_1 | \mathfrak{U}_0) \mathcal{R}_{\mathfrak{U}_0}(f_2 | \mathfrak{U}_0). \qquad \square$$

Using these two simple protocols repeatedly, we can compute any function of ν inputs that can be expressed as a multi-variate polynomial in ν variables. Note, however, that in general there are many ways in which we can compute a multi-variate polynomial, so instead we usually talk about an *algebraic circuit* as in Section 8.3, which is a directed graph where vertices with no incoming edges represent the inputs to the computation, the internal vertices represent either multiplications or fixed linear combinations, and the vertices with no outgoing edges represent the outputs of the computation.

More systematically, the circuit graph is a directed graph with no cycles. Each vertex is additionally labeled as

- input, either from a given player, or existing shares;
- multiplication, in which case there are exactly two incident edges;
- addition, where the label lists coefficients for incident edges; or
- output, either to a set of players, or as a sharing.

A player P gets as input s. No other player gets any input.

- Player P runs $\mathcal{D}$ on s, obtaining a sharing f. The player sends the share $f(P')$ privately to P' for all $P' \in \mathfrak{U}$.

The output of player P' is its share $f(P')$ of s.

Every player P gets as input their share of a sharing f of a value s. The list of share recipients is $\mathfrak{U}_0 \subseteq \mathfrak{U}$.

- Every player P sends its share $f(P)$ privately to P' for all $P' \in \mathfrak{U}_0$.
- When P has received shares from a set of players $\mathfrak{U}_1 \in \mathfrak{A}$, it computes $s \leftarrow \mathcal{R}(\mathfrak{U}_1, f|_{\mathfrak{U}_0})$.

The output of player $P' \in \mathfrak{U}_0$ is s.

FIGURE 12.3 Auxillary protocols: Share and reveal.

We order the vertices of the circuit graph such that a vertex precedes another vertex if there is a path from the first vertex to the second vertex.

To compute an algebraic circuit, we need two more auxilliary protocols, namely a sharing protocol and a reconstruction protocol, that players with inputs can distribute input shares to the other players, and a reconstruction protocol, where players can receive shares in order to reconstruct the output.

For any such algebraic circuit, we can use the protocols from Figures 12.1 and 12.2, together with the auxilliary protocols from Figure 12.3, to evaluate the circuit. Correctness follows from Propositions 12.7 and 12.8.

12.2.3 Non-adaptive Security

Our protocol is not secure for active adversaries. We shall not discuss adaptive adversaries, so we begin our analysis of security against passive, non-adaptive adversaries. Note that since the adversary is passive, we only have to care about confidentiality, and we know how the corrupted players will behave.

We prove security by showing that we can simulate the behaviour of the honest parties without knowledge of the honest parties inputs, until the end. We begin by noting that the linear combination protocol from Figure 12.1 does not involve any interaction, so it is trivial to simulate.

Next, we consider the multiplication protocol from Figure 12.2. In this protocol, the corrupted players receive shares from the honest players and then do a local computation. We need to simulate the shares without knowledge of the honest players' input shares. If our secret sharing scheme is statelessly simulatable, we simply run the simulator to generate the corrupted players'

The set of corrupted players is $\mathfrak{U}_0 \subseteq \mathfrak{U}$. A corrupted player P gets a share c_P of the input s.

$$f \leftarrow \mathcal{D}(s)$$
$$c_P \leftarrow f(P)$$

$$f|_{\mathfrak{U}_0} \leftarrow \mathcal{S}_1(\emptyset, \mathfrak{U}_0, \bot, \bot)$$
$$c_P \leftarrow f|_{\mathfrak{U}_0}(P).$$

FIGURE 12.4 Simulation for the sharing protocol from Figure 12.3. The adversary's real view on the left, the simulated view on the right.

The set of corrupted players is $\mathfrak{U}_0 \subseteq \mathfrak{U}$. A corrupted player P that should receive output gets for every honest player P' a share $c_{P,P'}$ of a sharing f of the input s. The corrupted players have $f|_{\mathfrak{U}_0}$.

$$c_{P,P'} \leftarrow f(P)$$

$$f \leftarrow \mathcal{S}_2(s, \mathfrak{U}_0, f|_{\mathfrak{U}_0}, \bot)$$
$$c_{P,P'} \leftarrow f(P).$$

FIGURE 12.5 Simulation for the revealing protocol from Figure 12.3. The adversary's real view on the left, the simulated view on the right.

shares for each of the honest players. The adversary cannot distinguish these shares from real shares, which means our simulation is done.

A useful way to formalise these arguments is in terms of the adversary's *view*, which contains everything the adversary sees. Figure 12.6 describes the adversary's view and how it is sampled in the real protocol. It also shows how we can simulate it. We have proved the following technical claim.

Proposition 12.9. *If the secret sharing scheme is simulatable, then the adversary view given by the simulator in Figure 12.6 is indistinguishable from the adversary view given by the multiplication protocol from Figure 12.2.*

Finally, we must consider the auxillary protocols. If the secret sharing

The set of corrupted players is $\mathfrak{U}_0 \subseteq \mathfrak{U}$. A corrupted player P gets a share $c_{P,P'}$ of the product share $f_1(P') \boxdot f_2(P')$ for every honest player $P' \notin \mathfrak{U}_0$.

$$h_{P'} \leftarrow \mathcal{D}(f_1(P') \boxdot f_2(P'))$$
$$c_{P,P'} \leftarrow h_{P'}(P)$$

$$h_{P'}|_{\mathfrak{U}_0} \leftarrow \mathcal{S}_1(\emptyset, \mathfrak{U}_0, \bot, \bot)$$
$$c_{P,P'} \leftarrow h_{P'}|_{\mathfrak{U}_0}(P)$$

FIGURE 12.6 Simulation for the multiplication protocol from Figure 12.2. The adversary's real view on the left, the simulated view on the right.

scheme is simulatable, the view given by the simulator in Figure 12.4 is indistinguishable from the view given by the sharing protocol from Figure 12.3.

The reconstruction protocol is different because we need to simulate shares that will not only give us a consistent reconstruction of the required shared secret, but the adversary's shares may not have been output by the dealer algorithm. This means that plain simulatability is not sufficient.

However, for our protocols we know that the adversary's shares will be linear combinations of shares that come from the dealer. It is therefore sufficient if the secret sharing scheme is linearly simulatable.

Proposition 12.10. *Suppose the secret sharing scheme is linearly simulatable and the input shares to the reveal protocol from Figure 12.3 were output by either the sharing protocol from Figure 12.3, the linear combination protocol from Figure 12.1 or the multiplication protocol from Figure 12.2. Then the adversary view given by the simulator from Figure 12.5 is indistinguishable from the adversary view given by the reveal protocol from Figure 12.3.*

12.2.4   Active Security

We shall only sketch a limited form of active security. Given the material we have already covered, the approach we sketch is the easiest approach. The idea is that the dealer publishes commitments to shares and reveals openings to the share owners. When the players create new shares, they must create new commitments and convince the other players that they are correct.

Let p be a large prime, and let G be a cyclic group of order p with a generators g, h. The Pedersen commitment scheme from Example 8.24 with commitment key (g, h) is group-homomorphic. This means that if we have commitments to two shares, the product of the commitments is a commitment to the sum of the shares under the sum of the openings. Similarily, we can recover commitments for any linear combination of shares.

The problem is multiplication of shares. The answer is to make a new commitment to the product and then use the multiplication argument from Section 11.4.3 to prove that the commitment can be opened to the product. Because zero knowledge is proven via simulation, it is very easy to simulate these multiplication arguments for commitments, so these multiplication proofs does not add a real burden when we want to simulate the protocol.

The multiparty computation protocols reveals shares of the result. Since each share now has an associated commitment, the natural thing would be to simply open the commitment. However, this causes problems when we want to simulate the computation. There are two easy options.

The first option is to cheat, so that we know $\log_g h$. Then we can open any commitment to any value. The second option is to cheat. We do not open the commitment, but instead prove that we know an opening of the commitment to the claimed share, using the argument from Section 11.4.1.

12.3   DISTRIBUTED DECRYPTION

Sometimes we need to decrypt ciphertexts, but we consider it too risky for a single party to have the decryption key. Distributed decryption allows us to mitigate this risk by sharing the decryption key among several parties and having them cooperate to compute the decryption. We also want robustness so that none of the parties can renege on their responsibilities.

The obvious approach is to use secret sharing and multi-party computation. We secret share the decryption key and express the decryption algorithm in terms of a algebraic circuit. However, we can do better.

The fundamental operation in many cryptosystems is exponentiation in a group G of prime order p. We have a secret exponent $b \in \{0, 1, 2, \ldots, p-1\}$ and want to compute x^b for some $x \in G$. We use a t-out-of-l Shamir secret sharing of b. For a set $\mathfrak{U}_0$ of at least t players and shares $f|_{\mathfrak{U}_0}$, we get

$$\sum_{P \in \mathfrak{U}_0} \rho_{P,\mathfrak{U}_0} f(P) \equiv b \pmod{p} \quad \text{and} \quad \prod_{P \in \mathfrak{U}_0} x^{\rho_{P,\mathfrak{U}_0} f(P)} = \prod_{P \in \mathfrak{U}_0} (x^{f(P)})^{\rho_{P,\mathfrak{U}_0}}.$$

If sufficiently many players compute $x^{f(P)}$, x^b can be reconstructed.

Definition 12.7. Let PKE $= (\mathcal{K}, \mathcal{E}, \mathcal{D})$ be a public key encryption scheme. A *distributed decryption* scheme for PKE and a set of players $\mathfrak{U}$ consists of three algorithms $(\mathcal{DK}, \mathcal{DD}, \mathcal{DR})$:

- The *key distribution* algorithm $\mathcal{DK}$ takes as input a decryption key dk and outputs a *decryption key sharing* d and a *public commitment* pc.

- The *distributed decryption* algorithm $\mathcal{DD}$ takes as input a player P, a decryption key share $d(P)$, associated data ad and a ciphertext c and outputs either a *decryption share* ms or the special symbol $\bot$.

- The *reconstruction* algorithm $\mathcal{DR}$ takes as input a public commitment pc, a set of players $\mathfrak{U}_0$, associated data ad, a ciphertext c and a set of decryption shares $\{(P, ms_P) \mid P \in \mathfrak{U}_0\}$. It outputs either a message m, or the special symbol $\bot$, or $\bot$ and $\mathfrak{U}_1 \subseteq \mathfrak{U}_0$.

We say that $(\mathcal{DK}, \mathcal{DD}, \mathcal{DR})$ is *correct for an access structure* $\mathfrak{A}$ if for any $\mathfrak{U}_0 \subseteq \mathfrak{U}$, any (ek, dk) output by $\mathcal{K}$, any decryption key sharing d and public commitment pc output by $\mathcal{DK}(dk)$, any associated data $ad \in \mathfrak{F}_{ek}$ and message $m \in \mathfrak{M}_{ek}$ and ciphertext c output by $\mathcal{E}(ek, ad, m)$, and any decryption shares ms_P output by $\mathcal{DD}(P, d(P), ad, c)$ for $P \in \mathfrak{U}_0$, we have that

$$\mathcal{DR}(pc, \mathfrak{U}_0, ad, c, \{(P, ms_P) \mid P \in \mathfrak{U}_0\}) = m.$$

Remark. A more general definition could be made by having $\mathcal{DK}$ replace the key generation algorithm $\mathcal{K}$. We do not discuss this further.

When the reconstruction algorithm outputs $(\bot, \mathfrak{U}_1)$, the intended interpretation is that decryption shares belonging to players in $\mathfrak{U}_1$ were invalid in some

The experiment for the integrity game proceeds as follows:

1. Compute $(ek, dk) \leftarrow \mathcal{K}$ and $(d, pc) \leftarrow \mathcal{DK}(dk)$. Send (ek, dk, d, pc) to $\mathcal{A}$.

2. When the adversary sends the query $(\mathfrak{U}_0, ad, c, \{(P, ms_P)\})$ (test), compute $m' \leftarrow \mathcal{DR}(pc, \mathfrak{U}_0, ad, c, \{(P, ms_P)\})$ and $m \leftarrow \mathcal{D}(dk, ad, c)$. If $m' \neq \perp$ and $m \neq m'$, send $\top$ to the adversary. Otherwise send $\perp$.

FIGURE 12.7 Integrity experiment for a distributed decryption scheme DDEC for a public key encryption scheme PKE with access structure $\mathfrak{A}$.

way. In many applications where a failure to decrypt has serious consequences, it is important to be able to find the responsible player.

Sometimes no single player is at fault, say if the ciphertext is invalid or if the set of players is not authorised. In this case, the reconstruction algorithm should only output $\perp$ and not blame any particular player.

Public key encryption schemes with distributed decryption are worthwhile to study on their own, and the combination is called a *threshold encryption* scheme. However, in our applications we find that it is easier to study the distributed decryption separately. Also, the security for threshold encryption are different than distributed decryption security.

Example 12.10 essentially describes distributed decryption for ElGamal, using Shamir's secret sharing. We shall expand on this example later.

E | *Exercise* 12.13. Write out the algorithms for the distributed decryption scheme for ElGamal from Example 12.10. What is the access structure? Show that the scheme is a distributed decryption scheme.

There are many ways to define security for distributed decryption, but there are two main properties we want: it should not ruin the confidentiality of the cryptosystem, and it should have some form of integrity.

Integrity   We try to give the adversary as much power as possible, but even for integrity we insist that the key material is honestly generated. This makes integrity easier to define, since the experiment will know the decryption key and therefore the correct decryption.

Remark. We could let the adversary generate the key material, but in order to know what the correct answer should be, the experiment would then have to generate the ciphertext, based on a message produced by the adversary. This is a different security notion. We shall not study if further.

D **Definition 12.8.** Let PKE be a public key encryption scheme and let DDEC be a distributed decryption scheme for PKE. An τ-*adversary against integrity* for DDEC is an algorithm that interacts with the experiment in Figure 12.7, and where the runtime of the adversary and the experiment is at most τ.

The *advantage* of $\mathcal{A}$ is

$$\mathbf{Adv}_{\mathrm{DDEC,PKE}}^{\mathrm{ddec\text{-}int}}(\mathcal{A}) = \Pr[E],$$

where E is the event that the experiment replied to a test query with $\top$.

E *Exercise* 12.14. Show that the distributed decryption from Example 12.10 and Exercise 12.13 does not have integrity. (That is, provide an adversary against integrity with trivial runtime and advantage 1.)

The natural way to provide integrity for the above distributed decryption scheme is to add some non-interactive argument that the decryption share has been computed correctly to the decryption share. The natural solution is to add a non-interactive argument for the equality of two discrete logarithms in the distributed decryption algorithm, for instance the argument from Example 11.4 made non-interactive using the construction from Figure 11.7. The reconstruction algorithm would then verify the arguments.

E *Example* 12.16. We extend Example 12.10 and Exercise 12.13 to a distributed decryption scheme for ElGamal with a t-out-of-l access structure, with the set of players being a subset of $\{1, 2, \ldots, p-1\}$. We use Shamir's secret sharing and a non-interactive argument of equality of discrete logarithms $(\mathcal{P}, \mathcal{V})$.

Let G be a group of prime order p with generator g.

- The key distribution algorithm $\mathcal{DK}$ takes b_0 as input, samples $b_1, \ldots, b_{t-1} \xleftarrow{r} \{0, 1, \ldots, p-1\}$ and computes the decryption key sharing $d(P) = \sum_i b_i P^i \bmod p$, and $y_P = g^{d(P)}$, for $P \in \mathfrak{U}$. It outputs d and the public commitment $\{(P, y_P)\}$.

- The distributed decryption algorithm $\mathcal{DD}$ takes as input a player P, a decryption key share $d(P)$ and a ciphertext $c = (x, w)$. It computes $x_P = x^{d(P)}$ and $arg_P \leftarrow \mathcal{P}(g, x, y_P, x_P, d(P))$, and outputs the decryption share (x_P, arg_P).

- The reconstruction algorithm $\mathcal{R}$ takes as input a public commitment $\{(P, y_P)\}$, a set of players $\mathfrak{U}_0$, a ciphertext $c = (x, w)$ and decryption shares $\{(P, x_P, arg_P) \mid P \in \mathfrak{U}_0\}$. If $\mathfrak{U}_0 \notin \mathfrak{A}$, it outputs $\bot$ and stops. If $\mathcal{V}(g, x, y_P, x_P, arg_P) = 0$ for any $P \in \mathfrak{U}_0$, output $\bot$ and the set of players for which verification failed. Otherwise, it computes and outputs

$$m = w \prod_{P \in \mathfrak{U}_0} x_P^{-\lambda_{P, \mathfrak{U}_0}}.$$

E| *Exercise* 12.15. Show that the above scheme has integrity.

Remark. When doing many arguments of essentially the same thing, such as when doing the distributed decryption of many ciphertexts, it is sometimes convenient to combine all the arguments into one *batched* argument, because this combined argument can sometimes be both proved and verified much faster. We shall not study batching further.

Confidentiality   Using distributed decryption should not compromise the confidentiality of the public key encryption scheme. We shall define this using simulation: given the decryption, we must simulate decryption shares.

Such a simulator must also simulate revealed decryption key shares and simulate consistent decryption shares. To keep complexity manageable, we go for non-adaptive corruption, where we define a set $\mathfrak{U}_1 \not\subseteq \mathfrak{A}$ and require that every reveal query is for a player in $\mathfrak{U}_1$.

D| **Definition 12.9.** A *non-adaptive simulator* for a distributed decryption scheme DDEC for a public key encryption scheme PKE is a pair of algorithms $(\mathcal{DS}_k, \mathcal{DS}_d)$. The algorithm $\mathcal{DS}_k$ takes as input of a set $\mathfrak{U}_1 \subseteq \mathfrak{U}$ and ek, and outputs $d|_{\mathfrak{U}_1}$ and pc. The algorithm $\mathcal{DS}_d$ takes as input $\mathfrak{U}_1$, $d|_{\mathfrak{U}_1}$, pc, associated data ad, a ciphertext c and a message m, and outputs decryption shares $\{(P, ms_P) \mid P \notin \mathfrak{U}_1\}$ such that

$$\mathcal{DR}(pc, \mathfrak{U} \setminus \mathfrak{U}_1, ad, c, \{(P, ms_P) \mid P \notin \mathfrak{U}_1\}) = m.$$

We shall also allow $\mathcal{DS}_d$ to output $\perp$, but if so the probability that it outputs $\perp$ must not be too large and must be independent of its input.

We say that DDEC is *non-adaptively simulatable* if there exists a non-adaptive simulator whose output distributions are indistinguishable from output distributions induced by the distributed decryption scheme algorithms.

We get computational, statistical and perfect variants of simulatable.

The simulatability approach gives us two interesting properties for a distributed decryption scheme. We can prove that the distributed decryption does not reduce security. Also, in certain situations we can use distributed decryption with cryptosystems that are not chosen-ciphertext secure.

First, we can modify the usual indistinguishability experiment for public key encryption to include distributed decryption. The public commitment is sent to the adversary along with the encryption key. The challenge query is unchanged. The chosen ciphertext query must be modified to indicate which decryption share to use, but will still refuse to process challenge ciphertexts. We also need a decryption key share reveal query.

Given a simulatable distributed decryption scheme, we can turn any adversary against the public key encryption scheme used with distributed decryption into an adversary against the simulator or against the public key encryption scheme. To see this, we change the game and use the simulator to generate the public commitment and the adversarial decryption key shares, instead of the key distribution algorithm. When the adversary makes a chosen ciphertext query, we use the decryption algorithm and the decryption key to decrypt the ciphertext, and then use the simulator with the now known decryption to simulate decryption shares. If the adversary's behaviour changes, we have a distinguisher for the simulator. If the behaviour does not change, we can build an adversary against the public key encryption scheme.

E │ *Exercise* 12.16. Write out the experiment for public key encryption with distributed decryption and use it to define the advantage of an adversary.

E │ *Exercise* 12.17. Bound the advantage of any adversary against public key encryption with distributed decryption using the advantages of a distinguisher for the simulator and an adversary against the public key encryption scheme.

Our ElGamal example would not be secure as a public key encryption scheme with distributed decryption under a chosen ciphertext attack, since ElGamal is not secure against a chosen ciphertext attack.

However, in certain cases we can use distributed decryption to good effect. The main example is when we (somehow) know what the decryption will be. In that event, we can replace the computation of the decryption shares with a simulation. This is an important step in many proofs, since we need to stop using the decryption key so that we can replace encryptions of messages by random encryptions. We shall continue with this in Chapter 14.

For simulating the distributed decryption from Example 12.16, we rely on the simulator for the non-interactive argument (this simulation will require reprogramming the random oracle) and the properties of Shamir's secret sharing, and in particular (12.3). If we have $t-1$ shares of the decryption key and the decryption key itself, we can recover any other share from this equation. Moreover, we can use this equation in the exponent, and then we do not even have to know the decryption key since we know the encryption key.

When we want to simulate the key distribution, we can sample $t-1$ shares and compute the remaining public commitments using the above strategy.

If the decryption of the ciphertext (x, w) under decryption key b_0 is m, then $x^{b_0} = wm^{-1}$. We recover $t-1$ decryption shares using the simulated decryption shares, and use our knowledge of x^{b_0} and the above approch to compute the other decryption shares. We then use the simulator for the non-interactive argument to simulate arguments for the simulated shares.

T │ **Proposition 12.11.** *Consider the non-adaptive simulator from Figure 12.8 for ElGamal with the distributed decryption scheme from Example 12.16. Suppose we use the argument from Exercise 11.4, made non-interactive using the construction from Figure 11.7 with the honest verifier simulator from Exercise 11.18. Then the non-adaptive simulator is perfect.*

Proof. The correct distribution of the public commitment and the decryption key sharing follows from the properties of Shamir's secret sharing.

The simulator $\mathcal{DS}_k$ takes as input $\mathfrak{U}_1 = \{P_1, \ldots, P_{t-1}\}$ and y:

1. It samples $b_1, \ldots, b_{t-1} \xleftarrow{r} \{0, 1, \ldots, p-1\}$ and computes $y_{P_i} \leftarrow g^{b_i}$ for $i = 1, 2, \ldots, t-1$, and

$$y_P = \left(y^{-1} \prod_{i=1}^{t-1} y_{P_i}^{\lambda_{P_i, \mathfrak{U}_P}} \right)^{-1/\lambda_{P, \mathfrak{U}_P}}$$

   for $P \notin \mathfrak{U}_1$, with $\mathfrak{U}_P = \mathfrak{U}_1 \cup \{P\}$.

2. It outputs the public commitment $\{(P, y_P)\}$ and the decryption sharing $d|_{\mathfrak{U}_1}$ given by $d|_{\mathfrak{U}_1}(P_i) = b_i$, $i = 1, 2, \ldots, t-1$.

The simulator $\mathcal{DS}_d$ takes as input $\mathfrak{U}_1$, $d|_{\mathfrak{U}_1}$, $\{(P, y_P)\}$, ad, $c = (x, w)$ and m:

1. It computes

$$ms_P = \left(mw^{-1} \prod_{i=1}^{t-1} x^{d|_{\mathfrak{U}_1}(P_i) \lambda_{P_i, \mathfrak{U}_P}} \right)^{-1/\lambda_{P, \mathfrak{U}_P}}$$

   for $P \notin \mathfrak{U}_1$, with $\mathfrak{U}_P = \mathfrak{U}_1 \cup \{P\}$.

2. It computes $arg_P \leftarrow \mathcal{S}_0(g, x, y_P, ms_P)$ for $P \notin \mathfrak{U}_1$.

3. It outputs decryption shares $\{(P, ms_P, arg_P) \mid P \notin \mathfrak{U}_1\}$.

FIGURE 12.8 Non-adaptive simulator $(\mathcal{DS}_k, \mathcal{DS}_d)$ for the distributed decryption scheme from Example 12.16 with an t-out-of-l access structure, using a non-interactive argument $(\mathcal{P}, \mathcal{V})$ with simulator $\mathcal{S}_0$.

The correct distribution of the decryption shares follows from the properties of Shamir's secret sharing and the perfect simulator for the argument. □

CHAPTER 13

Messaging Protocols

The original cryptographic problem is to have a secure conversation over an insecure channel, for reasonable values of secure, and we have briefly discussed the problem. We shall now formalise the problem and study solutions in detail.

The most extensive discussions were in Section 7.4 and Section 9.4. We build on these discussions and increase precision and the level of security we shall achieve. We shall also consider a greater variety of solutions.

We have previously distinguished between channels and messaging protocols, with the implicit understanding that channels are symmetric cryptosystems, while messaging protocols model multiple parties, though only two-party conversations. One goal for our work is to unify these treatments.

Unfortunately, there is a huge variety in messaging protocol design, and many of the variants have fairly subtle differences in security properties. Unlike the case for symmetric and public key encryption and digital signatures, it seems difficult to give an account of security notions for messaging protocols that is both exhaustive and of limited complexity. Unlike the case for key exchange, where we aimed for a somewhat comprehensive framework for security notions, we shall aim for a much simpler framework for messaging protocols, although there will be sufficient complexity left.

To guide our choices, we shall consider three reasonable classes of messaging protocols, for three distinct applications, and with concrete examples for each class. The first class is short-term, high-bandwidth, low-latency communications, such as establishing a short-term connection to a server. The second is long-term high-bandwidth, low-latency communications, such as establishing a long-term link between a remote sensor and a server. And the third is long-term, medium-bandwidth communications where latency is not crucial, such as text-based human-to-human messaging.

We shall make one crucial restriction on our messaging protocols. There should only be network traffic when establishing a conversation or when a message is sent, and each message sent should correspond to exactly one network message. This limits the types of security we can achieve. For instance, Alice cannot know if Bob has received a message until Bob responds with a message of his own. This choice is deliberate, with the assumption that applications can build such functionality and security on top of a messaging protocol.

13.1 MESSAGING PROTOCOLS

A messaging protocol must be able to deal with three processes: Establishing a state for a conversation, encrypting messages and decrypting messages. We shall use *stateful encryption*, something that we wanted to avoid in Chapters 7 and 8, but which now will be needed. Also, we must generate keys.

For convenience, the definition of messaging protocols comes in two parts: the algorithms and the correctness requirement, with some notation in between. Correctness essentially implies that if two parties running the protocol agree on the context for the conversation and what they have sent and received through the network, they will also agree on the messages sent and received.

Definition 13.1. A *basic messaging protocol* $\text{SM} = (\mathfrak{P}, \mathfrak{F}, \mathcal{K}, \mathcal{H}, \mathcal{E}_m, \mathcal{D}_m)$ consists of a set of plaintexts $\mathfrak{P}$, a set of associated data $\mathfrak{F}$, and four algorithms:

- The *key generation* algorithm $\mathcal{K}$ takes no input and outputs a public key pk and a secret key sk.

- The interactive *handshake* algorithm $\mathcal{H}$ takes as input a *role* $\rho \in \{0, 1\}$, associated data ad, a key pair (pk_ρ, sk_ρ) and a public key $pk_{1-\rho}$. It alternates between sending and receiving *signals*, initially sending if $\rho = 0$, otherwise initially receiving. Eventually, it either outputs a state st or the special symbol $\perp$ signifying failure.

- The *encryption* algorithm $\mathcal{E}_m$ takes as input a secret key sk, a state st, per-message associated data ad^m and a message $m \in \mathfrak{P}$, and outputs either the special symbol $\perp$, or a state st' and a *ciphertext* c.

- The *decryption* algorithm $\mathcal{D}_m$ takes as input a secret key sk, a state st, per-message associated data ad^m and c. It outputs either the special symbol $\perp$, or a state st' and a message $m \in \mathfrak{P}$.

The algorithms may also output $\perp$, and they will if given an input state $\perp$.

An *instance* of SM with *role* ρ, associated data ad, key pair (pk_ρ, sk_ρ) and $pk_{1-\rho}$ is described by signals $\hat{c}_1, \hat{c}_2, \ldots, \hat{c}_{\hat{j}}$, events $(\rho_i, ad_i^m, m_i, c_i)$, $i = 1, 2, \ldots, l$, and states $st_0, st_1, \ldots, st_l$, where

- $\mathcal{H}$ with input $(\rho, ad, pk_\rho, sk_\rho, pk_{1-\rho})$ sends/receives (receives/sends) the signals $\hat{c}_1, \hat{c}_2, \ldots, \hat{c}_{\hat{j}}$ and outputs the state st_0;

and for $i = 1, 2, \ldots, l$ we have either

- $\rho_i = \rho$ and $\mathcal{E}_m(sk, st_{i-1}, ad_i^m, m_i)$ output (st_i, c_i) (*encryption*); or
- $\rho_i = 1 - \rho$ and $\mathcal{D}_m(sk, st_{i-1}, ad_i^m, c_i)$ output (st_i, m_i) (*decryption*).

The public key of the instance is pk_ρ and $pk_{1-\rho}$ is its *partner key*. Two instances have *matching parameters* if they have the same associated data and public keys. Additionally, they have *matching key material* if either the public keys are distinct, or their public keys are both $\perp$ and the instances have identical secret keys. An instance is in its *handshake phase* until the handshake algorithm outputs a state, after which it is in its *messaging phase*.

Remark. As for key exchange schemes, we do not expect correctness if both instances use the same key pair. We shall explain the exception to this requirement when we discuss channels.

The *signalling transcript* tr_s is $(\hat{c}_1, \hat{c}_2, \ldots, \hat{c}_{\hat{j}})$. It must be easy to decide if a protocol has reached the messaging phase from its signalling transcript.

The *message transcript* tr_m is the list $((\rho_1, ad_1^m, m_1), \ldots, (\rho_l, ad_l^m, m_l))$. For $\rho' \in \{0, 1\}$ we define $tr_{m,\rho'}$ to be the list containing the tuples (ρ_i, ad_i^m, m_i) for which $\rho_i = \rho'$. In other words, $tr_{m,\rho}$ contains the messages encrypted by the role ρ, while $tr_{m,1-\rho}$ contains the messages decrypted by the role ρ.

The *ciphertext transcript* tr_c is the list $((\rho_1, ad_1^m, c_1), \ldots, (\rho_l, ad_l^m, c_l))$. For $\rho' \in \{0, 1\}$ we define $tr_{c,\rho'}$ to be the list containing the tuples (ρ_i, ad_i^m, c_i) for which $\rho_i = \rho'$. In other words, $tr_{c,\rho}$ contains the ciphertexts encrypted by the role ρ, while $tr_{c,1-\rho}$ contains the ciphertexts decrypted by the role ρ.

The *network transcript* tr_n is the pair (tr_s, tr_c).

We now have the concepts required to define correctness. The idea is that if two instances agree on the parameters and key material and which signals and ciphertexts were sent by which role with which associated data, then they should also agree on which messages were sent by which role. Note that as discussed we have only have a requirement on the internal order of messages sent by each role, not on the order of sent and received messages.

Definition 13.2. A basic messaging protocol SM is a *messaging protocol* if for any two instances with matching parameters, matching key material and transcripts (tr_m^0, tr_s^0, tr_c^0) and (tr_m^1, tr_s^1, tr_c^1) with $tr_s^0 = tr_s^1$, $tr_{c,0}^0 = tr_{c,0}^1$ and $tr_{c,1}^0 = tr_{c,1}^1$, we have that $tr_{m,0}^0 = tr_{m,0}^1$ and $tr_{m,1}^0 = tr_{m,1}^1$.

Our definition is intended as a cryptographic mechanism based on public keys. Section 7.4 introduced the concept of channels, which is very similar to messaging protocols. Since we are going to develop a somewhat substantial theory surrounding this definition, it would make sense if this theory could also apply to channels. And it turns out that it can, with fairly modest adaptions.

Definition 13.3. We say that SC $= (\mathfrak{P}, \mathfrak{F}, \mathfrak{K}_s, \mathcal{H}_c, \mathcal{E}_c, \mathcal{D}_c)$ is a *channel* protocol with key set $\mathfrak{K}_s$ if $(\mathfrak{P}, \mathfrak{F}, \mathcal{K}, \mathcal{H}_c, \mathcal{E}_c, \mathcal{D}_c)$ is a messaging protocol when $\mathcal{K}$ is the algorithm that samples its output from the uniform distribution on the set $\{(\perp, k) \mid k \in \mathfrak{K}_s\}$. SC is a *simple channel* if the signalling algorithm $\mathcal{H}_c$ sends and receives no signals.

Remark. This definition explains why the notion of matching key material has its one exception to distinct public keys. It also explains why we did not provide a formal definition of channel in Section 7.4.

As a matter of notation, we usually omit the special symbols $\perp$ marking the positions of the public keys in the invocation of $\mathcal{H}_c$.

We spent a significant amount of time discussing *closing* in Section 7.4. This is well-understood, but quite technical. Since closing adds significantly to the complexity and can easily be achieved by the application, we shall not discuss closing for messaging protocols.

A feature that is very similar to closing is *synchronisation*, which ensures that both parties have received everything that has been sent so far. This suggests that a more general signalling functionality could be useful for a messaging protocol. Another useful feature is *multiplexing*, where a single instance of a messaging protocol can pretend to be multiple instances. In fact, we could probably add an endless list of features to a messaging protocol.

The advantage of integrating such features into the messaging protocol is that it sometimes allows useful optimisations. For the general case, it probably saves on overall complexity if we build a protocol with more functionality on top of a messaging protocol. Such a protocol would be easier to analyse if the messaging protocol abstracts away the cryptographic details. We shall explore this general approach in further in Chapter 14.

We have included the notion of per-message associated data, which makes it easier to design such protocols on top of a messaging protocol. Note that as usual, we shall not require that per-message associated data is kept confidential, which has the usual security implications.

E *Exercise* 13.1. Propose techniques for achieving closing, synchronisation and multiplexing based on a messaging protocol. You may require the messaging protocol to support associated data sets of your choice.

One further thing that is remarkably useful, but which we have not modelled, is a *session identifier*. Being able to refer uniquely to a particular messaging protocol instance and its partner instance is actually of great value when building protocols on top of messaging protocols.

One particular use case is where some authentication or identification system external to the messaging protocol is used to authenticate something happening inside the messaging protocol, or identify either of the partners. In these cases, you sometimes get man-in-the-middle attacks if the authentication or identification is not tied to the particular messaging protocol session involved. Having a session identifier makes this fairly easy to defend against.

Another use of session identifiers is to get a compositional results for a messaging protocols similar to Theorem 7.8. This class of results are less useful than the previous result, and we shall not develop that strategy.

Adding a session identifier to a messaging protocol is actually somewhat non-trivial, but it is usually possible to do something that is sufficiently good for the application. In the simplest cases, randomness could be included in the associated data and hashed together with identifying information. Slightly more involved, the random nonces could be exchanged inside the messaging protocol and hashed together with the instance's belief about its partner. Or one could even include partial ciphertext transcripts in the hash.

There are two main reasons for us not to include session identifiers in the definition of messaging protocol. Most applications have sufficiently good alternatives, similar to the techniques used in Example 13.7. Also, modelling session identifiers would significantly complicate the security definition. We include this remark because this issue has real-world effects.

Conversation Structure We will need a more careful discussion of the structure of conversations in messaging protocols, since the network is asynchronous. Since there is a one-to-one correspondence between messages and ciphertexts, we shall instead define the structure on the ciphertexts.

Every ciphertext transcript tr_c induces a total order on the set of ciphertexts, namely the order in which the party saw the ciphertexts. This is the *temporal order*, which we denote by $\leq$. At any point in the messaging phase, the transcript also defines a *causal order* on the ciphertexts, which is a partial order on the set of ciphertexts in the transcript. The causal order must be consistent with both with the temporal order defined by the ciphertext transcript and the instance's belief about its partner instance's temporal order.

We denote the causal order by $\preceq_{tr_c}$. When the state and the transcript are clear from context, we shall just write $\preceq$.

The *minimal* causal order we shall consider is the union of the temporal order restricted to the ciphertexts of $tr_{c,0}$ and the temporal order restricted to the ciphertexts of $tr_{c,1}$. We could have modelled a less restrictive minimal causal order, for example sc-2 or sc-3 from Section 7.4, but we will not.

Example 13.1. Alice sends a ciphertext c_1 and receives two ciphertexts c_0, c_2, so that $c_1 \leq c_0 \leq c_2$. Alice did not see c_0 before she sent c_1. The causal order on tr_c cannot have c_0 preceding c_1, so $c_0 \not\preceq c_1$. It must have $c_0 \preceq c_2$.

If Bob sent c_0 before receiving c_1 and c_2 afterwards, then the messaging protocol may inform Alice of this fact, in which case $c_1 \preceq c_2$. The causal order must have $c_1 \not\preceq c_0$ in this case, leaving them uncomparable.

Some causal orders can be used to split the ciphertexts into *segments*. Given a transcript and a causal order on the transcript, two ciphertexts $c \preceq c'$ are in the same segment if they both originated with the same role and any c'' from the other role causally preceding c' also causally precedes c.

The segments naturally alternate between encryption and decryption, and we can consider successive segments as happening in parallel, in some sense.

For the minimal causal order, all the ciphertexts originating with a role belong to the same segment, so segments are not interesting. We shall study a causal order for which segments are more interesting later.

In principle, it could be difficult to determine the causal order based on a transcript. We shall only consider *public* causal orders, where it is easy to decide the causal order from the ciphertext transcript.

13.1.1 Examples

We shall consider four examples. The first two examples are channel protocols. We studied a variant of the first example in Section 7.4. The second example uses *session key evolution* to achieve stronger security in some contexts.

The third example is a natural combination of public key encryption and digital signatures, which we looked at in Section 9.4. It is our first messaging protocol. In some sense, this example is very similar to the first example.

The fourth example is another natural combination of a key exchange scheme and a channel protocol. Depending on which channel protocol we use, the composition is either suitable for short-lived or longer-lived instances.

The third and fourth examples use *compositional* constructions, where simpler constructions are composed to form new constructions. As previously mentioned, composition seems very natural, but can be quite subtle. Security properties do not compose nicely in general, but sometimes they do.

Channel As we saw in Section 7.4 there are a number of ways to realise a channel protocol. The following example is probably the simplest possible method. The channel protocol simply encrypts any message with the symmetric cryptosystem, encoding the sender and its order in the associated data. The recipient encodes its belief about the sender and the sender's order in the associated data it uses. It does not achieve any kind of security if more than one instance uses the same role, the same key and the same associated data.

Example 13.2. Let $\Sigma = (\mathfrak{K}_s, \mathfrak{P}, \mathfrak{F}_0, \mathfrak{C}, \mathcal{E}_s, \mathcal{D}_s)$ be a symmetric cryptosystem, with $\{0,1\} \times \mathbb{Z} \times \mathfrak{F} \times \mathfrak{F} \subseteq \mathfrak{F}_0$. The channel is SC-SYM $= (\mathfrak{P}, \mathfrak{F}, \mathfrak{K}_s, \mathcal{H}, \mathcal{E}_m, \mathcal{D}_m)$, where the algorithms all take the key k and state $(\rho, ad, j, j') \in \{0,1\} \times \mathfrak{F} \times \mathbb{Z} \times \mathbb{Z}$ as input, and work as follows:

- The *encryption* algorithm $\mathcal{E}_m$ also takes ad^m and m as input, computes $c \leftarrow \mathcal{E}_s(k, (\rho, j, ad, ad^m), m)$ and outputs $(\rho, ad, j+1, j')$ and c.

- The *decryption* algorithm $\mathcal{D}_m$ also takes ad^m and c as input and computes $m \leftarrow \mathcal{D}_s(k, (\rho, j', ad, ad^m), c)$. If $m = \bot$, the algorithm outputs $\bot$. Otherwise, it outputs $(\rho, ad, j, j'+1)$ and m.

This is a simple channel and $\mathcal{H}(\rho, ad, k)$ outputs the state $(\rho, ad, 0, 0)$.

This causal order of this channel is the minimal causal order, since the two directions are treated independently.

This channel protocol encodes the associated data in the ciphertexts, so many instances may share a key if they all have distinct associated data.

It is sometimes desirable to have a channel that can be used even if instances do not have distinct associated data. It is easy to use the handshake algorithm to establish a shared random value. The initiator sends a random value no_0, the responder responds with a random value no_1 and an encryption of some arbitrary message with the nonces and role included in the associated data. The initiator then responds with a similar encryption.

The same approach could also be used by the parties outside of the channel protocol. They exchange the nonces before they start the channel protocol. The only issue is that a party would not know if the partner actually exists until they receive a message from the partner. That is often not a problem, but some applications need to be careful about this.

As we shall see in the next example, some applications of such channel protocols ensure that the key is only ever used once by each party. In that case there is no need to ensure unique associated data.

Other applications ensure that each key is used at most once for each role, which again obviates the need for unique associated data for the channel. Also, we could get by with weaker encryption schemes. Using an AEAD scheme we could skip the random nonce. Or we could build an optimised version based on counter mode. We return to these ideas in Exercise 13.9.

Evolving Session Keys   The goal of this channel protocol is to evolve the session key, so that if the session key is at some point revealed, it does not reveal the decryption of earlier ciphertexts. When composed with a key exchange protocol as in Example 13.5 it is suitable for longer-term instances.

Example 13.3. Let $\Sigma = (\mathfrak{K}_s, \mathfrak{P}, \mathfrak{F}, \mathfrak{C}, \mathcal{E}_s, \mathcal{D}_s)$ be a symmetric cryptosystem, and let $kdf : \mathfrak{K}_s \to \mathfrak{K}_s^2$ be a key derivation function. The simple channel protocol SC-CHN = $(\mathfrak{P}, \mathfrak{F}, \mathfrak{K}_s^2, \mathcal{H}_c, \mathcal{E}_c, \mathcal{D}_c)$ works as follows:

- The *establishing* algorithm $\mathcal{H}_c$ takes as input $(\rho, ad, (kd, kd'))$ and outputs the state (ρ, ad, kd, kd').

- The *encryption* algorithm $\mathcal{E}_m$ takes as input sk, $st = (\rho, ad, kd, kd')$, ad^m and a message m. It computes $(kd'', k) \leftarrow kdf(kd)$ and $c \leftarrow \mathcal{E}_s(k, (\rho, ad, ad^m), m)$. Then it outputs (ρ, ad, kd'', kd') and c.

- The *decryption* algorithm $\mathcal{D}_m$ takes as input sk, $st = (\rho, ad, kd, kd')$, ad^m and a ciphertext c. It computes $(kd'', k) \leftarrow kdf(kd')$ and $m \leftarrow \mathcal{D}_s(k, (\rho, ad, ad^m), c)$. If $m = \bot$, it outputs $\bot$. Otherwise it outputs (ρ, kd, kd'') and m.

Again, the causal order is the minimal one.

Note that we have to be careful about the difference between the evolving session key in the state and the fixed long-term key. Revealing the long-term key will reveal every evolved session key. This is worrying, since revealing the long-term key is typically a more probable attack than revealing the state.

However, this channel protocol is not intended to be used as a stand-alone protocol. It is intended that a key exchange scheme is used to create the channel's long-term key, which is *erased* after the channel has been established.

Encryption and Signatures   Encryption and signatures combine in a natural way to build a messaging protocol. This protocol was essentially discussed in Section 9.4, but we now have a proper framework for its study.

Example 13.4. Let PKE $= (\mathcal{K}, \mathcal{E}, \mathcal{D})$ be a public key encryption scheme with plaintext space $\mathfrak{P}$, and let SIG $= (\mathcal{K}, \mathcal{S}, \mathcal{V})$ be a signature scheme. The simple messaging protocol SM-ENC-SIGN $= (\mathfrak{P}, \mathfrak{F}, \mathcal{K}, \mathcal{H}, \mathcal{E}_m, \mathcal{D}_m)$ works as follows:

- The *key generation* algorithm $\mathcal{K}$ computes $(ek, dk) \leftarrow \mathcal{K}$ and $(vk, sk) \leftarrow \mathcal{K}$, and outputs $pk = (ek, vk)$ and $sk = (dk, sk)$.

- The *establishing* algorithm $\mathcal{H}$ takes as input (ρ, ad, pk, pk') and outputs the state $st = (\rho, ad, pk, pk', 0, 0)$.

- The *encryption* algorithm $\mathcal{E}_m$ takes as input $sk = (dk, sk)$, $st = (\rho, ad, pk = (ek, vk), pk' = (ek', vk'), j, j)$, ad^m and m. It computes $c_0 \leftarrow \mathcal{E}(ek', (\rho, j, pk, pk', ad, ad^m), m)$, $\sigma_0 \leftarrow \mathcal{S}(sk, c_0)$, and outputs the state $(\rho, ad, pk, pk', j + 1, j')$ and $c = (c_0, \sigma_0)$.

- The *decryption* algorithm $\mathcal{D}_m$ takes as input $sk = (dk, sk)$, $st = (\rho, ad, pk = (ek, vk), pk' = (ek', vk'), j, j')$, ad^m and $c = (c_0, \sigma_0)$. It verifies that $\mathcal{V}(vk', c_0, \sigma_0) = 1$ and decrypts $m \leftarrow \mathcal{D}(dk, (1 - \rho, j', pk', pk, ad, ad^m), c_0)$. If either operation fails, it outputs $\perp$. Otherwise, it outputs the state $(\rho, ad, pk, pk', j, j' + 1)$ and the message m.

Again, the causal order is the minimal one.

This protocol is in general insecure if two instances have the same key material and the same associated data. As for Example 13.2, it is easy to use the establishing algorithm to agree on a shared random value.

Remark. It is interesting that the protocol's state only contains public values.

We have used the so-called encrypt-then-sign paradigm (compare with Section 1.4). It would also be possible to sign messages directly, and then encrypt both the message and the signature, the sign-then-encrypt paradigm. Since the signature may leak arbitrary information about the signed message, it is in general unsafe not to encrypt the signature, the so-called encrypt-and-sign paradigm. It is possible to have spirited arguments about which approach is best, a task we leave to the interested reader.

Remark. Public key encryption and digital signatures can be combined into something called *signcryption*. The direct composition as used in the above protocol is one way to do it, but there are other methods which can be more efficient. A signcryption scheme could be used in the above messaging protocol.

A Natural Composition The next example is a natural composition of cryptographic techniques, which already appears in Chapter 2. A key exchange scheme establishes a shared session key that is used by a channel protocol.

Example 13.5. This is a straight-forward composition of a key exchange protocol and a channel protocol. The messaging protocol begins with an establishing phase in which it runs the key exchange scheme until a session key is established. Then it enters a messaging phase in which the channel protocol is used with the session key to send the actual plaintexts.

Let KEX $= (\mathfrak{K}_s, \mathcal{K}, \mathcal{I}, \mathcal{R})$ be a key exchange protocol, and let SC $= (\mathfrak{P}, \{\perp\}, \mathfrak{K}_s, \mathcal{H}_c, \mathcal{E}_c, \mathcal{D}_c)$ be a simple channel. The messaging protocol is SM-KEX-SC $= (\mathfrak{P}, \mathfrak{F}, \mathcal{K}, \mathcal{H}, \mathcal{E}_m, \mathcal{D}_m)$, where the algorithms do:

- The *establishing* algorithm $\mathcal{H}$ on input of $(\rho, ad, (pk_\rho, sk), pk_{1-\rho})$ runs an instance of $\mathcal{I}(ad, pk_0, sk, pk_1)$ if $\rho = 0$, or $\mathcal{R}(ad, pk_1, sk, pk_0)$ if $\rho = 1$. When this instance sends a message, $\mathcal{H}$ sends the same message. When $\mathcal{H}$ receives a message, it sends the message to the instance. If the instance ever outputs $\perp$, $\mathcal{H}$ outputs $\perp$. When the instance outputs a session key k, $\mathcal{H}$ computes $st_0 \leftarrow \mathcal{H}_c(\rho, \perp, k)$ and outputs the state (k, st_0).

- The *encryption* algorithm $\mathcal{E}_m$ on input of sk, $st = (k, st_0)$, ad^m and a message m, computes $(st'_0, c) \leftarrow \mathcal{E}_c(k, st_0, ad^m, m)$. If the encryption fails, it outputs $\perp$. Otherwise it outputs (k, st'_0) and c.

- The *decryption* algorithm $\mathcal{D}_m$ on input of sk, $st = (k, st_0)$, ad^m and a ciphertext c, computes $(st'_0, m) \leftarrow \mathcal{D}_c(k, st_0, ad^m, c)$. If the decryption fails, it outputs $\perp$. Otherwise it outputs (k, st'_0) and m.

If the algorithms receive any input other than specified above, they output $\perp$. The causal order is inherited by the channel's causal order.

13.2 DEFINING SECURITY

We developed several integrity notions for channels in Section 7.4, suitable for a variety of underlying networks. These notions could all be suitable for messaging protocols, but since we want to add additional complications in the form of long-term key reveals and state reveals, combinatorial explosion would make general definitions hard to read. We therefore forego generality, and guarantee that plaintexts are received in the order they were sent, sc-4.

With respect to confidentiality, we shall model this in terms of a real-or-random game. We will model state reveals and long-term key reveals, as well as chosen ciphertexts, so managing trivially revealed messages is complex, and we will need some variants.

13.2.1 The Definition

The Experiment We shall model the security game in the usual fashion, as an interaction between an adversary and the experiment given in Figure 13.1. The experiment models honest users and the adversary's interaction with them. It does this by generating keys, starting instances of the messaging protocol with specified key pairs and partner keys and allowing the adversary to interact with these instances. It also allows the adversary to reveal various information about the instances.

We associate a *game transcript* to a game between the experiment and an adversary, which consists of a list of queries sent to the experiment and the corresponding responses. Note that all of the information encoded in the game transcript is known by the adversary. Also, the network transcript of every instance can be recovered from the game transcript, but in general the message transcript cannot be recovered.

The experiment models only honest instances, which will only run with a key pair generated through a key generation query. These are the *honestly generated* keys. The adversary is free to choose the partner public key, either an honestly generated public key or an adversary-generated key.

The adversary controls the network, modelled in the usual fashion, giving the ciphertexts and signals produced by the instances to the adversary, and letting the adversary give arbitrary ciphertexts and signals to the instances.

An adversary is *non-invasive* if it does no state reveal queries. It is *non-adaptive* if it does no state reveal queries or long-term key reveal queries. We could have defined a network-passive adversary, but this seems to be less interesting for messaging protocol. (Chapter 10 developed a theory for unauthenticated key exchange since this was a useful building block for stronger key exchange constructions. The analogous simpler messaging protocols are not the most useful building blocks for strong messaging protocol designs.)

An adversary is *key-respecting* if it does at most one execute query for each key-role pair. An adversary is *nonce-respecting* if it does at most one execute query for each key-role-associated data tuple.

Note that our modelling of the messaging protocol is different from how we model key exchange schemes. The former is modelled as a collection of algorithms producing or manipulating a state, while the latter is modelled as an interactive algorithm that takes input, produces output, and sends and receives messages. We could have used either modelling approach for both key exchange schemes and messaging protocols. The reason we use different modelling is because of the purpose of our state reveal queries.

For most key exchange protocols, revealing the random tape is the correct choice when we want to reveal the state. Other choices are possible and make sense in special cases or when we want to model different attacks.

But the design of many messaging protocols is predicated exactly on being able to erase the parts of the random tape and using that to achieve various

The experiment for a messaging protocol SM interacting with an adversary $\mathcal{A}$ proceeds as follows:

1. Sample $b, b'' \xleftarrow{r} \{0, 1\}$.

2. When the adversary sends the jth query $\perp$ (*key generation*), do:

   (a) Compute $(pk, sk) \leftarrow \mathcal{K}$, record (j, pk, sk) and send pk to $\mathcal{A}$.

3. When the adversary sends the ith query (ρ, j, pk', ad) (*execute*), do:

   (a) If (j, pk, sk) is not recorded, send $\perp$ to $\mathcal{A}$ and stop.

   (b) Start the ith establishing instance as $\mathcal{H}(\rho, ad, (pk, sk), pk')$.

   (c) If the instance sends a message $\hat{c}$, send $(i, \hat{c})$ to $\mathcal{A}$. If the instance outputs a state st, record $(i, 0, sk, st)$ and send $(i, \top)$ to $\mathcal{A}$. If the instance outputs $\perp$, record $(i, -1, \perp, \perp)$ and send $(i, \perp)$ to $\mathcal{A}$.

4. When the adversary sends $(i, \hat{c})$ (*chosen signalling*), do:

   (a) If $(i, \cdot, \cdot, \cdot)$ is recorded, send $\perp$ to $\mathcal{A}$ and stop.

   (b) Otherwise, send $\hat{c}$ to the ith instance and proceed as in Step 3c.

5. When the adversary sends the query (i, m) / (i, c) / (i, m_0) (*chosen plaintext / chosen ciphertext / challenge*), do:

   (a) If (i, ν, sk, st) is not recorded, send $\perp$ to $\mathcal{A}$.

   (b^{cc}) Compute $(st', m) \leftarrow \mathcal{D}_m(sk, st, c)$, record (i, ν, m) and send $\top$.

   (b^{cp}) Compute $(st', c) \leftarrow \mathcal{E}_m(sk, st, m)$ and send c to $\mathcal{A}$.

   (b^{ror}) Sample $m_1 \xleftarrow{r} \{m \in \mathfrak{P} \mid |m| = |m_0|\}$, compute $(st', c) \leftarrow \mathcal{E}_m(sk, st, m_b)$ and send c to $\mathcal{A}$.

   (c) If the computation failed, update the record to $(i, -1, sk, \perp)$ and send $\perp$ to $\mathcal{A}$. Otherwise update the record to $(i, \nu + 1, sk, st')$.

6. When the adversary sends the query (msg, i, ν) (*message reveal*), do:

   (a) If (i, ν, m) is recorded, send m to $\mathcal{A}$. Otherwise, send $\perp$.

7. When the adversary sends the query (state, i) (*state reveal*), do:

   (a) If $(i, \cdot, \cdot, st)$ is recorded, send st to $\mathcal{A}$. Otherwise, send the random tape of the ith $\mathcal{H}$ instance to $\mathcal{A}$.

8. When the adversary sends the query (ltk, j) (*long-term key reveal*), do:

   (a) If (j, pk, sk) is recorded, send sk to $\mathcal{A}$. Otherwise, send $\perp$.

FIGURE 13.1 Experiment $\mathbf{Exp}_{SM}(\mathcal{A})$ for a messaging protocol SM. The bit b'' is not used in the experiment but is used to simplify the calculation of advantage.

security goals. While one may quibble with this assumption, we cannot reasonably model such protocols with the approach used for key exchange.

One interesting effect of this modelling is that unlike for key exchange where an adversary can reveal the state even of instances that have output session keys, an adversary against a messaging protocol cannot "travel back in time". The adversary is only given access to an instance's current state, not its history. Nor does an instance's current state determine its future actions.

This choice prevents us from modelling some attacks, such as leakage from a faulty pseudo-random generator revealing previous pseudo-randomness.

Partnering Just as for key exchange, we need to decide which instances are sending messages to each other. There is no explicit correspondence between instances in the experiment because the adversary controls the network. As usual, we shall infer partners based on network transcripts. While we could define partnering notions in general, like we did for key exchange protocols, we shall avoid the added complexity of a general definition and instead define everything in terms of a relaxed version of matching conversations.

The idea is that the signals and ciphertexts that Alice send should be the signals and ciphertexts that Bob receives, and vice versa. The network is asynchronous, so it will under normal operations deliver network messages in the order they were sent, but only for messages in the same direction. The network cannot guarantee delivery, so we need to allow for the fact that some network messages may not have been delivered. We shall therefore compare ciphertexts in each direction separately. Signals, however, should match exactly.

Definition 13.4. Consider two instances with matching parameters, key material and distinct roles, where the instance with role ρ has network transcript $tr_n^\rho = (tr_s^\rho, tr_c^\rho)$ and causal order $\preceq_\rho$. The two instances are *partnered* if

- $tr_s^0 = tr_s^1$, $tr_{c,1}^0$ is a prefix of $tr_{c,1}^1$, $tr_{c,0}^1$ is a prefix of $tr_{c,0}^0$ and $\preceq_0$ and $\preceq_1$ agree on the ciphertexts in the prefixes; or

- for some ρ, $tr_s^{1-\rho}$ is a prefix of tr_s^ρ, which contains one more signal, and $tr_c^{1-\rho}$ and $tr_{c,1-\rho}^\rho$ are empty.

We say that the two instances are *partially partnered* if they are partnered; or

- $tr_s^0 = tr_s^1$ and both instances enter the messaging phase; or

- for some ρ, the ρ instance entered the messaging phase, sent the last signal, and $tr_s^{1-\rho}$ contains exactly the same signals as tr_s^ρ, except for the final signal, and $tr_c^{1-\rho}$ is empty.

This partnering notion ensures that the two instances agree on their signal transcripts, but either party may not (yet) have received the last few ciphertexts or the last signal sent by the other party. The two instances will also agree on the associated data, per-message associated data and the public keys involved, or the symmetric key in the channel case.

Since network transcripts are known by the adversary and causal orders are public, the adversary is always able to deduce the partnering of instances.

Partially partnered instances are intended to model a case where the parties start a conversation, but the adversary then does a state reveal on one or both instances and possibly corresponding long-term key reveals, and essentially impersonating one of the instances.

Integrity   The intuitive notion of integrity in the context of messaging protocols is that if Alice believes she has had a conversation with Bob, then she has had a conversation with Bob, and Bob agrees with her about their conversation and the causal order, possibly up to some messages that have not been received (yet). Our notion of integrity overlaps with authentication.

As usual, the intuitive notion is a bit too simple. Revealing both Alice' state and her long-term key must be sufficient to impersonate Alice to Bob and may also be sufficient to impersonate Bob to Alice. Revealing Bob's long-term key is trivially sufficient to impersonate Bob but perhaps not sufficient to impersonate Bob in an already ongoing conversation. An adversary that reveals Bob's state may learn enough to impersonate Bob, and it may even be sufficient to impersonate Alice to Bob. Again, we limit ourselves to four cases to avoid combinatorial explosion.

Definition 13.5. Consider an execution. A public key is *impersonatable* if it is not honestly generated or the long-term key has been revealed. We say that an instance is *viable* if its partner key was not impersonatable before the instance entered its messaging phase. An instance is *weakly impersonatable* if there is a state reveal for the instance after it entered its messaging phase. An instance is *impersonatable* if there is a long-term key reveal for its public key and a state reveal for the instance after it entered its messaging phase.

Integrity holds for an instance if it is not viable, or there is a unique partnered instance, or there is a unique partially partnered instance that is impersonatable. *Weak integrity* holds for the instance if integrity holds, or the partner key is impersonatable. *Implicit integrity* holds for the instance if integrity holds, or there is a unique partially partnered instance and either instance is impersonatable. *Weak implicit integrity* holds for the instance if integrity holds, or there is a unique partially partnered instance and either instance is weakly impersonatable.

Integrity (weak integrity, implicit integrity, weak implicit integrity) holds for the execution if it holds for every instance in the execution that has reached its messaging phase.

The game transcripts contain state and long-term key reveal queries, so network transcripts can be recreated from game transcripts. The game transcript is enough to decide whether the integrity notions holds.

Remark. We have that weak implicit integrity holds if implicit integrity holds, and implicit integrity and weak integrity hold if integrity holds. Against non-invasive adversaries no instance will be impersonatable or weakly impersonatable, so integrity, implicit integrity and weak implicit integrity coincide. Against non-adaptive adversaries no instance or public key will be impersonatable, so integrity, weak integrity and implicit integrity coincide.

We define integrity relative to partnering, which in turn is relative to the protocol's claimed causal order. This causal order must therefore be included in any comparison of security levels. Defining causal order generically requires significantly more complexity. Since we mostly consider the minimal causal order, the added complexity would not be worth it.

Freshness   When Alice has established a conversation with Bob, our intuitive notion says that the adversary should no longer be able to learn anything about their conversation, except possibly the length of their messages.

As for integrity, we need to expand the intuitive definition to capture all the queries an adversary can make. An adversary that reveals a chosen ciphertext query corresponding to a challenge query will trivially reveal the challenge bit. An adversary that reveals both the state and the long-term key of an instance just when it could decrypt a challenge ciphertext will also trivially be able to reveal the challenge bit. Sometimes, it is sufficient with just the state or just the long-term key. Sometimes, it is sufficient to reveal the state after the challenge query or before the challenge query. Sometimes, the state and long-term key of the instance is sufficient, while in other cases it is only the other party's state and long-term key that will work. Again, we limit ourselves to four cases to avoid combinatorial explotion.

We begin by defining trivial reveals, either directly through a revealed chosen ciphertext query or through revealing states or long-term keys.

Definition 13.6. Consider two partially partnered instances with network transcripts from some execution. The causal order is defined on the matching parts of the ciphertext transcripts, and we extend it to a partial order on the entire ciphertext transcripts by making the non-matching parts of the two transcripts non-comparable.

A state reveal query is *relevant* for a challenge query it is for the same instance and after the instance entered the messaging phase, and *partner-relevant* if it is for the partner instance and after the partner instance entered the messaging phase. A state reveal query *precedes* a challenge query if it is relevant and before the challenge query, or if it is partner-relevant and is not preceded by a chosen ciphertext query matching the challenge query. A state reveal query *immediately precedes* a challenge query if it precedes the challenge query and is subsequent to the start of the segment, or if it is partner-relevant, subsequent to the start of a matching segment and not preceded by a chosen ciphertext query matching the challenge query. A state reveal query is *adjacent* to a challenge query if it is in the same segment as the challenge query. A state reveal query is *partner-adjacent* to a challenge query if it is partner-relevant and in a segment matching the segment preceding the challenge query.

A challenge query is *directly revealed* if there is a matching chosen ciphertext query, and it has been revealed. A challenge query is *statically revealed* if it is directly revealed or the partner long-term key has been revealed. A challenge query is *weakly revealed* if it is direcly revealed or there is a state reveal query for either instance after entering the messaging phase. A challenge query is *implicitly revealed* if it is directly revealed, or there is a state reveal query that precedes the challenge query. A challenge query is *strongly revealed* if it is directly revealed, or there is a state reveal query that immediately precedes the challenge query, or the instance long-term key is revealed and there is an adjacent state reveal query, or the partner long-term key is revealed and there is a partner-adjacent state reveal query.

Now that we have defined how an adversary may have trivially revealed the contents of a challenge query, we are ready to say when the adversary's guess should count towards the advantage.

Definition 13.7. Consider an instance in an execution and a challenge query for that instance. The challenge query is *fresh* (*statically fresh, weakly fresh, implicitly fresh, strongly fresh*) if there is a partially partnered instance for which the challenge query is not directly (statically, weakly, implicitly, strongly) revealed, or the instance is viable and there is no partially partnered instance. We say that an execution is fresh (statically fresh, weakly fresh, implicitly fresh, strongly fresh) if every challenge query in the execution is fresh (statically fresh, weakly fresh, implicitly fresh, strongly fresh).

Remark. Weakly fresh and strongly fresh are variants of implicit fresh, where reveals for either partner may be sufficient to reveal a challenge query. This is unlike statically fresh, where the partner long-term key must be revealed.

The Definition   The security definition is in many ways a standard real-or-random definition. It combines the experiment from Figure 13.1 with partnering, an integrity notion and a freshness notion. As usual, we could mix and match, but we shall only consider five different combinations: integrity and fresh (ror), integrity and statically fresh (ror-s-f), implicit integrity and strongly fresh (ror-ii-sf), implicit integrity and implicitly fresh (ror-ii-if), and weak implicit integrity and weakly fresh (ror-wii-wf).

Definition 13.8. A $(\tau, l_c, l_u, l_s, l_e)$-*adversary* against a messaging protocol SM is an interactive algorithm $\mathcal{A}$ that interacts with the experiment in Figure 13.1 making at most l_c challenge queries, l_u key generation queries, l_s execute queries and l_e encryption queries, and where the runtime of the adversary and the experiment is at most τ.

The *advantage* of the adversary $\mathcal{A}$ is

$$\mathbf{Adv}_{\mathrm{SM}}^{\mathrm{ror}}(\mathcal{A}) = \max\{2|\Pr[E_d] - 1/2|, \Pr[E_{int}]\},$$

where E_d is the event that the adversary's guess equals b if the execution is fresh, or that the adversary's guess equals b'' if the execution is not fresh; and E_{int} is the event that integrity does not hold for some instance that enters its messaging phase. We define advantage in the same way with respect to integrity and statically fresh (ror-s-f), implicit integrity and strongly fresh (ror-ii-sf), implicit integrity and implicitly fresh (ror-ii-if) and weak implicit integrity and weakly fresh (ror-wii-wf).

Example: Encryption and Signatures   Example 13.4 is straight-forward to prove secure, using the multi-key variants of the usual security notions for public key encryption and signatures.

$\boxed{\text{T}}$ **Proposition 13.1.** *Let* SM-ENC-SIGN *be the construction from Example 13.4 based on a public key encryption scheme* PKE *and a signature scheme* SIG, *and let $\mathcal{A}$ be a non-adaptive, nonce-respecting $(\tau, l_c, l_u, l_s, l_e)$-adversary against* SM-ENC-SIGN. *Then there exists a l_u-multi-key (τ', l_c, l_e)-adversary $\mathcal{B}_1$ against* PKE *and a multi-key (τ', l_u, l_e)-adversary $\mathcal{B}_2$ against* SIG, *with τ' essentially equal to τ, such that*

$$\mathbf{Adv}_{\mathrm{SM\text{-}ENC\text{-}SIGN}}^{\mathrm{ror}}(\mathcal{A}) \leq \mathbf{Adv}_{\mathrm{PKE}}^{\mathrm{ror}}(\mathcal{B}_1) + \mathbf{Adv}_{\mathrm{SIG}}^{\mathrm{suf\text{-}cma}}(\mathcal{B}_2).$$

$\boxed{\text{E}}$ *Exercise* 13.2. Prove Proposition 13.1.

Since public key encryption and signatures are relatively slow, the messaging protocol from Example 13.4 is not suitable for all applications. It is, however, eminently suitable for written conversations between humans, where traffic is low-volume and infrequent, and the processing cost can be ignored. It also has a significant advantage that the state is essentially public, which means that only long-term, unchanging secret key material must be managed.

From a security point of view, the scheme is not secure against adaptive adversaries, so this scheme is not appropriate for some applications. However, this must be balanced against the simplicity of the scheme.

Remark. We could have proven security for statically fresh (ror-s-f) and non-invasive instead of fresh (ror) and non-adaptive, but we would have to restrict to a single challenge query to deal with corruptions. This shows that a seemingly stronger security notion may not give a stronger result overall.

Example: A Composition Theorem   Example 13.5 is a compositional construction: it uses a key exchange scheme to agree on a session key, which is then used for a channel protocol. Functionally, the composition is a messaging protocol. We show that the protocol is secure against non-invasive adversaries.

Proposition 13.2. *Let* SM-KEX-SC *be the construction from Example 13.5 based on a key exchange scheme* KEX *and a channel* SC, *and let* $\mathcal{A}$ *be a non-invasive* $(\tau, l_c, l_u, l_s, l_e)$-*adversary against* SM-KEX-SC. *Then there exists a non-invasive* $(\tau'_1, 0, l_u, l_s)$-*adversary* $\mathcal{B}_1$ *against* KEX, *two non-invasive* (τ'_2, l_s, l_u, l_s)-*adversaries* $\mathcal{B}_{2,0}$ *and* $\mathcal{B}_{2,1}$ *against* KEX, *and a non-adaptive, nonce-respecting* $(\tau'_3, l_c, l_s, l_s, l_e)$-*adversary* $\mathcal{B}_3$ *against* SC, *with* τ'_1, τ'_2 *and* τ'_3 *essentially equal to* τ, *such that*

$$\mathbf{Adv}^{\mathrm{ror}}_{\mathrm{SM\text{-}KEX\text{-}SC}}(\mathcal{A}) \leq 2\mathbf{Adv}^{\mathrm{kex\text{-}f\text{-}e}}_{\mathrm{KEX}}(\mathcal{B}_1) + 2\mathbf{Adv}^{\mathrm{kex\text{-}f\text{-}e}}_{\mathrm{KEX}}(\mathcal{B}_{2,0}) +$$
$$\mathbf{Adv}^{\mathrm{kex\text{-}f\text{-}e}}_{\mathrm{KEX}}(\mathcal{B}_{2,1}) + \mathbf{Adv}^{\mathrm{ror}}_{\mathrm{SC}}(\mathcal{B}_3).$$

The main complication is that there are three distinct ways in which the adversary can break the messaging protocol through the key exchange protocol: either break the authentication of the key exchange scheme so that there is no unique partner, or break the confidentiality of the key exchange scheme and then use knowledge of the key to either decrypt the challenge query directly or break integrity of the channel.

The proof is structured as a sequence of games. The initial game is the messaging experiment interacting with the adversary $\mathcal{A}$. We then modify the game so that if authentication fails for the key exchange scheme, we immediately halt the game. This ensures that any further integrity failures happen in the channel protocol. Then we modify the game so that when an instance enters the messaging phase, it uses a random session key, making sure that any partner for the key exchange uses the same key. If there is any change in the adversary's ability to guess the correct messaging experiment challenge bit, or any change in the adversary's ability to break the integrity of the messaging protocol, then the adversary is essentially capable of distinguishing the real session keys from random session keys.

Once we have made these changes, we can prove that the adversary in this game can be turned into an adversary against the channel with the same advantage. The key exchange adversaries are all constructed in the same manner, and we only write out two of them, leaving the third to the interested reader.

Proof of Proposition 13.2 In the following, let $\mathbf{GExp}_i$ be the experiment used in Game i. Let $E_{d,i}$ and $E_{int,i}$ be the events in Game i corresponding to E_d and E_{int} in Definition 13.8 for integrity and freshness. Let τ_i be the runtime cost of Game i. We also define F_i to be the event that for every viable instance that enters its messaging phase there is a unique partially partnered instance, but for some such instance there is no partnered instance. Define F'_i to be the event that there is a viable instance that enters its messaging phase and does not have a unique partially partnered instance. We see that $\Pr[F_i] + \Pr[F'_i] = \Pr[E_{int,i}]$. Since the adversary is non-invasive, no instance will be impersonatable.

Ⓖ **Game 0** The initial game experiment $\mathbf{GExp_0}$ is the messaging experiment $\mathbf{Exp_{SM}}$, which interacts with the adversary $\mathcal{A}$. Then

$$\mathbf{Adv}_{SM}^{ror}(\mathcal{A}) = \max\{2|\Pr[E_{d,0}] - 1/2|, \Pr[E_{int,0}]\}. \tag{13.1}$$

We have that $\tau_0 = \tau$.

Ⓖ **Game 1** In this game we modify the experiment so that whenever $\mathbf{GExp_0}$ would respond to a signalling query with $(\cdot, \top)$, the $\mathbf{GExp_1}$ experiment checks if the instance has some unique partial partner. If there is no such partial partner and the instance is viable, $\mathbf{GExp_1}$ samples $b' \xleftarrow{r} \{0, 1\}$ and the game terminates. If the instance is viable, it also verifies that the unique partial partner is viable and that it has a unique partial partner.

The game terminates because of the modification exactly when F_0' would happen in the previous game. Because of the termination, F_1' cannot happen in this game. Until the game terminates because of this modification, it behaves exactly as the previous game, so it follows that

$$|\Pr[E_{d,1}] - \Pr[E_{d,0}]| \leq \Pr[F_0'] \quad |\Pr[E_{int,1}] - \Pr[E_{int,1}]| = \Pr[F_0']. \tag{13.2}$$

The experiment $\mathbf{GExp_1}$ needs to determine if there are partners for each instance, but the extra runtime cost is small compared to the runtime cost of any interesting adversary. This means that τ_1 is essentially equal to τ_0.

Ⓣ **Lemma 13.3.** *There exists a* $(\tau', 0, l_u, l_s)$*-adversary* $\mathcal{B}_1$ *against* KEX, *with* τ_1' *essentially equal to* τ_1, *such that*

$$\mathbf{Adv}_{KEX}^{kex\text{-}f\text{-}e}(\mathcal{B}_1) = \Pr[F_0'].$$

Ⓔ *Exercise* 13.3. Prove the above lemma.

Ⓖ **Game 2** In this game we modify the experiment so that whenever it is about to respond to a signalling query with $(\cdot, \top)$, $\mathbf{GExp_2}$ checks if the instance has some partial partner that has entered the messaging phase. If so, $\mathbf{GExp_2}$ replaces the session key in the state with the session key from the partner's state. If the partial partner has not entered its messaging phase, $\mathbf{GExp_2}$ replaces the session key in the instance's state with a random session key.

If $\mathcal{A}$ exceeds its bounds the game samples $b' \xleftarrow{r} \{0, 1\}$ and terminates.

The accounting needed in $\mathbf{GExp_2}$ had already been included in the previous game. Sampling the runtime keys is cheap, so because bounds on $\mathcal{A}$ are enforced we get that τ_2 is essentially equal to τ_1.

Lemma 13.4. *There exists two (τ_2', l_s, l_u, l_s)-adversaries $\mathcal{B}_{2,0}$ and $\mathcal{B}_{2,1}$, against* KEX, *with τ_2' essentially equal to τ_2, such that*

$$|\Pr[E_{d,2}] - \Pr[E_{d,1}]| \leq \mathbf{Adv}_{\mathrm{KEX}}^{\mathrm{kex\text{-}f\text{-}e}}(\mathcal{B}_{2,0})$$
$$|\Pr[E_{int,2}] - \Pr[E_{int,1}]| \leq \mathbf{Adv}_{\mathrm{KEX}}^{\mathrm{kex\text{-}f\text{-}e}}(\mathcal{B}_{2,1}).$$

Proof. The experiment $\mathbf{GExp}_2$ can be simulated by a composition of the key exchange experiment $\mathbf{Exp}_{\mathrm{KEX}}^{\mathrm{kex}}$ and a simulator $\mathbf{Sim}$ that runs a copy of $\mathbf{GExp}_2$ modified as follows:

- When $\mathbf{GExp}_2$ would run the key exchange key generation algorithm, $\mathbf{Sim}$ sends a key generation query to $\mathbf{Exp}_{\mathrm{KEX}}^{\mathrm{kex}}$ and forwards the response.

- Any long-term key reveal queries are forwarded to $\mathbf{Exp}_{\mathrm{KEX}}^{\mathrm{kex}}$.

- When $\mathbf{GExp}_2$ would start a KEX instance, $\mathbf{Sim}$ sends a corresponding execute query to $\mathbf{Exp}_{\mathrm{KEX}}^{\mathrm{kex}}$ with the correct keys and associated data.

- When $\mathbf{GExp}_2$ receives a signal that would be handled by a key exchange instance, it sends this signal to $\mathbf{Exp}_{\mathrm{KEX}}^{\mathrm{kex}}$ and outputs the response.

- When an instance in $\mathbf{Exp}_{\mathrm{KEX}}^{\mathrm{kex}}$ outputs a session key and the corresponding messaging instance would enter its messaging phase, $\mathbf{Sim}$ determines if the key exchange instance has a unique partner. It does not and it is not viable, $\mathbf{Sim}$ does a session key reveal for the key exchange instance and uses the response as the session key for this messaging instance. Otherwise, if there is a partner and it has entered its messaging phase, $\mathbf{Sim}$ uses the session key from the partner's state as this instance's session key. Otherwise $\mathbf{Sim}$ sends a test query to $\mathbf{Exp}_{\mathrm{KEX}}^{\mathrm{kex}}$ for the key exchange instance and uses the response as this instance's session key.

By inspection, if the challenge bit in $\mathbf{Exp}_{\mathrm{KEX}}^{\mathrm{kex}}$ is 0, then $\mathbf{Exp}_{\mathrm{KEX}}^{\mathrm{kex}}$ together with $\mathbf{Sim}$ perfectly simulate $\mathbf{GExp}_1$. Also by inspection, if the challenge bit in $\mathbf{Exp}_{\mathrm{KEX}}^{\mathrm{kex}}$ is 1, then the $\mathbf{Exp}_{\mathrm{KEX}}^{\mathrm{kex}}$ together with $\mathbf{Sim}$ perfectly simulate $\mathbf{GExp}_2$.

Furthermore, by careful design, every instance for which $\mathbf{Sim}$ makes a test query will be fresh, so any execution with $\mathbf{Exp}_{\mathrm{KEX}}^{\mathrm{kex}}$ will be fresh.

Our adversaries will be the composition of $\mathbf{Sim}$ and $\mathcal{A}$. All that remains is to specify which bit our adversary will output: if the messaging experiment simulated by $\mathbf{Sim}$ is fresh, $\mathcal{B}_{2,0}$ outputs 1 if the $\mathcal{A}$'s guess b' equals the messaging experiment's challenge bit and 0 otherwise; if the messaging experiment is not fresh, $\mathcal{B}_{2,0}$ outputs a random bit; $\mathcal{B}_{2,1}$ outputs 0 if integrity holds for the messaging experiment, and 1 otherwise. □

We must output a random bit when the messaging experiment is not fresh, even though the underlying key exchange experiment is fresh. The reason is that the adversary's output when the execution is not fresh does not count when evaluating $E_{d,i}$, so we cannot then allow it to affect the output of our key exchange adversary. In principle, it could cancel out the advantage.

While it would be possible to combine the two adversaries into a single adversary, this requires careful work. We must also know the exact probabilities involved in the advantage calculations for the various games. The reason is that the adversary's advantage come from two components: guessing the challenge bit and breaking integrity, and these components are somehow independent. This means that when we make changes to the game experiment, these components may move in opposite directions.

T **Lemma 13.5.** *There exists a non-adaptive, nonce-respecting $(\tau_3', l_c, l_s, l_s, l_e)$-adversary $\mathcal{A}_3$ against* SC, *with τ_3' essentially equal to τ_2, such that*

$$\max\{|2|\Pr[E_{d,2}] - 1/2|, \Pr[E_{int,2}]\} = \mathbf{Adv}_{\mathrm{SC}}^{\mathrm{ror}}(\mathcal{B}_3).$$

Proof. The experiment $\mathbf{GExp}_2$ can be simulated by a composition of $\mathbf{Exp}_{\mathrm{SC}}$ and a simulator $\mathbf{Sim}$ that runs a copy of $\mathbf{GExp}_2$ modified as follows:

- Whenever $\mathbf{GExp}_2$ samples a session key at random, $\mathbf{Sim}$ makes a key generation query to $\mathbf{Exp}_{\mathrm{SC}}$.

- When $\mathbf{GExp}_2$ starts an instance of SC with a particular session key, $\mathbf{Sim}$ makes an execute query to $\mathbf{Exp}_{\mathrm{SC}}$ for this session key.

- Whenever $\mathbf{GExp}_2$ would have done a computation with a channel instance, $\mathbf{Sim}$ forwards the query to $\mathbf{Exp}_{\mathrm{SC}}$ and outputs the response. In particular, challenge queries are handled by $\mathbf{Exp}_{\mathrm{SC}}$.

- Whenever the adversary makes a message reveal query, $\mathbf{Sim}$ forwards the query to $\mathbf{Exp}_{\mathrm{SC}}$, with an appropriate event count shift.

By inspection, we see that $\mathbf{Exp}_{\mathrm{SC}}$ together with $\mathbf{Sim}$ perfectly simulate $\mathbf{GExp}_2$, with the challenge bit and random bit of $\mathbf{Exp}_{\mathrm{SC}}$ identified with the challenge bit and random bit in $\mathbf{GExp}_2$. Also, the event $F_2 = E_{int,2}$ corresponds exactly with the event E_{int} for the game with $\mathbf{Exp}_{\mathrm{SC}}$.

Our adversary $\mathcal{A}_3$ runs a copy of $\mathbf{Sim}$ and $\mathcal{A}$, and the above argument shows that its advantage is as claimed. The simulator $\mathbf{Sim}$ does forward some queries and their responses, but this cost is small. It follows that the time cost of $\mathcal{A}_3$ interacting with $\mathbf{Exp}_{\mathrm{SC}}$ is essentially equal to τ_2. □

Proposition 13.2 follows from (13.1) and (13.2), and the three lemmas.

13.2.2 Fewer Queries Suffice – Maybe

We shall now prove several theorems about the number of queries needed. First, we prove that we only need a single challenge query. Second, we prove that for messaging protocols in general, we only need two key generation queries. For channels, this reduces to a single key generation query. We then use these theorems to prove results about our examples.

Challenge queries   Multiple challenge queries are more convenient for application security proofs. However, multiple challenge queries makes our proof obligation much more complicated when analysing the messaging protocol. We therefore need the following result.

Ⓣ **Proposition 13.6.** *Let $\mathcal{A}$ be a $(\tau, l_c, l_u, l_s, l_e)$-adversary against a messaging protocol* SM. *Then there exists a $(\tau', 1, l_u, l_s, l_e)$-adversary $\mathcal{B}$ against* SM, *where τ' is essentially equal to τ, and*

$$\mathbf{Adv}_{\mathrm{SM}}(\mathcal{A}) \leq l_c \mathbf{Adv}_{\mathrm{SM}}(\mathcal{B}).$$

The proof is a standard hybrid argument, where we replace all but one challenge query with chosen plaintext queries, and where we deliberately randomise the plaintext for some of these queries. Note that we must output a random bit if a simulated challenge query is somehow revealed.

Ⓔ *Exercise* 13.4. Prove Proposition 13.6.

Key generation queries   The analysis of messaging protocols is essentially the same as we saw for key exchange protocols in Section 10.7. The idea is that as long as there is a single challenge query, we can guess which key pairs are involved. Likewise, we can also guess the key pairs involved in the first instance that breaks integrity. These techniques work since the experiment handles every instance independently.

Ⓣ **Proposition 13.7.** *Let $\mathcal{A}$ be a $(\tau, 1, l_u, l_s, l_e)$-adversary against a messaging protocol* SM. *Then there exists a $(\tau', 1, 2, l_s, l_e)$-adversary $\mathcal{B}$ against* SM, *with τ' essentially equal to τ, such that*

$$\mathbf{Adv}_{\mathrm{SM}}(\mathcal{A}) \leq l_u^2 \mathbf{Adv}_{\mathrm{SM}}(\mathcal{B}).$$

The proof is essentially identical to the proof of Proposition 10.7. First, we guess which key pairs will be involved. The experiment is then split into two subexperiments, where one handles the two guessed key pairs and the other handles the remaining key pairs. Because instances are independent, the two experiments are indistinguishable from a single experiment. The first experiment then becomes the messaging experiment, while the second is simulated. If we have guessed correctly, the adversary's success is preserved. If we have guessed incorrectly, we simply abort.

Ⓔ *Exercise* 13.5. Prove Proposition 13.7.

The situation is even simpler for channels, since a single key suffices.

Ⓣ **Proposition 13.8.** *Let $\mathcal{A}$ be a $(\tau, 1, l_u, l_s, l_e)$-adversary against a channel protocol* SC. *Then there exists a $(\tau', 1, 1, l_s, l_e)$-adversary $\mathcal{B}$ against* SC, *with τ'*

essentially equal to τ, *such that*

$$\mathbf{Adv}_{\mathrm{SM}}(\mathcal{A}) \leq l_u \mathbf{Adv}_{\mathrm{SM}}(\mathcal{B}).$$

Again, the proof is essentially the same, though with a single guess.

E | *Exercise* 13.6. Prove Proposition 13.8.

Example: Channel   Analysing the channel from Example 13.2 is now easy.

T | **Proposition 13.9.** *Let* $\mathcal{A}$ *be a nonce-respecting* $(\tau, 1, 1, l_s, l_e)$-*adversary against* SC-SYM *from Example 13.2. Then there exists a* $(\tau', 1, l_e, l_e)$-*adversary* $\mathcal{B}_1$ *real-or-random for* Σ *and a* (τ', l_e, l_e)-*adversary* $\mathcal{B}_2$ *against integrity for* Σ, *with* τ' *essentially equal to* τ, *such that*

$$\mathbf{Adv}^{\mathrm{ror}}_{\mathrm{SC\text{-}SYM}}(\mathcal{A}) \leq \mathbf{Adv}^{\mathrm{ror\text{-}cca}}_{\Sigma}(\mathcal{B}_1) + \mathbf{Adv}^{\mathrm{int\text{-}ctxt}}_{\Sigma}(\mathcal{B}_2).$$

E | *Exercise* 13.7. Prove Proposition 13.9.

V | *Remark.* It is worthwhile to note that we now have a solid cryptographic solution to the original cryptographic problem of private conversations.

Given a good block cipher, a good hash function and a good key encapsulation mechanism, we instantiate the compositional construction from Example 13.5 with the channel from Example 13.2 and the key exchange protocol from Example 10.7. We can instantiate the channel with a symmetric cryptosystem constructed using Examples 7.1, 7.2, 7.4 and 7.5. We would instantiate the key derivation functions using the hash function. The key exchange scheme can be instantiated with the key encapsulation mechanism. We could use signatures in addition to the key encapsulation mechanism to instantiate the key exchange protocol.

This messaging protocol is usable for most applications including real-time low-latency high-volume traffic. There are a number of possible optimisations, both with respect to the construction and the theorems used to prove it secure.

From a security point of view, the messaging protocol has one downside if the conversations are long-lasting. The session key should then be thought of as a long-term key and assuming a non-invasive adversary becomes unrealistic.

E | *Exercise* 13.8. Proposition 13.9 is not optimal for use with Proposition 13.2. We can provide a stronger result if we rely on multi-key notions for the symmetric cryptosystem, as discussed in Section 7.1.7. State and prove a better result for multi-key notions.

E | *Exercise* 13.9. The channel in Example 13.2 is suboptimal for the compositional construction in Example 13.5, since it allows for keys to be reused, while the compositional construction will use each session key once. Design a deterministic channel protocol that is secure against key-respecting adversaries.

Hint: Consider Example 7.6 and the comments following Example 13.2.

Example: Evolving Session Keys The analysis of the channel from Example 13.3 is significantly simplified by the above results, since we only have to consider a single key. It is even further simplified in that we are looking at key-respecting adversaries, so there will be at most two instances.

The potential difficulty is that using the leaf key k_ρ, to encrypt reveals information about the leaf key, which reveals information about the derivation key kd_ρ. The answer is to use random choices for k_ρ, and kd'_ρ until after the challenge query. By freshness and the security of the key derivation function, this cannot be noticed. (This explains why we use a different leaf key for the message encryption, which makes kd'_ρ independent of the message encryption.)

T **Proposition 13.10.** *Let* SC-CHN *be the construction from Example 13.3 based on a key exchange scheme* KEX, *a key derivation function* $kdf : \mathfrak{K}_s \to \mathfrak{K}_s^2$ *and a symmetric cryptosystem* $\Sigma = (\mathfrak{K}_s, \mathfrak{P}, \mathfrak{F}, \mathfrak{C}, \mathcal{E}_s, \mathcal{D}_s)$, *and let* $\mathcal{A}$ *be a key-respecting* $(\tau, 1, 1, 2, l_e)$-*adversary against* SC-CHN *that makes no long-term key reveal queries. Then there exists a* τ'_1-*adversary* $\mathcal{B}_1$ *against* kdf *and* $(\tau'_2, 1, 1)$-*adversaries* $\mathcal{B}_2$ *and* $\mathcal{B}_3$ *against* Σ, *with* τ'_1 *and* τ'_2 *essentially equal to* τ, *such that*

$$\mathbf{Adv}_{\text{SC-CHN}}^{\text{ror-wii-wf}}(\mathcal{A}) \leq l_e \mathbf{Adv}_{kdf}^{\text{prf}}(\mathcal{B}_1) + \mathbf{Adv}_{\Sigma}^{\text{ind-cca}}(\mathcal{B}_2) + l_e \mathbf{Adv}_{\Sigma}^{\text{int-ctxt}}(\mathcal{B}_3).$$

Remark. The requirement that the adversary does no long-term key reveal queries is not significant, as discussed after Example 13.3.

The proof is structured as a sequence of two games. The initial game is the messaging experiment interacting with the adversary $\mathcal{A}$. We then modify the game so that each instance simply chooses a random keys as k_ρ, and kd'_ρ. We then build a real-or-random adversary against the symmetric cryptosystem simply by forwarding the challenge query plaintext and associated data as a challenge query to the real-or-random symmetric cryptosystem experiment. We build a ciphertext integrity adversary against the symmetric cryptosystem in the same way, but we must guess when the first integrity failure happens.

E *Exercise* 13.10. Prove Proposition 13.10.

It is tempting to apply Propositions 13.2, 13.8 and 13.10 to SM-KEX-SC from Example 13.5 when instantiated with SC-CHN. But the proof of Proposition 13.2 does not work for adversaries making state reveal queries.

The reason is that encrypting a message with a given symmetric key in some sense commits to the key. For non-invasive adversaries, that is not much of a problem since we know that an instance will be fresh when it outputs a session key. But when the adversary is allowed to reveal states, we do not know if an instance will be fresh until the adversary no longer is allowed to reveal the state. In fact, we may very well have to encrypt messages with a given symmetric key before we know whether the instance will be fresh or not, committing us to a particular symmetric key.

We solve this problem by using a single challenge query and guessing an instance. If the first integrity failure or the challenge query does not involve this instance, we abort the game.

Proposition 13.11. *Let* SM-KEX-SC *be the construction from Example 13.5 based on a key exchange scheme* KEX *and the channel* SC-CHN *from Example 13.3, and let* $\mathcal{A}$ *be a* $(\tau, 1, l_u, l_s, l_e)$*-adversary against* SM-KEX-SC. *Then there exists a* $(\tau_1', 0, l_u, l_s)$*-adversary* $\mathcal{B}_1$ *against* KEX, *two* $(\tau_2', 1, l_u, l_s)$*-adversaries* $\mathcal{B}_{2,0}$ *and* $\mathcal{B}_{2,1}$ *against* KEX, *and a non-adaptive, nonce-respecting* $(\tau_3', 1, 1, 2, l_e)$*-adversary against* SC-CHN, *with* τ_1', τ_2' *and* τ_3' *essentially equal to* τ, *such that*

$$\mathbf{Adv}_{\text{SM-KEX-SC}}^{\text{ror-wii-wf}}(\mathcal{A}) \leq 2\mathbf{Adv}_{\text{KEX}}^{\text{kex-fe-e}}(\mathcal{B}_1) + 2l_s\mathbf{Adv}_{\text{KEX}}^{\text{kex-fe-e}}(\mathcal{B}_{2,0}) +$$
$$2l_s\mathbf{Adv}_{\text{KEX}}^{\text{kex-fe-e}}(\mathcal{B}_{2,1}) + l_s\mathbf{Adv}_{\text{SC-CHN}}^{\text{ror-wii-wf}}(\mathcal{B}_3).$$

As usual, the proof is structured as a sequence of games. The first modification is to stop the game if some instance enters its messaging phase without a unique partner. We then have the game choose an instance at random and stop the game if the first integrity failure or the challenge query involves some other instance. Finally, we have the chosen instance (and its partner) use a random key instead of the key from the key exchange scheme. At this point, the adversary is essentially a key-respecting adversary against the channel that makes no long-term key reveal queries.

Exercise 13.11. Prove Proposition 13.11.

13.3   INVASIVE ADVERSARIES

So far we have shown how to deal with non-invasive adversaries and proven a fairly weak form of security against invasive adversaries. Our goal in this section is to show how to provide stronger security against invasive adversaries.

This protocol combines the idea of continuous key exchange first introduced in Section 10.6 with evolving session keys. The goal is to achieve continuous authentication as well as an ability to recover from state compromise. Therefore, each exchanged session key is used to key one channel, and each channel is either used to encrypt or decrypt, never both. A single session key is used for all the ciphertexts in a segment.

We already injected the needed definitions into our definitions, so there is no need for further definitional work.

Example 13.6. Let $\text{KEX}_{rs} = (\mathfrak{K}_s^2, \mathfrak{N}_{rs}, \mathcal{K}, \mathcal{MG}, \mathcal{SK})$ be a role-symmetric key exchange scheme and let SC-CHN be the channel from Example 13.3. Let ex be the function that maps a tuple $(\rho, \cdot, k_{0,}, k_{1,})$ to $k_{\rho,}$. The messaging protocol SM-CNT $= (1, \mathfrak{P}, \mathfrak{F}, \mathcal{K}, \mathcal{H}, \mathcal{E}_m, \mathcal{D}_m)$ works as follows.

In the messaging phase, the state consists of a tuple $(\rho, ad, pk, pk', hik, hic,$ $hic', \hat{c}, r, st_c, st_c')$ listing the role ρ, the associated data ad, the public keys pk, pk', histories hik, hic, hic' of key exchange messages and ciphertexts encrypted and decrypted, the last key exchange message received $\hat{c}$, the last nonce used r, and the state of the encrypting and decrypting channels st_c, st_c'. The *establishing* algorithm $\mathcal{H}$ does:

- On input of $(0, ad, pk_0, sk_0, pk_1)$, it computes $(r_0, \hat{c}_0) \leftarrow \mathcal{MG}(sk_0)$ and sends $\hat{c}_0$. When it receives $\hat{c}_1$ and t', it computes $ks_1 \leftarrow \mathcal{SK}(0, (ad, \bot, \bot), (pk_0, sk_0), pk_1, \hat{c}_1, r_0)$, $(kd_{1,0}, t') \leftarrow kdf(ks_1)$, $(k, k') \leftarrow kdf(kd_{1,0})$, $st_c \leftarrow \mathcal{H}_c(0, \bot, k, \bot)$ and $st_c' \leftarrow \mathcal{H}_c(0, \bot, k', \bot)$. If $t \neq t'$, the algorithm outputs $\bot$. Otherwise it outputs the state $st = (0, ad, pk_0, pk_1, (\hat{c}_0, \hat{c}_1), \bot, \bot, \hat{c}_1, \bot, st_c, st_c')$.

- On input of $(1, ad, pk_1, sk_1, pk_0)$, it waits to receive a signal $\hat{c}_0$, computes $(r_1, \hat{c}_1) \leftarrow \mathcal{MG}(sk_1)$, $ks \leftarrow \mathcal{SK}(1, (ad, \bot, \bot), (pk_1, sk_1), pk_0, \hat{c}_0, r_1)$, $(kd_1, t) \leftarrow kdf(ks)$, $(k, k') \leftarrow kdf(kd_1)$, $st_c \leftarrow \mathcal{H}_c(0, \bot, k, \bot)$ and $st_c' \leftarrow \mathcal{H}_c(0, \bot, k', \bot)$. It sends $(\hat{c}_1, t)$ and outputs the state $st = (1, ad, pk_0, pk_1, (\hat{c}_0, \hat{c}_1), \bot, \bot, \bot, r_1, st_c', st_c)$.

The *encryption* algorithm $\mathcal{E}_m$ on input of sk, st, ad^m and m does:

- (responder not ready) If $\rho = 1$ and $hic = hic' = \bot$, output $\bot$.

- (start new segment) If $r = \bot$, compute $(r', \hat{c}') \leftarrow \mathcal{MG}(sk)$, $ks \leftarrow \mathcal{SK}(\rho, (ad, hik, hic), (pk, sk), pk', \hat{c}, r')$, $(kd, k) \leftarrow kdf(ks, ex(st_c))$ and $(st_c'', c) \leftarrow \mathcal{E}_c(\mathcal{H}_c(0, \bot, (k, \bot)), ad^m, m)$, and output $st' = (\rho, ad, pk, pk', hik||\hat{c}', hic||(ad^m, c), hic', \bot, r', st_c'', st_c')$ and $(\hat{c}', c)$.

- (continue segment) If $r \neq \bot$, compute $(st_c'', c) \leftarrow \mathcal{E}_c(st_c, ad^m, m)$, and output $st' = (\rho, ad, pk, pk', hik, hic||(ad^m, c), hic', \bot, r, st_c'', st_c')$ and c.

The *decryption* algorithm $\mathcal{D}_m$ on input of sk, st, ad^m and a ciphertext does:

- (start new segment) If the ciphertext is of the form $(\hat{c}', c)$, compute $ks \leftarrow \mathcal{SK}(\rho, (ad, hik, hic'), (pk, sk), pk', \hat{c}', r)$, $(kd, k) \leftarrow kdf(ks, ex(st_c'))$ and $(st_c'', m) \leftarrow \mathcal{D}_c(\mathcal{H}_c(1, \bot, (\bot, k)), ad^m, c)$, and output $st' = (\rho, ad, pk, pk', hik||\hat{c}', hic, hic'||(ad^m, c), \hat{c}', \bot, st_c, st_c'')$ and m.

- (continue segment) If the ciphertext is of the form c, compute $(st_c'', m) \leftarrow \mathcal{D}_c(st_c', ad^m, c)$, and output $st' = (\rho, ad, pk, pk', hik, hic, hic'||(ad^m, c), \hat{c}', r, st_c, st_c'')$ and m.

If any computation fails in any algorithm, the algorithm stops and outputs $\bot$.

In the above, hik contains the key exchange messages sent and received, in order; hic contains the ciphertexts encrypted; and hic' contains the ciphertexts decrypted. Both ciphertext lists also contain the per-message associated data.

The causal order is the minimal partial order that contains the temporal order restricted to the set of ciphertexts of the form $(\hat{c}, c)$ that start new segments. In other words, the partners agree on the temporal order of these ciphertexts, but may not agree on the temporal order of other ciphertexts.

Remark. As discussed in Section 10.6, the use of a role-symmetric key exchange scheme in this messaging protocol is not essential, but it is a nice optimisation.

The establishing algorithm uses the round-symmetric key exchange scheme to do an implicitly authenticated key exchange, and then explicitly authenticate the responder by revealing a value derived from the session key.

The encryption algorithm either begins sending a message segment, or continues sending a segment. When a segment begins, the encryption also includes a key exchange message. The encryption algorithm will not encrypt messages for the responder until the initiator's first ciphertext has been decrypted.

The decryption algorithm either begins decrypting a new segment, or continues decrypting the current segment. When a segment begins, the decryption algorithm completes the key exchange started by the encryption algorithm. The decryption algorithm will also be able to bring the responder properly into the messaging phase when it receives the first encrypted message.

The initiator's first message brings the responder properly into the messaging phase, and until that happens the responder will not be able to send any messages. This is the intended functionality.

If the recipient should be allowed to send the first message, we could have the initiator send a signal explicitly authenticating the key exchange phase. In practice, this signal may be sent together with the encryption of the first message, in the event that the initiator has a first message to send. The downside is that this makes the design and analysis slightly more complicated.

We could allow the responder to encrypt messages immediately, but without knowing if the initiator is present. A responder entering the messaging phase without an initiator would usually be considered an attack, but if we accept that flaw, we could model this in a definition of integrity.

Some deployed schemes actually follow this approach, but with a twist. The initiator precomputes many of its initial handshake signals and publishes them. The responder (somehow) chooses an unused initial signal, computes the handshake signal and sends its together with its first message. This works since the first signal is independent of the associated data or partner public key. It also makes the responder the actual initiator of the conversation, and allows the conversation to start without interaction, which is convenient.

The histories hik, hic and hic' as we have defined them are fundamentally impractical, since they include every ciphertext with per-message associated data. Since messaging protocols can be used to send a lot of data, requiring the cryptographic subsystem to store the histories would require too much storage. In practice, we would therefore compute these histories as the *iterated hash* of the terms it is built from, instead of concatenating or appending. From a security point of view this would be equivalent if the hash function is collision resistant. And from a functional point of view, the histories could now be computed with fixed-size storage. However, including the hash would needlessly complicate the design and its analysis.

This fact explains another design decision. From a security point of view, we should want to include more information than just hic when we encrypt. In particular, we should include hic', the list of ciphertexts decrypted at the time of encryption. This would allow each partner to share its temporal order on the messages with the other partner, which would be good. As we have defined the histories, it would be sufficient to include the number of decrypted messages j' in the clear (or perhaps inside the encryption).

But when the messaging protocol uses a chain of hashes instead of storing everything, this imposes potentially large storage requirements on the protocol. The reason is that we do not know when the partner starts a new segment until we receive the key exchange term marking a new segment, so we do not know which of the iterated hash values it will use. This means that we need to record every potential hash value as we encrypt messages. For some applications, there could be a large number of such terms and this approach would be impractical. For some applications, this does not impose undue storage requirements, in which case this approach would be feasible.

Remark. In some sense, this protocol achieves near-optimal security. If everything has been revealed in one party, new randomness must be generated to recover. Also, if the partner has had everything revealed, we must wait for it to encrypt something to regain security. Recovering for every encryption would therefore be costly. We could also have given the protocol non-implicit security by using temporary signature key pairs, updated with every key exchange.

Remark. The use of implicit integrity in the protocol is interesting in that in some cases it can provide *deniability*. The idea is that even though Alice has a complete record of a conversation with Bob, this does not prove that there was a conversation with Bob, since Alice could have made the whole thing up by impersonating Bob herself. It must be admitted that deniability has limited applicability and power, but it is sometimes a requested feature.

The security of the protocol follows from the security of the role-symmetric key exchange scheme and the channel. The idea is that we guess which instance and which segment where integrity first fails or the challenge query happens. In this case, the role-symmetric key exchange instance must be fresh, which means that we can replace the session key with a random key. Once this is done, we get an adversary against the channel scheme.

E *Exercise* 13.12. State a claim similar to Proposition 13.11 for the scheme SM-CNT from Example 13.6, and prove it.

13.4   SOMEWHAT ANONYMOUS MESSAGING

In all our work so far there is an implicit assumption that the initiator and responder somehow agree not only to communicate but also on what the associated data should be. We have not discussed how this happens, but it typically involves transmitting the associated data in the clear. Also, we often want this process to run in parallel with establishing the conversation, since latency can be a significant issue for applications. But there can be other reasons for not agreeing on the associated data in the clear, besides efficiency.

The goal of this section is to achieve a measure of *anonymity* against certain adversaries. Since the associated data typically identifies the parties involved or their keys, the associated data may be sensitive. We cannot therefore just assume that the parties agree on the associated data somehow, but we have to specify how this is done in the protocol while preserving the secrecy of the associated data. We choose to consider the associated data to be a pair, where the initiator provides the first half and the responder provides the second half.

Remark. Recall that none of our earlier cryptosystems guarantee confidentiality for associated data, which is generally assumed to be public.

The parties first use unauthenticated key exchange to establish a session key for a channel. Then they agree on the associated data and authenticate each other *inside the channel* before entering the messaging phase.

Remark. In order to fit into our framework, we have to give both parties the complete associated data as input. This makes no sense, given what we just said, but we want the protocol to have the usual integrity and confidentiality properties, so we want to fit it into our framework. Also note that the protocol does not use the partner's associated data until it has been received.

This particular protocol uses digital signatures for authentication, but other techniques could be used. In order to slightly simplify the exposition, we also use a key encapsulation mechanism to do key exchange, which as we saw in Section 10.3 is more or less equivalent to a two-round key exchange scheme.

E *Example* 13.7. Let KEM $= (\mathfrak{K}_s^2, \mathcal{KK}, \mathcal{KE}, \mathcal{KD})$ be a key encapsulation mechanism, SIG $= (\mathcal{K}, \mathcal{S}, \mathcal{V})$ a signature scheme, $\Sigma = (\mathfrak{K}_s, \mathfrak{P}, \mathfrak{F}_0, \mathfrak{C}, \mathcal{E}_s, \mathcal{D}_s)$ a symmetric cryptosystem, and SC $= (\mathfrak{P}, \mathfrak{F}, \mathfrak{K}_s, \mathcal{H}_c, \mathcal{E}_c, \mathcal{D}_c)$ a simple channel.

The messaging protocol SM-ANON $= (\mathfrak{P}, \mathfrak{F}, \mathcal{K}, \mathcal{H}, \mathcal{E}_m, \mathcal{D}_m)$ works as follows:

- The *establishing* algorithm $\mathcal{H}$ on input of $(0, ad, (vk_0, sk), vk_1)$ does:

  1. Compute $(ek, dk) \leftarrow \mathcal{KK}$. Send ek. Wait for $(\hat{c}_1, \hat{c}_1')$.

2. Compute $k \leftarrow \mathcal{KD}(dk, \hat{c}_1)$, $st_c \leftarrow \mathcal{H}_c(0, k)$, and $(st'_c, (ad_1, vk_1) \leftarrow \mathcal{D}_c(k, st_c, 1, \hat{c}'_1))$.

3. Compute $\sigma_2 \leftarrow \mathcal{S}(sk, m_2)$, with $m_2 = (ad_0, ad_1, vk_0, vk_1, ek, \hat{c}_1, \hat{c}'_1)$, and $(st''_c, \hat{c}_2) \leftarrow \mathcal{E}_c(k, st'_c, 2, (ad_0, vk_0, \sigma_2))$. Send $\hat{c}_2$. Wait for $\hat{c}_3$.

4. Compute $(st'''_c, \sigma_3) \leftarrow \mathcal{D}_c(k, st''_c, 3, \hat{c}_3)$ and verify $\mathcal{V}(vk_1, m_3, \sigma_3) = 1$, with $m_3 = (ad_0, ad_1, vk_0, vk_1, ek, \hat{c}_1, \hat{c}'_1, \hat{c}_2)$.

5. If any decapsulation, decryption or verification fails, the algorithm immediately stops with output $\bot$. Otherwise, it outputs $(0, k, st'''_c)$.

- The *establishing* algorithm $\mathcal{H}$ on input of $(1, ad, (vk_1, sk), vk_0)$ does:

  1. Wait for ek. Compute $(\hat{c}_1, k) \leftarrow \mathcal{KE}(ek)$, $st_c \leftarrow \mathcal{H}_c(1, k)$ and $(st'_c, \hat{c}'_1) \leftarrow \mathcal{E}_c(k, st_c, 1, (ad_1, vk_1))$. Send $(\hat{c}_1, \hat{c}'_1)$. Wait for $\hat{c}_2$.

  2. Compute $(st''_c, (ad_0, vk_0, \sigma_2)) \leftarrow \mathcal{D}_c(k, st'_c, 2, \hat{c}_2)$, and verify that $\mathcal{V}(vk_0, m_2, \sigma_2) = 1$, with m_2 as above.

  3. Compute $\sigma_3 \leftarrow \mathcal{S}(sk, m_3)$, with m_3 as above, and $(st'''_c, \hat{c}_3) \leftarrow \mathcal{E}_c(k, st''_c, 3, \sigma_3)$. Send $\hat{c}_3$.

  4. If any decapsulation, decryption or verification fails, the algorithm immediately stops with output $\bot$. Otherwise, it outputs $(1, k, st'''_c)$.

- The *encryption* algorithm $\mathcal{E}_m$ on input of sk, $st = (\rho, k, st_c)$, ad^m and a message m, computes $(st'_c, c) \leftarrow \mathcal{E}_c(k, st_c, ad^m, m)$, and outputs (ρ, k, st'_c). If the encryption fails, the algorithm outputs $\bot$.

- The *decryption* algorithm $\mathcal{D}_m$ on input of sk, $st = (\rho, k, st_c)$, ad^m and a ciphertext c, computes $(st'_c, m) \leftarrow \mathcal{D}_c(k, st_c, ad^m, c)$, and outputs (ρ, k, st'_c). If the decryption fails, the algorithm outputs $\bot$.

On any other input than specified above, the algorithms output $\bot$.

Recall that a symmetric cryptosystem does not hide the length of the messages. In order to avoid trivial attacks, we must use fixed-length encodings of verification keys and signatures for the signature scheme. Users of the scheme will typically also want to ensure that both parts of the associated data used is given an encoding of fixed length, or that care is taken so that the length of the associated data does not reveal too much information.

There is actually no need for the constant associated data 2, 3 and so forth in the protocol, since the channel will take care of roles and order. This is why we do not have to worry about collisions between these constants and the per-message associated data used for encryption and decryption.

If either party using the messaging protocol does not know the other party's associated data or public key before the establishing protocol is finished, the protocol would also need to provide a way for the user to extract this information after establishment. We could easily store this information inside the state, for instance.

In order to properly model the security of this protocol, we need to modify the experiment. The idea is to add a *challenge execute query*, which takes as input two sets of public key references and two sets of equal-length associated data (that is, the first parts of the associated data have equal length, as does the second parts). It starts two matching instances with opposite roles and input according to the challenge bit.

If the two instances cannot be each other's potential partner, the adversary wins trivially. The execution will not be fresh unless they are.

The adversary will still be allowed to make ordinary execute and challenge queries, in order to capture the fact that the messaging protocol must still preserve confidentiality and integrity.

E | *Exercise* 13.13. Write out the experiment for somewhat anonymous messaging as sketched above.

Multiple challenge execute and challenge queries is as usual a complicating factor, but we do not have to deal with more than one such query, which will either be a challenge execute query or a challenge query. In particular, this means that we can consider the anonymity and confidentiality separately, which simplifies our work.

E | *Exercise* 13.14. Show that a single challenge execute query *or* a single challenge query suffices.

In principle, we could reduce the number of key generation queries to four (no lower, because a challenge execute query could involve four public keys), but this is less useful.

In order to prove the scheme secure with a challenge query, we first of all forbid integrity failures during the handshake. We can do that, since any integrity failure will lead to an attack on the signature scheme or the channel. (This is actually a fairly complicated step: the signature ensure partnering for the key exchange, which ensures confidentiality for the signatures and integrity for the signals.)

Next, we guess which instance will receive the challenge query or have the first integrity failure and use a random key for that particular channel. We can do that since we have forbidden integrity failures during the handshake, which means that the unauthenticated key exchange has a partner, and we can therefore rely on its security. Now, any integrity failure of attack reduces to an attack on the channel.

The argument for anonymity is very similar.

E | *Exercise* 13.15. State a claim similar to Proposition 13.11 and prove it.

Remark. We have only claimed that this scheme is "somewhat" anonymous. It will not be anonymous against active adversaries, but an active adversary will prevent establishment which the users will notice.

Noticing the attack is little comfort against certain adversaries. However, passive adversaries against anonymity are fairly common, so providing security against them may be worthwhile.

CHAPTER 14

Cryptographic Voting

We began the discussion of voting in Section 8.5, with two very simple schemes based on public key encryption. The main issue with these schemes is the lack of integrity and privacy if the party doing the counting is dishonest.

The most important property of an election is integrity, that the result reflects the ballots the voters wanted to cast. There are many elections with no privacy at all, but an election without integrity does not mean much. If there are no requirements for privacy, integrity is usually trivial to achieve.

On the other hand, integrity cannot be achieved without privacy in many elections, because social mechanisms will constrain voter choice unless that choice is private. While this is often subtle, an extreme variant is where a voter is *coerced* to cast a particular ballot.

We begin this chapter by expanding the definition of voting scheme to first of all distribute the counting operation and make it verifiable. Second, we add a shuffle step, which rerandomises and shuffles a collection of ciphertexts.

We should immediately remark that there is a huge number of choices possible for the general strategy of a voting scheme. The choices we make are quite popular and general, but there are reasonable alternatives.

For most of the cryptographic schemes in this book, it has been obvious how the scheme is used. This is less obvious for cryptographic voting schemes, so we must discuss how to build a voting protocol based on a cryptographic voting scheme. Again, given a particular cryptographic voting scheme, there are many ways to organise an election, typically based on some *trust* structure.

One important notion with respect to integrity is that of a *verifiable* election, where in addition to the election result the voting scheme provides arguments for each operation. We say that protocol is verifiable if the result is consistent with the cast ballots whenever these arguments are all accepted.

The threat landscape for voting is varied and interesting. Our main definition is mostly geared towards protecting a voter with trusted computational ability from the election authorities.

Since voters cannot do non-trivial computations, they will be using some ballot casting device. If this device is not under the voter's sole control, it

DOI: 10.1201/9781003149422-14

is a threat to both privacy and integrity. Our second topic in this chapter is to study how to achieve integrity in the presence of a compromised ballot casting device, ensuring that the ballot was *cast as intended*. There are generic approaches to solving this problem, but it is unclear how well they work. We shall study more specific methods that may work better in practice.

The third topic in this chapter is to study coercion resistance, which is a form of privacy, but under a different threat model than we usually study. Achieving coercion resistance on its own is not very difficult under reasonable coercion models. The challenge is that we want to achieve both coercion resistance and verifiability, two notions that are in some sense contradictory.

14.1 DEFINITIONS

We will now augment Definition 8.13 to include functionality to distribute counting, shuffle encrypted ballots and verify all operations.

Distributing the counting requires changing key generation, so that the decryption key is replaced by a collection of key shares. There must be an access structure for the key shares, which is part of the scheme definition. The distributed counting operation produces a share of the result, as well as a proof of correct counting. The result shares must be combined into a result, and the algorithm to do so should also verify the proofs of correct counting.

Shuffling encrypted ballots requires a shuffle operation, which outputs a collection of encrypted ballots and a proof of correct shuffling. We need a separate algorithm to verify the shuffles.

There may be a limit on how many times encrypted ballots can be shuffled. The encrypted ballots output by the shuffle operation may also be easily distinguishable from the encrypted ballots output by the casting algorithm.

Remark. In order to simplify notation, we shall adopt the convention of writing a tuple $(c_1, \ldots, c_l)$ as $\mathbf{c}$, similar to the notation in Section 11.4.4. We denote the length l of the tuple by $|\mathbf{c}|$. When we have multiple tuples we denote the jth tuple by $\mathbf{c}_j$ and the ith entry in the jth tuple by $c_{j,i}$.

Definition 14.1. A *cryptographic voting scheme* VOTE consists of a (totally ordered) set of ballots $\mathfrak{P}$, a set of associated data $\mathfrak{F}$, an access structure $\mathfrak{A}$ on $\mathfrak{U}$ and five algorithms that work as follows:

- The *setup* algorithm $\mathcal{VS}$ runs with no input and outputs a *ballot casting key* bk and a *counting key sharing* ck.

- The *casting* algorithm $\mathcal{CB}$ takes as input a ballot casting key bk, associated data $ad \in \mathfrak{F}$ and a ballot $v \in \mathfrak{P}$, and outputs an encrypted ballot c and a *ballot argument* arg^c.

- The *shuffle* algorithm $\mathcal{SH}$ takes as input a ballot casting key and encrypted ballots. It outputs the same number of encrypted ballots and a *shuffle argument* arg^s.

- The *verification* algorithm $\mathcal{VV}$ either takes as input a ballot casting key, associated data, an encrypted ballot and a ballot argument, or a ballot casting key, two equal-length sequences of encrypted ballots and a shuffle argument. In either case, the algorithm outputs 0 or 1.

- The *counting* algorithm $\mathcal{VC}$ takes as input a counting key share ck_P and l encrypted ballots $\mathbf{c}$ and outputs a result share.

- The *reconstruction* algorithm $\mathcal{VR}$ takes as input a ballot casting key, a sequence of encrypted ballots and a collection of result shares, and outputs either a result or the special symbol $\perp$.

A voting scheme is (l_0, ν_0)-*correct for a counting function* f if for any $l < l_0$, $\nu < \nu_0$, any (bk, ck) output by $\mathcal{VS}$, any set $\mathfrak{U}_0 \in \mathfrak{A}$, any list of distinct associated data ad and any list of ballots $\mathbf{v}$, and with

$$(c_{0,i}, arg_i^c) \leftarrow \mathcal{CB}(bk, ad_i, v_i) \text{ for } i = 1, 2, \dots, l,$$
$$(\mathbf{c}_j, arg_j^s) \leftarrow \mathcal{SH}(bk, \mathbf{c}_{j-1}) \text{ for } j = 1, 2, \dots, \nu,$$
$$res_P \leftarrow \mathcal{VC}(ck_P, \mathbf{c}_\nu) \text{ for } P \in \mathfrak{U}_0, \text{ and}$$
$$res \leftarrow \mathcal{VR}(bk, \mathbf{c}_\nu, \{(P, res_P) \mid P \in \mathfrak{U}_0\}),$$

then $res = f(\mathbf{v})$. Also,

$$\mathcal{VV}(bk, ad_i, c_i, arg_i^c) = 1 \text{ for } i = 1, 2, \dots, l \text{ and}$$
$$\mathcal{VV}(bk, \mathbf{c}_{j-1}, \mathbf{c}_j, arg_j^s) = 1 \text{ for } j = 1, 2, \dots, \nu.$$

The scheme is *correct for a counting function* f if it is (l_0, ν_0)-correct for any l_0 and ν_0.

The indirect choice of encrypted ballots to count in Figure 8.6 simplified much of the accounting in Section 8.5, but this does not work in the more complex setting. The accounting in this game is therefore quite different.

The experiment from Figure 8.6 must be modified to take into account these extra algorithms. We also want to model integrity for a voting system.

Before we can define privacy and integrity, we must first define a few important concepts with respect to the game between the experiment and the adversary. Consider a count or test query with a ciphertext list $\mathbf{c}$ of length l. A *path* for the query is a sequence of ciphertext lists $\mathbf{c}_0, \mathbf{c}_1, \dots, \mathbf{c}_\nu$ such that $\mathbf{c}_\nu = \mathbf{c}$, there are tuples $(\mathbf{c}_{j-1}, \mathbf{c}_j, \cdot)$ in L_S for $j = 1, 2, \dots, \nu$, and there are tuples $(ad_i, v_{0,i}, v_{1,i}, c_{0,i}, \cdot)$ in L_C for $i = 1, 2, \dots, l$. The ciphertext list $\mathbf{c}_0$ is an *origin* for the query, and the ballot list $v_{b,1} v_{b,2} \dots v_{b,l}$ are *originating ballots* for the query. A count or test query is *proper* if it has a path. A priori, a count or test query can have many paths and origins.

A proper test query is *wrong* if there are no originating ballots $v_{b,1} v_{b,2} \dots v_{b,l}$, $\perp$ appearing exactly l' times, and ballots $v_1' v_2' \dots v_{l'}'$ such that the result equals $f(v_{b,1} v_{b,2} \dots v_{b,l} v_1' v_2' \dots v_{l'}')$. (Recall that we extended the definition of count functions to strings that contained the symbol $\perp$.) A proper

The experiment for the voting scheme security game with an adversary $\mathcal{A}$ proceeds as follows:

1. Sample $b, b'' \xleftarrow{r} \{0, 1\}$, and let L_C and L_S be empty lists.

2. Compute $(bk, ck_1, \ldots, ck_{l_s}) \leftarrow \mathcal{VS}$ and send bk to $\mathcal{A}$.

3. When the adversary sends a query (ad, v_0, v_1) (*challenge*), do:

    (a) Compute $(c, arg^c) \leftarrow \mathcal{CB}(bk, ad, v_b)$.

    (b) Append (ad, v_0, v_1, c, arg^c) to L_C. Send (c, arg^c) to $\mathcal{A}$.

4. When the adversary sends a query (ad, c, arg^c) (*chosen ciphertext*), do:

    (a) If some tuple $(ad, \cdot, \cdot, c, arg^c)$ is in L_C or $\mathcal{VV}(bk, ad, c, arg^c) \neq 1$, send $\perp$ to the adversary and stop.

    (b) Append $(ad, \perp, \perp, c, arg^c)$ to L_C. Send $\top$ to $\mathcal{A}$.

5. When the adversary sends a query $\mathbf{c}$ (*shuffle*), do:

    (a) Compute $(\mathbf{c}', arg^s) \leftarrow \mathcal{SH}(bk, \mathbf{c})$.

    (b) Append $(\mathbf{c}, \mathbf{c}', arg^s)$ to L_S. Send $(\mathbf{c}', arg^s)$ to $\mathcal{A}$.

6. When the adversary sends a query $(\mathbf{c}, \mathbf{c}', arg^s)$ (*chosen shuffle*), do:

    (a) If $(\mathbf{c}, \mathbf{c}', arg^s)$ is recorded in L_S, $|\mathbf{c}| \neq |\mathbf{c}'|$ or $\mathcal{VV}(bk, \mathbf{c}, \mathbf{c}', arg^s) \neq 1$, send $\perp$ to the adversary and stop.

    (b) Append $(\mathbf{c}, \mathbf{c}', arg^s)$ to L_S. Send $\top$ to $\mathcal{A}$.

7. When the adversary sends a query $(j, \mathbf{c})$ (*count*), do:

    (a) Compute $res_j \leftarrow \mathcal{VC}(ck_j, \mathbf{c})$ and send res_j to $\mathcal{A}$.

8. When the adversary sends a query $(\mathfrak{U}_0, \mathbf{c}, \{(P, res_P) \mid P \in \mathfrak{U}_0\})$ (*test*), do:

    (a) Compute $res \leftarrow \mathcal{VR}(bk, \mathbf{c}, \{(P, res_P) \mid P \in \mathfrak{U}_0\})$. Send res to $\mathcal{A}$.

9. When the adversary sends a query P (*reveal*), do:

    (a) Send ck_P to $\mathcal{A}$.

Eventually, the adversary outputs $b' \in \{0, 1\}$.

FIGURE 14.1 The experiment $\mathbf{Exp}_{\text{VOTE}}^{\text{vote}}(\mathcal{A})$ for the security game for a voting scheme VOTE with adversary $\mathcal{A}$. The bit b'' is not used in the experiment but is used to simplify the calculation of advantage.

test query $(\mathfrak{U}_1, \mathbf{c}, S)$ *fails* if it outputs $\perp$, but there is some subset $\mathfrak{U}_0 \subseteq \mathfrak{U}_1$ such that $\mathfrak{U}_0 \in \mathfrak{A}$ and for any $P \in \mathfrak{U}_0$, there was a count query $(P, \mathbf{c})$ that resulted in $resp$ and $(P, resp) \in S$. Two proper test queries that can have identical origin (up to order) are *inconsistent* if they return distinct results.

A proper count query is *balanced* if $f(v_{0,1}v_{0,2} \cdots v_{0,l}) = f(v_{1,1}v_{1,2} \cdots v_{1,l})$.

An execution is *unsound* if there are proper test queries that are inconsistent, are wrong or fail. An execution is *fresh* if every count query is both proper and balanced, and the set of revealed count key shares is not in the voting scheme's access structure.

Definition 14.2. A $(\tau, l_v, l_c, l_d, l_s)$-*adversary against a voting scheme* VOTE is an interactive algorithm $\mathcal{A}$ that interacts with the experiment in Figure 14.1 making count queries with at most l_v encrypted ballots, l_c challenge queries and l_d chosen ciphertext queries, l_s shuffle/chosen shuffle queries, and where the runtime of the adversary and the experiment is at most τ.

An adversary $\mathcal{A}$ is *non-adaptive* if there is a set $\mathfrak{U}_{\mathcal{A}} \notin \mathfrak{A}$ such that $\mathcal{A}$ never makes a reveal query for $P \notin \mathfrak{U}_{\mathcal{A}}$.

The *advantage* of this adversary is defined to be

$$\mathbf{Adv}^{\text{vote}}_{\text{VOTE}}(\mathcal{A}) = \max\{2|\Pr[E_d] - 1/2|, \Pr[E_{int}]\},$$

where the *confidentiality* event E_d is the event that $b' = b$ given that the execution is fresh, or that $b' = b''$ if the execution is not fresh; and the *integrity* event E_{int} is the event the execution is unsound.

We discussed the event E_d in Section 8.5. The idea is that for a list of encrypted ballots, the result should be uniquely determined and should be based on the ballots that were cast, even with adversarially cast ballots.

The experiment only knows the ballots that have been submitted through challenge queries, not the ballots submitted through chosen ciphertext queries. Our notion of wrong test query on its own gives the adversary wide latitude in how to count these encrypted ballots. It is the notion of inconsistent test queries that prevent the adversary from using chosen ciphertext queries to introduce unwanted flexibility in the count.

Our definition of voting scheme does not provide an algorithm for determining which ballot is inside a particular encrypted ballot. This could have simplified the integrity definition slightly. While we could add such functionality, our definition works and avoids superfluous synthetic machinery.

14.1.1 Example: An ElGamal-based Voting Scheme

This scheme is in some sense a natural evolution of the scheme from Section 8.5.1, augmented with a shuffle, distributed decryption and some extra work during encryption. The idea is that one or more infrastructure players rerandomise and shuffle the encrypted ballots before a collection of infrastructure players jointly decrypt.

One thing worth noting is that we no longer encode the associated data in the encrypted ballot, but instead encode it in the ballot argument. This is because we need to separate the decryption from the associated data, which typically contains information identifying the voter who cast the ballot.

Concretely, the scheme is based on ElGamal from Section 3.2.1, using the equality of discrete logarithms argument from Example 11.4 with the shuffle argument from Section 11.4.4 and the distributed decryption scheme from Section 12.3. We add an extra term to the initial ElGamal encryption that allows efficient ballot extraction in our security proofs.

Remark. It is fairly easy to generalise this example to use a generic public key encryption schemes, given appropriate primitives. Of course, these primitives may be non-trivial or expensive. In particular, the extra term added to the ElGamal encryption does not easily generalise, but it can often be replaced by a commitment to the ballot and an argument that the encryption and the commitment contain the same ballot.

Example 14.1. Let G be a cyclic group of prime order p with generator g, let $(\mathcal{P}_e, \mathcal{V}_e)$ be an argument for equal discrete logarithms for G, let $(\mathcal{P}_s, \mathcal{V}_s)$ be a shuffle argument for ElGamal over G, and let $\text{DDEC} = (\mathcal{DK}, \mathcal{DD}, \mathcal{DR})$ be a distributed decryption scheme for ElGamal over G.

The voting scheme $\text{EGVOTE} = (G, \mathfrak{F}, \mathfrak{A}, \mathcal{VS}, \mathcal{CB}, \mathcal{SH}, \mathcal{VV}, \mathcal{VC}, \mathcal{VR})$ works as follows:

- The *setup* algorithm $\mathcal{VS}$ samples $h \xleftarrow{r} G$, $b \xleftarrow{r} \{0, 1, \ldots, p-1\}$, computes $y = g^b$ and $(d, pc) \leftarrow \mathcal{DK}(b)$. It outputs the ballot casting key $bk = (y, h, pc)$ and the counting key sharing $ck = d$.

- The *casting* algorithm $\mathcal{CB}$ takes as input a ballot casting key $bk = (y, h, \cdot)$, associated data ad and a ballot $v \in G$. It samples $r \xleftarrow{r} \{0, 1, \ldots, p-1\}$ and computes $x \leftarrow g^r$, $\hat{x} \leftarrow h^r$, $w \leftarrow y^r v$ and $arg^e \leftarrow \mathcal{P}_e(g, h, x, \hat{x}, r, (bk, ad, w))$. It outputs $c = (x, w)$ and $arg^c = (\hat{x}, arg^e)$.

- The *shuffle* algorithm $\mathcal{SH}$ takes as input $bk = (y, \cdot, \cdot)$ and encrypted ballots $c_1 = (x_1, w_1), \ldots, c_l = (x_l, w_l)$. It samples $r_1, \ldots, r_l \xleftarrow{r} \{0, 1, \ldots, p\}$ and π from the set of permutations on $\{0, 1, \ldots, p\}$, and computes $x_i' \leftarrow x_{\pi(i)} g^{r_{\pi(i)}}$ and $w_i' \leftarrow w_{\pi(i)} y^{r_{\pi(i)}}$ for $i = 1, 2, \ldots, l$, letting $c_i' = (x_i', w_i')$. It then computes $arg^s \leftarrow \mathcal{P}_s(y, c_1, \ldots, c_l, c_1', \ldots, c_l', \pi, r_1, \ldots, r_l)$ and outputs $c_1', \ldots, c_l'$ and arg^s.

- The *verification* algorithm $\mathcal{VV}$ with input $bk = (y, h, \cdot)$, ad, $c = (x, w)$ and $arg^c = (\hat{x}, arg^e)$ outputs the result of $\mathcal{V}_e(g, h, x, \hat{x}, arg^e, (bk, ad, w))$.

- The *verification* algorithm $\mathcal{VV}$ with input $(y, \cdot, \cdot)$, $c_1, \ldots, c_l$, $c_1', \ldots, c_l'$ and arg^s outputs the result of $\mathcal{V}_s(y, c_1, \ldots, c_l, c_1', \ldots, c_l', arg^s)$.

- The *counting* algorithm $\mathcal{VC}$ takes as input P, ck_P, encrypted ballots $c_1, \ldots, c_l$ and computes $ms_{P,i} \leftarrow \mathcal{DD}(P, ck_P, c_i)$ for $i = 1, 2, \ldots, l$. It outputs $(ms_{P,1}, \ldots, ms_{P,l})$.

- The *reconstruction* algorithm $\mathcal{VR}$ takes as input $(\cdot, \cdot, pc)$, $\mathfrak{U}_0$, encrypted ballots $c_1, \ldots, c_l$ and $\{(P, (ms_{P,1}, \ldots, ms_{P,l})) \mid P \in \mathfrak{U}_0\}$. It outputs $\bot$ and stops if $\mathfrak{U}_0 \not\subseteq \mathfrak{A}$. Otherwise, it computes $v_i \leftarrow \mathcal{DR}(pc, \mathfrak{U}_0, c_i, \{(P, ms_{P,i}) \mid P \in \mathfrak{U}_0\})$ for $i = 1, 2, \ldots, l$. If any computation fails, it outputs $\bot$ and stops. Otherwise it outputs $f(v_1, \ldots, v_l)$.

E | *Exercise 14.1.* Show that EGVOTE voting scheme is correct for f.

The scheme is correct for any counting function, but it is not secure for any counting function, since the ballots can be derived from the count shares.

T | **Theorem 14.1.** *Let f be the counting function that returns a sorted list of ballots. Let $\mathcal{A}$ be a non-adaptive $(\tau, l_v, l_c, l_d, l_s)$-adversary against the EGVOTE voting scheme making at most l_h random oracle queries. Then there exists a τ_1'-adversary $\mathcal{B}_1$ against integrity for DDEC, a τ_2'-adversary $\mathcal{B}_2$ against Decision Diffie-Hellman, τ_3'-adversaries $\mathcal{B}_{3,1}$ and $\mathcal{B}_{3,2}$ against Decision Diffie-Hellman and a τ_4'-adversary $\mathcal{B}_4$ against Decision Diffie-Hellman, with τ_1', τ_2', τ_3' and τ_4' all essentially equal to at most a small multiple of τ, such that*

$$\mathbf{Adv}_{\text{vote}}^{\text{EGVOTE}}(\mathcal{A}) \leq \mathbf{Adv}_{\text{DDEC,PKE}}^{\text{ddec-int}}(\mathcal{B}_1) + 2\mathbf{Adv}_{G,g}^{\text{DDH}}(\mathcal{B}_2)$$
$$+ 4l_s(\mathbf{Adv}_{G,g}^{\text{DDH}}(\mathcal{B}_{3,1}) + \mathbf{Adv}_{G,g}^{\text{DDH}}(\mathcal{B}_{3,2})) + 2\mathbf{Adv}_{G,g}^{\text{DDH}}(\mathcal{B}_2)$$
$$+ \frac{2l_h l_c}{p} + \frac{2l_h(l_c + l_d)}{|\mathfrak{B}| - (l_c + l_d)} + \frac{2l_h}{|\mathfrak{B}|} + \frac{2l_h l_v}{p}.$$

The integrity part of the claim follows directly from the soundness of the arguments used, as well as integrity for the distributed decryption.

The remainder of the proof of Theorem 14.1 is structured as a sequence of games. We first simulate all the honestly generated non-interactive arguments during ballot casting. This allows us to randomise the check value $\hat{x}$ in honestly generated ballot arguments, so that we can afterwards embed a trapdoor in h. This trapdoor allows us to extract ballots from adversarially generated encrypted ballots. This in turn allows us to use the known ballots for chosen ciphertexts to simulate the distributed decryption. In order for this to work, we have to extract the permutations from the chosen shuffles, so that we know the correct order for the ballots in the distributed decryption. Finally, we simulate the shuffle arguments and randomise both challenge ciphertexts and shuffle ciphertexts, which makes the ciphertexts independent of the challenge ballots.

Remark. We must be careful with the shuffle argument simulation, since it is computational zero knowledge, not perfect zero knowledge. In particular, the simulation depends on the Decision Diffie-Hellman problem with our decryption key as one of the exponents. This means that we have to stop using the decryption key before we can simulate the shuffle arguments.

Proof of Theorem 14.1   We begin with integrity.

T **Lemma 14.2.** *Under the conditions of Theorem 14.1, there exists an τ_1'-adversary $\mathcal{B}_1$ against integrity for the distributed decryption scheme* DDEC, *with τ_1' essentially equal to τ, such that*

$$\Pr[E_{int}] \leq \mathbf{Adv}_{\mathrm{DDEC,PKE}}^{\mathrm{ddec\text{-}int}}(\mathcal{B}_1) + \frac{l_h(l_c + l_d)}{|\mathfrak{B}| - (l_c + l_d)} + \frac{l_h}{|\mathfrak{B}|}.$$

Proof. If a shuffle argument holds, we know that the decryptions of the shuffled ciphertexts equal the decryptions of the original ciphertexts, up to order. It follows that a proper test query cannot be wrong unless a shuffle argument fails or the distributed decryption is wrong. Since the decryption of ciphertexts is uniquely defined, we cannot have two proper test queries that are inconsistent unless a shuffle argument fails or the distributed decryption is wrong.

Exercise 11.58 tells us the soundness error of the non-interactive shuffle argument. It follows that we can use $\mathcal{A}$ to build an adversary against integrity for the distributed decryption, and the claim follows. □

In the following, let $E_{d,i}$ be the event in Game i that the adversary's guess b' equals the experiment's challenge bit b given that the execution is fresh.

G **Game 0** The initial game is $\mathcal{A}$ interacting with the voting experiment $\mathbf{Exp}_{\mathrm{EGVOTE}}^{\mathsf{vote}}(\mathcal{A})$. Then

$$\mathbf{Adv}_{\mathrm{EGVOTE}}^{\mathsf{vote}}(\mathcal{A}) = \max\{2|\Pr[E_{d,0}] - 1/2|, \Pr[E_{int}]\}. \tag{14.1}$$

Let $\tau_0 = \tau$.

G **Game 1** In this game, when responding to a challenge query we simulate the equality of discrete logarithm argument. This is only noticable if the reprogramming of the random oracle fails. We get that

$$|\Pr[E_{d,1}] - \Pr[E_{d,0}]| \leq \frac{l_h l_c}{p}. \tag{14.2}$$

The runtime τ_1 is essentially equal to τ_0.

G **Game 2** In this game, when responding to a challenge query we sample $\hat{x} \xleftarrow{r} G$ instead of computing it. If the adversary exceeds its bounds, we sample $b' \xleftarrow{r} \{0,1\}$ and stop.

The runtime τ_2 is essentially equal to τ_1.

E *Exercise 14.2.* Show that there exists an τ_2'-distinguisher $\mathcal{B}_2$ for Decision Diffie-Hellman, with τ_2' essentially equal to τ_2, such that

$$|\Pr[E_{d,2}] - \Pr[E_{d,1}]| \leq \mathbf{Adv}_{G,g}^{\mathsf{DDH}}(\mathcal{B}_2). \tag{14.3}$$

Hint: With ElGamal, we can encrypt using the decryption key, so that we do not have to know the randomness used to encrypt.

Game 3 This game is about bookkeeping. We sample $b_0 \xleftarrow{r} \{0, 1, \ldots, p - 1\}$ and compute $h \leftarrow y^{b_0}$. For a hash query for the non-interactive shuffle argument involving $\xi \in G$ to get (ζ, β), we sample $w \xleftarrow{r} \{0, 1, \ldots, p - 1\}$, compute the hash response as $\zeta \leftarrow g^w$ and record w together with h.

For a chosen ciphertext query $(ad, x, w, \hat{x}, arg^e)$ that is accepted, we compute $v \leftarrow w\hat{x}^{-b_0^{-1}}$ and append (ad, v, v, x, w, arg^e) to L_C.

When the adversary makes a shuffle query $\mathbf{c}$, we append $(\mathbf{c}, \mathbf{c}', arg^s, \pi)$ to L_S, where π is the permutation used by $\mathcal{SH}$.

For a chosen shuffle query $(\mathbf{c}, \mathbf{c}', arg^s)$ that is accepted, we use the value w that we recorded for the relevant hash query together with Exercise 11.58 to extract the permutation π used. If the extraction fails, we set π to be the identity permutation. Then we append $(\mathbf{c}, \mathbf{c}', arg^s, \pi)$ to L_S

These changes are not observable, so

$$\Pr[E_{d,3}] = \Pr[E_{d,2}]. \tag{14.4}$$

Embedding the trapdoor in ζ requires one exponentiation, potentially for each of the l_h hash queries. This means that τ_3 is at most a small multiple of τ_2.

Game 4 In this game, we simulate the distributed decryption and the sharing of the decryption key. To get the correct decryption we find some origin for the ciphertext list, and extract a ballot list $\mathbf{v}$ from L_C. Then we compose the permutations recorded in L_S from the query's path into a single permutation π, and use $\pi\mathbf{v}$ as the input to the distributed decryption simulator.

While the distributed decryption simulator is perfect by Proposition 12.11, it may fail if reprogramming random oracles fails. Therefore, we run the simulator until we get a success, with the maximal number of runs set so that the probability that any query response fails is at most $1/p$.

If the adversary exceeds its bounds, we sample $b' \xleftarrow{r} \{0, 1\}$ and stop.

This change is only observable if we recover incorrect ballots from a chosen ciphertext query or an incorrect permutation from a chosen shuffle query. In either case, if this happens we must have accepted some unsound argument. The maximal soundness error for the arguments used is $(l_c + l_d)/(|\mathfrak{B}| - (l_c + l_d)) + 1/|\mathfrak{B}|$. It follows that

$$|\Pr[E_{d,4}] - \Pr[E_{d,3}]| \leq \frac{l_h(l_c + l_d)}{|\mathfrak{B}| - (l_c + l_d)} + \frac{l_h}{|\mathfrak{B}|} + \frac{l_h l_v}{p}. \tag{14.5}$$

We do some extra work for each decryption share query, but since the number of decryption share queries is small compared to τ, it follows that τ_4 is essentially equal to τ_3.

At this point, the decryption key shares are no longer used.

Game 5 In this game, we simulate every honest shuffle argument. If the adversary exceeds its bounds, we sample $b' \xleftarrow{r} \{0, 1\}$ and stop the game.

Again, the number of honest shuffle queries is small, so the runtime τ_5 is essentially equal to τ_4.

E | *Exercise* 14.3. Use Proposition 11.20 and a standard hybrid argument to show that τ_3'-distinguishers $\mathcal{B}_{3,1}$ and $\mathcal{B}_{3,2}$ exist, with τ_3' essentially equal to τ_5, such that

$$|\Pr[E_{d,5}] - \Pr[E_{d,4}]| \le 2l_s(\mathbf{Adv}_{G,g}^{\mathsf{DDH}}(\mathcal{B}_{3,1}) + \mathbf{Adv}_{G,g}^{\mathsf{DDH}}(\mathcal{B}_{3,2})). \tag{14.6}$$

Remark. We could probably write a tighter reduction for this step, getting rid of the factor l_s, but the number of shuffles will typically be very small, so there would be little gain for significant extra work.

G | **Game 6**   In this game, we encrypt random messages in challenge queries. We also let the output ciphertexts of shuffle queries be encryptions of random messages, instead of rerandomisations of the input ciphertexts.

For each challenge query and shuffle query, we actually do slightly less work, so the runtime τ_6 is essentially equal to τ_5.

E | *Exercise* 14.4. Show that a τ_4'-distinguisher $\mathcal{B}_4$ exist, with τ_4' essentially equal to τ_6, such that

$$|\Pr[E_{d,6}] - \Pr[E_{d,5}]| \le \mathbf{Adv}_{G,g}^{\mathsf{DDH}}(\mathcal{B}_4). \tag{14.7}$$

Everything the adversary now sees is independent of b, so

$$\Pr[E_{d,6}] = 1/2. \tag{14.8}$$

The theorem follows from Lemma 14.2 and equations (14.1)–(14.8).

We use non-adaptive corruption because adaptive corruption will significantly complicate the construction and the security proof, and seems not add much to the practical security, as discussed in Chapter 12.2.

The extra term $\hat{x}$ in the ballot argument is required because we need to extract the decryptions of the chosen ciphertexts without using the decryption key. Practically, this term and the ballot argument defends against an attack where an adversary copies and resubmits someone else's encrypted ballot, after which the result will reveal the challenge bit.

The traditional way to defend against this attack is to include an argument of knowledge of the ballot in the ballot argument, so that it is hard to construct an encrypted ballot with an argument without also knowing its decryption. In particular, the traditional solution has been the Schnorr argument from Exercise 11.5 made non-interactive with Fiat-Shamir from Section 11.3.3.

Extraction via rewinding is tricky to use in these contexts, because we need to use the extracted value to complete the game. In general the adversary may structure the generation of the non-interactive arguments such that rewinding does not work. We can reasonably avoid this situation by forbidding the adversary from making chosen ciphertext queries after the first count query.

A more important problem is that extractor analysed in Section 11.14 has a very low success rate. In order to improve the success rate, we would need to run the adversary many times, which would cause a significant increase in runtime, which may force us to use a large group. As usual, it is somewhat unclear how we should interpret that result.

A less important problem is that using an extractor would significantly complicate the presentation of the security proof.

This explains why we do not use a Schnorr argument, even though that would seem to be less computationally expensive than our chosen solution.

Remark. The attack where an adversary copies and resubmits someone else's encrypted ballot is blocked because of the requirement that every encrypted ballot in the origin must have distinct associated data, which means that the equality of discrete logarithms argument cannot be directly copied.

But we never proved that our arguments cannot be copied like this, and we will not prove it directly. Indeed, it would be tricky to prove, since it *would not contradict soundness*. Instead, it follows as a side effect of the general proof strategy, where we first simulate the arguments and later randomise the check value. At that point, the statement is no longer true. And since the adversary must use different associated data, the adversary must essentially create a new argument for an untrue statement, which would contradict soundness.

This effect is subtle. When we apply the simulator to untrue statements, the simulator is under no obligation to provide anything useful at all. But if so, it would trivially turn into a Decision Diffie-Hellman distinguisher. Since we do not believe the simulator is a good distinguisher, we must also accept that it can simulate arguments for untrue statements.

Remark. It is worthwhile to note that we now have a solid cryptographic solution to the highly non-trivial cryptographic problem of voting. It is also worth noting that this solution required almost all of the cryptographic theory we have covered in this book.

14.1.2 Example: Homomorphic Counting

Our next system is essentially based on the scheme from Section 8.5.2, with the added ingredients from the previous example. This voting scheme only supports a yes/no election, but it actually has stronger security properties, and we do not have to use shuffles to achieve security.

The scheme is again based on ElGamal. We encode 0 as the identity element and 1 as g. We count the ballots by multiplying the ciphertexts, decrypting the product and then computing the discrete logarithm of the decryption.

Remark. Note that the discrete logarithm is very small, so exhaustive search will typically be sufficiently fast. In the unlikely event that something faster is needed, Shanks' method from Section 2.2.4 suitably tuned will work well.

As discussed in Section 8.5.2, we must ensure that every cast ballot is 0 or 1 to get a correct count, so we use the argument from Exercise 11.36.

Remark. In order to make the scheme fit the definition, we have to define a dummy shuffle algorithm and a trivial shuffle verification algorithm.

Example 14.2. Let G be a cyclic group of prime order p with generator g, let $(\mathcal{P}_e, \mathcal{V}_e)$ be an argument for equal discrete logarithms for G, let $(\mathcal{P}_v, \mathcal{V}_v)$ be an argument for ElGamal encryptions of 1 or g, and let $\text{DDEC} = (\mathcal{DK}, \mathcal{DD}, \mathcal{DR})$ be a distributed decryption scheme for ElGamal over G.

The voting scheme HOMVOTE $= (\{0,1\}, \mathfrak{F}, \mathfrak{A}, \mathcal{VS}, \mathcal{CB}, \mathcal{SH}, \mathcal{VV}, \mathcal{VC}, \mathcal{VR})$ works as follows:

- The *setup* algorithm $\mathcal{VS}$ samples $h \xleftarrow{r} G$, $b \xleftarrow{r} \{0, 1, \ldots, p-1\}$, computes $y = g^b$ and $(d, pc) \leftarrow \mathcal{DK}(b)$. It outputs the ballot casting key $bk = (y, h, pc)$ and the counting key sharing $ck = d$.

- The *casting* algorithm $\mathcal{CB}$ takes as input a ballot casting key $bk = (y, h, \cdot)$, associated data ad and a ballot $v \in \{0, 1\}$. It samples $r \xleftarrow{r} \{0, 1, \ldots, p-1\}$ and computes $x \leftarrow g^r$, $\hat{x} \leftarrow h^r$, $w \leftarrow y^r g^v$, $arg^e \leftarrow \mathcal{P}_e(g, h, x, \hat{x}, r, (bk, ad, w))$ and $arg_v \leftarrow \mathcal{P}_v(ad, \{1, g\}, y, x, w, g^v, r)$. It outputs $c = (x, w)$ and $arg^c = (\hat{x}, arg^e, arg_v)$.

- The dummy *shuffle* algorithm $\mathcal{SH}$ takes as input $bk = (y, \cdot, \cdot)$ and encrypted ballots $\mathbf{c}$ and outputs $\mathbf{c}$ and an empty shuffle argument.

- The *verification* algorithm $\mathcal{VV}$ with input $bk = (y, h, \cdot)$, ad, $c = (x, w)$ and $arg^c = (\hat{x}, arg^e, arg_v)$ outputs 1 if $\mathcal{V}_e(g, h, x, \hat{x}, arg^e, (bk, ad, w))$ and $arg_v(ad, \{1, g\}, y, x, w, arg_v)$ both output 1, and otherwise outputs 0.

- The *verification* algorithm $\mathcal{VV}$ with input $(\cdot, \cdot, \cdot)$, $\mathbf{c}$, $\mathbf{c}'$, and an empty argument outputs 1 if $\mathbf{c} = \mathbf{c}'$, and otherwise 0.

- The *counting* algorithm $\mathcal{VC}$ takes as input P, ck_P, encrypted ballots $c_1 = (x_1, w_1), \ldots, c_l = (x_l, w_l)$, computes $ms_P \leftarrow \mathcal{DD}(P, ck_P, (\prod_i x_i, \prod_i w_i))$, and outputs ms_P.

- The *reconstruction* algorithm $\mathcal{VR}$ takes as input $(\cdot, \cdot, pc)$, $\mathfrak{U}_0$, encrypted ballots $c_1 = (x_1, w_1), \ldots, c_l = (x_l, w_l)$ and $\{(P, ms_P) \mid P \in \mathfrak{U}_0\}$. It outputs $\perp$ and stops if $\mathfrak{U}_0 \notin \mathfrak{A}$. Otherwise, it computes $m \leftarrow \mathcal{DR}(pc, \mathfrak{U}_0, (\prod_i x_i, \prod_i w_i), \{(P, ms_{P,i}) \mid P \in \mathfrak{U}_0\})$. If the computation fails, it outputs $\perp$ and stops. Otherwise it outputs $\log_g m$.

Exercise 14.5. Show that the HOMVOTE voting scheme is correct for the counting function that counts the number of 1 ballots.

Remark. Note the similarity of HOMVOTE and EGVOTE, where the cast algorithms are essentially the same, up to one scheme enforcing the smaller ballot set. The real difference is in shuffle and counting algorithms. Informally, it sometimes makes sense to consider a voting scheme as a composition of a ballot casting scheme and a counting scheme. Unfortunately, doing so causes a lot of problems, so we shall not try to formalise this.

Theorem 14.3. *Let f be the counting function that counts the number of 1 ballots. Let $\mathcal{A}$ be a non-adaptive $(\tau, l_v, l_c, l_d, l_s)$-adversary against the* HOMVOTE *voting scheme making at most l_h random oracle queries. Then there exists a τ_1'-adversary $\mathcal{B}_1$ against integrity for* DDEC, *a τ_2'-distinguisher $\mathcal{B}_2$ for Decision Diffie-Hellman and a τ_3'-distinguisher for Decision Diffie-Hellman, with τ_1', τ_2' and τ_3' all essentially equal to τ, such that*

$$\mathbf{Adv}^{\text{vote}}_{\text{HOMVOTE}}(\mathcal{A}) \le \mathbf{Adv}^{\text{ddec-int}}_{\text{DDEC,PKE}}(\mathcal{B}_1) + 2\mathbf{Adv}^{\text{DDH}}_{G,g}(\mathcal{B}_2) + 2\mathbf{Adv}^{\text{DDH}}_{G,g}(\mathcal{B}_3) +$$
$$\frac{2l_h(l_c + l_v) + 1}{p}.$$

Again, integrity follows from the soundness of the arguments used as well as integrity for the distributed decryption. The remainder of the proof could then be structured as a sequence of games, which we proceed to sketch.

We begin with the initial game between the experiment and the adversary which defines the advantage. The first change is to simulate the equality of discrete logarithm argument and the encryption of 1 or g argument when responding to challenge queries. The simulators are perfect, so this is only noticable if the reprogramming of the random oracles fail.

Next, we randomise the check value $\hat{x}$. If this is noticable we get a distinguisher for Decision Diffie-Hellman.

Then we embed a trapdoor in h and introduce bookkeeping to extract ballots from chosen ciphertexts. This change is not observable.

Now we simulate the distributed decryption and the sharing of the decryption key. Again, we recover the correct decryption by finding the origin for the count query and recovering ballots from L_C. Again, we have to account for random oracle reprogramming failures as well as extracting an incorrect ballot from a chosen ciphertext query.

Finally, we can randomise the challenge ciphertexts. If this is noticable, we get another distinguisher for Decision Diffie-Hellman.

Exercise 14.6. Write out the proof of Theorem 14.3 carefully.

The advantage of HOMVOTE over EGVOTE is that it is simpler and it avoids fairly expensive shuffles. Also, the arguments involved in counting are very small. But it just supports yes/no elections, which is a significant functional restriction. As we shall see in the next section, there are some practical security differences between the two schemes.

If we replace ElGamal by the Paillier public key encryption scheme (and also modify somewhat the arguments), we get a scheme that can be extended to more complicated counting functions, for instance 1-out-of-l choices elections. We do not investigate this further than we did in Section 8.5.

14.2   HOW TO USE A VOTING SCHEME

While we have not followed this convention, it is sometimes convenient to distinguish between a *scheme* and a *protocol* in the following sense: The former is essentially a collection of (interactive) algorithms, perhaps with a few sets attached, while the latter is a collection of interactive algorithms attached to *roles*, along with a description of user sets and infrastructures the protocol relies on, such as networks and public key infrastructures, again perhaps with a few sets attached. For the former, the adversary is allowed to reveal key material, while for the latter the adversary *corrupts* players.

The objects we have studied so far in this book would all be cryptographic schemes under this classification. But when relevant there is mostly a trivial and obvious cryptographic protocol built on top of a public key infrastructure and some suitable network. It is also mostly obvious how the security properties of the cryptographic scheme imply security for the application. We have studied schemes instead of protocols because schemes are easier to analyse.

For voting schemes, it is far less obvious how to build a protocol, and in fact a single voting scheme can be used in different ways to achieve different election security goals. This section gives a brief overview of how we can build and analyse cryptographic voting protocols. We note that a proper discussion of these topics is no longer pure cryptography, but a multi-disciplinary study involving psychology, human-computer-interaction, game theory and other fields. This is beyond the scope of this book.

14.2.1   A Cryptographic Protocol

Design   Figure 14.2 describes one possible way to use a cryptographic voting scheme to achieve useful election security properties. The election proceeds in three *phases*: setup, casting and counting.

During the *setup phase*, the setup algorithm is run by a trusted party, which distributes the ballot casting key to every player and one counting key share to each counter. This is not shown in the figure.

Remark. We do not want a trusted party to run the setup algorithm. Since this process happens before the election, it is usually not very time critical. Standard multiparty computation will usually be sufficient to distribute setup. Faster and more convenient methods are often available.

In the *casting phase*, the voters input their ballots to their ballot casting devices, which run the casting algorithm with appropriate associated data. The output and the associated data is given to the voter and the ballot box. The voters accept their ballots as cast when they receive the device output.

The *ballot box* serves as storage and conduit for encrypted ballots and their corresponding associated data and ballot proofs. When the *counting phase* starts, the ballot box selects the encrypted ballots to be counted and sends them to the first shuffle server. The ballot box will follow some for now

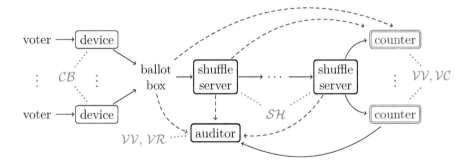

FIGURE 14.2 One configuration for an election using a cryptographic voting scheme. Players enclosed by double lines have secret key material. The solid arrows denote the flow of (encrypted) ballots and result shares. The dashed arrows denote the flow of audit data. The dotted lines connect players with the algorithms they run. (Many arrows to and from the counters were omitted to aid legibility.)

unspecified rule to select the ballots, but in particular it will never select an encrypted ballot with a ballot argument that does not verify, nor will the ballot box select two ballots with identical associated data.

The *shuffle servers* simply apply the shuffle algorithm to the encrypted ballots they receive, passing the output encrypted ballots to the next shuffle server. The final shuffle server sends the output to the counters.

The *counters* run the counting algorithm on their counting key share and the encrypted ballots they receive, sending result shares to the auditor. The *auditor* runs the result reconstruction algorithm and announces the result.

In order to ensure that the encrypted ballots have been properly processed, the ballot box and the shuffle servers all send their inputs and outputs to all the counters, who individually verify this information before running the counting algorithm with their counting key shares. The counters include this information in the associated data they use. The auditor also receives and verifies this information, before running the reconstruction algorithm and making the result public.

The auditor also outputs a *certificate* for the election, which contains the ballot casting key, the contents of the ballot box, the output of every shuffle server and the output of the counters. With the certificate, anyone can verify that the auditor output the correct result. Any voter can also verify that their cast ballot was included in the certificate.

Remark. A popular approach is to replace the ballot box and the auditor with a *(web) bulletin board*. A bulletin board is a write-only, authenticated public database. The voters post their ballots to the bulletin board. The shuffle servers and counters get their input from and post their output ton the board. Finally, anyone can recover the result from the bulletin board.

This is in some sense a much simpler design. But a bulletin board is an idealisation. How to realise a bulletin board in the real world is somewhat complicated, and any realisation will not be perfect. This means that the simpler design is analysed in a slightly imperfect model of the real world, which is awkward. This can be handled, but we will not study this approach.

14.2.2   Voting Protocol Security

To properly analyse security, we must consider an *execution model* for the election as a cryptographic protocol. This is described as one of our usual experiments, but we only sketch the technicalities.

The players in the protocol are the *infrastructure* players (ballot box, shuffle servers, counters and auditor), the *voters* and their *devices*. All players are modelled as interactive algorithms, managed by an experiment. The adversary may start pairs of voter and device instances. The adversary will specify the associated data the device uses. The adversary may have one voter cast multiple ballots in parallel. (This is required to model certain real-world attacks.)

The players communicate via a network that is authenticated, which means that the adversary may schedule the delivery of network messages, but cannot modify messages sent between parties, nor change the delivery order. Also, the voters and their devices communicate privately, meaning that the adversary cannot see the contents of network messages sent between voters and devices.

Remark. For voter-device communication, this models physical reality. For other communication, this is a very simplified model of messaging protocols as discussed in Chapter 13. We could easily include a suitable messaging protocol and public key infrastructure into the protocol, but for our purposes this would significantly complicate analysis at no gain to understanding.

The setup phase begins by starting the infrastructure players. Eventually, the adversary signals for setup to end, at which point the setup algorithm is run and the key material is distributed to the players.

The casting phase then starts. In this phase, the adversary provides a voter identity, associated data and a pair of ballots to the experiment, which starts an instance of the voter and a paired ballot casting device instance with the appropriate associated data and inputs either the left or the right ballot to the voter instance according to the challenge bit. The adversary then schedules network messages for delivery. The only infrastructure player active during the casting phase is the ballot box.

Finally, the adversary signals the end of the casting phase. Any network message that has not been delivered at the start of the phase is erased and any remaining voter or ballot casting device instances are halted. In the counting phase, the adversary schedules network messages for delivery. If the auditor publishes a result and a certificate, it sends this to the adversary. Eventually, the adversary outputs its guess for the challenge bit.

The adversary may *corrupt* players, in which case the adversary receives a copy of the player's state and random tape, as well as any player key material. The player then stops working, but the adversary may send and receive messages on the player's behalf. Any key material received by the player is received by the adversary. Any player that has not been corrupted is *honest*.

In general, the adversary may adaptively corrupt a player at any point in time. However, we shall restrict attention to *static* corruption, where no player is corrupted after the setup algorithm has been run. We shall also assume that a voter's ballot casting device is corrupted if and only if the voter has been corrupted, a requirement we relax in Section 14.3.

Privacy   *Privacy* means that who cast which ballot is kept secret. We measure the advantage of the adversary in the usual way. We define *freshness* of the execution similarly to how we defined balanced for a count query.

The intention is that the design preserves the confidentiality of the cryptographic voting scheme under the following *trust* structure: There must be an honest shuffle server and the set of honest counters must be in the access structure. The idea is that freshness for the execution and the actions of the honest players make sure we can simulate a protocol execution using an execution of the voting scheme experiment.

Exercise 14.7. Explain how an adversary $\mathcal{A}$ against privacy for an election can be turned into an adversary $\mathcal{B}$ against the cryptographic voting scheme. Show that under the above trust assumptions the resulting execution of the game between $\mathbf{Exp}^{\text{vote}}_{\text{VOTE}}(\mathcal{B})$ and $\mathcal{B}$ will be fresh, if the protocol execution is fresh. Conclude that $\mathcal{A}$'s advantage equals $\mathcal{B}$'s advantage.

Integrity   *Integrity* means that if a result is published, it is consistent with the cast ballots (subject to a selection rule). A variant is δ-*integrity*, where if a result is published, it is consistent with a $(1-\delta)$-fraction of the cast ballots.

The integrity of an election does not follow from the integrity of the cryptographic voting scheme in the same way that privacy follows, since the former requires all cast ballots to be counted, while the latter does not. Indeed, in the latter the adversary chooses which ballots to count.

There are a number of ways to ensure the integrity of an election. One way is to have one or more observers of the ballot box. The ballot box shares every ballot it receives with the observers, who return their signatures on the ballot. These signatures are returned to the voter's device, which accepts the ballot as cast only if the observer's signatures are correct. At the end of the election, the observers send a copy of the ballots they have seen to the auditor, which compares the list to the data received from the ballot box. Care must be taken to preserve robustness, as discussed below. Like privacy, this requires trusting at least one of the observers.

The ballot box could pretend to have received ballots that the voter did not send, in order to exploit the selection rule. The voter's ballot casting device may sign the encrypted ballot on behalf of the voter.

Verifiability We can achieve integrity without trust by exploiting *verifiability*. The auditor publishes a certificate of the election, which an individual voter can use to verify two things: That their ballot was included in the count and that the count was correctly computed. If so, we say that the voter *accepts* the result. By integrity of the voting scheme, the result must be consistent with the ballots of the voters that accept the result.

In practice, only a very small minority of voters will bother verifying an election certificate. This is sufficient to conclude that the count was correctly computed, but it is not sufficient to conclude that every cast ballot was included in the count. To conclude that almost every ballot was included in the count, we may apply a statistical argument.

E *Exercise 14.8.* Let S be a finite set. A distribution $\mathcal{X}$ on 2^S is (ϵ, δ)-*unavoidable* if for any $V \subseteq S$ with $|V|/|S| \leq \delta$, then if $T \xleftarrow{r} \mathcal{X}$ we have $\Pr[V \cap T = \emptyset] \leq \epsilon$.

(a) Suppose $|S| = 10\,000$ and $\mathcal{X}$ samples subsets with 100 elements with equal probability. For which values of ϵ and δ is $\mathcal{X}$ (ϵ, δ)-unavoidable?

(b) Suppose we have an election with $10\,000$ voters and that 100 voters verify and accept the election certificate. Suppose further that the set of verifying voters was chosen uniformly at random from the sets of 100 voters, and it was chosen after the certificate was published. What is the probability that 5 cast votes were left uncounted. 10? 100? 1000?

Remark. The distribution of voters that verify an election will almost certainly not look uniform to any serious adversary. There will be a small number of voters that will almost always verify, and a large number of voters that will almost certainly never verify. But it may still be sufficiently unpredictable to provide security. How to determine what this distribution is for a given election is not a cryptographic question, and beyond the scope of this book.

With the above discussion, we get δ-integrity for the election under the assumption that a (ϵ, δ)-unavoidable set of voters verify the election.

E *Exercise 14.9.* Show that any adversary against δ-integrity with advantage ϵ_0 for the election when a (ϵ, δ)-unavoidable set of voters verify the election gives an adversary against integrity for the voting scheme with advantage at least $\epsilon_0 - \epsilon$. (If $\epsilon \geq \epsilon_0$, this is obviously an empty statement.)

Robustness   *Robustness* is the property that it is hard for an adversary to prevent an accurate count being published. It is not enough for us to detect that something has gone wrong. We must also be able to recover, somehow. Robustness follows directly from the integrity and design of the scheme, as well as some weak trust assumptions.

If a shuffle server misbehaves, the server can be removed and the shuffles rerun by the others. Sometimes, new servers could be recruited.

As long as some set of counters in the access structure behave correctly, the counters' output will yield a result. In some sense we balance robustness against privacy, with the access structure representing our choice of balance.

Exercise 14.10. Suppose we use a t-out-of-l access structure. Describe the trade-off between privacy and robustness encoded by t and l.

The auditor can obviously refuse to publish any information, but any faulty auditor can easily be replaced. Also, there is no reason why there should only be one auditor. The same applies to shuffle servers.

Fairness and Eligibility   *Fairness* is the property that no voter casts more than one ballot. *Eligibility* is the property that every counted ballot was cast by an eligible voter (however defined).

We get fairness by encoding the voter identity into the associated data, having the ballot box reject ballots with duplicate voter identities, and rejecting any certificates with duplicate voter identities.

The obvious way to ensure eligibility is to have the voter's device sign the ballot on the voter's behalf with a signing key that is known to belong to an eligible voter. This typically relies on a pre-existing public key infrastructure.

Exercise 14.11. A public key infrastructure can be modelled as a public database of pairs of verification keys and voter identities. The adversary creates the contents of the database. The adversary may ask for the database to generate a key pair using the signature key generation algorithm and add the verification key with an adversary-chosen voter identity. The adversary may ask for the database to add a particular adversary-chosen pair to the database. We require that no two pairs may be added to the database with the same voter identity. The adversary may reveal signing keys.

Describe how to use a pre-existing public key infrastructure to ensure that only ballots cast by eligible voters are counted.

We could have integrated signatures into the cryptographic voting scheme by adding a registration algorithm that generates per-voter key material, and modifying the casting and verification algorithms. This would increase the complexity of the definition and analysis. In practice, it also turns out to be difficult to organise an ad-hoc public key infrastructure in the context of an election. Even maintaining a list of eligible voters is surprisingly non-trivial in some situations. Furthermore, as we shall see, there are other methods to ensure eligibility. Therefore, we do not integrate digital signatures into the cryptographic voting scheme, but deal with it at the voting protocol level.

Remark. Using a pre-existing public key infrastructure instead of a dedicated key structure means that another application could generate signatures that confuses the election. Great care must be taken to ensure cryptographic separation when cryptographic key material is used in more than one application. This is easy for signatures if no two applications can sign the same message, which we can guarantee by giving each application a unique message prefix.

This is a special case of a more general technique. It applies to encryption where we can use a unique prefix for associated data. And this technique applies to voting schemes, where we can reuse key material for more than one election by careful use of associated data. Key material reuse may not be desirable for other reasons, though.

Another way to ensure eligibility is to rely on a pre-existing identification system and use a trusted player, often called an authentication server. One approach is for the voter to identify to the authentication server, which then issues a voting credential (perhaps the server's signature on a verification key generated by the voter's device). The credential is somehow used to tie the encrypted ballot (and its ballot proof) to a voter identity. The downside of this approach is that we have to trust the authentication server, but this trust requirement can often be mitigated, perhaps by making logs public or audited.

Remark. In this context we do not consider the identification schemes from Section 9.3, but instead something built on some form of authenticated key exchange, perhaps using passwords and one-time codes or authentication tokens. These systems are typically quite complex and difficult to analyse, but are often easier to deploy and use than classical public key infrastructures.

Accountability *Accountability* is the property that if anything goes wrong in an election, we can determine where the fault lies. This is typically easy to achieve by letting the ballot box, auditor, shuffle servers and counters all have a signature key pair and let them sign their outputs. If any player's outputs later turn out to be faulty, the signature proves that the player was faulty.

The ballot box should also sign the voter's encrypted ballots. If any encrypted ballot has been left out of the count, the voter can prove that it should have been included and that the ballot box is faulty.

Accountability requires that the $\mathcal{VR}$ algorithm outputs the set of faulty result shares. This is often easy (see Section 12.3).

E | *Exercise* 14.12. Explain how to modify the $\mathcal{VR}$ algorithm of the above two examples such that it outputs the set of faulty input shares.

The above is a useful list of election security properties. It is only a partial list, and depending on the election there may be other desirable properties.

14.3   CAST AS INTENDED

So far, we have assumed that a voter's ballot casting device is corrupted if and only if the voter has been corrupted. In some settings the focus is on the voter defending against corrupt election authorities (or rather, not wanting to trust election authorities), which makes it reasonable to assume that the honest voter has some uncorruptible computational ability, in particular because it is hard to achieve anything otherwise. In many other settings we will find that the adversary is not the election authorities, but rather some external attacker (which may compromise one or more infrastructure players, of course). It is completely unreasonable to assume that these attackers will not corrupt an honest voter's ballot casting device.

We have also assumed a *personal* ballot casting device. This need not be the case, and groups of voters may share a single physical ballot casting device. In this case corrupting a ballot casting device will have much greater effect, which means that adversaries will spend more effort doing so.

It follows that we may want to defend against a corrupt ballot casting device. This is often known as ensuring that the ballot was *cast as intended*, that is, that the encrypted ballot really is an encryption of the intended ballot.

Remark. There are concepts known as *stored as cast* and *counted as stored* which through composition ensure that the ballot was *counted as intended*.

Stored as cast essentially means that we should be able to check that the correct encrypted ballot was stored in the ballot box, which is in some sense trivial. Counted as stored would be ensured by the cryptographic machinery.

This is sometimes a useful mental model for thinking about some security properties. It is a reasonable approach for informal discussions of voting protocols with non-experts. It does not replace proper cryptographic definitions.

Cryptography in general is not able to force anyone to do anything. The entire goal of cryptography is instead to detect when someone is not doing what they are supposed to be doing. Given this detection ability, there are other mechanisms that incentivise or enforce proper behaviour.

Therefore, our goal is not forcing a device to cast the intended ballot, but to detect misbehaviour by corrupted devices. The detection need not be perfect, but the detection rate should be fairly robust. This also applies to an adversary trying to avoid detection. The avoidance need not be perfect and a relatively small detection rate can be survivable for a practical adversary.

14.3.1   Benaloh Challenges

Recall the discussion of the Schnorr identification scheme in Section 4.4. The verifier wants to check that the prover behaved correctly. It either asks the prover to reveal its randomness or some other computation. A rational actor will usually behave correctly, since otherwise the detection risk is too large.

We can use a similar idea to detect misbehaviour by the ballot casting device. The device first computes the encrypted ballot as usual. It then commits to the encrypted ballot and asks the voter if they want to cast the ballot or if they want to challenge the encrypted ballot. If the voter decides to cast the ballot, the device forwards it to the ballot box as usual. If the voter decides to challenge, the device reveals the randomness used to encrypt the ballot. The voter can use this randomness and the commitment to the encrypted ballot to check that the device behaved honestly. This is called a *Benaloh challenge*.

As usual, there are a number of practical issues with this idea. Some voters may find Benaloh challenges hard to understand. Unless the scheme is very carefully designed, there is a non-trivial risk that they end up not casting a ballot. Also, a more complicated voting process may cause some fraction of all voters not to vote. These issues can be mitigated, but mitigation will often be a trade-off between potential harm rate and the challenge rate.

The human voter cannot do any non-trivial calculation. This means that the voter needs a second device in order to check the computations of the ballot casting device. We cannot assume that the second device is uncorrupted (otherwise it should have cast the ballot in the first place), but the adversary must now corrupt two devices to guarantee success without detection.

We cannot require that voters have a second device, nor can we require that voters challenge. This may allow an adversary to adaptively select target voters. For instance, if the ballot casting device can detect candidate second devices nearby, it can desist from tampering. The voters who challenge (with significant probability) may also be distinguishable from voters who do not challenge, allowing the adversary to desist from tampering with certain voters.

Some voters may decide to challenge the ballot before they enter it into the ballot casting device. This may change their behaviour, something that may be detectable by the ballot casting device. For instance, if the ballot casting device knows what ballot the voter intends to cast, any other ballot suggests that the voter has decided to challenge the device.

Ensuring safe and effective deployment of Benaloh challenges is non-trivial.

14.3.2   Return Codes

Another approach for detecting tampering is for the ballot box to return some human-verifiable information related to the cast ballot that the ballot casting device cannot fake. If the set of possible ballots is small, this is possible.

Suppose we have a small set of possible ballots $\mathfrak{P}$. For each ballot $v \in \mathfrak{P}$, sample a human-readable code t from some suitable set $\mathfrak{T}$. This process defines a random function $\mu : \mathfrak{P} \to \mathfrak{T}$, which can be presented in a human-readable form (for instance as a small table). The idea is that the human voter has μ, the ballot box somehow computes a *return code* $t \leftarrow \mu(v)$ using the encrypted ballot sent by the ballot casting device, and then sends t to the ballot casting device, which in turn presents t to the voter. The voter may use the human-readable form of μ to verify that the return code presented by the ballot casting device equals $\mu(v)$. Or the voter may choose to ignore the return code.

There is some anecdotal evidence that voters do check return codes in practice. Laboratory studies suggest that return codes are vulnerable to an attack where the ballot casting device simply omits displaying the return code. One plausible explanation of this seeming contradiction is that displaying the return code reminds many voters to verify it, something they do not remember (or care about) if the return code is not displayed.

How would the ballot box compute $\mu(v)$? First, we may use a pseudo-random function instead of a random function. Second, the ballot casting device should not be able to compute the function itself, though it may participate in computing it. Third, the ballot box should not be able to learn anything about the ballot v from $\mu(v)$. Fourth, while we could use any protocol for two-party computation (a slightly more restricted version of multi-party computation), we would prefer a two-move protocol where the ballot box computes a response to the encrypted ballot and the ballot argument.

One candidate function that works well with our ElGamal-based cryptosystems is the function $\mu : \mathfrak{K}_s \times \{0, 1, \ldots, p-1\} \times \mathfrak{P} \to \mathfrak{T}$ given by

$$\mu(v) = f(k, v^d)$$

where $f : \mathfrak{K}_s \times G \to \mathfrak{T}$ is a suitable pseudo-random function. This function is a composition of $f(k, \cdot)$ and $v \mapsto v^d$, so it is a natural two-stage computation. Since $\mathfrak{P} \subseteq G$, the function $v \mapsto v^d$ is a group homomorphism that matches ElGamal as a homomorphic encryptiongroup-homomorphic cryptosystem, so the first stage can be computed on encrypted ballots.

There are many ways to organise the computation, but one way is to give d to the ballot casting device and k to the ballot box. The ballot box should also have an ElGamal key pair (y_r, b_r), and the ballot casting device should know y_r. The idea is that the ballot casting device creates an encryption of v^d under y_r, which the ballot box can decrypt and use to finish the computation of t. The obstacle is that the ballot casting device must convince the ballot box that its encryption contains v^d, when the encrypted ballot contains v.

In order to simplify the argument, we shall construct this encryption in a very particular way. Let $x = g^r$ and $w = y^r v$. Compute $\bar{x} = x^d$, $\bar{w} = w^d$ and $\hat{w} = (y_r/y)^{rd}$. Then $(\bar{x}, \bar{w})$ is an encryption of v^d under y, while $\hat{w}$ is a *key shift* so that $(\bar{x}, \bar{w}\hat{w})$ is an encryption of v^d under y_r.

Remark. The order of the key shift is important. If we do it first, the ballot box would get an encryption of the ballot under (y_r, b_r).

E *Exercise* 14.13. Suppose $\mathfrak{P}$ is a randomly chosen subset of G. Explain why we expect the above technique to hide v from an adversary that knows b_r.

Hint: First, use a collection of DDH tuples together with knowledge of b and b_r to simulate the set $\mathfrak{P}$, the pairs (v, v^d) and the relationship between

(x, w), $(\bar{x}, \bar{w})$ and $\hat{w}$. Then assume $\bar{x}$ and $\bar{w}$ are random and use a different collection of DDH tuples together with knowledge of b_r to simulate (x, w).

The ballot casting device will also have to argue for correctness, but given how we have constructed the encryption this can be done using a series of equality of discrete logarithms arguments. To verify these arguments, the ballot box needs a a public commitment $e = g^d$ to d.

Since the adversary may corrupt voters, both d and k may only be used for one voter. In other words, we need per-voter key material. Note that as usual, we do not want to prescribe how voter identities and authentication is handled, so instead of assigning key material to voters, we tag the material with *voter associated data* ad^v, which is distinct from data associated to casting a ballot. This is because key material and cast ballots have different contexts.

Modified Voting Scheme   We shall sketch the changes to the definition of a voting scheme. It turns out that the two parts of the return code function are best understood as belonging to different layers of abstraction, so we shall only model the map $v \mapsto v^d$ here. The pseudo-random function will be part of the cryptographic protocol to be discussed later. We need to modify the setup, cast and verification algorithms, and also add algorithms for generating per-voter key material (*register*) and return code decryption.

- The *setup* algorithm $\mathcal{VS}$ also outputs a return code decryption key dk_r.
- The *register* algorithm $\mathcal{VU}$ takes as input a ballot casting key bk, ad^v and outputs a set $S = \{(v, \hat{t}) \mid v \in \mathfrak{P}\}$, a *voter casting key* vck and a public commitment pc_v.
- The *cast* algorithm also takes as input ad^v and a voter casting key.
- The *verification* algorithm also takes as input ad^v and a public commitment to a voter casting key.
- The *return code decryption* algorithm $\mathcal{VD}$ takes as input a return code decryption key dk_r, public commitment pc_v, associated data ad and ad^v, an encrypted ballot c and a ballot argument arg^c, and outputs $\bot$ or $\hat{t}$.

In the context of the usual correctness requirements, we also require that when $(S, vck, pc_v) \leftarrow \mathcal{VU}(bk, ad^v)$, $(c, arg^c) \leftarrow \mathcal{CB}(bk, vck, ad, ad^v, v)$ and $\hat{t} \leftarrow \mathcal{VD}(dk_r, pc_v, ad, ad^v, c, arg^c)$, then $(v, \hat{t}) \in S$.

Remark. Return code decryption overlaps with verification of a cast ballot. But in a voting protocol these are distinct tasks done by distinct parties.

E *Exercise* 14.14. Write out carefully the definition of a voting scheme with return codes, as sketched above.

E | *Exercise* 14.15. Adapt the EGVOTE from Example 14.1 to use return codes as sketched above. Show that the scheme is correct.

E | *Exercise* 14.16. Adapt the HOMVOTE from Example 14.2 to use return codes as sketched above. Show that the scheme is correct.

Hint: The direct adaption is actually not secure, since $\{1, g\}$ is not a random subset of G. Make the appropriate modification (see Example 8.32).

We must also modify the experiment from Figure 14.1 to adapt the cast and chosen ciphertext queries to the modified algorithms, add a *register query* that allows the adversary to generate per-voter key material, and add a *return code decrypt query*. We must also adapt the reveal query to allow the adversary to reveal per-voter key material (vck) and the return code decryption key (dk_r).

We must also adapt the definition of fresh. If the adversary reveals some vck and makes a return code decrypt query for some challenge ciphertext, this may reveal the challenge bit. We could attempt to make rules for which combinations of queries are allowed, but this is tricky because of chosen ciphertexts. Instead of going for a complex and optimal solution, we opt for a simple rule: If a challenge query is made for a particular ad^v, no return code decrypt queries may be made for that particular ad^v. If dk_r has been revealed, at most one challenge query per ad^v may be made. With this adaption to freshness, the definition of the confidentiality event E_d works.

We must also adapt the definition of the integrity event. The idea is that the adversary can also break integrity by somehow causing the computation of a $\hat{t}$ that is inconsistent with the decryption of the ballot. We formalise this by requiring that for any test query, the ballots replacing $\bot$ among the originating ballots must also be consistent with any return code decryption query for any ciphertext in the origin. Here, consistency is with respect to the set $S = \{(v, \hat{t})\}$ output by $\mathcal{VU}$ for the particular ad^v.

E | *Exercise* 14.17. Write out carefully the experiment for voting schemes with return codes and the definitions of freshness and integrity, as sketched above.

E | *Exercise* 14.18. State and prove a claim similar to Theorem 14.1 for the scheme from Exercise 14.15 and the security notion from Exercise 14.17.

E | *Exercise* 14.19. State and prove a claim similar to Theorem 14.3 for the scheme from Exercise 14.16 and the security notion from Exercise 14.17.

Modified Voting Protocol   When we build a cryptographic voting protocol on top of a voting scheme with return codes, we must modify the setup phase and the casting phase. One possible configuration is shown in Figure 14.3. During the setup phase, we must generate appropriate key material and distribute it to the ballot casting devices and the shuffle server. We must also allow voters to register with appropriate voter associated data, and distribute

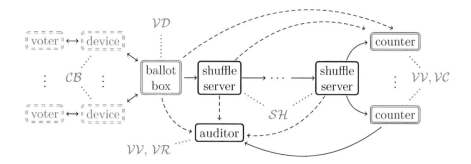

FIGURE 14.3 One configuration for an election using a cryptographic voting scheme with return codes. Compare with Figure 14.2. Players enclosed by dashed double lines have per-voter secret key material.

the resulting keys to the voter, their ballot casting device and the shuffle server. (Alternatively, if we have observers monitoring the ballot box, we could have one of these observers deal with return codes instead of the ballot box.)

Generating per-voter key material and distributing it in a practical fashion is a difficult problem. The ballot-return code table must usually be printed on paper, but should only be known by the voter. There are physical processes that can achieve reasonable security, but some trust seems essential.

During the casting phase, the ballot box will have to respond to the encrypted ballot with the return code, which the ballot casting device presents to the voter. The voter uses μ to check that the return code matches the ballot and accepts their ballot as cast if so.

Again, the design preserves the confidentiality of the cryptographic voting scheme under reasonable trust assumptions: In addition to the previous requirements, we require that the voter's ballot casting device is not corrupt, and that either the voter casts just one ballot, or the ballot box is not corrupted.

The usual techniques work well for integrity, with the trust assumption that the ballot casting device and the ballot box cannot both be corrupt.

Previously, verifiability relied on the voter's ballot casting device not being corrupted, which was true by assumption. Now, verifiability relies on a relaxed assumption, namely that either the ballot casting device is not corrupted (and the voter casts exactly one ballot) or the ballot box is not corrupted.

As for Benaloh challenges, return codes must be carefully evaluated to ensure effective deployment, but safe deployment is somewhat easier.

14.4   COERCION RESISTANCE

Influencing what ballots other voters decide to cast through discussion and argument is desired behaviour for most elections. Coercion is when the undesirable methods are used to influence other voters' decisions. Unchecked coercion is a serious problem in an election, first of all because it may affect the outcome, but even if the practical effect on the election result is small, the perceived effect on the election outcome may be significant. This compromises legitimacy. Defending against coercion is a non-trivial problem.

Coercion takes many forms. The obvious is a direct threat to life or limb. There may be a promise of a reward in some form, though for some elections that may be legitimate. Or there may be a more subtle threat of social consequences. Practically, the coercer can essentially cast the ballot on the voter's behalf using their secrets, or the coercer can observe the voter while the voter casts the desired ballot. Or the coercer can expect the voter to provide proof that the desired ballot was cast and counted, as for verifiability. One important point to keep in mind about coercion is that a voter may not be willing to admit that they were coerced, even to themself. Some coercers may also honestly believe that they are simply being helpful.

The natural goal for coercion is to get the voter to cast a desired ballot. Alternatively, the coercer may also want the voter to abstain, which may have a desired (if weaker) effect on the outcome. A variant of abstention is for the voter to cast a random ballot. The effect comes about because a coercer may know (to some degree) the voters' preferences, so may know the effect on the election outcome of abstention or casting a random ballot.

14.4.1 Definitions

A *coercion-resistant* voting system is a voting system that allows individual voters to cast their desired ballot in the presence of coercion. The techniques we shall study rely on a *recovery step* being performed after coercion.

Remark. Coercion resistance can only be defined for voting protocols, not for voting schemes. Voting schemes may support coercion resistance, though.

We shall model coercion by considering two types of voters: those that accept coercion and those that resist coercion and use the recovery step. We measure the coercer's power by its ability to distinguish between the two voter types after attempting to coerce.

We first discuss how to model the coercer and its powers. We shall consider two types of coercers: an active coercer that reveals all the voter key material and then effectively replaces the voter and its ballot casting device, and a passive coercer that reveals all the voter key material and then receives a copy of the conversation between the voter and the ballot casting device. Neither coercer is allowed to see messages between other players in the system. In some sense, this is the opposite of the usual adversary.

If a coercer learns the voter's key material, then from a cryptographic point of view the coercer is indistinguishable from the voter. There are two limiting factors. First of all, when a coercer reveals the voter's key material, the voter is allowed to *lie*. Second, our execution model considers only authenticated communication, so after the conversation with the ballot casting device ends, infrastructure players can distinguish between the coercer using the voter's key material and the voter and their ballot casting device acting honestly.

The latter point seems strange, but may be a realistic model for some voting systems. A voting system may rely on some general public key infrastructure that the voter is unwilling to reveal key material for, even though the voter is willing to be coerced. The active coercer is simply allowed to use the voter's key material once. The passive coercer would not be. For some voting systems and some realistic coercers, this is not a realistic model.

Remark. A third type of coercer also exists, who expects the voter to provide some form of evidence for the fact that the voter cast the desired ballot, often called a *receipt*. We shall not discuss such coercers further.

Next, we shall model the voter and three distinct cases: a voter that accepts coercion, a voter that resists coercion, and a voter that uses the recovery step.

- A *meek voter* accepts coercion, sends its key material to the coercer and then follows its usual behaviour.

- A *resisting voter* behaves as a meek voter, but may generate and use fake key material instead of its real key material.

- A *recovering voter* employs the recovery step, follows its procedures and may signal to the ballot casting device act accordingly.

The ballot casting device follows its programming regardless.

Remark. Note that it is the *voter* that may generate new, fake key material. This means, in particular, that the voter needs to keep a state, so that if it is coerced multiple times, it will present consistent key material to the coercer. In practice, voters may find this difficult, something that must be taken into account when evaluating the security and effectiveness of such schemes.

Our execution model allows the coercer to control what ballots were cast by non-coerced voters. This means that the coercer can predict the result, both for accepting and resisting voters. Since the coercer essentially tries to affect the result, the two results will differ and the coercer can trivially decide wether the voters accepted or resisted coercion. We want to exclude any advantage gained purely by looking at the result from the adversary's advantage.

Remark. We immediately clarify that this does not mean that a coercer cannot succeed by looking at the result. There are supposedly real-world examples where coercers have done this. In particular, if the set of ballots is very large, but many ballots have approximately the same effect, a coercer can ask the voters to cast unique ballots with the desired effect on the result. If the cast ballots are public, which they are in schemes such as EGVOTE from Example 14.1, the coercer can check if the unique ballot is present or not. This is called an *Italian attack*. We cannot defend against such attacks without changing the voting system. But we can avoid making the problem worse.

One possible approach is to let the non-coerced voters cast random ballots. This introduces randomness into the result. The adversary's coercion will have an effect on the output, but we can bound the advantage gleaned from the result by the statistical distance between the two distributions.

However, resistance followed by the recovery step will typically cause some noticable effect on the election certificate. Hiding signs of resistance is hard, so instead we will make sure this behaviour always appears. One idea is to introduce a *compensating voter*. The idea is that the coercer knows the possible effects of the coercion attempt. It must then ask the experiment, wich knows the actual effect, to cause the opposite effect for some other voter.

Remark. What if an active coercer lies to the experiment? This would allow the coercer to cause some noticable effect. In general, assuming particular adversarial behaviour often leads to modelling mistakes which may hide real security problems. But it is not unreasonable to claim that we are not interested in "lying coercers". Furthermore, the coercer's claims are testable. We shall therefore assume that an active coercer does not lie to the experiment.

If we use a compensating voter, we may as well require this voter to also campensate the electoral outcome. So we abandon the random ballot idea.

We change the experiment for the voting protocol as follows.

- The adversary may make coerce queries, identifying a voter, the voter's intended ballot and the coercer's desired behaviour. The experiment starts instances of the voter and the ballot casting device, where the voter is either meek or resisting, according to the challenge bit.

- The adversary may make compensate queries, identifying a coerce query and a compensating voter. The experiment starts instances of the compensating voter and its ballot casting device under passive corruption with the coercer's desired behaviour, and where the compensating voter is either resisting or meek, according to the challenge bit.

  Once the compensating voter finishes, the experiment immediately starts a recovering voter and a ballot casting device, either for the coerced voter if the coerced voter was resisting, or for the compensating voter.

The challenge queries made by the adversary must all have $v_0 = v_1$. The adversary may corrupt other voters and cast ballots on their behalf, as usual. The adversary will only see messages sent and received by coerced voters and their ballot casting devices. The adversary will not see messages sent and received by infrastructure players or non-coerced or compensating voters and their ballot casting devices. Eventually, the adversary sees the certificate output by the auditor and outputs a guess for the challenge bit.

Remark. The voting system must be secure against conventional adversaries.

14.4.2 Voter Secrets

One way to protect against coercion is to include a secret known only to the voter in the ballot argument when casting a ballot, but include no other identifying information. The resisting voter can then lie about the secret to any coercer, which would result in the ballot argument containing (almost certainly) an invalid secret. The recovering voter simply casts a ballot as usual, with the correct secret. If the infrastructure players know the set of voter secrets, they can remove any encrypted ballots cast with invalid secrets.

The above approach is simple. As long as lies are indistinguishable from secrets, the coercer should not be able to distinguish meek and resisting voters.

The problem is that we also want verifiability, which means that the certificate must somehow convince voters that their honestly cast ballot was not removed. Since the coercer learns the certificate, the certificate must not allow the coercer to learn if a particular encrypted ballot was removed.

One idea is to make encryptions of all the voter secrets available, and then include an independent encryption of the voter secret in the ballot argument.

The first *filtering* step is to remove duplicates (two encrypted ballots are *duplicate* if they contain the same voter secret). We can prove that two ciphertexts have the same decryption (Example 11.9), so it is easy to prove that something is a duplicate. We want to allow one of the duplicates to remain, determined by some selection rule.

The second filtering step removes ballots with invalid voter secrets and proves that there are no duplicates remaining. We can use shuffles to do this. We shuffle pairs of encrypted ballots and encrypted voter secrets (to decorrelate the duplicates) and then indicate which encrypted ballots to keep. We give a correspondence between a subset of the encryptions of valid secrets (which we also rerandomise and reorder using shuffles) and the (rerandomised) encrypted voter secrets belonging to encrypted ballots we will keep and prove that corresponding ciphertexts contain identical voter secrets. Finally, for each discarded encrypted ballot, we prove that the secret is not a valid secret.

For ElGamal, one possibility is to raise the ciphertexts to a fixed random power and then reveal decryptions of the resulting ciphertexts. Under Decision Diffie-Hellman, this reveals no information about the invalid secret or the valid secrets. It is, however, important that distinct random powers are used for each discarded ciphertext, otherwise the coercer may encode relations among secrets that reveal when a particular encrypted ballot is discarded.

The number of operations required for the second filtering step equals the product of the number of voters and the number of discarded ballots. The number of discarded ballots may be large, which means that the second filtering step may become inconveniently expensive.

The ballot box may know who submits which encrypted ballot, so it can prevent an excessive number of encrypted ballot submissions, although one must use rate limiting rather than a maximum number of submissions.

The filtering steps define a *filter function* ϕ from equal-length lists of ballots and voter secrets to lists of ballots. This function essentially describes how to count duplicates, encoding the selection rule previously used by the ballot box. Typically, we interpret the order of the ballots as a temporal order and count (say) the last ballot for any given voter secret. We extend the function to allow $\perp$ instead of ballots or voter secrets, but $\perp$ is ignored, as is any ballot for which the corresponding voter secret is $\perp$.

Modified Voting Scheme   In order to analyse such a scheme with voter secrets, we would again have to change the definition of a voting scheme, modifying the setup, cast and verification algorithms, and also add a new filtering algorithm. As for return codes, we abstract the above concrete sketches.

- The *setup* algorithm $\mathcal{VS}$ also outputs a filter key fk.

- The *register* algorithm $\mathcal{VU}$ takes as input a ballot casting key bk and outputs a *voter secret* vs and a public commitment evs. The voter secret must be sampled from the uniform distribution on some set.

- The *cast* algorithm $\mathcal{CB}$ also takes as input vs.

- The *filter* algorithm $\mathcal{VF}$ takes as input a ballot casting key bk, a *filter key* fk and l encrypted ballots $\mathbf{c}$ with corresponding ballot arguments $\mathbf{arg^c}$, and outputs (possibly fewer) encrypted ballots $\mathbf{c}'$ and a *filter argument* arg_{vf}. We require that the output subset is completely determined by bk, fk, $\mathbf{c}$ and $\mathbf{arg^c}$.

- The *verification* algorithm may, in addition to the existing variants, also take as input either a ballot casting key bk, l encrypted ballots $\mathbf{c}$ with corresponding ballot arguments $\mathbf{arg^c}$, a subset of $\{1, 2, \ldots, l\}$ and arg_{vf}.

The usual correctness requirement must be adapted to use voter secrets, and correctness is with respect to a given filter function ϕ.

Exercise 14.20. Write out carefully the definition of a voting scheme with voter secrets, as sketched above.

Exercise 14.21. Adapt the EGVOTE scheme from Example 14.1 to use voter secrets as sketched above. Show that the scheme is correct.

Exercise 14.22. Adapt the HOMVOTE scheme from Example 14.2 to use voter secrets as sketched above. Show that the scheme is correct.

We must modify the experiment from Figure 14.1 to add a *register* query to generate per-voter secrets and corresponding public commitments. The challenge query must include a reference to a register query, allowing the adversary to specify which per-voter secret to use. This information is also recorded in L_C. The chosen ciphertext query does not include such a reference. The adversary must be allowed to make (chosen) filter queries and reveal per-voter secrets. Finally, we must add *coerce* and *compensate* queries, which supports coercion resistance. We sketch the new queries in some detail, since the details are highly technical and somewhat tricky.

The register query takes as input voter associated data (supposed to identify a unique voter) and returns a public commitment. The register queries in an execution are *proper* if every query has distinct voter associated data.

The coerce and compensate queries give the adversary a voter secret and two honestly generated ciphertexts and let the adversary specify a ciphertext. This is sufficient to simulate the voting protocol behaviour. (The three ciphertexts would usually be created in a different order, so great care must be taken if global state or time stamps are used with associated data.)

The coerce query takes as input ballots v_0 and v_1, and voter associated data ad_0^v, ad_1^v corresponding to distinct register queries with voter secrets vs_{00} and vs_{11}. It samples two secrets vs_{01} and vs_{10}, computes $(c_0, arg_0^c) \leftarrow \mathcal{CB}(bk, vs_{1b}, v_0)$ and $(c_1, arg_1^c) \leftarrow \mathcal{CB}(bk, vs_{1-b,1-b}, v_1)$, and outputs $(vs_{0b}, c_0, arg_0^c, c_1, arg_1^c)$. If $b = 0$, it records $(\perp, \perp, \perp, \perp, c_0, arg_0^c)$ and $(\perp, v_1, v_1, ad_1^v, c_1, arg_1^c)$ in L_C. Otherwise, it records $(\perp, v_0, v_0, ad_1^v, c_0, arg_0^c)$ and $(\perp, v_1, v_1, ad_0^v, c_0, arg_0^c)$ in L_C.

The compensate query takes as input an encrypted ballot c, a ballot argument arg^c and a reference to a previous coerce query $(v_0, v_1, ad_0^v, ad_1^v)$ with response $(vs, \cdot, \cdot, \cdot, \cdot)$. If $b = 0$, it records $(\perp, v_0, v_0, ad_0^v, c, arg^c)$ in L_C. Otherwise, it records $(\perp, \perp, \perp, \perp, c, arg^c)$.

The coerce queries in an execution are *proper* if every coerce query is referenced by exactly one compensate query.

Remark. For challenge bit $b = 0$, the coercion query will provide one encrypted ballot with a random secret and one encrypted ballot with vs_{11}. In this case, the compensate query will provide an encrypted ballot with vs_{00}.

For challenge bit $b = 1$, the coercion query will provide encrypted ballots for both vs_{11} and vs_{00}, while the compensate query will provide an encrypted ballot with a random secret.

Note also that after the queries, ad_0^v and ad_1^v have both been recorded exactly one more time in L_C. Also note that the encrypted ballot with a random voter secret is recorded in the same way as a chosen ciphertext query.

A filter query for a ciphertext list $\mathbf{c}$ with ballot arguments arg^c is *proper* if there are tuples $(\cdot, v_{0,i}, v_{1,i}, ad_i^v, c_i, arg_i^c)$ in L_C for $i = 1, 2, \ldots, l$. A count or test query is *proper* if there are intermediate shuffles in L_S connecting the query to the output of a proper filter query. We define *path*, *origin* $\mathbf{c} = (c_1, \ldots, c_l)$ and *originating ballots* $\mathbf{v}_0, \mathbf{v}_1$ as before for proper count, test and filter queries. We also define *originating identities* to be the list $ad^v = (ad_1^v, \ldots, ad_l^v)$. (Note that the voter associated data will be $\perp$ for chosen ciphertext queries.)

A proper filter or count query is *balanced* if $f(\phi(\mathbf{v}_0, ad^v)) = f(\phi(\mathbf{v}_1, ad^v))$.

A proper filter or count query is *compensated* if every encrypted ballot from any coerce or compensate query is included in the origin.

An execution is *fresh* if the register queries are proper and there are either challenge queries or coerce queries, but not both. If there are challenge queries, then freshness requires that every count and filter query is both proper and balanced, and that no register query is referenced in both a challenge query and a reveal query, and that set of revealed decryption shares is not in the access structure. If there are coerce queries, then freshness requires that every count and filter query is both proper and compensated, that the coerce queries are proper, and that no register query is referenced by more than one coerce query, or by both a coerce query and a reveal query. Also, no reveal query should be made for the filter key or a decryption key share.

The *integrity* event E_{int} is the event that a proper test query is wrong, that two proper test queries are inconsistent, or that a proper test query fails. Additionally, E_{int} includes the event that the number of encrypted ballots output by a proper filter query is larger than the number of register queries.

Exercise 14.23. Write out carefully the experiment for a voting scheme with voting secrets, the definition of fresh and the definition of integrity, as sketched above.

Exercise 14.24. State a claim similar to Theorem 14.1 for the scheme from Exercise 14.21 and the security notion from Exercise 14.23, and prove it.

Remark. Voter secrets are in some sense orthogonal to the rest of the voting scheme. It should be possible to add the voter secrets and the filter algorithms as extensions to a standard voting scheme as in Definition 14.1. This approach could simplify the security analysis and the design. We leave the development of this approach to the interested reader.

Modified Voting Protocol   We describe a voting protocol on top of a cryptographic voting scheme with voter secrets. The configuration is shown in Figure 14.4. We modify the setup, casting and counting phases.

During the setup phase, we must distribute the key material to the appropriate players. We must also allow voters to register with appropriate voter associated data, and distribute the resulting keys to the voters. The public commitments are made public, which essentially means that every single player receives a copy, which is included in the auditor's certificate.

During the casting phase, the voter gives the voter secret to the ballot casting device along with the ballot, and receives the ciphertext in return.

During the counting phase, we have a filter player that removes duplicates and invalid encrypted ballots. The scheme then proceeds as usual.

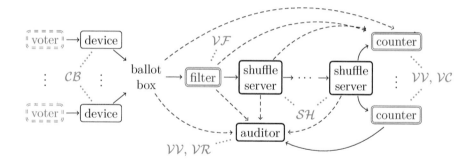

FIGURE 14.4 One configuration for an election using a cryptographic voting scheme with voter secrets. Compare with Figure 14.2. Players enclosed by dashed double lines have per-voter secret key material.

Again, we see that the design preserves the privacy of the cryptographic scheme under reasonable trust assumptions. The protocol preserves the integrity of the scheme under the trust assumption that voter secrets are only known by the voters and that the ballot casting device is not corrupt. This assumption also gives verifiability.

Key generation for ElGamal is easy to distribute using multiparty computation. The generation of voter secrets presents a more difficult problem.

Ideally, the voter should generate the voter secret. But in this case, the voter could trivially generate the voter secret and the commitment in such a way that the voter cannot lie about the voter secret. Which would break coercion resistance. Typically, this also happens if the voter participates in a protocol that verifiable generates the secret, because the voter can derandomise its behaviour in a way that cannot be simulated.

The most common alternative is to have a *registrar* that simply runs the register algorithm and is trusted. In principle, one could have multiple registrars and trust all of them. But of course, this also increases the risk that one of the registrars cannot be trusted, after which coercion resistance is lost since choosing the untrusted registrar may be part of the coercion process.

Remark. This protocol has no security against a corrupt ballot casting device. Voter secrets cannot easily be combined with return codes, since a corrupt ballot casting device may leave the ballot untouched, but change the voter secret. Benaloh challenges could be used to protect against a corrupt ballot casting device, but the discussion in Section 14.3 still applies.

Remark. Voter secrets are inadvisable in practice, because voters may misremember their voter secrets, effectively self-cancelling their own ballots. It seems hard to find a mitigation strategy that preserves coercion resistance.

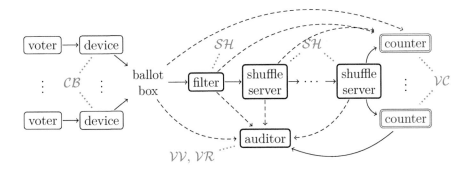

FIGURE 14.5 One possible configuration for an election using a cryptographic voting scheme. Compare with Figure 14.2.

For more limited coercion scenarios, for instance when the coercer is not present when the voter casts the ballot, other strategies could be possible.

14.4.3 Re-voting

Achieving both coercion-resistance and verifiability is hard because the goals are in some sense in opposition to each other. A different possibility is to remember that integrity is the actual goal we want for a voting protocol. Verifiability is just a tool we use to achieve integrity. Could we achieve integrity otherwise? We already saw some ideas on how to achieve integrity in Section 14.2.1, and we can combine those with the ideas in verifiability. This will allow us to use an extremely simple coercion recovery strategy: The voter simply casts another ballot.

We shall use a standard voting scheme as in Definition 14.1 and an additional mechanism to hide who cast which ballot. The configuration is shown in Figure 14.5.

Instead of giving the associated data to the cast algorithm, the ballot casting device commits to the associated data and gives the commitment to the cast algorithm instead of the associated data. It will then send the encrypted ballot, the ballot argument, the commitment and an opening of the commitment to the associated data to the ballot box.

The filter player first decides which encrypted ballots should be counted. It then shuffles all the encrypted ballots, and outputs only the shuffled encrypted ballots that correspond to the encrypted ballots that should be counted. In addition, it sends the randomness used to shuffle to the counters and the auditor, who can then verify the selection of encrypted ballots simply by recomputing the shuffle.

The auditor verifies everything. It includes in the certificate the encrypted ballots with ballot arguments and commitments, the filter shuffle, shuffle argument and discarded ballots, every subsequent shuffle with shuffle arguments, and the decryption shares, and excludes the associated data, the opening of the commitments and the randomness for the filter player's shuffle.

The certificate reveals no information about the associated data of the cast ballots, because the commitment scheme is hiding, the shuffle argument is zero knowledge, and the shuffle properly rerandomises ballots. Since three encrypted ballots are added to the ballot box for every coercion and compensate query pair, and one of the encrypted ballots is discarded, regardless of the challenge bit, we get coercion resistance.

Confidentiality follows under the same trust assumptions as in Section 14.2.1, and by essentially the same argument.

A voter can recognise when their encrypted ballot has not been included in the count, but they cannot know if the encrypted ballot was discarded by the filter or not. This can happen for two reasons: Either the filter is cheating, or someone has revoted on the voter's behalf. The former cannot happen if at least one honest player is included in the certificate. With regard to the latter, the strategies we discussed under eligibility in Section 14.2.1 will still work, and now have the side effect of guaranteeing integrity. It follows that we can in practice ensure integrity under relatively mild trust assumptions.

Unlike voter secrets, re-voting as a concept is fairly easy to understand and the recovery strategy seems easy to follow, though in practice it is advisable to test this conjecture before deployment.

This protocol has no security against a corrupt ballot casting device. Unlike the case with voter secrets, this protocol can easily be based on a voting scheme with return codes, which gives some measure of protection against a corrupt ballot casting device. Depending on the circumstances, such a solution may provide a practical compromise between the various security properties.

Benaloh challenges could be used when practical with respect to the voters.

Sometimes, a cryptographic voting protocol is used together with traditional paper voting. In this case, it is possible to use paper voting to provide coercion resistance. The idea is to count a paper ballot in preference to any encrypted ballot. Essentially, the filter (and any other infrastructure player) receives a list of voters who cast ballots on paper and use this list to discard encrypted ballots. This allows us to use an even simpler coercion strategy: cast a paper ballot. Again, we need to trust that the list is correct, but in many cases this is plausible.

Index